SAXON Math™
HOMESCHOOL
7/6

Saxon Publishers gratefully acknowledges the contributions of the following individuals in the completion of this project:

Authors: Stephen Hake, John Saxon

Editorial: Chris Braun, Dana Nixon, Bo Björn Johnson, Brian E. Rice

Editorial Support Services: Christopher Davey, Jay Allman, Jenifer Sparks, Shelley Turner, Jean Van Vleck, Darlene Terry

Production: Alicia Britt, Karen Hammond, Donna Jarrel, Brenda Lopez, Adriana Maxwell, Cristi D. Whiddon

Project Management: Angela Johnson, Becky Cavnar

Printed in the United States of America

ISBN: 1-59141-319-2

2 3 4 5 6 7 8   062   12 11 10 09 08 07 06

SAXON Math™
HOMESCHOOL
7/6

Stephen Hake
John Saxon

SAXON™
PUBLISHERS

CONTENTS

Dear Student,

We study mathematics because of its importance to our lives. Our study schedule, our trip to the store, the preparation of our meals, and many of the games we play involve mathematics. You will find that the word problems in this book are often drawn from everyday experiences.

As you grow into adulthood, mathematics will become even more important. In fact, your future in the adult world may depend on the mathematics you have learned. This book was written to help you learn mathematics and to learn it well. For this to happen, you must use the book properly. As you work through the pages, you will see that similar problems are presented over and over again. **Solving each problem day after day is the secret to success.**

Your book is made up of daily lessons and investigations. Each lesson has four parts. The first part is a Warm-Up that includes practice of basic facts and mental math. These exercises improve your speed, accuracy, and ability to do math "in your head." The Warm-Up also includes a problem-solving exercise to familiarize you with strategies for solving complicated problems. The second part of the lesson is the New Concept. This section introduces a new mathematical concept and presents examples that use the concept. In the next section, the Lesson Practice, you have a chance to solve problems involving the new concept. The problems are lettered a, b, c, and so on. The final part of the lesson is the Mixed Practice. This problem set reviews previously taught concepts and prepares you for concepts that will be taught in later lessons. Solving these problems helps you remember skills and concepts for a long time.

Investigations are variations of the daily lesson that often involve activities. Investigations contain their own set of questions instead of a problem set.

Remember, solve every problem in every practice set, every problem set, and every investigation. Do not skip problems. With honest effort, you will experience success and true learning that will stay with you and serve you well in the future.

Stephen Hake
Temple City, California

PREFACE

Dear Parent-Teacher,

Congratulations on your decision to use *Saxon Math 7/6—Homeschool*! Proven results, including higher test scores, have made Saxon Math™ the hands-down favorite for homeschoolers. Only Saxon helps you teach the way your child learns best—step-by-step. With Saxon, each new skill builds on those already taught, daily reviews of earlier material increase understanding, and frequent, cumulative assessments ensure that your child masters each skill before new ones are added. The result? More confidence, more willingness to learn, more success!

SAXON PHILOSOPHY

The unique structure of Saxon Math™ promotes student success through the sound, proven educational practices of *incremental development* and *continual review*. Consider how most other mathematics programs are structured: content is organized into topical chapters, and topics are developed rapidly to prepare students for end-of-chapter tests. Once a chapter is completed, the topic changes, and often practice on the topic ends as well. Many students struggle to absorb the large blocks of content and often forget the content after practice on it ends. Chapter organization might be good for reference, but it is not the best organization for learning. Incremental development and continual review are structural designs that improve student learning.

Incremental development With incremental development, topics are developed in small steps spread over time. One facet of a concept is taught and practiced before the next facet is introduced. Both facets are then practiced together until it is time for the third to be introduced. Instead of being organized in chapters that rapidly develop a topic and then move on to the next strand, Saxon Math™ is organized in a series of lessons that gradually develop concepts. This approach gives students the time to develop a deeper understanding of concepts and how to apply them.

Continual review Through continual review, previously presented concepts are practiced frequently and extensively throughout the year. Saxon's cumulative daily practice strengthens students' grasp of concepts and improves their long-term retention of concepts.

John Saxon often said, "Mathematics is not difficult. Mathematics is just different, and time is the elixir that turns things different into things familiar." This program provides the time and experiences students need to learn the skills and concepts necessary for success in mathematics, whether those skills are applied in quantitative disciplines or in the mathematical demands of everyday life.

PROGRAM COMPONENTS

Saxon Math 7/6—Homeschool consists of three components: 1) textbook, 2) Tests and Worksheets, and 3) Solutions Manual. **Before using the program, please ensure that you have each component.**

Textbook   The *Saxon Math 7/6—Homeschool* textbook is divided into 120 lessons and 12 investigations. The textbook also contains an appendix topic that can be presented at the teacher's discretion, supplemental practice problems for remediation, an illustrated glossary, and a comprehensive index.

Tests and Worksheets   The *Saxon Math 7/6—Homeschool Tests and Worksheets* booklet provides all the worksheets and tests needed by one student to complete the program. It also contains the following recording forms for students to show their work and for parents to track student progress:

- Recording Form A: Facts Practice
- Recording Form B: Lesson Worksheet
- Recording Form C: Mixed Practice Solutions
- Recording Form D: Scorecard
- Recording Form E: Test Solutions

Directions for using the recording forms are provided in the Program Overview (below), as well as in the introduction to the Tests and Worksheets booklet.

Note: The recording forms are blackline masters that should be photocopied, as they may be used more than once.

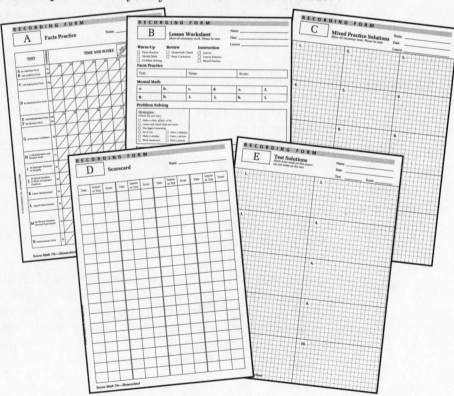

Solutions Manual   The *Saxon Math 7/6—Homeschool Solutions Manual* contains step-by-step solutions to all textbook and test exercises.

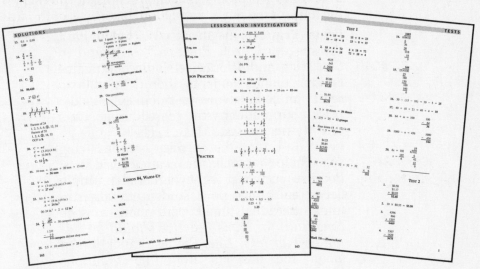

PROGRAM OVERVIEW

Saxon Math 7/6—Homeschool contains three types of math "sessions": lessons, investigations, and tests. Concepts are introduced and reviewed in a carefully planned sequence. **It is therefore crucial to complete all the lessons and investigations in *Saxon Math 7/6—Homeschool* in the given order.** If lessons are skipped or presented out of sequence, students will encounter problems on the tests and in the problem sets that they might not be equipped to solve.

By completing one lesson, investigation, or test per day, you can finish the entire program in thirty-one or thirty-two weeks. However, faster or slower paces may be appropriate, depending on students' individual learning styles.

Lessons   Each of the program's 120 lessons is divided into four sections: Warm-Up, New Concept(s), Lesson Practice, and Mixed Practice. Below we show a lesson from the textbook.

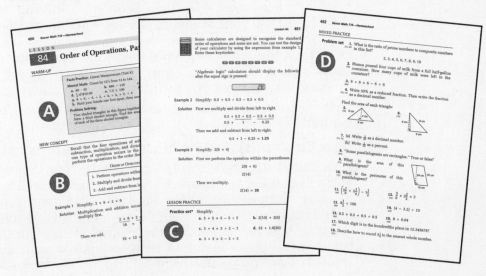

Ⓐ Warm-Up (10–15 minutes)

The Warm-Up promotes mental math and problem-solving skills and sets the tone for the day's instruction. It consists of three activities:

Facts Practice: Begin the Warm-Up with the suggested Facts Practice Test (found in the Tests and Worksheets). Facts Practice covers content students should be able to recall immediately or to calculate quickly. Have your student write his or her answers directly on the test. Make Facts Practice an event by timing the exercise—emphasizing speed helps automate the recall of basic facts. Because each test is encountered multiple times, encourage your student to improve upon previous timed performances. *The time limit for Facts Practice should be five minutes or less.* After the test, quickly read aloud the answers from the Solutions Manual as your student checks his or her answers. If desired, Facts Practice scores and times can be tracked on Recording Form A (from the Tests and Worksheets) or in a math notebook. The time invested in Facts Practice is repaid in students' ability to work more quickly.

Mental Math: Follow Facts Practice with Mental Math. Read the problems aloud while your student follows along. Have your student perform the calculations mentally and write the answers on a copy of Recording Form B or on blank paper. (*Note:* Students should **not** use pencil and paper to perform the calculations.) Mental math ability pays lifelong benefits and improves markedly with practice. *Complete the Mental Math activity in two to three minutes.* Mental Math answers are provided in the Solutions Manual.

Problem Solving: Finish the Warm-Up with the daily Problem Solving exercise. Problem Solving promotes critical-thinking skills and offers opportunities for students to use such strategies as drawing diagrams and pictures, making lists, acting out situations, and working backward. If the Problem Solving exercise presents difficulties for your student, you are encouraged to suggest strategies for tackling the problem, referring to the Solutions Manual as necessary. Students may write their answers on Recording Form B and check off the strategies they used in solving the exercise. *Most Problem Solving exercises can be solved in a few minutes.*

Note: Because of time considerations, the Warm-Up to Lesson 104 is omitted.

Ⓑ New Concept(s) (5–15 minutes for most lessons)

After completing the Warm-Up, present the New Concept(s). In this section you will find the new instructional increment as well as example problems to work through with your student. Important vocabulary terms are highlighted in color, and each of these terms is defined in the textbook's glossary. It is recommended that you read through the New Concept(s) before presenting a lesson to become familiar with the content. Because students learn most effectively by actually working math problems, keep the presentation of the New Concept(s) brief. This will maximize the time your student has to solve problems in the Lesson Practice and Mixed Practice problem sets (which are described in the next sections).

Some lessons involve activities that require the use of household items. Refer to page xxi for a list of necessary materials and the lessons in which they are used. Certain lessons also call for students to use Activity Sheets (see example at right). Activity Sheets are referenced in the textbook and can be found in the Tests and Worksheets booklet.

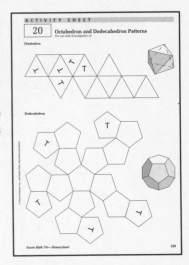

C Lesson Practice (5–10 minutes)

The Lesson Practice provides practice on the New Concept. Closely monitor student work on the Lesson Practice problems, providing immediate feedback as appropriate. Have your student solve **all** the problems in the Lesson Practice before proceeding to the next section of the lesson. Answers may be written on Recording Form B or on blank paper. If your student has difficulty with the Lesson Practice, you may wish to reteach the relevant examples in the New Concept section in order to identify the particular aspect of the concept that is causing problems.

Some Lesson Practice sets are marked with an asterisk (see example at right). An asterisk indicates that additional problems on the lesson's concept appear in the Supplemental Practice section of the textbook's appendix. Supplemental Practice problems are intended for remediation. Assign your student these problems only if he or she has difficulty with a concept several lessons after it is presented.

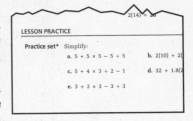

D Mixed Practice (20–40 minutes)

The Mixed Practice is the fourth and most important component of the lesson. This section contains twenty-five to thirty problems that prepare students for upcoming lessons, allow them to work with several strands of mathematics concurrently, and provide them with the distributed practice that promotes long-term retention of concepts. **Have your student work independently on the Mixed Practice, and ensure that no problem is skipped.** Students may show their work on a copy of Recording Form C or on blank paper.

> If your student encounters difficulty with Mixed Practice problems, have him or her refer to the Lesson Reference Numbers that appear in parentheses below each problem number. Lesson Reference Numbers indicate which lessons explain concepts relevant to the problems they label. Because many problems involve multiple concepts, more than one reference number might be given for a problem.

At the end of the math period, check your student's work, referring to the Solutions Manual as necessary. If there are incorrect answers, help your student identify which solution steps led to the errors. Then have the student rework the problems to achieve the correct answers. If desired, track the completion of your student's daily assignments on Recording Form D.

Investigations   Following every tenth lesson is an investigation. Investigations are in-depth treatments of concepts that often involve activities. Because of the length of investigations, no Warm-Up or Mixed Practice is included. As with lessons, investigations might call for students to use Activity Sheets, which can be found in the Tests and Worksheets booklet.

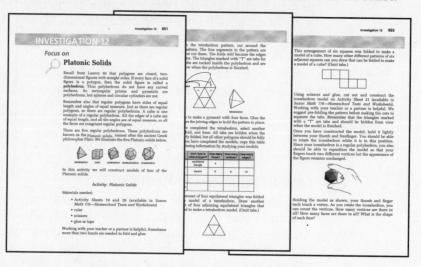

Tests   Twenty-three cumulative tests are provided in *Saxon Math 7/6—Homeschool*. The problems on the tests are similar to those in the textbook, and the tests are scheduled so that students have about five days to practice concepts before being assessed on them. For detailed information regarding when to give each test, refer to the Testing Schedule in the Tests and Worksheets booklet.

Begin each test day with Facts Practice. The appropriate Facts Practice Test is specified at the top of the scheduled cumulative test (see example at right). After the Facts Practice, administer the cumulative test. Have your student show his or her work and record his or her answers on a copy of Recording Form E. The textbook should not be used during the test.

After the test, compare each student answer to the one given in the Solutions Manual. Note any incorrect answers, and review the test with your student. Determine whether errors were caused by computational mistakes or conceptual misunderstandings. If necessary, stress to your student that computational errors can be prevented by writing neatly and by checking the work. If he or she misunderstands a concept, be sure to address the misunderstanding promptly. Work through textbook examples that demonstrate the concept (identify the appropriate lesson by referring to the Lesson Reference Numbers on the test), and assign additional practice problems for the student to solve. (Check for additional practice problems in the textbook's appendix.)

HOW TO GET HELP

If you need help implementing your homeschool program, you can call our Parent Support Line at (405) 217-1717, and our veteran teachers will counsel you on how to set up and teach Saxon Homeschool Math. If you and your student need help with a specific math problem, please e-mail our tutors at mathhelp@saxonhomeschool.com, and they will respond promptly.

We encourage you to visit the Saxon Homeschool Web site, www.saxonhomeschool.com, for descriptions of Saxon's math and phonics programs and for downloadable documents such as our homeschool catalog, placement tests, and state-standards correlations. The Web site also provides online math and phonics activities for your student.

We wish you success and enjoyment in the coming year, and please remember to contact us with any questions or comments!

The following materials are used throughout *Saxon Math 7/6—Homeschool*. We suggest you acquire these materials before beginning the program.

- inch/centimeter ruler

  (*Note:* a ruler that shows both customary and metric scales is preferred. However, separate customary and metric rulers are acceptable.)

- protractor
- compass (for drawing circles)
- calculator with memory functions
- graph paper (grid paper)
- plain, unlined paper
- scissors

Certain lessons and investigations contain activities that call for additional materials. Refer to the following list before beginning the specified lessons/investigations.

Lesson 7

- 6-by-1 inch strip of tagboard

Lesson 17 (Warm-Up)

- dot cube

Investigation 2

- envelope or zip-top plastic bag
- colored pencils or markers (optional)

Lesson 47

- 2-4 circular objects for measurement (e.g., dinner plates, pie pans, flying disks, bicycle tires, can tops or bottoms, trash cans)
- string or masking tape
- cloth tape measure

Investigation 9

- 4 red and 2 white marbles (If marbles are unavailable, use colored plastic chips, craft sticks, or slips of paper.)
- small, opaque bag

Investigation 10

- plastic straws or wooden skewers

Investigation 12

- glue or tape

1 Adding Whole Numbers and Money • Subtracting Whole Numbers and Money • Fact Families, Part 1

WARM-UP†

NEW CONCEPTS

Adding whole numbers and money

To combine two or more numbers, we add. The numbers that are added together are called **addends.** The answer is called the **sum.** Changing the order of the addends does not change the sum. For example,

$$3 + 5 = 5 + 3$$

This property of addition is called the **commutative property.** When adding numbers, we add digits that have the same place value.

Example 1 Add: 345 + 67

Solution When we add whole numbers on paper, we write the numbers so that the place values are aligned. Then we add the digits by column.

$$\begin{array}{r} \overset{1\,1}{345} \\ +67 \\ \hline 412 \end{array}$$ addend
addend
sum

Changing the order of the addends does not change the sum. One way to check an addition answer is to change the order of the addends and add again.

$$\begin{array}{r} \overset{1\,1}{67} \\ +345 \\ \hline 412 \end{array}$$ check

---

†For instructions on how to use the Warm-up, please consult the preface.

Example 2   Add: $1.25 + $12.50 + $5

Solution   When we add money, we write the numbers so that the decimal points are aligned. We write $5 as $5.00 and add the digits in each column.

$$\begin{array}{r} \$1.25 \\ \$12.50 \\ + \quad \$5.00 \\ \hline \mathbf{\$18.75} \end{array}$$

If one of two addends is zero, the sum of the addends is identical to the nonzero addend. This property of addition is called the **identity property of addition.**

$$5 + 0 = 5$$

Subtracting whole numbers and money   We subtract one number from another number to find the **difference** between the two numbers. In a subtraction problem, the **subtrahend** is taken from the **minuend.**

$$5 - 3 = 2$$

In the problem above, 5 is the minuend and 3 is the subtrahend. The difference between 5 and 3 is 2.

The commutative property does not apply to subtraction; for example, 2 − 4 does not equal 4 − 2.

Example 3   Subtract: 345 − 67

Solution   When we subtract whole numbers, we align the digits by place value. We subtract the bottom number from the top number and regroup when necessary.

$$\begin{array}{r} \overset{2}{\cancel{3}}\,\overset{13}{\cancel{4}}\,\overset{1}{5} \\ - \quad 6\,7 \\ \hline \mathbf{2\,7\,8} \end{array}$$
difference

Example 4   Jim spent $1.25 for a hamburger. He paid for it with a five-dollar bill. Find how much change he should get back by subtracting $1.25 from $5.

Solution   Order matters when we subtract. The starting amount is put on top. We write $5 as $5.00. We line up the decimal points to align the place values. Then we subtract. Jim should get back **$3.75.**

$$\begin{array}{r} \$\overset{4}{\cancel{5}}.\overset{9}{\cancel{0}}\,\overset{1}{0} \\ - \quad \$1.2\,5 \\ \hline \$3.7\,5 \end{array}$$

We can check the answer to a subtraction problem by adding. If we add the answer (difference) to the amount subtracted, the total should equal the starting amount. We do not need to rewrite the problem. We just add the two bottom numbers to see whether their sum equals the top number.

SUBTRACT DOWN	$5.00	ADD UP
To find the	− $1.25	To check
difference	$3.75	the answer

Fact families, part 1
Addition and subtraction are called **inverse operations**. We can "undo" an addition by subtracting one addend from the sum. The three numbers that form an addition fact also form a subtraction fact. For example,

$$4 + 5 = 9 \qquad 9 - 5 = 4$$

The numbers 4, 5, and 9 are a **fact family**. They can be arranged to form the two addition facts and two subtraction facts shown below.

$$\begin{array}{cccc} 4 & 5 & 9 & 9 \\ +\,5 & +\,4 & -\,5 & -\,4 \\ \hline 9 & 9 & 4 & 5 \end{array}$$

Example 5 Rearrange the numbers in this addition fact to form another addition fact and two subtraction facts.

$$11 + 14 = 25$$

Solution We form another addition fact by reversing the addends.

$$14 + 11 = 25$$

We form two subtraction facts by making the sum, 25, the first number of each subtraction fact. Then each remaining number is subtracted from 25.

$$25 - 11 = 14$$

$$25 - 14 = 11$$

Example 6 Rearrange the numbers in this subtraction fact to form another subtraction fact and two addition facts.

$$\begin{array}{r} 11 \\ -\ 6 \\ \hline 5 \end{array}$$

Solution The commutative property does not apply to subtraction, so we may not reverse the first two numbers of a subtraction problem. However, we may reverse the last two numbers.

$$\begin{array}{r} 11 \\ -\ 6 \\ \hline 5 \end{array} \times \begin{array}{r} 11 \\ -\ 5 \\ \hline 6 \end{array}$$

For the two addition facts, 11 is the sum.

$$\begin{array}{r} 5 \\ +\ 6 \\ \hline 11 \end{array} \qquad \begin{array}{r} 6 \\ +\ 5 \\ \hline 11 \end{array}$$

LESSON PRACTICE

Practice set Simplify:

a. $3675 + 426 + 1357$

b. $\$6.25 + \$8.23 + \$12$

c. $5374 - 168$

d. $\$5 - \$1.35$

e. Arrange the numbers 6, 8, and 14 to form two addition facts and two subtraction facts.

f. Rearrange the numbers in this subtraction fact to form another subtraction fact and two addition facts.

$$25 - 10 = 15$$

MIXED PRACTICE

Problem set 1. What is the sum of 25 and 40?

2. Johnny had 137 apple seeds in one pocket and 89 in another. He found 9 more seeds in his cuff. Find how many seeds he had in all by adding 137, 89, and 9.

3. What is the difference when 93 is subtracted from 387?

4. Keisha paid $5 for a movie ticket that cost $3.75. Find how much change Keisha should get back by subtracting $3.75 from $5.

5. Tatiana had $5.22 and earned $1.15 more. Find how much money Tatiana had in all by adding $1.15 to $5.22.

6. The hamburger cost $1.25, the fries cost $0.70, and the drink cost $0.60. To find the total price of the lunch, add $1.25, $0.70, and $0.60.

7.
```
  63
  47
+ 50
 166
```
8.
```
  632
   57
+ 198
  887
```
9.
```
  78
   9
+ 967
 105.4
```
10.
```
  432
  579
+ 3604
  4615
```

11. 345 − 67

12. 678 − 416

13. 3764 − 96

14. 875 + 1086 + 980

15. 10 + 156 + 8 + 27

16.
```
  $3.47
− $0.92
  2.55
```
17.
```
  $24.15
−  $1.45
  22.70
```
18.
```
  $0.75
+ $0.75
  1.50
```
19.
```
  $0.12
  $0.46
+ $0.50
  1.08
```

20. What is the name for the answer when we add?

21. What is the name for the answer when we subtract?

22. The numbers 5, 6, and 11 are a fact family. Form two addition facts and two subtraction facts with these three numbers.

23. Rearrange the numbers in this addition fact to form another addition fact and two subtraction facts.

$$27 + 16 = 43$$

24. Rearrange the numbers in this subtraction fact to form another subtraction fact and two addition facts.

$$50 − 21 = 29$$

25. Describe a way to check the correctness of a subtraction answer.

LESSON 2

Multiplying Whole Numbers and Money • Dividing Whole Numbers and Money • Fact Families, Part 2

WARM-UP

Facts Practice: 64 Addition Facts (Test A)

Mental Math: Count by 5's from 5 to 100 and from 100 to 0.
Count by 2's from 2 to 20 and from 20 to 2.

a. 500 + 40   **b.** 60 + 200   **c.** 30 + 200 + 40
d. 70 + 300 + 400   **e.** 400 + 50 + 30   **f.** 60 + 20 + 400

Problem Solving:

Robert formed three triangle patterns using 3 coins, 6 coins, and 10 coins. If he continues forming larger triangle patterns, how many coins will he need to form each of the next two triangle patterns?

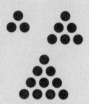

NEW CONCEPTS

Multiplying whole numbers and money

When we add the same number several times, we get a sum. We can get the same result by multiplying.

$$\underbrace{67 + 67 + 67 + 67 + 67}_{\text{Five 67's equal 335.}} = 335$$

$$5 \times 67 = 335$$

Numbers that are multiplied together are called **factors**. The answer is called the **product**.

To indicate multiplication, we can use a times sign or a dot. We also can write factors side by side without a sign. Each of these expressions means that l and w are multiplied:

(1) $l \times w$ (2) $l \cdot w$ (3) lw

Notice that the multiplication dot in form (2) is elevated and is not in the position of a decimal point. Form (3) can be used to show the multiplication of two or more letters or of a number and letters, as we show below.

$$lwh \qquad 4s \qquad 4st$$

Form (3) can also be used to show the multiplication of two or more numbers. To prevent confusion, however, we use parentheses to separate the numbers in the multiplication. Each of the following is a correct use of parentheses to indicate "3 times 5," although the first form is most commonly used. Without the parentheses, we would read each of these simply as the number 35.

$$3(5) \qquad (3)(5) \qquad (3)5$$

When we multiply by a two-digit number on paper, we multiply twice. To multiply 28 by 14, we first multiply 28 by 4. Then we multiply 28 by 10. For each multiplication we write a partial product. We add the partial products to find the final product.

```
    28   factor
  × 14   factor
  ─────
   112   partial product (28 × 4)
   280   partial product (28 × 10)
  ─────
   392   product (14 × 28)
```

When multiplying dollars and cents by a whole number, the answer will have cents places, that is, two places after the decimal point.

```
   $1.35
 ×     6
 ───────
   $8.10
```

Example 1 Find the cost of two dozen pencils at 35¢ each.

Solution Two dozen is two 12's, which is 24. To find the cost of 24 pencils, we multiply 35¢ by 24.

```
    35¢
  × 24
  ─────
   140
   700
  ─────
   840¢
```

The cost of two dozen pencils is 840¢, which is **$8.40.**

The **commutative property** applies to multiplication as well as addition, so changing the order of the factors does not change the product. For example,

$$4 \times 2 = 2 \times 4$$

One way to check multiplication is to reverse the order of factors and multiply again.

$$
\begin{array}{r}
23 \\
\times\ 14 \\
\hline
92 \\
230 \\
\hline
322
\end{array}
\qquad
\begin{array}{r}
14 \\
\times\ 23 \\
\hline
42 \\
280 \\
\hline
322
\end{array}
\ \text{check}
$$

The **identity property of multiplication** states that if one of two factors is 1, the product equals the other factor. The **zero property of multiplication** states that if zero is a factor of a multiplication, the product is zero.

Example 2   Multiply:
$$
\begin{array}{r}
400 \\
\times\ 874 \\
\end{array}
$$

Solution   To simplify the multiplication, we reverse the order of the factors and write trailing zeros so that they "hang out" to the right.

$$
\begin{array}{r}
\overset{2\,1}{874} \\
\times\quad 400 \\
\hline
\mathbf{349{,}600}
\end{array}
$$

Dividing whole numbers and money   When a number is to be separated into a certain number of equal parts, we divide. Each of these expressions means "24 divided by 2":

$$
24 \div 2 \qquad 2\overline{)24} \qquad \frac{24}{2}
$$

The answer to a division problem is the **quotient**. The number that is divided is the **dividend**. The number by which the dividend is divided is the **divisor**. We can indicate division with a division symbol ($\div$), a division box ($\overline{)\ \ }$), or a division bar ($-$), as shown below.

dividend ÷ divisor = quotient
$\mathrm{divisor}\overline{)\mathrm{dividend}}^{\ \ \text{quotient}}$
$\dfrac{\text{dividend}}{\text{divisor}} = \text{quotient}$

When the dividend is zero, the quotient is zero. The divisor may not be zero. When the dividend and divisor are equal (and not zero), the quotient is 1.

Example 3   Divide: 3456 ÷ 7

Solution   We show both the long-division and short-division methods.

LONG DIVISION

$$
\begin{array}{r}
493 \text{ R } 5 \\
7\,\overline{)3456} \\
\underline{28} \\
65 \\
\underline{63} \\
26 \\
\underline{21} \\
5
\end{array}
$$

SHORT DIVISION

$$
\begin{array}{r}
4\;9\;3 \text{ R } 5 \\
7\,\overline{)3\;4^6 5^2 6}
\end{array}
$$

Using the short-division method, we perform the multiplication and subtraction steps mentally, recording only the result of each subtraction.

To check our work, we multiply the quotient by the divisor. Then we add the remainder to this answer. The result should be the dividend. For this example we multiply 493 by 7. Then we add 5.

$$
\begin{array}{r}
^{6\,2} \\
493 \\
\times \quad 7 \\
\hline
3451 \\
+ \quad 5 \\
\hline
3456
\end{array}
$$

When dividing dollars and cents, cents will be included in the answer. Notice that the decimal point in the quotient is directly above the decimal point in the division box, separating the dollars from the cents.

$$
\begin{array}{r}
\$1.60 \\
3\,\overline{)\$4.80} \\
\underline{3} \\
1\,8 \\
\underline{1\,8} \\
00 \\
\underline{0} \\
0
\end{array}
$$

Fact families, part 2   Multiplication and division are inverse operations, so there are multiplication and division fact families just as there are addition and subtraction fact families. The numbers 5, 6, and 30 are a fact family. We can form two multiplication facts and two division facts with these numbers.

$$5 \times 6 = 30 \qquad 30 \div 5 = 6$$

$$6 \times 5 = 30 \qquad 30 \div 6 = 5$$

Example 4   Rearrange the numbers in this multiplication fact to form another multiplication fact and two division facts.

$$5 \times 12 = 60$$

Solution   By reversing the factors, we form another multiplication fact.

$$12 \times 5 = 60$$

By making 60 the dividend, we can form two division facts.

$$60 \div 5 = 12$$

$$60 \div 12 = 5$$

LESSON PRACTICE

Practice set*†   **a.** $20 \times 37¢$ **b.** $37 \cdot 0$ **c.** $407(37)$

d. $5\overline{)\$8.40}$ **e.** $200 \div 12$ **f.** $\dfrac{234}{3}$

g. Which numbers are the divisors in problems **d, e,** and **f**?

h. Use the numbers 8, 9, and 72 to form two multiplication facts and two division facts.

MIXED PRACTICE

Problem set   **††1.** If the factors are 7 and 11, what is the product?

(2)

2. What is the difference of 97 and 79?

(1)

3. If the addends are 170 and 130, what is the sum?

(1)

4. If 36 is the dividend and 4 is the divisor, what is the quotient?

(2)

5. Find the sum of 386, 98, and 1734.

(1)

6. Fatima spent $2.25 for a hamburger. She paid for it with a five-dollar bill. Find how much change she should get back by subtracting $2.25 from $5.

(1)

---

†The asterisk after "Practice set" indicates that additional practice problems intended for remediation are available in the appendix.

††The italicized numbers within parentheses underneath each problem number are called *lesson reference numbers.* These numbers refer to the lesson(s) in which the major concept of that particular problem is introduced. If additional assistance is needed, refer to the discussion, examples, or practice problems of that lesson.

7. Luke wants to buy a $70.00 radio for his car. He has
(1) $47.50. Find how much more money he needs by
subtracting $47.50 from $70.00.

8. Each energy bar costs 75¢. Find the cost of one dozen
(2) energy bars by multiplying 75¢ by 12.

9. (1)	**10.** (1)	**11.** (1)	**12.** (1)
312 − 86	4106 + 1398	4000 − 1357	$10.00 − $2.83

13. 405(8)   **14.** 25 ✗ 25   **15.** $\dfrac{288}{6}$ **16.** $\dfrac{225}{15}$
(2)   (2)   (2)   (2)

17. $1.25 × 8   **18.** 400 × 50
(2)   (2)

19. 1000 ÷ 8   **20.** $45.00 ÷ 20
(2)   (2)

21. Use the numbers 6, 8, and 48 to form two multiplication
(2) facts and two division facts.

22. Rearrange the numbers in this division fact to form
(2) another division fact and two multiplication facts.

$$4\overline{)36}^{\,9}$$

23. Rearrange the numbers in this addition fact to form
(1) another addition fact and two subtraction facts.

$$12 + 24 = 36$$

24. (a) Find the sum of 9 and 6.
(1)
   (b) Find the difference between 9 and 6.

25. The divisor, dividend, and quotient are in these positions
(2) when we use a division sign:

Dividend ÷ divisor = quotient

On your paper, draw a division box and show the
positions of the divisor, dividend, and quotient.

26. Multiply to find the answer to this addition problem:
(2)
   39¢ + 39¢ + 39¢ + 39¢ + 39¢ + 39¢

27. 365 × 0   **28.** 0 ÷ 50   **29.** 365 ÷ 365
(2)   (2)   (2)

30. Describe a way to check the correctness of a division
(2) answer that has no remainder.

3

Missing Numbers in Addition •
Missing Numbers in Subtraction

WARM-UP

Facts Practice: 100 Addition Facts (Test B)

Mental Math: Count by 5's from 5 to 100 and from 100 to 0.
           Count by 50's from 50 to 1000 and from 1000 to 0.

a. 3000 + 4000   **b.** 600 + 2000   **c.** 20 + 3000
d. 600 + 300 + 20   **e.** 4000 + 300 + 200 **f.** 70 + 300 + 4000

Problem Solving:

Shoes in a typical shoe box cannot "get out," because they are closed in by a number of flat surfaces. How many surfaces enclose a pair of shoes in a closed shoe box?

NEW CONCEPTS

Missing numbers in addition   Below is an addition fact with three numbers. If one of the addends were missing, we could use the other addend and the sum to find the missing number.

$$\begin{array}{r} 4 \\ + 3 \\ \hline 7 \end{array}$$

Cover the 4 with your finger. How can you use the 7 and the 3 to find that the number under your finger is 4?

Now cover the 3 instead of the 4. How can you use the other two numbers to find that the number under your finger is 3?

Notice that we can find a missing addend by subtracting the known addend from the sum. We will use a letter to stand for a missing number. The letter may be lowercase or uppercase.

Example 1   Find the number for m:

$$\begin{array}{r} 12 \\ + m \\ \hline 31 \end{array}$$

Solution   One of the addends is missing. The known addend is 12. The sum is 31. If we subtract 12 from 31, we find that the missing addend is **19**. We check our answer by using 19 in place of *m* in the original problem.

$$\begin{array}{r} \overset{2}{\cancel{3}}\overset{1}{1} \\ -\ 1\ 2 \\ \hline 1\ 9 \end{array} \quad \begin{array}{l}\text{Use 19 in}\\ \text{place of } m.\end{array} \longrightarrow \quad \begin{array}{r} \overset{1}{1}2 \\ +\ 19 \\ \hline 31 \end{array} \quad \text{check}$$

Example 2   Find the number for *n*:

$$36 + 17 + 5 + n = 64$$

Solution   First we add all the known addends.

$$\underbrace{36 + 17 + 5}_{58} + n = 64$$
$$58 \qquad + n = 64$$

Then we find *n* by subtracting 58 from 64.

$$64 - 58 = 6 \quad \text{So } n \text{ is } \mathbf{6.}$$

We check our work by using 6 in place of *n* in the original problem.

$$36 + 17 + 5 + 6 = 64 \quad \text{The answer checks.}$$

Missing numbers in subtraction   Below is a subtraction fact. Cover the 8 with your finger. How can you use the other two numbers to find that the number under your finger is 8?

$$\begin{array}{r} 8 \\ -\ 3 \\ \hline 5 \end{array}$$

Now cover the 3 instead of the 8. How can you use the other two numbers to find that the covered number is 3?

As we will show below, we can find a missing minuend (first number in a subtraction problem) by adding the other two numbers. We can find a missing subtrahend (second number in a subtraction problem) by subtracting the difference from the minuend.

Example 3   Find the number for *W*:

$$\begin{array}{r} W \\ -\ 16 \\ \hline 24 \end{array}$$

Solution   We can find the first number of a subtraction problem by adding the other two numbers. We add 16 and 24 to get **40.** We check our answer by using 40 in place of *W*.

$$\begin{array}{r} \overset{1}{1}6 \\ +\ 24 \\ \hline 40 \end{array} \longrightarrow \begin{array}{r} \text{Use 40 in} \\ \text{place of } W. \\ \overset{3}{\cancel{4}}\overset{1}{0} \\ -\ 1\ 6 \\ \hline 2\ 4 \quad \text{check} \end{array}$$

Example 4   Find the number for *y*:

$$236 - y = 152$$

Solution   One way to determine how to find a missing number is to think of a simpler problem that is similar. Here is a simpler subtraction fact:

$$5 - 3 = 2$$

In the problem, *y* is in the same position as the 3 in the simpler subtraction fact. Just as we can find 3 by subtracting 2 from 5, we can find *y* by subtracting 152 from 236.

$$\begin{array}{r} \overset{1}{2}\overset{1}{3}6 \\ -\ 1\ 5\ 2 \\ \hline 8\ 4 \end{array}$$

We find that *y* is **84.** Now we check our answer by using 84 in place of *y* in the original problem.

$$\begin{array}{r} \overset{1}{2}\overset{1}{3}6 \\ -\quad 8\ 4 \quad \leftarrow \text{Use 84 in place of } y. \\ \hline 1\ 5\ 2 \quad \leftarrow \text{The answer checks.} \end{array}$$

LESSON PRACTICE

Practice set*   Find the missing number in each problem. Check your work by using your answer in place of the letter in the original problem.

a.	b.	c.	d.
$\begin{array}{r} A \\ +\ 12 \\ \hline 45 \end{array}$	$\begin{array}{r} 32 \\ +\ B \\ \hline 60 \end{array}$	$\begin{array}{r} C \\ -\ 15 \\ \hline 24 \end{array}$	$\begin{array}{r} 38 \\ -\ D \\ \hline 29 \end{array}$

e. $e + 24 = 52$　　　　　　f. $29 + f = 70$

g. $g - 67 = 43$　　　　　　h. $80 - h = 36$

i. $36 + 14 + n + 8 = 75$

58

MIXED PRACTICE

Problem set

1. If the two factors are 25 and 12, what is the product?
(2)

2. If the addends are 25 and 12, what is the sum?
(1)

3. What is the difference of 25 and 12?
(1)

4. Each of the 31 families took 75 aluminum cans to the
(2) recycling bin. Find how many cans the families collected by multiplying 31 by 75.

5. Find the total price of one dozen pepperoni pizzas at
(2) $7.85 each by multiplying $7.85 by 12.

6. The basketball team scored 63 of its 102 points in the
(1) first half of the game. Find how many points the team scored in the second half by subtracting 63 points from 102 points.

7.　　$3.68　　**8.**　　407　　**9.**　　28¢　　**10.**　　370
(2)　　$\times$　　9　　(2)　$\times$　80　　(2)　$\times$　14　　(2)　$\times$ 140

11. $100 \cdot 100$　　　**12.** $144 \div 12$　　**13.** $(12)(5)$
(2)　　　　　　　　(2)　　　　　　　　(2)

14.　　3627　　**15.**　　5010　　**16.**　　$10.00
(1)　　　598　　(1)　$-$ 1376　　(1)　$-$　$0.26
　　　$+$ 4881

Find the missing number in each problem:

17.　　A　　**18.**　　23　　**19.**　　C　　**20.**　　42
(3)　$+$ 16　　(3)　$+$　B　　(3)　$-$ 17　　(3)　$-$　D
　　　48　　　　　52　　　　　31　　　　　25

21. $x + 38 = 75$　　　　**22.** $x - 38 = 75$
(3)　　　　　　　　　　(3)

23. $75 - y = 38$　　　　**24.** $6 + 8 + w + 5 = 32$
(3)　　　　　　　　　　(3)

25. Rearrange the numbers in this addition fact to form
(1) another addition fact and two subtraction facts.

$$24 + 48 = 72$$

26. Rearrange the numbers in this multiplication fact to form
(2) another multiplication fact and two division facts.

$$6 \times 15 = 90$$

27. Find the quotient when the divisor is 20 and the
(2) dividend is 200.

28. Multiply to find the answer to this addition problem:
(2)

$$15 + 15 + 15 + 15 + 15 + 15 + 15 + 15$$

29. $144 \div 144$
(2)

30. Describe how to find a missing addend in an addition
(3) problem.

L E S S O N

4

Missing Numbers in Multiplication • Missing Numbers in Division

WARM-UP

Facts Practice: 64 Addition Facts (Test A)

Mental Math: Count up and down by 5's between 5 and 100.
Count up and down by 50's between 50 and 1000.

a. 600 + 2000 + 300 + 20   **b.** 3000 + 20 + 400 + 5000
c. 7000 + 200 + 40 + 500   **d.** 700 + 2000 + 50 + 100
e. 60 + 400 + 30 + 1000   **f.** 900 + 8000 + 100 + 50

Problem Solving:

The digits 1, 3, and 5 can be arranged to make six different three-digit numbers. Two of the six numbers are 135 and 153. What are the other four numbers?

NEW CONCEPTS

Missing numbers in multiplication

This multiplication fact has three numbers. If one of the factors were missing, we could use the other factor and the product to figure out the missing factor.

$$\begin{array}{r} 4 \\ \times\ 3 \\ \hline 12 \end{array}$$

With your finger, cover the factors in this multiplication fact one at a time. How can you use the two uncovered numbers to find the covered number? Notice that we can find a missing factor by dividing the product by the known factor.

Example 1   Find the missing number:
$$\begin{array}{r} A \\ \times\ 6 \\ \hline 72 \end{array}$$

Solution   The missing number is a factor. The product is 72. The factor that we know is 6. Dividing 72 by 6, we find that the missing factor is **12.** We check our work by using 12 in the original problem.

$$\begin{array}{r} 1\,2 \\ 6\overline{)7^12} \end{array} \longrightarrow \begin{array}{r} {}^{1}\ \\ 12 \\ \times\ 6 \\ \hline 72 \end{array} \text{ check}$$

Example 2 Find the missing number: $6w = 84$

Solution When a number and a letter are written side by side, it means that they are to be multiplied. In this problem $6w$ means "6 times w." We divide 84 by 6 and find that the missing factor is **14**. We check our work by multiplying.

$$
\begin{array}{r}
1\,4 \\
6\overline{)8\,^24}
\end{array}
\longrightarrow
\begin{array}{r}
\overset{2}{1}4 \\
\times\ \ 6 \\
\hline
84 \quad \text{check}
\end{array}
$$

Missing numbers in division This division fact has three numbers. If one of the numbers were missing, we could figure out the third number.

$$
6\overline{)24}^{\,4}
$$

Cover each of the numbers with your finger. How can you use the other two numbers to find the covered number? Notice that we can find the dividend (the number inside the division box) by multiplying the other two numbers. We can find either the divisor or quotient (the numbers outside of the box) by dividing.

Example 3 Find the missing number:

$$
\frac{k}{6} = 15
$$

Solution The letter k is in the position of the dividend. If we rewrite this problem with a division box, it looks like this:

$$
6\overline{)k}^{\,15}
$$

We find a missing dividend by multiplying the divisor and quotient. We multiply 15 by 6 and find that the missing number is **90**. Then we check our work.

$$
\begin{array}{r}
\overset{3}{1}5 \\
\times\ \ 6 \\
\hline
90
\end{array}
\qquad
\begin{array}{r}
1\,5 \\
6\overline{)9\,^30}
\end{array} \quad \text{check}
$$

Example 4 Find the missing number: $126 \div m = 7$

Solution The letter m is in the position of the divisor. If we were to rewrite the problem with a division box, it would look like this:

$$m\overline{)126}^{\,7}$$

We can find m by dividing 126 by 7.

$$7\overline{)1\,2^{5}6}^{\,1\;8}$$

We find that m is **18**. We can check our division by multiplying as follows:

$$\begin{array}{r} \overset{5}{18} \\ \times\ \ 7 \\ \hline 126 \end{array}$$

In the original equation we can replace the letter with our answer and test the truth of the resulting equation.

$$126 \div 18 = 7$$
$$7 = 7$$

LESSON PRACTICE

Practice set* Find each missing number. Check your work by using your answer in place of the letter in the original problem.

a.
$$\begin{array}{r} A \\ \times\ 7 \\ \hline 91 \end{array}$$

b.
$$\begin{array}{r} 20 \\ \times\ B \\ \hline 440 \end{array}$$

c. $7\overline{)C}^{\,15}$

d. $D\overline{)144}^{\,8}$

e. $7w = 84$

f. $112 = 8m$

g. $\dfrac{360}{x} = 30$

h. $\dfrac{n}{5} = 60$

MIXED PRACTICE

Problem set

1. Five dozen carrot sticks are to be divided evenly among
(2) 15 children. Find how many carrot sticks each child should receive by dividing 60 by 15.

2. Matt separated 100 pennies into 4 equal piles. Find how
(2) many pennies were in each pile by dividing 100 by 4.

3. Sandra put 100 pennies into stacks of 5 pennies each.
(2) Find how many stacks she formed by dividing 100 by 5.

4. For the upcoming season, 294 players signed up for
(2) soccer. Find the number of 14-player soccer teams that
can be formed by dividing 294 by 14.

5. Angela is reading a 280-page book. She has just finished
(1) page 156. Find how many pages she still has to read by
subtracting 156 from 280.

6. Each month Bill earns $0.75 per customer for delivering
(2) newspapers. Find how much money he would earn in a
month in which he had 42 customers by multiplying
$0.75 by 42.

Find each missing number. Check your work.

7.
(4)
$$\begin{array}{r} J \\ \times\ 5 \\ \hline 60 \end{array}$$

8.
(3)
$$\begin{array}{r} 27 \\ +\ K \\ \hline 72 \end{array}$$

9.
(3)
$$\begin{array}{r} L \\ +\ 36 \\ \hline 37 \end{array}$$

10.
(3)
$$\begin{array}{r} 64 \\ -\ M \\ \hline 46 \end{array}$$

11. $n - 48 = 84$
(3)

12. $7p = 91$
(4)

13. $q \div 7 = 0$
(4)

14. $144 \div r = 6$
(4)

15. $6\overline{)\$12.36}$
(2)

16. $\dfrac{5760}{8}$
(2)

17. $526 \div 18$
(2)

18. $563 + 563 + 563 + 563$
(1)

19. $\$3.75 \cdot 16$
(2)

20. $\$3 + \$2.86 + \$0.98$
(1)

21. $\$10 - \$6.43$
(1)

22. If the divisor is 3 and the quotient is 12, what is the
(4) dividend?

23. If the product is 100 and one factor is 5, what is the other
(4) factor?

24. Rearrange the numbers in this subtraction fact to form
(1) another subtraction fact and two addition facts.

$$17 - 9 = 8$$

25. Rearrange the numbers in this division fact to form
(2) another division fact and two multiplication facts.

$$72 \div 8 = 9$$

26. $w + 6 + 8 + 10 = 40$
(3)

27. Find the answer to this addition problem by multiplying:
(2)

$$23¢ + 23¢ + 23¢ + 23¢ + 23¢ + 23¢ + 23¢$$

28. $25m = 25$ **29.** $15n = 0$
(4)   (4)

30. Describe how to find a missing factor in a multiplication
(4) problem.

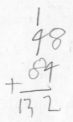
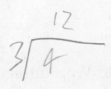

LESSON

5   Order of Operations, Part 1

WARM-UP

Facts Practice: 100 Addition Facts (Test B)

Mental Math: Count by 25's from 25 to 1000.

    a. 560 + 200   **b.** 840 + 30   **c.** 5200 + 2000
    d. 650 + 140   **e.** 3800 + 2000   **f.** 440 + 200

Problem Solving:

Copy this addition problem and fill in the five missing digits:

$$\begin{array}{r} 1\_5 \\ +\ \_3\_ \\ \hline \_\_93 \end{array}$$

NEW CONCEPT

When there is more than one addition or subtraction step within a problem, we take the steps in order from left to right. In this problem we first subtract 4 from 9. Then we add 3.

$$9 - 4 + 3 = 8$$

If a different order of steps is desired, parentheses are used to show which step should be taken first. In the problem below, we first add 4 and 3 to get 7. Then we subtract 7 from 9.

$$9 - (4 + 3) = 2$$

Example 1 (a) $18 - 6 - 3$ (b) $18 - (6 - 3)$

Solution (a) We subtract in order from left to right.

         $\underline{18 - 6} - 3$ First subtract 6 from 18.
            $12\ \ \ - 3$ Then subtract 3 from 12.
                9 The answer is 9.

  (b) We subtract within the parentheses first.

         $18 - \underline{(6 - 3)}$ First subtract 3 from 6.
         $18 - \ \ \ 3$ Then subtract 3 from 18.
            15 The answer is 15.

When there is more than one multiplication or division step within a problem, we take the steps in order from left to right. In this problem we divide 24 by 6 and then multiply by 2.

$$24 \div 6 \times 2 = 8$$

If there are parentheses, then we first do the work within the parentheses. In the problem below, we first multiply 6 by 2 and get 12. Then we divide 24 by 12.

$$24 \div (6 \times 2) = 2$$

Example 2   (a) $18 \div 6 \div 3$ (b) $18 \div (6 \div 3)$

Solution   (a) We take the steps in order from left to right.

$$\underbrace{18 \div 6} \div 3 \quad \text{First divide 18 by 6.}$$
$$3 \quad \div 3 \quad \text{Then divide 3 by 3.}$$
$$1 \quad \text{The answer is 1.}$$

(b) We divide within the parentheses first.

$$18 \div \underbrace{(6 \div 3)} \quad \text{First divide 6 by 3.}$$
$$18 \div \quad 2 \quad\;\; \text{Then divide 18 by 2.}$$
$$9 \quad \text{The answer is 9.}$$

Only two numbers are involved in each step of a calculation. If three numbers are added (or multiplied), changing the two numbers selected for the first addition (or first multiplication) does not change the final sum (or product).

$$(2 + 3) + 4 = 2 + (3 + 4) \qquad (2 \times 3) \times 4 = 2 \times (3 \times 4)$$

This property applies to addition and multiplication and is called the **associative property.** As shown by examples 1 and 2, the associative property does not apply to subtraction or to division.

Example 3 $\dfrac{5 + 7}{1 + 2}$

Solution   Before dividing we perform the operations above the bar and below the bar. Then we divide 12 by 3.

$$\frac{5 + 7}{1 + 2} = \frac{12}{3} = 4$$

LESSON PRACTICE

Practice set **a.** $16 - 3 + 4$ **b.** $16 - (3 + 4)$

c. $24 \div (4 \times 3)$ **d.** $24 \div 4 \times 3$

e. $24 \div 6 \div 2$ **f.** $24 \div (6 \div 2)$

g. $\dfrac{6 + 9}{3}$ **h.** $\dfrac{12 + 8}{12 - 8}$

MIXED PRACTICE

Problem set **1.** Jack paid $5 for a hamburger that cost $1.25 and a drink
(1) that cost $0.60. How much change should he get back?

2. In one day the elephant ate 82 pounds of straw, 8 pounds
(1) of apples, and 12 pounds of peanuts. How many pounds
of food did it eat in all?

3. What is the difference of 110 and 25?
(1)

4. What is the total price of one dozen apples that cost
(2) 25¢ each?

5. What number must be added to 149 to total 516?
(3)

6. Judy plans to read a 235-page book in 5 days. Describe
(2) how to find the average number of pages she needs to
read each day.

7. $5 + (3 \times 4)$ **8.** $(5 + 3) \times 4$
(5)   (5)

9. $800 - (450 - 125)$ **10.** $600 \div (20 \div 5)$
(5)   (5)

11. $800 - 450 - 125$ **12.** $600 \div 20 \div 5$
(5)   (5)

13. $144 \div (8 \times 6)$ **14.** $144 \div 8 \times 6$
(5)   (5)

15. $\$5 - (\$1.25 + \$0.60)$
(5)

16. Use the numbers 63, 7, and 9 to form two multiplication
(2) facts and two division facts.

17. If the quotient is 12 and the dividend is 288, what is the
(4) divisor?

18. 25)$10.00
(2)

19. (378)(64)
(2)

20.   506
(2) × 370

21. $10.10
(1) − $9.89

Find each missing number. Check your work.

22. $n - 63 = 36$
(3)

23. $63 - p = 36$
(3)

24. $56 + m = 432$
(3)

25. $8w = 480$
(4)

26. $5 + 12 + 27 + y = 50$
(3)

27. $36 \div a = 4$
(4)

28. $x \div 4 = 8$
(4)

29. Use the numbers 7, 11, and 18 to form two addition facts
(1) and two subtraction facts.

30. $3 \cdot 4 \cdot 5$
(5)

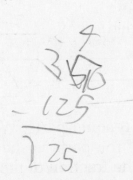

LESSON

6   Fractional Parts

WARM-UP

Facts Practice: 100 Subtraction Facts (Test C)

Mental Math: Count up and down by 25's between 25 and 1000.

a. 2500 + 400   b. 6000 + 2400   c. 370 + 400

d. 9500 + 240   e. 360 + 1200   f. 480 + 2500

Problem Solving:

In his pocket Alex had seven coins totaling exactly one dollar. Name a possible combination of coins in his pocket. How many different combinations of coins are possible?

NEW CONCEPT

As young children we learned to count objects using whole numbers. As we grew older, we discovered that there are parts of wholes—like parts of a candy bar—that cannot be named with whole numbers. We can name these parts with **fractions**. A common fraction is written with two numbers and a fraction bar. The "bottom" number is the **denominator**. The denominator shows the number of equal parts in the whole. The "top" number, the **numerator**, shows the number of the parts that are being represented.

We see that this whole circle has been divided into 4 equal parts; 1 part is shaded. The fraction of the circle that is shaded is 1 out of 4 parts. We call this part "one fourth" and write it as $\frac{1}{4}$.

Example 1   What fraction of this circle is shaded?

Solution   The circle has been divided into 6 equal parts. We use 6 for the bottom of the fraction. One of the parts is shaded, so we use 1 for the top of the fraction. The fraction of the circle that is shaded is one sixth, which we write as $\frac{1}{6}$.

We can also use a fraction to name a part of a group. There are 6 members in this group. We can divide this group in half by dividing it into two equal groups with 3 in each half. We write that $\frac{1}{2}$ of 6 is 3.

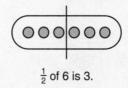

$\frac{1}{2}$ of 6 is 3.

We can divide this group into thirds by dividing the 6 members into three equal groups. We write that $\frac{1}{3}$ of 6 is 2.

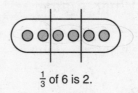

$\frac{1}{3}$ of 6 is 2.

Example 2   (a) What number is $\frac{1}{2}$ of 450?

(b) What number is $\frac{1}{3}$ of 450?

(c) How much money is $\frac{1}{5}$ of $4.50?

Solution   (a) To find $\frac{1}{2}$ of 450, we divide 450 into two equal parts and find the amount in one of the parts. We find that $\frac{1}{2}$ of 450 is **225.**

$$2\overline{)450}^{\,225} \quad \longrightarrow \quad \tfrac{1}{2} \text{ of } 450 \text{ is } 225.$$

(b) To find $\frac{1}{3}$ of 450, we divide 450 into three equal parts. Since each part is 150, we find that $\frac{1}{3}$ of 450 is **150.**

$$3\overline{)450}^{\,150} \quad \longrightarrow \quad \tfrac{1}{3} \text{ of } 450 \text{ is } 150.$$

(c) To find $\frac{1}{5}$ of $4.50, we divide $4.50 by 5. We find that $\frac{1}{5}$ of $4.50 is **$0.90.**

$$5\overline{)\$4.50}^{\,\$0.90} \quad \longrightarrow \quad \tfrac{1}{5} \text{ of } \$4.50 \text{ is } \$0.90.$$

Example 3   Copy the figure at right, and shade $\frac{1}{3}$ of it:

Solution   The rectangle has six parts of equal size. Since $\frac{1}{3}$ of 6 is 2, we shade any two of the parts.

LESSON PRACTICE

Practice set Use both words and digits to write the fraction that is shaded in problems **a–c**.

a.   **b.**  **c.**

d. What number is $\frac{1}{2}$ of 72?

e. What number is $\frac{1}{2}$ of 1000?

f. What number is $\frac{1}{3}$ of 180?

g. How much money is $\frac{1}{3}$ of $3.60?

h. Copy this figure and shade one half of it.

MIXED PRACTICE

Problem set **1.** What number is $\frac{1}{2}$ of 540?
   (6)

2. What number is $\frac{1}{3}$ of 540?
   (6)

3. In four days of sight-seeing the Richmonds drove 346 miles,
   (1) 417 miles, 289 miles, and 360 miles. How many miles did they drive in all?

4. Willis paid $20 for a book that cost $12.08. How much
   (1) money should he get back?

5. How many days are in 52 weeks?
   (2)

6. How many $20 bills would it take to make $1000?
   (2)

7. Use words and digits to write the
   (6) fraction of this circle that is shaded.

8.
(1)
$$3604$$
$$5186$$
$$+ 7145$$

9.
(1)
$$\$30.01$$
$$- \$15.76$$

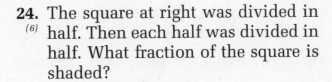

10.
(2)
$$376$$
$$\times \ 87$$

11.
(2)
$$470$$
$$\times \ 203$$

12. $20 - \$11.98
(1)

13. $596 - (400 - 129)$
(5)

14. $32 \div (8 \times 4)$
(5)

15. $8\overline{)4016}$
(2)

16. $15\overline{)6009}$
(2)

17. $36\overline{)9000}$
(2)

Find each missing number. Check your work.

18. $8w = 480$
(4)

19. $x - 64 = 46$
(3)

20. $\dfrac{49}{N} = 7$
(4)

21. $\dfrac{M}{7} = 15$
(4)

22. $365 + P = 653$
(3)

23. $36¢ + 25¢ + m = 99¢$
(3)

24. The square at right was divided in
(6) half. Then each half was divided in
half. What fraction of the square is
shaded?

25. Copy this figure on your paper, and
(6) shade one fourth of it.

26. 6.35×12
(2)

27. Use the numbers 2, 4, and 6 to form two addition facts
(1) and two subtraction facts.

28. Write two multiplication facts and two division facts
(2) using the numbers 2, 4, and 8.

29. Multiply to find the answer to this addition problem:
(2)

$$38 + 38 + 38 + 38 + 38 + 38 + 38 + 38 + 38 + 38$$

30. Make up a fractional-part question about money, as in
(6) example 2(c). Then find the answer.

LESSON

7

Lines, Segments, and Rays • Linear Measure

WARM-UP

Facts Practice: 100 Subtraction Facts (Test C)

Mental Math: Count up and down by $\frac{1}{2}$'s between $\frac{1}{2}$ and 10.
Count up and down by 2's between 2 and 40.

a. $800 - 300$ b. $3000 - 2000$ c. $450 - 100$
d. $2500 - 300$ e. $480 - 80$ f. $750 - 250$

Problem Solving:

Sharon made three square patterns using 4 coins, 9 coins, and 16 coins. If she continues forming larger square patterns, how many coins will she need for each of the next two square patterns?

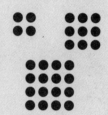

NEW CONCEPTS

Lines, segments, and rays

In everyday language the following figure is often referred to as a line:

_____

However, using mathematical terminology, we say that the figure represents a **segment**, or line segment. A segment is part of a line and has two **endpoints**. A mathematical **line** has no endpoints. To represent a line, we use arrowheads to indicate a line's unending quality.

◄————————————►

A **ray** has one endpoint. We represent a ray with one arrowhead.

•————————————►

A ray is roughly represented by a beam of sunlight. The beam begins at the sun (which represents the endpoint of the ray) and continues across billions of light years of space.

Linear measure Line segments have length. In the United States we have two systems of units that we use to measure length. One system is the **U.S. Customary System**. Some of the units in this system are inches (in.), feet (ft), yards (yd), and miles (mi). The other system is the **metric system**. Some of the units in the metric system are millimeters (mm), centimeters (cm), meters (m), and kilometers (km). In this lesson we will practice measuring line segments with an inch ruler and with a centimeter ruler.

Activity: *Inch Ruler*

Materials needed:

- inch ruler
- narrow strip of tagboard about 6 inches long and 1 inch wide
- pencil

Use your pencil and ruler to draw inch marks on the strip of tagboard. Number the inch marks. When you are finished, the tagboard strip should look like this:

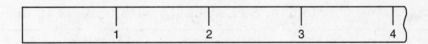

Now set aside your ruler. We will use estimation to make the rest of the marks on the tagboard strip. Estimate the halfway point between inch marks, and make the half-inch marks slightly shorter than the inch marks, as shown below.

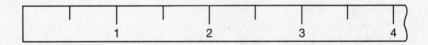

Now show every quarter inch on your tagboard ruler. To do this, estimate the halfway point between each mark on the ruler, and make the quarter-inch marks slightly shorter than the half-inch marks, as shown below.

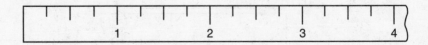

Save your tagboard ruler. We will be making more marks on it in Lesson 17.

A metric ruler is divided into centimeters. There are 100 centimeters in a meter. Each centimeter is divided into 10 millimeters. So 1 centimeter equals 10 millimeters, and 2 centimeters equals 20 millimeters.

By comparing an inch ruler with a centimeter ruler, we see that an inch is about $2\frac{1}{2}$ centimeters.

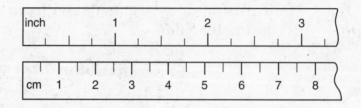

A stick of gum 3 inches long is about $7\frac{1}{2}$ cm long. A foot-long ruler is about 30 cm long.

Example 1   How long is the line segment?

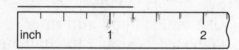

Solution   The line is one whole inch plus a fraction. The fraction is one fourth. So the length of the line is **$1\frac{1}{4}$ in.**

> *Note:* In this book the abbreviation for "inches" ends with a period. The abbreviations for other units do not end with periods.

Example 2   How long is the line segment?

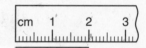

Solution   We simply read the scale to see that the line is **2 cm** long. The segment is also **20 mm** long.

LESSON PRACTICE

Practice set   How long is each line segment?

a.

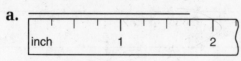

b.

c. Measure the following segment twice, once with an inch ruler and once with a centimeter ruler.

_____

Use the words *line, segment,* or *ray* to describe each of these figures:

d. •—————————————→

e. ←————————————→

f. •————————————•

MIXED PRACTICE

Problem set

1. To earn money for gifts, Debbie sold decorated pinecones.
(2) If she sold 100 pinecones at $0.25 each, how much money did she earn?

2. There are 365 days in a common year. April 1 is the 91st
(1) day. How many days are left in the year after April 1?

3. The Smiths are planning to complete a 1890-mile trip in 3
(5) days. If they drive 596 miles the first day and 612 miles the second day, how far must they travel the third day? (*Hint:* This is a two-step problem. First find how far the Smiths traveled the first two days.)

4. What number is $\frac{1}{2}$ of 234?
(6)

5. How much money is $\frac{1}{3}$ of $2.34?
(6)

6. Use words and digits to write the
(6) fraction of this circle that is shaded.

7. 3654	**8.** $41.01	**9.** 28¢	**10.** 906
(1) 2893	*(1)* − $15.76	*(2)* × 74	*(2)* × 47
+ 5614			

11. 6)5000
(2)

12. 800 ÷ 16
(2)

13. 60)3174
(2)

14. 3 + 6 + 5 + w + 4 = 30
(3)

15. 300 − 30 + 3
(5)

16. 300 − (30 + 3)
(5)

17. $4.32 · 20
(2)

18. 24(48¢)
(2)

19. $8.75 ÷ 25
(2)

Find each missing number. Check your work.

20. $W ÷ 6 = 7$
(4)

21. $6n = 96$
(4)

22. $58 + r = 213$
(3)

23. Rearrange the numbers in this subtraction fact to form
(1) another subtraction fact and two addition facts.

$$60 − 24 = 36$$

24. How long is the line segment below?
(7)

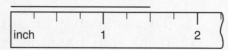

25. Find the length, in centimeters and in millimeters, of the
(7) line segment below.

26. Use the numbers 9, 10, and 90 to form two multiplication
(2) facts and two division facts.

27. Describe how to find a missing dividend in a division
(4) problem.

28. $w − 12 = 8$
(3)

29. $12 − x = 8$
(3)

30. A meterstick is 100 centimeters long. One hundred
(7) centimeters is how many millimeters?

LESSON

8 Perimeter

WARM-UP

Facts Practice: 64 Addition Facts (Test A)

Mental Math: Count up by $\frac{1}{4}$'s from $\frac{1}{4}$ to 10.

a. 400 + 2400   b. 750 + 36   c. 8400 + 520

d. 980 − 60   e. 4400 − 2000   f. 480 − 120

Problem Solving:

As you sit in your chair facing forward, you can describe the locations of people and objects in the room compared to your position. Perhaps another chair is four feet in front of you and two feet to the left. Perhaps a door is about 6 feet directly to your right. Describe the location of a doorway, a window, and another object (or person) of your choice.

NEW CONCEPT

The distance around a shape is its **perimeter**. The perimeter of a square is the distance around it. The perimeter of a room is the distance around the room.

Activity: *Perimeter*

Walk the perimeter of a room in your home. Start at a point along a wall of the room, and, staying close to the walls, walk around the room until you return to your starting point. Count your steps as you travel around the room. How many of your steps is the perimeter of the room? When you are finished, have another person walk the perimeter, and then answer these questions:

a. Does the other person count the same number of steps?

b. Does the perimeter depend upon who is measuring it?

c. Which of these is the better physical example of perimeter?

1. The tile or carpet that covers the floor.

2. The molding along the base of the wall.

Here we show a rectangle that is 3 cm long and 2 cm wide.

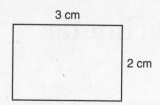

If we were to start at one corner and trace the perimeter of the rectangle, our pencil would travel 3 cm, then 2 cm, then 3 cm, and then 2 cm to get all the way around. We add these lengths to find the perimeter of the rectangle.

$$3 \text{ cm} + 2 \text{ cm} + 3 \text{ cm} + 2 \text{ cm} = \textbf{10 cm}$$

Example 1 What is the perimeter of this triangle?

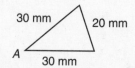

Solution The perimeter of a shape is the distance around the shape. If we trace around the triangle from point A, the point of the pencil would travel 30 mm, then 20 mm, and then 30 mm. Adding these distances, we find that the perimeter is **80 mm.**

Example 2 The perimeter of a square is 20 cm. What is the length of each side?

Solution The four sides of a square are equal in length. So we divide the perimeter by four to find the length of each side. We find that the length of each side is **5 cm.**

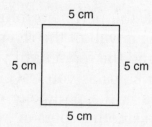

LESSON PRACTICE

Practice set What is the perimeter of each shape?

a. square

12 mm

b. rectangle

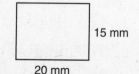

15 mm

20 mm

c. pentagon

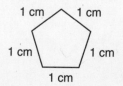

d. equilateral triangle

2 cm

e. trapezoid

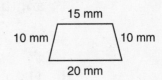

15 mm

10 mm   10 mm

20 mm

f. The perimeter of a square is 60 cm. How long is each side of the square?

MIXED PRACTICE

Problem set

1. In an auditorium there are 25 rows of chairs with 18 chairs
(2) in each row. How many chairs are in the auditorium?

2. All the king's horses numbered 765. All the king's men
(1) numbered 1750. Find how many fewer horses than men the king had by subtracting 765 from 1750.

3. Susan B. Anthony divided 140 of her suffragettes into 5
(2) equal groups. How many suffragettes were in each group?

4. What is the perimeter of this
(8) triangle?

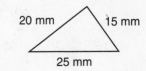

20 mm   15 mm

25 mm

5. How much money is $\frac{1}{2}$ of $6.54?
(6)

6. What number is $\frac{1}{3}$ of 654?
(6)

7. What fraction of this rectangle is
(6) shaded?

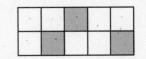

8. 4)$9.00   **9.** 10)373   **10.** 12)1500   **11.** 39)800
(2)   *(2)*   *(2)*   *(2)*

12. 400 ÷ 20 ÷ 4   **13.** 400 ÷ (20 ÷ 4)
(5)   *(5)*

14. Use the numbers 240, 20, and 12 to form two
(2) multiplication facts and two division facts.

15. Rearrange the numbers in this addition fact to form
(1) another addition fact and two subtraction facts.

$$60 + 80 = 140$$

16. The ceiling tiles used in many buildings have sides that
(8) are 12 inches long. What is the perimeter of a square tile
with sides 12 inches long?

17. (a) Find the sum of 6 and 4.
(1, 2)

(b) Find the product of 6 and 4.

18. $5 − M = $1.48
(3)

19. $10 \times 20 \times 30$
(5)

20. $825 \div 8$
(2)

Find each missing number. Check your work.

21. $w − 63 = 36$
(3)

22. $150 + 165 + a = 397$
(3)

23. $12w = 120$
(4)

24. If the divisor is 8 and the quotient is 24, what is the
(4) dividend?

25. (a) Estimate the length of the line segment below in
(7)   centimeters.

(b) Measure the length of the segment in millimeters.

_____

26. Use a ruler to draw a line segment that is $2\frac{3}{4}$ in. long.
(7)

27. $w − 27 = 18$
(3)

28. $27 − x = 18$
(3)

29. Multiply to find the answer to this addition problem:
(2)

$$35 + 35 + 35 + 35$$

30. Describe how to calculate the perimeter of a rectangle.
(8)

LESSON
9

The Number Line:
Ordering and Comparing

WARM-UP

Facts Practice: 100 Subtraction Facts (Test C)

Mental Math: Count up and down by 25's between 25 and 1000.

a. 48 + 120   b. 76 + 10 + 3   c. 7400 + 320

d. 860 − 50   e. 960 − 600   f. 365 − 200

Problem Solving:

The digits 2, 4, and 6 can be arranged to form six different three-digit numbers. List the six numbers in order from least to greatest.

NEW CONCEPT

A **number line** is a way to show numbers in order.

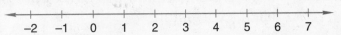

The arrowheads show that the line continues without end and that the numbers continue without end. The small marks crossing the horizontal line are called **tick marks.** Number lines may be labeled with various types of numbers. The numbers we say when we count (1, 2, 3, 4, and so on) are called **counting numbers.** All the counting numbers along with the number zero make up the **whole numbers.** To the left of zero on this number line are **negative numbers,** which will be described in later lessons. As we move to the right on this number line, the numbers are greater and greater in value. As we move to the left, the numbers are less and less in value.

Example 1   Arrange these numbers in order from least to greatest:

$$121 \qquad 112 \qquad 211$$

Solution   On a number line, these three numbers appear in order from least (on the left) to greatest (on the right).

For our answer, we write

$$\textbf{112} \qquad \textbf{121} \qquad \textbf{211}$$

When we **compare** two numbers, we decide whether the numbers are equal; if they are not equal, we determine which number is greater and which is less. We show a comparison with symbols. If the numbers are equal, the comparison symbol we use is the **equal sign** (=).

$$1 + 1 = 2$$

If the numbers are not equal, we use one of the **greater than/less than symbols** (> or <). When properly placed between two numbers, the small end of the symbol points to the smaller number.

Example 2   Compare: 5012 ◯ 5102

Solution   In place of the circle we should write =, >, or < to make the statement true. Since 5012 is less than 5102, we point the small end to the 5012.

$$5012 < 5102$$

Example 3   Compare: $16 \div 8 \div 2$ ◯ $16 \div (8 \div 2)$

Solution   Before we compare the two expressions, we find the value of each expression.

$$\underbrace{16 \div 8 \div 2}_{1} \bigcirc \underbrace{16 \div (8 \div 2)}_{4}$$

Since 1 is less than 4, the comparison symbol points to the left.

$$16 \div 8 \div 2 < 16 \div (8 \div 2)$$

Example 4   Use digits and other symbols to write this comparison:

One fourth is less than one half.

Solution   We write the numbers in the order stated.

$$\frac{1}{4} < \frac{1}{2}$$

LESSON PRACTICE

Practice set   **a.** Arrange these amounts of money in order from least to greatest:

12¢ $12 $1.20

b. Compare: $16 - 8 - 2$ ◯ $16 - (8 - 2)$

c. Compare: $8 \div 4 \times 2 \bigcirc 8 \div (4 \times 2)$

d. $2 \times 3 \bigcirc 2 + 3$ **e.** $1 \times 1 \times 1 \bigcirc 1 + 1 + 1$

f. Use digits and other symbols to write this comparison:

One half is greater than one fourth.

MIXED PRACTICE

Problem set

1. Tamara arranged 144 books into 8 equal stacks. How many
(2) books were in each stack?

2. Find how many years there were from 1492 to 1603 by
(1) subtracting 1492 from 1603.

3. Martin is carrying groceries in from the car. If he can
(2) carry 2 bags at a time, how many trips will it take him to
carry in 9 bags?

4. Use a centimeter ruler to measure the length and width of
(7, 8) the rectangle below. Then calculate the perimeter of the
rectangle.

5. How much money is $\frac{1}{2}$ of $5.80?
(6)

6. How many cents is $\frac{1}{4}$ of a dollar?
(6)

7. Use words and digits to name the
(6) fraction of this triangle that is
shaded.

Zelda

8. Compare: $5012 \bigcirc 5120$
(9)

9. Arrange these numbers in order from least to greatest:
(9)

$$1, 0, \tfrac{1}{2}$$

10. Compare: $100 - 50 - 25 \bigcirc 100 - (50 - 25)$
(9)

11. 478
(1) 3692
+ 45

12. $50.00
(1) − $31.76

13. $4.20
(2) × 60

14. 78
(2) × 36

15. 9)7227
(2)

16. 25)7600
(2)

17. 20)8014
(2)

18. 7136 ÷ 100
(2)

19. 736 ÷ 736
(2)

Find each missing number. Check your work.

20. 165 + a = 300
(3)

21. b − 68 = 86
(3)

22. 9c = 144
(4)

23. $\dfrac{d}{15}$ = 7
(4)

24. Use an inch ruler to draw a line segment two inches long.
(7) Then use a centimeter ruler to find the length of the segment to the nearest centimeter.

25. Which of the figures below represents a ray?
(7)

A. •————————————•

B. ◄————————————►

C. •————————————►

26. Use digits and symbols to write this comparison:
(9)

One half is greater than one third.

27. Arrange the numbers 9, 11, and 99 to form two
(2) multiplication facts and two division facts.

28. Compare: 25 + 0 ◯ 25 × 0
(9)

29. 100 = 20 + 30 + 40 + x
(3)

30. Describe how you chose the positions of the small and
(9) large ends of the greater than/less than symbol that you used in problem 8.

LESSON

10   Sequences • Scales

WARM-UP

> **Facts Practice:** 100 Subtraction Facts (Test C)
>
> **Mental Math:** Count by $\frac{1}{4}$'s from $\frac{1}{4}$ to 10.
>
> | | | |
> |---|---|---|
> | **a.** 43 + 20 + 5 | **b.** 670 + 200 | **c.** 254 + 20 + 5 |
> | **d.** 100 − 50 | **e.** 300 − 50 | **f.** 3600 − 400 |
>
> **Problem Solving:**
>
> Use the digits 5, 6, 7, and 8 to complete this addition problem. There are two possible arrangements.
>
> $$\begin{array}{r} \overline{}\ \overline{} \\ +\ \ 9 \\ \hline \overline{}\ \overline{} \end{array}$$

NEW CONCEPTS

Sequences   A **sequence** is an ordered list of numbers, called **terms**, that follows a certain rule. Here are two different sequences:

(a) 5, 10, 15, 20, 25, …

(b) 5, 10, 20, 40, 80, …

Sequence (a) is an **addition sequence** because the same number is added to each term of the sequence to get the next term. In this case, we add 5 to the value of a term to find the next term. Sequence (b) is a **multiplication sequence** because each term of the sequence is multiplied by the same number to get the next term. In (b) we find the value of a term by multiplying the preceding term by 2. When we are asked to find missing numbers in a sequence, we inspect the numbers to discover the rule for the sequence. Then we use the rule to find other numbers in the sequence.

Example 1   Describe the following sequence as an addition sequence or a multiplication sequence. State the rule of the sequence, and find the next term.

1, 3, 9, 27, _____, …

Solution   The sequence is a **multiplication sequence** because each **term in the sequence can be multiplied by 3 to find the next term.** Multiplying 27 by 3, we find that the term that follows 27 in the sequence is **81.**

The numbers ..., 0, 2, 4, 6, 8, ... form a special sequence called **even numbers.** We say even numbers when we "count by twos." Notice that zero is an even number. Any whole number with a last digit of 0, 2, 4, 6, or 8 is an even number. Whole numbers that are not even numbers are **odd numbers.** The odd numbers form the sequence ..., 1, 3, 5, 7, 9, ... An even number of objects can be divided into two equal groups. An odd number of objects cannot be divided into two equal groups.

Example 2   Think of a whole number. Double that number. Is the answer even or odd?

Solution   The answer is **even.** Doubling any whole number—odd or even—results in an even number.

Scales   Numerical information is often presented to us in the form of a **scale.** A scale is a display of numbers with an indicator to show the value of a certain measure. The "trick" to reading a scale is to discover the value of the marks on the scale. Marks on a scale may show divisions of one unit. Scales may also show divisions of other numbers of units, such as two, five, ten, or one fourth (as with the inch ruler from Lesson 7). We study the scale to find the value of the units before we try to read the indicated number.

Two commonly used scales on thermometers are the **Fahrenheit scale** and the **Celsius scale.** The temperature at which water freezes under standard conditions is 32 degrees Fahrenheit (abbreviated 32°F) and zero degrees Celsius (0°C). The boiling temperature of water is 212°F, which is 100°C. Normal body temperature is 98.6°F (37°C). A cool room may be 68°F (20°C).

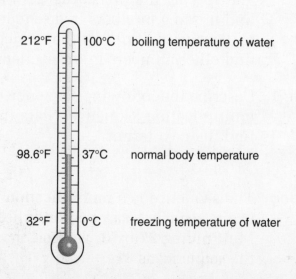

212°F — 100°C   boiling temperature of water

98.6°F — 37°C   normal body temperature

32°F — 0°C   freezing temperature of water

Example 3   What temperature is shown on this thermometer?

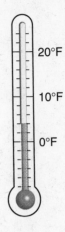

Solution   As we study the scale on this Fahrenheit thermometer, we see that the tick marks divide the distance from 0°F to 10°F into five equal sections. So the number of degrees from one tick mark to the next must be 2°F. Since the fluid in the thermometer is two marks above 0°F, the temperature shown is **4°F.**

LESSON PRACTICE

Practice set   For the sequence in problems **a** and **b**, determine whether the sequence is an addition sequence or a multiplication sequence, state the rule of the sequence, and find the next three terms.

a. ..., 18, 27, 36, 45, _____, _____, _____, ...

b. 1, 2, 4, 8, _____, _____, _____, ...

c. Think of a whole number. Double that number. Then add 1 to the answer. Is the final number even or odd?

d. This thermometer indicates a comfortable room temperature. Find the temperature indicated on this thermometer to the nearest degree Fahrenheit and to the nearest degree Celsius.

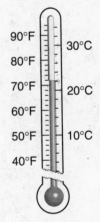

MIXED PRACTICE

Problem set   **1.** State the rule of the following sequence. Then find the
(10)   next three terms.

16, 24, 32, _____, _____, _____, ...

2. Find how many years there were from 1620 to 1776 by
(1)   subtracting 1620 from 1776.

3. Is the number 1492 even or odd? How can you tell?
(10)

4. What weight is indicated on this scale?
(10)

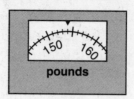

pounds

5. If the perimeter of a square is 40 mm, how long is each side
(8) of the square?

6. How much money is $\frac{1}{2}$ of $6.50?
(6)

7. Compare: $4 \times 3 + 2 \bigcirc 4 \times (3 + 2)$
(9)

8. Use words and digits to write the
(6) fraction of this circle that is **not** shaded.

9. What is the
(1, 2)

   (a) product of 100 and 100?

   (b) sum of 100 and 100?

10. (2)	**11.** (2)	**12.** (2)	**13.** (2)
365 $\times 100$	146 $\times 240$	78¢ $\times 48$	907 $\times 36$

14. $\dfrac{4260}{10}$ **15.** $\dfrac{4260}{20}$ **16.** $\dfrac{4260}{15}$
(2)   (2)   (2)

17. $28{,}347 - 9{,}637$ **18.** $8 + w = 11.49$
(1)   (3)

19. $10 - 0.75   **20.** $0.56 \times 60   **21.** $6.20 \div 4
(1)   (2)   (2)

Find each missing number. Check your work.

22. $56 + 28 + 37 + n = 200$
(3)

23. $a - 67 = 49$ **24.** $67 - b = 49$
(3)   (3)

25. $8c = 120$
(4)

26. $\dfrac{d}{8} = 24$
(4)

27. Here are three ways to write "12 divided by 4."
(2)

$$4\overline{)12} \qquad 12 \div 4 \qquad \dfrac{12}{4}$$

Show three ways to write "20 divided by 5."

28. What number is one third of 36?
(6)

29. Arrange the numbers 346, 463, and 809 to form two
(1) addition equations and two subtraction equations.

30. At what temperature on the Fahrenheit scale does water
(10) freeze?

INVESTIGATION 1

Focus on

Frequency Tables •
Histograms • Surveys

Frequency tables   Hannah made a frequency table to record her test scores throughout the year. For each test she made a tally mark in the row showing the number of correct answers on the test.

Frequency Table

Number Correct	Tally	Frequency
18–20	卌 IIII	9
16–17	卌 II	7
14–15	IIII	4
12–13	II	2

When Hannah finished recording the test scores, she counted the number of tally marks in each row and then recorded the count in the frequency column. For example, the table shows nine tests with 18, 19, or 20 correct answers. A **frequency table** is a way of pairing selected data, in this case specified test scores, with the number of times the selected data occur.

1. Can you tell from this frequency table how many of Hannah's tests had 20 correct answers? Why or why not?

2. Hannah tallied the number of scores in each interval. How wide is each interval? Suggest a reason why Hannah arranged the scores in such intervals.

3. Show how to make a tally for 12.

4. Create a frequency table to track your test scores throughout the year. After each test update your frequency table. At the end of the year, you will have a record of all your test scores.

Histograms   Using the information in the frequency table, Hannah created a histogram to display the scores on her tests.

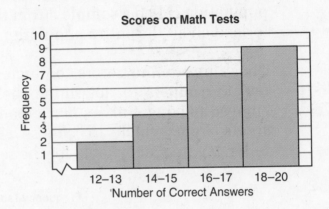

Bar graphs display numerical information with shaded rectangles (bars) of various lengths. Bar graphs are often used to show comparisons.

A **histogram** is a special type of bar graph. This histogram displays the data (test scores) in intervals (ranges of scores). There are no spaces between the bars. The break in the horizontal scale ($\sim$) indicates that the portion of the scale between 0 and 13 has been omitted. The height of each bar indicates the number of test scores in each interval.

Refer to the histogram to answer problems 5–7.

   5. Which interval had the lowest frequency of scores?

   6. Which interval had the highest frequency of scores?

   7. Which interval had exactly twice as many scores as the 12–13 interval?

   8. Make a frequency table and a histogram for the following set of test scores. (Use 50–59, 60–69, 70–79, 80–89, and 90–99 for the intervals.)

       63, 75, 58, 89, 92, 84, 95, 63, 78, 88,

       96, 67, 59, 70, 83, 89, 76, 85, 94, 80

Surveys A **survey** is a way of collecting data about a population. Rather than collecting data from every member of a population, a survey might focus on only a small part of the population called a **sample.** From the sample, conclusions are formed about the entire population.

The camp administrators conducted a survey of 100 male and female campers to determine the sport that campers most enjoyed playing. Survey participants were given six different sports from which to choose. The surveyors made the following frequency table for the responses:

Frequency Table

Sport	Tally	Frequency
Basketball	JHT JHT JHT I	16
Bowling	JHT JHT II	12
Football	JHT JHT JHT	15
Softball	JHT JHT JHT JHT JHT I	26
Table Tennis	JHT JHT II	12
Volleyball	JHT JHT JHT IIII	19

From the frequency table, the camp administrators constructed a bar graph to display the results.

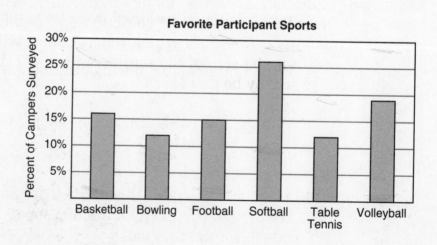

Since 16 out of 100 campers selected basketball as their favorite sport to play, basketball was the choice of 16% (which means "16 out of 100") of the campers surveyed.

Refer to the frequency table and bar graph for this survey to answer problems 9–12.

9. Which sport was the favorite sport of about $\frac{1}{4}$ of the campers surveyed?

10. Which sport was the favorite sport of the girls who were surveyed?

   A. softball

   B. volleyball

   C. basketball

   D. cannot be determined from information provided

11. How might changing the sample group change the results of the survey? For example, what if the camp administrators surveyed only girls or only boys?

12. How might changing the survey question—the choice of sports—change the results of the survey?

This survey was a **closed-option survey** because the responses were limited to the six choices offered. An **open-option survey** does not limit the choices. An example of an open-option survey question is "What is your favorite sport?"

Extension Conduct a survey to determine the favorite foods of your family and friends. If you choose to conduct a closed-option survey, determine which food choices will be offered. What will be the size of the sample? How will the data gathered by the survey be displayed?

LESSON

11

Problems About Combining • Problems About Separating

WARM-UP

Facts Practice: 64 Multiplication Facts (Test D)

Mental Math: Count up and down by $\frac{1}{2}$'s between $\frac{1}{2}$ and 12.

a. 3×40	**b.** 3×400	**c.** $\$4.50 + \$1.25$
d. $451 + 240$	**e.** $4500 - 400$	**f.** $\$5.00 - \$1.50$

g. Start with 10. Add 2; divide by 2; add 2; divide by 2; then subtract 2. What is the answer?

Problem Solving:

Zuna has 2¢ stamps, 3¢ stamps, 10¢ stamps, and 37¢ stamps. She wants to mail a package that requires $1.29 postage. In order to pay the exact postage, what is the smallest number of stamps Zuna can use?

NEW CONCEPTS

Like stories in your reading books, many of the stories we analyze in mathematics have plots. We can use the plot of a math story to write an equation for the story. Stories with the same plot are represented by the same equation. That is why we say there are **patterns** for certain plots.

Problems about combining

Many math stories are about **combining.** Here is an example:

Before he went to work, Pham had $24.50. He earned $12.50 more putting up a fence. Then Pham had $37.00. (Plot: Pham had some money and then he earned some more money.)

Stories about combining have an **addition pattern.**

Some + some more = total

$$S + M = T$$

There are three numbers in the pattern. In a story problem one of the numbers is missing, as in the story below.

Katya had 734 stamps in her collection. Then her uncle gave her some more stamps. Now Katya has 813 stamps. How many stamps did Katya's uncle give her?

This story has a plot similar to the previous story. (Plot: Katya had some stamps and was given some more stamps.) In this story, however, one of the numbers is missing, and we are asked to find the missing number.

Many story problems can be solved by using the following four-step process.

Step 1: Read the problem and identify its pattern.

Step 2: Write an equation for the given information.

Step 3: Find the number that solves the equation and check the answer.

Step 4: Review the question and write the answer.

In this story Katya's uncle gave her some stamps, but the story does not say how many. We use the four-step process to solve the problem.

Step 1: Since this story is about combining, it has an addition pattern.

Step 2: We use the pattern to set up an equation.

PATTERN: Some + some more = total

EQUATION: 734 stamps + B = 813 stamps

Step 3: The answer to the question is the missing number in the equation. Since the missing number is an addend, we subtract 734 from 813 to find the number.

Find answer:	Check answer:
813 stamps	813 stamps
− 734 stamps	− 79 stamps
79 stamps	734 stamps

Step 4: We review the question and write the answer. Katya's uncle gave her 79 stamps.

Example 1 Jenny rode her bike on a trip with her bicycling club. After the first day Jenny's trip odometer showed that she had traveled 86 miles. After the second day the trip odometer showed that she had traveled a total of 163 miles. How far did Jenny ride the second day?

Solution **Step 1:** Jenny rode some miles and then rode some more miles. The distances from the two days combine to give a total. Since this is a story about combining, it has an **addition pattern.**

Step 2: The trip odometer showed how far she traveled the first day and the total of the first two days. We record the information in the pattern. We use a letter in place of the missing number. This time we show that the equation may be written vertically.

$$\begin{array}{r} \text{Some} \\ + \text{ Some more} \\ \hline \text{Total} \end{array} \qquad \begin{array}{r} 86 \text{ miles} \\ + \; m \text{ miles} \\ \hline 163 \text{ miles} \end{array}$$

Step 3: We solve the equation by finding the missing number. From Lesson 3 we know that we can find the missing addend by subtracting 86 miles from 163 miles. We check the answer.

Find answer: Check answer:

$$\begin{array}{r} 163 \text{ miles} \\ - \;\; 86 \text{ miles} \\ \hline 77 \text{ miles} \end{array} \qquad \begin{array}{r} 86 \text{ miles} \\ + \; 77 \text{ miles} \\ \hline 163 \text{ miles} \end{array}$$

Step 4: We review the question and write the answer. Jenny rode **77 miles** on the second day of the trip.

Problems about separating   Another common plot in math stories is **separating.** There is a beginning amount, then some goes away, and some remains. Stories about separating have a **subtraction pattern.**

Beginning amount − some went away = what remains

$$B - A = R$$

Here is an example:

> *Waverly took $37.00 to the music store. She bought headphones for $26.17. Then Waverly had $10.83. (Plot: Waverly had some money, but some of her money went away when she spent it.)*

This is a story about separating. Thus it has a subtraction pattern.

PATTERN:

Beginning amount − some went away = what remains

EQUATION:

$37.00 − $26.17 = $10.83

Example 2   Nancy counted 47 prairie dogs standing in the field. When a hawk flew over the field, some of the prairie dogs ducked into their burrows. Then Nancy counted 29 prairie dogs. How many prairie dogs ducked into their burrows when the hawk flew over the field?

Solution   **Step 1:** There were 47 prairie dogs. Then some went away. This story has a **subtraction pattern.**

Step 2: We use the pattern to write an equation for the given information. We show the equation written vertically.

$$
\begin{array}{ll}
\quad\text{Beginning amount} & 47 \text{ prairie dogs} \\
-\ \text{Some went away} & -\ d \text{ prairie dogs} \\
\hline
\quad\text{What remains} & 29 \text{ prairie dogs}
\end{array}
$$

Step 3: We find the missing number by subtracting 29 from 47. We check the answer.

$$
\begin{array}{ll}
\quad\text{Find answer:} & \text{Check answer:} \\
\quad 47 \text{ prairie dogs} & 47 \text{ prairie dogs} \\
-\ 29 \text{ prairie dogs} & -\ 18 \text{ prairie dogs} \\
\hline
\quad 18 \text{ prairie dogs} & 29 \text{ prairie dogs}
\end{array}
$$

Step 4: We review the question and write the answer. When the hawk flew over the field, **18 prairie dogs** ducked into their burrows.

LESSON PRACTICE

Practice set   Follow the four-step method to solve each problem. Along with each answer, include the equation you use to solve the problem.

a. When Tim finished page 129 of a 314-page book, how many pages did he still have to read?

b. The football team scored 19 points in the first half of the game and 42 points by the end of the game. How many points did the team score in the second half of the game?

MIXED PRACTICE

Problem set   **1.** John ran 8 laps and rested. Then he ran some more laps.
(11)   If John ran 21 laps in all, how many laps did he run after he rested? (Write an equation and solve the problem.)

2. (a) Find the product of 8 and 4.
(1, 2)
(b) Find the sum of 8 and 4.

3. The expression below means "the product of 6 and 4
(5) divided by the difference of 8 and 5." What is the
quotient?

$$(6 \times 4) \div (8 - 5)$$

4. Marcia went to the store with $20.00 and returned home
(11) with $7.75. How much money did Marcia spend at the
store? (Write an equation and solve the problem.)

5. When Jack went to bed at night, the beanstalk was one
(11) meter tall. When he woke up in the morning, the beanstalk
was one thousand meters tall. How many meters had the
beanstalk grown during the night? (Write an equation and
solve the problem.)

6. $0.65 + $0.40
(1)

Find each missing number. Check your work.

7. $87 + w = 155$ **8.** $1000 - x = 386$
(3)   (3)

9. $y - 1000 = 386$ **10.** $42 + 596 + m = 700$
(3)   (3)

11. Compare: $1000 - (100 - 10) \bigcirc 1000 - 100 - 10$
(9)

12. $8\overline{)1000}$ **13.** $10\overline{)987}$ **14.** $12\overline{)w}^{\,35}$
(2)   (2)   (4)

15. 600×300 **16.** $365w = 365$
(2)   (4)

17. What are the next three numbers in the following
(10) sequence?

$$2, 6, 10, \underline{\qquad}, \underline{\qquad}, \underline{\qquad}, \ldots$$

18. $2 \times 3 \times 4 \times 5$
(5)

19. What number is $\frac{1}{2}$ of 360?
(6)

20. What number is $\frac{1}{4}$ of 360?
(6)

21. What is the product of eight and one hundred twenty-five?
(2)

22. How long is the line segment below?
(7)

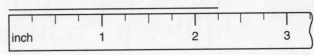

23. What fraction of the circle at right
(6) is not shaded?

24. What is the perimeter of the square
(8) shown?

9 mm

25. What is the sum of the first five odd numbers greater
(10) than zero?

26. Here are three ways to write "24 divided by 4":
(2)

$$4\overline{)24} \qquad 24 \div 4 \qquad \frac{24}{4}$$

Show three ways to write "30 divided by 6."

27. Seventeen of the 30 cars in the lot are sedans. So $\frac{17}{30}$ of the
(6) cars in the lot are sedans. The rest of the 30 cars are
coupes. What fraction of the cars in the lot are coupes?

28. At what temperature on the Celsius scale does water freeze?
(10)

29. Use the numbers 24, 6, and 4 to write two multiplication
(2) facts and two division facts.

30. In the second paragraph of this lesson there is a story
(11) with an addition pattern. Rewrite the same story as a
problem by removing one of the numbers from the story
and asking a question instead.

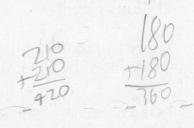

LESSON

12

Place Value Through Trillions • Multistep Problems

WARM-UP

Facts Practice: 64 Multiplication Facts (Test D)

Mental Math: Count by $\frac{1}{4}$'s from $\frac{1}{4}$ to 12.

    a. 6 × 40   **b.** 6 × 400   **c.** $12.50 + $5.00

    d. 451 + 24   **e.** 7500 − 5000   **f.** $10.00 − $2.50

    g. Start with 12. Divide by 2; subtract 2; divide by 2; then subtract 2. What is the answer?

Problem Solving:

    A dot cube has six faces (surfaces). The faces are marked with 1, 2, 3, 4, 5, and 6 dots. Altogether, how many dots are on a dot cube?

NEW CONCEPTS

Place value through trillions

In our number system the value of a digit depends upon its position. The value of each position is called its **place value**.

Whole-Number Place Values

hundred trillions	ten trillions	trillions		hundred billions	ten billions	billions		hundred millions	ten millions	millions		hundred thousands	ten thousands	thousands		hundreds	tens	ones
—	—	—	,	—	—	—	,	—	—	—	,	—	—	—	,	—	—	—

Example 1 In the number 123,456,789,000, which digit is in the ten-millions place?

Solution Either by counting places or looking at the chart, we find that the digit in the ten-millions place is **5**.

Example 2 In the number 5,764,283, what is the place value of the digit 4?

Solution   By counting places or looking at the chart, we can see that the place value of 4 is **thousands.**

Large numbers are easy to read and write if we use commas to group the digits. To place commas, we begin at the right and move to the left, writing a comma after every three digits.

Putting commas in 1234567890, we get 1,234,567,890.

Commas help us read large numbers by marking the end of the trillions, billions, millions, and thousands. We read the three-digit number in front of each comma and then say "trillion," "billion," "million," or "thousand" when we reach the comma.

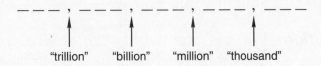

"trillion"   "billion"   "million"   "thousand"

Example 3   Use words to write the number 1024305.

Solution   First we insert commas.

$$1,024,305$$

We write **one million, twenty-four thousand, three hundred five.**

Note: We write commas after the words *trillion, billion, million,* and *thousand.* We hyphenate compound numbers from 21 through 99. We do not say or write "and" when naming whole numbers.

Example 4   Use digits to write the number one trillion, two hundred fifty billion.

Solution   When writing large numbers, it may help to sketch the pattern before writing the digits.

___ , ___ , ___ , ___ , ___

↑   ↑   ↑   ↑
"trillion"   "billion"   "million"   "thousand"

We write a 1 to the left of the trillions comma and 250 in the three places to the left of the billions comma. The remaining places are filled with zeros.

1,250,000,000,000

Multistep problems   The **operations of arithmetic** are addition, subtraction, multiplication, and division. In this table we list the terms for the answers we get when we perform these operations:

Sum	The answer when we add
Difference	The answer when we subtract
Product	The answer when we multiply
Quotient	The answer when we divide

We will use these terms in problems that have several steps.

Example 5   What is the difference between the product of 6 and 4 and the sum of 6 and 4?

Solution   We see the words *difference, product,* and *sum* in this question. We first look for phrases such as "the product of 6 and 4." We will rewrite the question, emphasizing these phrases.

What is the difference between the <u>product of 6 and 4</u> and the <u>sum of 6 and 4</u>?

For each phrase we find one number. "The product of 6 and 4" is 24, and "the sum of 6 and 4" is 10. So we can replace the two phrases with the numbers 24 and 10 to get this question:

What is the difference between 24 and 10?

We find this answer by subtracting 10 from 24. The difference between 24 and 10 is **14.**

LESSON PRACTICE

Practice set*   **a.** Which digit is in the millions place in 123,456,789?

b. What is the place value of the 1 in 12,453,000,000?

c. Use words to write 21,350,608.

d. Use digits to write four billion, five hundred twenty million.

e. When the product of 6 and 4 is divided by the difference of 6 and 4, what is the quotient?

MIXED PRACTICE

Problem set   **1.** What is the difference between the product of 1, 2, and 3
(12)   and the sum of 1, 2, and 3?

2. Earth is about ninety-three million miles from the Sun.
(12)   Use digits to write that distance.

3. Gilbert and Sullivan cooked 342 pancakes for the
(11)   pancake breakfast. If Gilbert cooked 167 pancakes, how
many pancakes did Sullivan cook? (Write an equation
and solve the problem.)

4. The two teams scored a total of 102 points in the
(11)   basketball game. If the winning team scored 59 points,
how many points did the losing team score? (Write an
equation and solve the problem.)

5. What is the perimeter of the
(8)   rectangle at right?

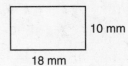

10 mm

18 mm

6. $6m = 60$
(4)

7. (a) What number is $\frac{1}{2}$ of 100?
(6)

(b) What number is $\frac{1}{4}$ of 100?

8. Compare: $300 \times 1 \bigcirc 300 \div 1$
(9)

9. $(3 \times 3) - (3 + 3)$
(5)

10. What are the next three numbers in the following
(10)   sequence?

$$1, 2, 4, 8, \underline{\quad}, \underline{\quad}, \underline{\quad}, \dots$$

11. $1 + m + 456 = 480$ **12.** $1010 - n = 101$
(3)   (3)

13. $1234 \div 10$ **14.** $1234 \div 12$
(2)   (2)

15. What is the sum of the first five even numbers greater
(10)   than zero?

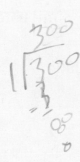

16. How many millimeters long is the line segment below?
(7)

mm 10   20   30   40

17. In the number 123,456,789,000, which digit is in the ten-
(12) billions place?

18. In the number 5,764,283,000, what is the place value of
(12) the digit 4?

19. Which digit is in the hundred-thousands place in the
(12) number 987,654,321?

20. $1 \times 10 \times 100 \times 1000$
(5)

21. $\$3.75 \times 3$ **22.** $22y = 0$
(2)   (4)

23. $100 + 200 + 300 + 400 + w = 2000$
(3)

24. 24×26 **25.** $m\overline{)625}$
(2)   (4)

with quotient 25

26. If the divisor is 4 and the quotient is 8, what is the
(4) dividend?

27. Show three ways to write "27 divided by 3."
(2)

28. Seven of the ten marbles in a bag are red. So $\frac{7}{10}$ of the
(6) marbles are red. What fraction of the marbles are not red?

29. Use digits to write four trillion.
(12)

30. Using different numbers, make up a question similar to
(12) example 5 in this lesson. Then find the answer.

LESSON
13 Problems About Comparing • Elapsed-Time Problems

Facts Practice: 64 Addition Facts (Test A)

Mental Math: Count up and down by 25's between 25 and 1000.
Count up and down by 2's between 2 and 40.

a. 5 × 300
b. 5 × 3000
c. $7.50 + $1.75
d. 3600 + 230
e. 4500 − 500
f. $20.00 − $5.00

Problem Solving:

Tom followed the directions on the treasure map. Starting at the big tree, he walked 5 steps north. Then he turned right and walked 7 steps. He turned right again and walked 9 steps. Then he turned left and walked 3 steps. Finally, he turned left and walked 4 steps. In what direction was he facing? How many steps was he from the big tree?

NEW CONCEPTS

We practiced using patterns to solve story problems in Lesson 11. If things combined, we used an addition pattern. If things separated, we used a subtraction pattern. In this lesson we will look at two other kinds of math story problems.

Problems about comparing

Some math stories are about **comparing** the size of two groups. They usually ask questions such as "How many more are in the first group" and "How many fewer are in the second group?" Comparison problems such as these have a **subtraction pattern.** We write the numbers in the equation in this order:

$$\text{Larger} - \text{smaller} = \text{difference}$$

In place of the words, we can use letters. We use the first letter of each word.

$$L - S = D$$

Example 1   There were 324 girls and 289 boys in the contest. How many fewer boys than girls were there in the contest?

Solution   Again we use the four-step process to solve the problem.

Step 1: We are asked to compare the number of boys to the number of girls. The question asks "how many fewer?" This problem has a **subtraction pattern.**

Step 2: We use the pattern to write an equation. There are more girls than boys, so the number of girls replaces "larger" and the number of boys replaces "smaller."

PATTERN: Larger − smaller = difference

EQUATION: $324 - 289 = D$

Step 3: We find the missing number by subtracting.

324 girls
− 289 boys
————————
35 fewer boys

Step 4: We review the question and write the answer. There were **35 fewer boys** than girls in the contest. We can also state that there were 35 more girls than boys in the contest.

Elapsed-time problems   **Elapsed time** is the length of time between two events. We illustrate this on the ray below.

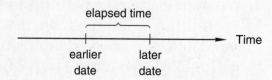

The time that has elapsed since the moment you were born until now is your age. Subtracting your birth date from today's date gives your age.

Today's date   (later)
− Your birth date (earlier)
————————————
Your age   (difference)

Elapsed-time problems are like comparison problems. They have a **subtraction pattern.**

Later − earlier = difference

We use the first letter of each word to represent the word.

$L - E = D$

Example 2 How many years were there from 1492 to 1620?

Solution **Step 1:** This is an **elapsed-time** problem. It has a **subtraction pattern.** We use *L, E,* and *D* to stand for "later," "earlier," and "difference."

$$L - E = D$$

Step 2: The later year is 1620. The earlier year is 1492.

$$1620 - 1492 = D$$

Step 3: We find the missing number by subtracting. We can check the answer by adding.

$$
\begin{array}{r}
1620 \\
- 1492 \\
\hline
128
\end{array}
\qquad
\begin{array}{r}
1492 \\
+ \quad 128 \\
\hline
1620
\end{array}
$$

Step 4: We review the question and write the answer. There were **128 years** from 1492 to 1620.

Example 3 Abraham Lincoln was born in 1809 and died in 1865. How many years did he live?

Solution **Step 1:** This is an **elapsed-time** problem. It has a **subtraction pattern.** We use *L* for the later time, *E* for the earlier time, and *D* for the difference of the times.

$$L - E = D$$

Step 2: We write an equation using 1809 for the earlier year and 1865 for the later year.

$$1865 - 1809 = D$$

Step 3: We find the missing number by subtracting. We may add his age to the year of his birth to check the answer.

$$
\begin{array}{r}
1865 \\
- 1809 \\
\hline
56
\end{array}
\qquad
\begin{array}{r}
1809 \\
+ \quad 56 \\
\hline
1865
\end{array}
$$

We also note that 56 is a reasonable age, so our computation makes sense.

Step 4: We review the question and write the answer. Abraham Lincoln lived **56 years.**

LESSON PRACTICE

Practice set   Follow the four-step method to solve each problem. Along with each answer, include the equation you use to solve the problem.

    a. The population of Castor is 26,290. The population of Weston is 18,962. How many more people live in Castor than live in Weston?

    b. How many years were there from 1066 to 1215?

MIXED PRACTICE

Problem set   **1.** When the sum of 8 and 5 is subtracted from the product
(12) of 8 and 5, what is the difference?

2. The Moon is about two hundred fifty thousand miles
(12) from the Earth. Use digits to write that distance.

3. Use words to write 521,000,000,000.
(12)

4. Use digits to write five million, two hundred thousand.
(12)

5. Robin Hood roamed Sherwood Forest with sevenscore
(2) merry men. A score is twenty, so sevenscore is seven twenties. Find how many merry men roamed with Robin.

6. The beanstalk was 1000 meters tall. The giant had climbed
(11) down 487 meters before Jack could chop down the beanstalk. How far did the giant fall? (Write an equation and solve the problem.)

7. At Big River Summer Camp there are 503 girls and
(13) 478 boys. How many more girls than boys attend Big River Summer Camp? (Write an equation and solve the problem.)

8. 99 + 100 + 101   **9.** 9 × 10 × 11
(1)   (5)

10. Which digit is in the thousands place in 54,321?
(12)

11. What is the place value of the 1 in 1,234,567,890?
(12)

12. The three sides of an equilateral
(8) triangle are equal in length. What is the perimeter of the equilateral triangle shown?

18 mm

BOO!

↓

EXTIRMINATE

FAIL

2
18
18
18
+18
―――
54

13. 5432 ÷ 100
(2)

14. $\dfrac{60,000}{30}$
(2)

15. 1000 ÷ 7
(2)

16. $4.56 ÷ 3
(2)

17. Compare: 3 + 2 + 1 + 0 ◯ 3 × 2 × 1 × 0
(9)

18. The rule for the sequence below is different from the rules for addition sequences or multiplication sequences. What is the next number in the sequence?
(10)

1, 4, 3, 6, 5, 8, __7__ , ... 9 10 9 12

19. What is $\frac{1}{2}$ of 5280?
(6)

20. 365 ÷ w = 365
(4)

21. (5 + 6 + 7) ÷ 3
(5)

22. Use a ruler to find the length in inches of the rectangle below.
(7)

width

length

23. Describe two ways to find the perimeter of a square: one way by adding and the other way by multiplying.
(8)

24. Multiply to find the answer to this addition problem:
(2)

125 + 125 + 125 + 125 + 125 + 125

25. At what temperature on the Fahrenheit scale does water boil?
(10)

26. Show three ways to write "21 divided by 7."
(2)

Find each missing number. Check your work.

27. 8a = 816
(4)

28. $\dfrac{b}{4}$ = 12
(4)

29. $\dfrac{12}{c}$ = 4
(4)

30. d − 16 = 61
(3)

LESSON
14
The Number Line: Negative Numbers

Facts Practice: 64 Multiplication Facts (Test D)

Mental Math: Count up and down by $\frac{1}{4}$'s between $\frac{1}{4}$ and 6.

a. 8 × 400   b. 6 × 3000   c. $7.50 + $7.50

d. 360 + 230   e. 1250 − 1000   f. $10.00 − $7.50

g. Start with 10. Add 2; divide by 3; multiply by 4; then subtract 5. What is the answer?

Problem Solving:

Andy, Bob, and Carol stood side by side for a picture. Then they changed their order for another picture. Then they changed their order again. List all the possible side-by-side arrangements.

NEW CONCEPT

We have seen that a number line can be used to arrange numbers in order.

On the number line above, the points to the right of zero represent **positive numbers.** The points to the left of zero represent **negative numbers.** Zero is neither positive nor negative.

Negative numbers are used in various ways. A temperature of five degrees below zero Fahrenheit may be written as −5°F. An elevation of 100 feet below sea level may be indicated as "elev. −100 ft." The change in a stock's price from $23.00 to $21.50 may be shown in a newspaper as −1.50.

Example 1 Arrange these numbers in order from least to greatest:

$$0, 1, -2$$

Solution All negative numbers are less than zero. All positive numbers are greater than zero.

$$-2, 0, 1$$

Example 2 Compare: $-3 \bigcirc -4$

Solution Negative three is three less than zero, and negative four is four less than zero. So

$$-3 > -4$$

The number -5 is read "negative five." Notice that the points on the number line marked 5 and -5 are the same distance from zero but are on opposite sides of zero. We say that 5 and -5 are **opposites**. Other opposite pairs include -2 and 2, -3 and 3, and -4 and 4. The tick marks show the location of numbers called **integers**. Integers include all of the counting numbers and their opposites, as well as the number zero.

If you subtract a larger number from a smaller number (for example, $2 - 3$), the answer will be a negative number. One way to find the answer to such questions is to use the number line. We start at 2 and count back (to the left) three integers. Maybe you can figure out a faster way to find the answer.

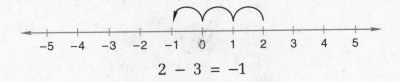

$$2 - 3 = -1$$

Example 3 Subtract 5 from 2.

Solution **Order matters in subtraction.** Start at 2 and count to the left 5 integers. You should end up at **-3**. Try this problem with a calculator by entering ⬚2⬚ ⬚−⬚ ⬚5⬚ ⬚=⬚. What number is displayed after the ⬚=⬚ is pressed?

Example 4 Arrange these four numbers in order from least to greatest:

$$1, -2, 0, -1$$

Solution A number line shows numbers in order. By arranging these numbers in the order they appear on a number line, we arrange them in order from least to greatest.

$$-2, -1, 0, 1$$

Example 5 What number is 7 less than 3?

Solution The phrase "7 less than 3" means to start with 3 and subtract 7.

$$3 - 7$$

We count to the left 7 integers from 3. The answer is **-4**.

LESSON PRACTICE

Practice set   **a.** Compare: −8 ◯ −6

b. Use words to write this number: −8.

c. What number is the opposite of 3?

d. Arrange these numbers in order from least to greatest:

0, −1, 2, −3

e. What number is 5 less than 0?

f. What number is 10 less than 5?

g. 5 − 8   **h.** 1 − 5

i. All five of the numbers below are integers. True or false?

−3, 0, 2, −10, 50

j. The temperature was twelve degrees below zero Fahrenheit. Use a negative number to write the temperature.

k. The desert floor was 186 feet below sea level. Use a negative number to indicate that elevation.

l. The stock's price dropped from $18.50 to $16.25. Use a negative number to express the change in the stock's value.

MIXED PRACTICE

Problem set   **1.** What is the quotient when the sum of 15 and 12 is
(12)   divided by the difference of 15 and 12?

2. What is the place value of the 7 in 987,654,321,000?
(12)

3. Light travels at a speed of about one hundred eighty-six
(12)   thousand miles per second. Use digits to write that speed.

4. What number is three integers to the left of 2 on the
(14)   number line?

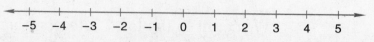

5. Arrange these numbers in order from least to greatest:
(14)

$$5, -3, 1, 0, -2$$

6. What number is halfway between -4 and 0 on the
(14) number line?

7. Seventy-two of the 140 merry men remained in
(11) Sherwood Forest while the rest rode out with Robin. How
many of the merry men rode out of the forest with Robin?
(Write an equation and solve the problem.)

8. Compare: $1 + 2 + 3 + 4 \bigcirc 1 \times 2 \times 3 \times 4$
(9)

9. What is the perimeter of this right
(8) triangle?

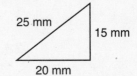

25 mm

15 mm

20 mm

10. What are the next two numbers in the following
(10) sequence?

$$\ldots, 16, 8, 4, \underline{\qquad}, \underline{\qquad}, \ldots$$

11. There are 365 days in a common year. How much less
(13) than 500 is 365? (Write an equation and solve the
problem.)

12. What number is 8 less than 6?
(14)

13. $1020 \div 100$ **14.** $\dfrac{36,180}{12}$ **15.** $18\overline{)564}$
(2)   (2)   (2)

16. $1234 + 567 + 89$ **17.** $n - 310 = 186$
(1)   (3)

18. $10 \cdot 11 \cdot 12$ **19.** $\$3.05 - m = \$2.98$
(5)   (3)

20. Estimate the length of this nail in centimeters. Then use a
(7) centimeter ruler to find its length to the nearest
centimeter and to the nearest millimeter.

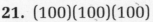

21. $(100)(100)(100)$
(5)

22. What digit in 123,456,789 is in the ten-thousands place?
(12)

23. If you know the length of an object in centimeters, how
(7)   can you figure out the length of the object in millimeters
without remeasuring?

24. Use the numbers 19, 21, and 399 to write two
(2)   multiplication facts and two division facts.

25. Compare: $12 \div 6 \times 2 \bigcirc 12 \div (6 \times 2)$
(9)

26. Show three ways to write "60 divided by 6."
(2)

27. The human brain has about nine trillion nerve cells. Use
(12)   digits to write that number of nerve cells.

28. One third of the 12 eggs in the carton were cracked. How
(6)   many eggs were cracked?

29. What number is the opposite of 10?
(14)

30. Arrange these numbers in order from least to greatest:
(14)

$$1, 0, -1, \tfrac{1}{2}$$

LESSON
15 Problems About Equal Groups

WARM-UP

Facts Practice: 100 Subtraction Facts (Test C)

Mental Math: Count up and down by $\frac{1}{4}$'s between $\frac{1}{4}$ and 10.

  a. 7 × 4000   **b.** 8 × 300   **c.** $12.50 + $12.50

  d. 80 + 12   **e.** 6250 − 150   **f.** $20.00 − $2.50

  g. Start with a dozen. Subtract 3; divide by 3; subtract 3; then multiply by 3. What is the answer?

Problem Solving:

Copy this subtraction problem and fill in the missing digits:

$$\begin{array}{r} 4\_7 \\ -\ \_9\_ \\ \hline 21 \end{array}$$

NEW CONCEPT

We have studied several mathematical story plots. Stories about combining have an addition pattern. Stories about separating, stories about comparing, and elapsed-time problems have subtraction patterns. Another type of mathematical story is the **equal groups** story. Here is an example:

> *In the auditorium there were 15 rows of chairs with 20 chairs in each row. Altogether, there were 300 chairs in the auditorium.*

The chairs were arranged in 15 groups (rows) with 20 chairs in each group. Here is how we write the pattern:

15 rows × 20 chairs in each row = 300 chairs

Number of groups × number in group = total

$$N \times G = T$$

In a problem about equal groups, any one of the numbers might be missing. We multiply to find the unknown total. We divide to find an unknown factor.

Example   At Buddy's Bookstore there were 232 biographies on eight shelves. If there were the same number of biographies on each shelf, how many biographies would be on each of the eight shelves?

Solution   **Step 1:** A number of biographies is divided into equal groups (shelves). This is a problem about **equal groups.** The words *in each* often appear in "equal groups" stories.

Step 2: We draw the pattern and record the numbers, writing a letter in place of the missing number.

PATTERN	EQUATION
Number in each group	n on each shelf
× Number of groups	× 8 shelves
Number in all groups	232 on all shelves

Step 3: We find the missing factor by dividing. Then we check our work.

$$\begin{array}{r} 29 \\ 8\overline{)232} \end{array} \qquad \begin{array}{r} 29 \\ \times\ \ 8 \\ \hline 232 \end{array}$$

Step 4: We review the question and write the answer. If there were the same number of biographies on each shelf, there would be **29 biographies** on each of the eight shelves.

LESSON PRACTICE

Practice set   Follow the four-step method to solve each problem. Along with each answer, include the equation you use to solve the problem.

a. Marcie collected $4.50 selling lemonade at 25¢ for each cup. How many cups of lemonade did Marcie sell? (*Hint:* Record $4.50 as 450¢.)

b. In the store parking lot there were 18 parking spaces in each row, and there were 12 rows of parking spaces. Altogether, how many parking spaces were in the parking lot?

MIXED PRACTICE

Problem set   **1.** The second paragraph of this lesson contains an "equal
                  (15)   groups" story. Rewrite the story as a problem by
                         removing one of the numbers and writing a question.

2. On the Fahrenheit scale, water freezes at 32°F and boils at
(13) 212°F. How many degrees difference is there between the
     freezing and boiling points of water? (Write an equation
     and solve the problem.)

3. There are about three hundred twenty little O's of cereal
(15) in an ounce. About how many little O's are there in a one-
     pound box? Write an equation and solve the problem
     (1 pound = 16 ounces).

4. There are 31 days in August. How many days are left in
(11) August after August 3? (Write an equation and solve the
     problem.)

5. Compare: $3 - 1 \bigcirc 1 - 3$
(14)

6. Subtract 5 from 2. Use words to write the answer.
(14)

7. The stock's value dropped from $28.00 to $25.50. Use a
(14) negative number to show the change in the stock's value.

8. What are the next three numbers in the following
(10) sequence?

$$\ldots, 6, 4, 2, 0, \underline{\quad\quad}, \underline{\quad\quad}, \underline{\quad\quad}, \ldots$$

9. What is the temperature reading on
(10, 14) this thermometer? Write the answer
         twice, once with digits and an
         abbreviation and once with words.

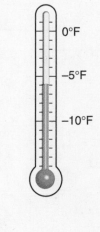

10. $10 − 10¢
(1)

11. How much money is $\frac{1}{2}$ of $3.50?
(6)

12. To which hundred is 587 closest?
(9)

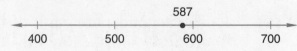

13. 9 + 87 + 654 + 3210
(1)

14. 574 × 76
(2)

15. $\dfrac{4320}{9}$
(2)

16. $36\overline{)493}$
(2)

Find each missing number. Check your work.

17. 1200 ÷ w = 300
(4)

18. 63w = 63
(4)

19. $\dfrac{76}{m}$ = 1
(4)

20. w + \$65 = \$1000
(3)

21. 3 + n + 12 + 27 = 50
(3)

22. There are 10 millimeters in 1 centimeter. How many
(7) millimeters long is this paper clip?

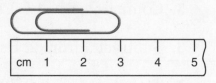

23. (8 + 9 + 16) ÷ 3
(5)

24. What is the place value of the 5 in 12,345,678?
(12)

25. Which digit occupies the ten-billions place in
(12) 123,456,789,000?

26. Use the numbers 19, 21, and 40 to write two addition
(1) facts and two subtraction facts.

27. Arrange these numbers in order from least to greatest:
(14)

$$0, -1, 2, -3$$

28. Susan sold seven of her seventeen seashells down by the
(6) seashore. What fraction of her seashells did she sell?

29. Susan sold seashells for 75¢ each. How much money did
(15) Susan receive selling seven seashells?

30. What number is neither positive nor negative?
(14)

LESSON
16 Rounding Whole Numbers • Estimating

WARM-UP

Facts Practice: 64 Multiplication Facts (Test D)

Mental Math: Count by 3's from 3 to 60.

a. 3 × 30 plus 3 × 2   **b.** 4 × 20 plus 4 × 3
c. 150 + 20   **d.** 75 + 9
e. 800 − 50   **f.** 8000 − 500
g. Start with 1. Add 2; multiply by 3; subtract 4; then divide by 5. What is the answer?

Problem Solving:

Fran has 8 coins that total exactly $1.00. If at least one of the coins is a dime, what are Fran's 8 coins?

NEW CONCEPTS

Rounding whole numbers

When we **round** a whole number, we are finding another whole number, usually ending in zero, that is close to the number we are rounding. The number line can help us visualize rounding.

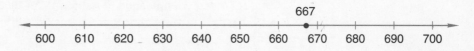

In order to round 667 to the nearest ten, we recognize that 667 is closer to 670 than it is to 660. In order to round 667 to the nearest hundred, we recognize that 667 is closer to 700 than to 600.

Example 1 Round 6789 to the nearest thousand.

Solution The number we are rounding is between 6000 and 7000. It is closer to **7000.**

Example 2 Round 550 to the nearest hundred.

Solution The number we are to round is halfway between 500 and 600. When the number we are rounding is halfway between two round numbers, we round **up.** So 550 rounds to **600.**

Estimating Rounding can help us **estimate** the answer to a problem. Estimating is a quick way to "get close" to the answer. It can also help us decide whether an answer is reasonable. To estimate, we round the numbers before we add, subtract, multiply, or divide.

Example 3 Estimate the sum of 467 and 312.

Solution Estimating is a skill we can learn to do "in our head." First we round each number. Since both numbers are in the hundreds, we will round each number to the nearest hundred.

467 rounds to 500

312 rounds to 300

To estimate the sum, we add the rounded numbers.

$$\begin{array}{r} 500 \\ + \ 300 \\ \hline 800 \end{array}$$

We estimate the sum of 467 and 312 to be **800.**

Example 4 According to this graph, about how many more people lived in Ashton in 2000 than in 1980?

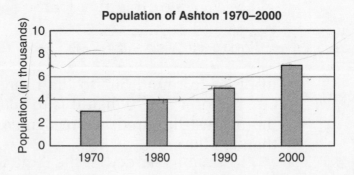

Population of Ashton 1970–2000

Solution We often need to use estimation skills when reading graphs. The numbers along the left side of the graph (the vertical axis) indicate the population in thousands. The bar for the year 2000 is about halfway between the 6000 and 8000 levels, so the population was about 7000. In 1980 the population was about 4000. This problem has a subtraction pattern. We subtract and find that about **3000 more people** lived in Ashton in 2000 than in 1980.

LESSON PRACTICE

Practice set* Round each of these numbers to the nearest ten:

    a. 57   **b.** 63   **c.** 45

Round each of these numbers to the nearest hundred:

    d. 282   **e.** 350   **f.** 426

Round each of these numbers to the nearest thousand:

    g. 4387   **h.** 7500   **i.** 6750

Use rounded numbers to estimate each answer.

    j. 397 + 206   **k.** 703 − 598

    l. 29 × 31   **m.** 29$\overline{)591}$

Use the graph in example 4 to answer problems **n** and **o.**

    n. About how many fewer people lived in Ashton in 1980 than in 1990?

    o. The graph shows an upward trend in the population of Ashton. If the population grows the same amount from 2000 to 2010 as it did from 1990 to 2000, what would be a reasonable projection for the population in 2010?

MIXED PRACTICE

Problem set   **1.** What is the difference between the product of 20 and 5 and the sum of 20 and 5?
    (12)

  2. Columbus landed in the Americas in 1492. The Pilgrims landed in 1620. How many years after Columbus did the Pilgrims land in America? (Write an equation and solve the problem.)
    (13)

  3. Robin Hood separated his 140 merry men into 5 equal groups. He sent one group north, one south, one east, and one west. The remaining group stayed in camp. How many merry men stayed in camp? (Write an equation and solve the problem.)
    (15)

4. Which digit in 159,342,876 is in the hundred-thousands
(12) place?

5. In the 2000 U.S. presidental election, 105,396,641 votes
(12) were tallied for president. Use words to write that
number of votes.

6. What number is halfway between 5 and 11 on the
(9) number line?

7. Round 56,789 to the nearest thousand.
(16)

8. Round 550 to the nearest hundred.
(16)

9. Estimate the product of 295 and 406 by rounding each
(16) number to the nearest hundred before multiplying.

10. 45 + 5643 + 287   **11.** 40,312 − 14,908
(1)   (1)

12. $\dfrac{7308}{12}$ **13.** $100\overline{)5367}$
(2)   (2)

14. (5 + 11) ÷ 2
(5)

15. How much money is $\frac{1}{2}$ of \$5?
(6)

16. How much money is $\frac{1}{4}$ of \$5?
(6)

17. \$0.25 × 10   **18.** 325(324 − 323)
(2)   (5)

19. Compare: 1 + (2 + 3) $\bigcirc$ (1 + 2) + 3
(9)

20. *Wind chill* describes the effect of temperature and wind
(14) combining to make it feel colder outside. At 3 p.m. in
Minneapolis, Minnesota, the wind chill was −10°
Fahrenheit. At 11 p.m. the wind chill was −3° Fahrenheit.
At which time did it feel colder outside, 3 p.m. or 11 p.m.?
Explain how you arrived at your answer.

21. Your heart beats about 72 times per minute. At that rate,
(15) how many times will it beat in one hour? (Write an
equation and solve the problem.)

22. The distance between bases on a major league baseball
(8) diamond is 90 feet. A player who runs around the
diamond runs about how many feet?

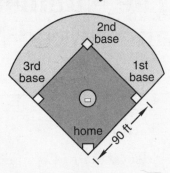

Refer to the bar graph shown below to answer problems 23–26.

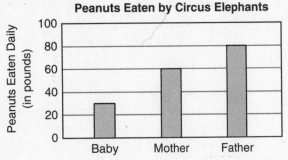

23. How many more pounds of peanuts does the father
(16) elephant eat each day than the baby elephant?

24. Altogether, how many pounds do the three elephants eat
(16) each day?

25. How many pounds would the mother elephant eat in
(16) one week?

26. Using the information in this graph, write a comparison
(16) story problem.

Find each missing number. Check your work.

27. $6w = 66$ **28.** $m - 60 = 37$ **29.** $60 - n = 37$
(4)   (3)   (3)

30. Each day Chico, Fuji, and Rolo eat 6, 8, and 9 bananas
(Inv. 1) respectively. Draw a bar graph to illustrate this
information.

LESSON

17

The Number Line: Fractions and Mixed Numbers

WARM-UP

Facts Practice: 100 Multiplication Facts (Test E)

Mental Math: Count up and down by $\frac{1}{4}$'s between $\frac{1}{4}$ and 12.

  a. 5 × 30 plus 5 × 4   **b.** 4 × 60 plus 4 × 4

  c. 180 + 12   **d.** 64 + 9

  e. 3000 − 1000 − 100   **f.** $10.00 − $7.50

  g. Start with 5. Multiply by 4; add 1; divide by 3; then
     subtract 2. What is the answer?

Problem Solving:

If you pick up a dot cube with two fingers by holding your fingers against opposite faces, your fingers will cover a total of how many dots? (Use a dot cube to find out.)

NEW CONCEPT

On this number line the tick marks show the location of the integers:

There are points on the number line between the integers that can be named with fractions or **mixed numbers**. A mixed number is a whole number plus a fraction. Halfway between 0 and 1 is $\frac{1}{2}$. Halfway between 1 and 2 is $1\frac{1}{2}$. Halfway between −1 and −2 is $-1\frac{1}{2}$.

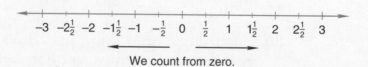

We count from zero.

The distance between consecutive integers on a number line may be divided into halves, thirds, fourths, fifths, or any other number of equal divisions. To determine which fraction or mixed number is represented by a point on the number line, we follow the steps described in the next example.

Example 1 Point *A* represents what mixed number on this number line?

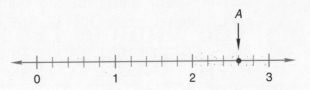

Solution We see that point *A* represents a number greater than 2 but less than 3. So point *A* represents a mixed number, which is a whole number plus a fraction. To find the fraction, we first notice that the segment from 2 to 3 has been divided into five smaller segments. The distance from 2 to point *A* crosses three of the five segments. Thus, point *A* represents the mixed number $2\frac{3}{5}$.

Note: It is important to focus on the *number of segments* and not on the number of vertical tick marks. The four vertical tick marks divide the space between 2 and 3 into five segments, just as four cuts divide a candy bar into five pieces.

Activity: *Inch Ruler to Sixteenths*

Materials needed:

  • inch ruler made in Lesson 7

In Lesson 7 we made an inch ruler divided into fourths. In this activity we will divide the ruler into eighths and sixteenths. First we will review what we did in Lesson 7.

We used a ruler to make one-inch divisions on a strip of tagboard.

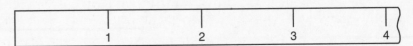

Then we estimated the halfway point between inch marks and drew new marks. The new marks were half-inch divisions. Then we estimated the halfway point between the half-inch marks and made quarter-inch divisions.

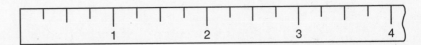

We made the half-inch marks a little shorter than the inch marks and the quarter-inch marks a little shorter than the half-inch marks.

Now divide your ruler into eighths of an inch by estimating the halfway point between the quarter-inch marks. Make these eighth-inch marks a little shorter than the quarter-inch marks.

Finally, divide your ruler into sixteenths by estimating the halfway point between the eighth-inch marks. Make these marks the shortest marks on the ruler.

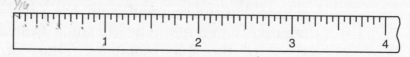

Example 2   Use your ruler to find the length of this line segment to the nearest sixteenth of an inch.

Solution   The ruler has been divided into sixteenths. We align the zero mark (or end of the ruler) with one end of the line segment. Then we find the mark on the ruler closest to the other end of the line segment and read this mark. We will enlarge a portion of a ruler to show how each mark is read.

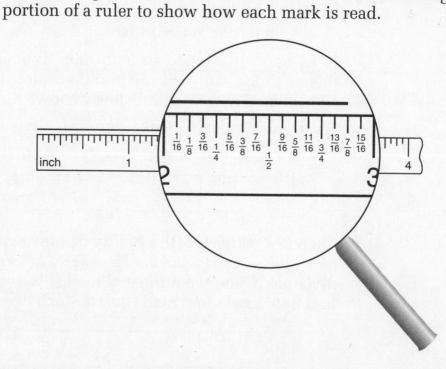

We find that the line segment is about $2\frac{7}{8}$ **inches long.** This is the nearest sixteenth because the end of the segment aligns more closely to the $\frac{7}{8}$ mark $\left(\text{which equals } \frac{14}{16}\right)$ than it does to the $\frac{13}{16}$ mark or to the $\frac{15}{16}$ mark.

LESSON PRACTICE

Practice set   **a.** Continue this sequence to $1\frac{1}{2}$:

$$\frac{1}{16}, \frac{1}{8}, \frac{3}{16}, \frac{1}{4}, \frac{5}{16}, \frac{3}{8}, \frac{7}{16}, \frac{1}{2}, \ldots$$

b. What number is halfway between −2 and −3?

c. What number is halfway between 2 and 5?

d. Point A represents what mixed number on this number line?

Use your ruler to find the length of each of these line segments to the nearest sixteenth of an inch:

e. ———————

f. —————————————

g. ———————————————————

MIXED PRACTICE

Problem set   **1.** What is the sum of twelve thousand, five hundred and
$^{(12)}$ ten thousand, six hundred ten?

2. In 1903 the Wright brothers made the first powered
$^{(13)}$ airplane flight. In 1969 Americans first landed on the
moon. How many years was it from the first powered
airplane flight to the first moon landing? (Write an
equation and solve the problem.)

3. Captain Hook often ran from the sound of ticking clocks.
$^{(15)}$ If he could run 6 yards in one second, how far could he
run in 12 seconds? (Write an equation and solve the
problem.)

4. Aladdin found two dozen gold coins. If the value of
$^{(15)}$ each coin was $1000, what was the value of all the
coins Aladdin found? (Write an equation and solve the
problem.)

5. Estimate the sum of 5280 and 1760 by rounding each
(16) number to the nearest thousand before adding.

6. $\dfrac{480}{3}$
(2)

7. $\dfrac{6 - 6}{3}$
(5)

8. The letters a, b, and c represent three different numbers.
(1) The sum of a and b is c.

$$a + b = c$$

Rearrange the letters to form another addition fact and two subtraction facts.

9. Rewrite $2 \div 3$ with a division bar, but do not divide.
(2)

10. A square has sides 10 cm long. Describe how to find its
(8) perimeter.

11. Use a ruler to find the length of the line segment below to
(17) the nearest sixteenth of an inch.

_____

Find each missing number. Check your work.

12. $\$3 - y = \$1.75$
(3)

13. $m - 20 = 30$
(3)

14. $12n = 0$
(4)

15. $16 + 14 = 14 + w$
(3)

16. Compare: $19 \times 21 \bigcirc 20 \times 20$
(9)

17. $100 - (50 - 25)$
(5)

18. $\dfrac{5280}{44}$
(2)

19. $365 + 4576 + 50{,}287$
(1)

20. What number is missing in the following sequence?
(10)

$$5, 10, \underline{\qquad}, 20, 25, \ldots$$

21. Which digit in 987,654,321 is in the hundred-millions
(12) place?

22. $250{,}000 \div 100$
(2)

23. $\$3.75 \times 10$
(2)

24. The magician pulled 38 rabbits out of a hat. Half of the
(15) rabbits were white. How many rabbits were not white?
(Write an equation and solve the problem.)

25. What is the sum of the first six positive odd numbers?
(10)

26. Describe how to find $\frac{1}{4}$ of 52.
(6)

27. A quarter is $\frac{1}{4}$ of a dollar.
(6)
   (a) How many quarters are in one dollar?

   (b) How many quarters are in three dollars?

28. On an inch ruler, which mark is halfway between the
(17) $\frac{1}{4}$-inch mark and the $\frac{1}{2}$-inch mark?

29. Point A represents what mixed number on the number
(17) line below?

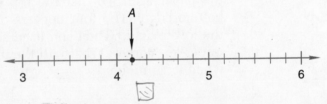

30. A segment that is $\frac{1}{2}$ of an inch long is how many
(17) sixteenths of an inch long?

LESSON

18   Average • Line Graphs

WARM-UP

Facts Practice: 100 Addition Facts (Test B)

Mental Math: Count by 3's from 3 to 60.

a. 4 × 23 equals 4 × 20 plus 4 × 3. Find 4 × 23.

b. 4 × 32   c. 3 × 42   d. 3 × 24

e. Start with a half dozen. Add 2; multiply by 3; divide by 4; then subtract 5. What is the answer?

Problem Solving:

Jeana folded a square piece of paper in half so that the left edge aligned with the right edge. Then she folded the paper again so that the top edge aligned with the bottom edge. (The four corners of the paper aligned at the lower right.) Then she used scissors and cut off the upper left corner. What will the piece of paper look like when it is unfolded?

cut

4 corners

NEW CONCEPTS

Average   Here we show three stacks of books; the stacks contain 8 books, 7 books, and 3 books respectively. Altogether there are 18 books, but the number of books in each stack is not equal.

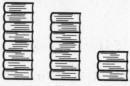

If we move some of the books from the taller stacks to the shortest stack, we can make the three stacks the same height. Then there will be 6 books in each stack.

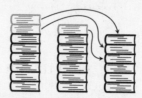

By making the stacks equal, we have found the **average** number of books in the three stacks. Notice that the average number of books in each stack is greater than the number in the smallest stack and less than the number in the largest stack. One way we can find an average is by making equal groups.

Example 1   In four aquariums there were 28 goldfish, 27 goldfish, 26 goldfish, and 31 goldfish respectively. What was the average number of goldfish in each aquarium?

Solution   The average number of goldfish in each aquarium is how many goldfish there would be in each aquarium if we made the numbers equal. So we will take the total number of goldfish and make four equal groups. To find the total number of goldfish, we add the numbers in each aquarium.

$$
\begin{array}{r}
28 \text{ goldfish} \\
27 \text{ goldfish} \\
26 \text{ goldfish} \\
+\ 31 \text{ goldfish} \\
\hline
112 \text{ goldfish in all}
\end{array}
$$

We make four equal groups by dividing the total number of goldfish by four.

$$
\begin{array}{r}
2\,8 \text{ goldfish} \\
4\,\overline{)11^32 \text{ goldfish}}
\end{array}
$$

If the groups were equal, there would be 28 goldfish in each aquarium. The average number of goldfish in each aquarium was **28**.

Notice that an average problem is a "combining" problem and an "equal groups" problem. First we found the total number of goldfish in all the aquariums ("combining" problem). Then we found the number of goldfish that would be in each group if the groups were equal ("equal groups" problem).

Example 2   What is the average of 3, 7, and 8?

Solution   This question does not tell us whether the numbers 3, 7, and 8 refer to books or goldfish or coins or quiz scores. Still, we can find the average of these numbers by combining and then making equal groups. Since there are three numbers, there will be three groups. First we find the total.

$$3 + 7 + 8 = 18$$

Then we divide the total into three equal groups.

$$18 \div 3 = 6$$

The average of 3, 7, and 8 is **6**.

Example 3 What number is halfway between 27 and 81?

Solution The number halfway between two numbers is also the average of the two numbers. For example, the average of 7 and 9 is 8, and 8 is halfway between 7 and 9. So the average of 27 and 81 will be the number halfway between 27 and 81. We add 27 and 81 and divide by 2.

$$\text{Average of 27 and 81} = \frac{27 + 81}{2}$$

$$= \frac{108}{2}$$

$$= 54$$

The number halfway between 27 and 81 is **54.**

Line graphs **Line graphs** display numerical information as points connected by line segments. Whereas bar graphs often display comparisons, line graphs often show how a measurement changes over time.

Example 4 This line graph shows Margie's height in inches from her eighth birthday to her fourteenth birthday. During which year did Margie grow the most?

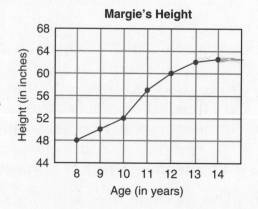

Solution From Margie's eighth birthday to her ninth birthday, she grew about two inches. She also grew about two inches from her ninth to her tenth birthday. From her tenth to her eleventh birthday, Margie grew about five inches. Notice that this is the steepest part of the growth line. So the year Margie grew the most was **the year she was ten.**

Your teacher might ask you to keep a line graph of your math test scores. Activity Sheet 1 in *Saxon Math 7/6—Homeschool Tests and Worksheets* can be used for this purpose.

LESSON PRACTICE

Practice set* **a.** There were 26 books on the first shelf, 36 books on the second shelf, and 43 books on the third shelf. Vijay rearranged the books so that there were the same number of books on each shelf. After Vijay rearranged the books, how many were on the first shelf?

b. What is the average of 96, 44, 68, and 100?

c. What number is halfway between 28 and 82?

d. What number is halfway between 86 and 102?

e. Find the average of 3, 6, 9, 12, and 15.

Use the information in the graph in example 4 to answer these questions:

f. How many inches did Margie grow from her eighth to her twelfth birthday?

g. During which year did Margie grow the least?

h. Based on the information in the graph, would you predict that Margie will grow to be 68 inches tall?

MIXED PRACTICE

Problem set **1.** Jumbo ate two thousand, sixty-eight peanuts in the
(11)   morning and three thousand, nine hundred forty in the afternoon. How many peanuts did Jumbo eat in all? What kind of pattern did you use?

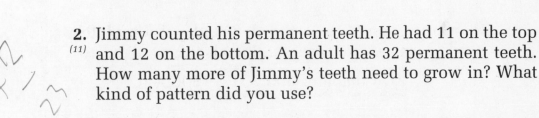

2. Jimmy counted his permanent teeth. He had 11 on the top
(11)   and 12 on the bottom. An adult has 32 permanent teeth. How many more of Jimmy's teeth need to grow in? What kind of pattern did you use?

3. Olive bought one dozen cans of spinach as a birthday
(15)   present for her boyfriend. The spinach cost 53¢ per can. How much did Olive spend on spinach? What kind of pattern did you use?

4. Estimate the difference of 5035 and 1987 by rounding each
(16) number to the nearest thousand before subtracting.

5. Find the average of 9, 7, and 8.
(18)

6. What number is halfway between 59 and 81?
(18)

7. What number is 6 less than 2?
(14)

8. $0.35 × 100   **9.** 10,010 ÷ 10   **10.** 34,180 ÷ 17
(2)   (2)   (2)

11. $3.64 + $94.28 + 87¢   **12.** 41,375 − 13,576
(1)   (1)

13. 125 × 16   **14.** 4 · 3 · 2 · 1 · 0
(2)   (5)

Find each missing number. Check your work.

15. $w − 84 = 48$ **16.** $\dfrac{234}{n} = 6$
(3)   (4)

17. $(1 + 2) × 3 = (1 × 2) + m$
(3, 5)

18. Draw a rectangle 5 cm long and 3 cm wide. What is its
(8) perimeter?

19. What is the sum of the first six positive even numbers?
(10)

20. Describe the rule of the following sequence. Then find
(10) the missing term.

$$1, 2, 4, \underline{\hspace{1cm}}, 16, 32, 64, \ldots$$

21. Compare: 500 × 1 ◯ 500 ÷ 1
(9)

22. What number is $\frac{1}{2}$ of 1110?
(6)

23. What is the place value of the 7 in 987,654,321?
(12)

Refer to the line graph shown below to answer problems 24–26.

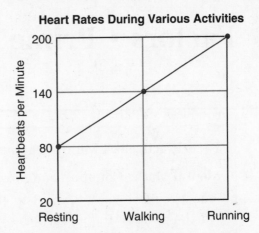

Heart Rates During Various Activities

48
125
125
125
125
125
125
125
125
125
125
125
125
12 5
12 5
125
+ 125
———
2005

24. Running increases a resting person's heart rate by about
(18) how many heartbeats per minute?

25. About how many times would a person's heart beat
(18) during a 10-minute run?

26. Using the information in the line graph, write a story
(18) problem about comparing. Then answer the problem.

27. In three stores there are 24, 27, and 33 customers
(18) respectively. How many customers will be in each store
if some customers move so that the number of customers
in each store is equal?

28. A dime is $\frac{1}{10}$ of a dollar.
(6) (a) How many dimes are in a dollar?

(b) How many dimes are in three dollars?

29. Use a ruler to draw a rectangle that is $2\frac{1}{4}$ inches long and
(17) $1\frac{3}{4}$ inches wide.

30. Story problems about finding an average include which
(18) two types of stories? (Select two.)

combining   separating   comparing   equal groups

LESSON

19

Factors • Prime Numbers

WARM-UP

Facts Practice: 64 Multiplication Facts (Test D)

Mental Math: Count up and down by $\frac{1}{4}$'s between $\frac{1}{4}$ and 12.

a. 3×64
b. 3×46
c. $120 + 18$
d. $34 + 40 + 9$
e. $34 + 50 - 1$
f. $\$20.00 - \$12.50$
g. Start with 100. Divide by 2; subtract 1; divide by 7; then add 3. What is the answer?

Problem Solving:

Alex wanted to make a 40¢ phone call at a pay phone. He had a nickel, a dime, and a quarter. He could put in the nickel, then the dime, then the quarter. Or he could put in the quarter, then the dime, then the nickel. What are the other possible orders of coin drops for his phone call?

NEW CONCEPTS

Factors   Recall from Lesson 2 that a factor is one of the numbers multiplied to form a product.

$$2 \times 3 = 6 \qquad \text{Both 2 and 3 are factors.}$$

$$1 \times 6 = 6 \qquad \text{Both 1 and 6 are factors.}$$

We see that each of the numbers 1, 2, 3, and 6 can serve as a factor of 6. Notice that when we divide 6 by 1, 2, 3, or 6, the resulting quotient has no remainder (that is, it has a remainder of zero). We say that 6 is "divisible by" 1, 2, 3, and 6.

$$
\begin{array}{cccc}
6 & 3 & 2 & 1 \\
1\overline{)6} & 2\overline{)6} & 3\overline{)6} & 6\overline{)6} \\
\underline{6} & \underline{6} & \underline{6} & \underline{6} \\
0 & 0 & 0 & 0
\end{array}
$$

This leads us to another definition of *factor*.

> The **factors** of a given number are the whole numbers that divide the given number without a remainder.

We can illustrate the factors of 6 by arranging 6 tiles to form rectangles. With 6 tiles we can make a 1-by-6 rectangle. We can also make a 2-by-3 rectangle.

The number of tiles along the sides of these two rectangles (1, 6, 2, 3) are the four factors of 6.

Example 1 What are the factors of 10?

Solution The factors of 10 are all the numbers that divide 10 evenly (with no remainder). They are **1, 2, 5,** and **10.**

$$1)\overline{10}^{\,10} \qquad 2)\overline{10}^{\,5} \qquad 5)\overline{10}^{\,2} \qquad 10)\overline{10}^{\,1}$$

We can illustrate the factors of 10 with two rectangular arrays of tiles.

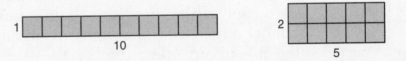

The number of tiles along the sides of the two rectangles (1, 10, 2, 5) are the factors of 10.

Example 2 How many different whole numbers are factors of 12?

Solution Twelve can be divided evenly by 1, 2, 3, 4, 6, and 12. The question asked "How many?" Counting factors, we find that 12 has **6** different whole-number factors.

Twelve tiles can be arranged in three rectangular arrays to illustrate that the six factors of 12 are 1, 12, 2, 6, 3, and 4.

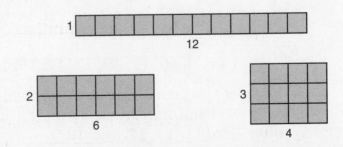

Prime numbers   Here we list the first ten counting numbers and their factors. Which of the numbers have exactly two factors?

NUMBER	FACTORS
1	1
2	1, 2
3	1, 3
4	1, 2, 4
5	1, 5
6	1, 2, 3, 6
7	1, 7
8	1, 2, 4, 8
9	1, 3, 9
10	1, 2, 5, 10

Numbers that have exactly two factors are **prime numbers**. The first four prime numbers are 2, 3, 5, and 7. The number 1 is not a prime number, because it has only one factor, itself. The only factors of a prime number are the number itself and 1. Therefore, to determine whether a number is prime, we may ask ourselves the question, "Is this number divisible by any number other than the number itself and 1?" If the number is divisible by any other number, the number is not prime.

Example 3   The first four prime numbers are 2, 3, 5, and 7. What are the next four prime numbers?

Solution   We will consider the next several numbers and eliminate those that are not prime.

8, 9, 10, 11, 12, 13, 14, 15, 16, 17, 18, 19, 20

All even numbers have 2 as a factor. So no even numbers greater than two are prime numbers. We can eliminate the even numbers from the list.

8̸, 9, 1̸0̸, 11, 1̸2̸, 13, 1̸4̸, 15, 1̸6̸, 17, 1̸8̸, 19, 2̸0̸

Since 9 is divisible by 3, and 15 is divisible by 3 and by 5, we can eliminate 9 and 15 from the list.

8̸, 9̸, 1̸0̸, 11, 1̸2̸, 13, 1̸4̸, 1̸5̸, 1̸6̸, 17, 1̸8̸, 19, 2̸0̸

Each of the remaining four numbers on the list is divisible only by itself and by 1. Thus the next four prime numbers after 7 are **11, 13, 17,** and **19.**

Activity: *Prime Numbers*

List the counting numbers from 1 to 100 (or use the hundred number chart, Activity Sheet 2, from *Saxon Math 7/6— Homeschool Tests and Worksheets*). Then follow these directions:

Step 1: Draw a line through the number 1. The number 1 is not a prime number.

Step 2: Circle the prime number 2. Draw a line through all the other multiples of 2 (4, 6, 8, etc.).

Step 3: Circle the prime number 3. Draw a line through all the other multiples of 3 (6, 9, 12, etc.).

Step 4: Circle the prime number 5. Draw a line through all the other multiples of 5 (10, 15, 20, etc.).

Step 5: Circle the prime number 7. Draw a line through all the other multiples of 7 (14, 21, 28, etc.).

Step 6: Circle all remaining numbers on your list (the numbers that do not have a line drawn through them).

When you have finished, all the prime numbers from 1 to 100 will be circled on your list.

LESSON PRACTICE

Practice set List the factors of the following numbers:

a. 14

b. 15

c. 16

d. 17

Which number in each group is a prime number?

e. 21, 23, 25

f. 31, 32, 33

g. 43, 44, 45

Which number in each group is not a prime number?

h. 41, 42, 43

i. 31, 41, 51

j. 23, 33, 43

Prime numbers can be multiplied to make whole numbers that are not prime. For example, $2 \cdot 2 \cdot 3$ equals 12 and $3 \cdot 5$ equals 15. (Neither 12 nor 15 are prime.) Show which prime numbers we multiply to make these products:

k. 16

l. 18

MIXED PRACTICE

Problem set

1. If two hundred fifty-two is the dividend and six is the quotient, what is the divisor?
(4)

2. Lincoln began his speech, "Fourscore and seven years ago ...." A score is twenty. How many years is fourscore and seven?
(11, 15)

3. Overnight the temperature dropped from 4°C to −3°C. This was a drop of how many degrees?
(14)

4. If 203 turnips are to be shared equally among seven rabbits, how many should each receive? Write an equation and solve the problem.
(15)

5. What is the average of 1, 2, 4, and 9?
(18)

6. What is the next number in the following sequence?
(10)

$$1, 4, 9, 16, 25, \underline{\hspace{1cm}}, \ldots$$

7. A regular hexagon has six sides of equal length. If each side of a hexagon is 25 mm, what is the perimeter?
(8)

8. One centimeter equals ten millimeters. How many millimeters long is the line segment below?
(7)

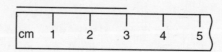

9. What are the whole-number factors of 20?
(19)

10. How many different whole numbers are factors of 15?
(19)

11. Which of the numbers below is a prime number?
(19)

A. 25   B. 27   C. 29

12. 250,000 ÷ 100
(2)

13. 1234 ÷ 60
(2)

14. $\dfrac{6 + 18 + 9}{3}$
(5)

15. $3.45 × 10
(2)

Find each missing number. Check your work.

16. $\$10.00 - w = \$1.93$ **17.** $\dfrac{w}{3} = 4$

(3)   (4)

18. The letters a, b, and c represent three different numbers.
(2) The product of a and b is c.

$$ab = c$$

Rearrange the letters to form another multiplication fact and two division facts.

19. Arrange these numbers in order from least to greatest:
(17)

$$3, -2, 1, \tfrac{1}{2}, 0$$

20. Compare: $123 \div 1 \bigcirc 123 - 1$
(9)

21. Which digit in 135,792,468,000 is in the ten-millions place?
(12)

22. Round 123,456,789 to the nearest million.
(16)

23. How much money is $\tfrac{1}{2}$ of $\$11.00$?
(6)

24. If a square has a perimeter of 48 inches, how long is each
(8) side of the square?

25. $(51 + 49) \times (51 - 49)$
(5)

26. Which of the numbers below is a prime number?
(19)

A. 2   B. 22   C. 222

27. Prime numbers can be multiplied to make whole
(19) numbers that are not prime. To make 18, we perform the multiplication $2 \cdot 3 \cdot 3$. Show which prime numbers we multiply to make 20.

28. The dictionaries are placed in three stacks. There are
(18) 6 dictionaries in one stack and 12 dictionaries in each of the other two stacks. How many dictionaries will be in each stack if some dictionaries are moved from the taller stacks to the shortest stack so that there are the same number of dictionaries in each stack?

29. Draw a square with sides that are $1\tfrac{3}{8}$ inches long.
(17)

30. Describe how to use the concepts of "even" and "odd"
(10, 19) numbers to determine whether a number is divisible by 2.

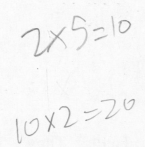
$2 \times 5 = 10$

$10 \times 2 = 20$

LESSON
20   Greatest Common Factor (GCF)

WARM-UP

Facts Practice: 100 Subtraction Facts (Test C)

Mental Math: Count up and down by 3's between 3 and 60.

a. 6×23 b. 6×32 c. $640 + 1200$

d. $63 + 20 + 9$ e. $63 + 30 - 1$ f. $\$100.00 - \$75.00$

g. Start with 10. Multiply by 10; subtract 1; divide by 9; then add 1. What is the answer?

Problem Solving:

Use the digits 5, 6, 7, and 8 to complete this subtraction problem. There are two possible arrangements.

$$\begin{array}{r} \overline{} \\ -\ 9 \\ \hline \overline{} \end{array}$$

NEW CONCEPT

The factors of 8 are

1, 2, 4, and 8

The factors of 12 are

1, 2, 3, 4, 6, and 12

We see that 8 and 12 have some of the same factors. They have three factors in common. These common factors are 1, 2, and 4. Their **greatest common factor**—the largest factor that they both have—is 4. *Greatest common factor* is often abbreviated **GCF.**

Example 1 Find the greatest common factor of 12 and 18.

Solution The factors of 12 are

1, 2, 3, 4, 6, and 12

The factors of 18 are

1, 2, 3, 6, 9, and 18

We see that 12 and 18 share four common factors: 1, 2, 3, and 6. The greatest of these is **6.**

Example 2 Find the GCF of 6, 9, and 15.

Solution The factors of 6 are

1, 2, 3, and 6

The factors of 9 are

1, 3, and 9

The factors of 15 are

1, 3, 5, and 15

The GCF of 6, 9, and 15 is **3**.

Note: The search for the greatest common factor of two or more numbers is a search for the **largest** number that evenly divides each of them. In this problem we can quickly determine that 3 is the largest number that evenly divides 6, 9, and 15. A complete listing of the factors might be helpful but is not required.

LESSON PRACTICE

Practice set* Find the greatest common factor (GCF) of the following:

a. 10 and 15   **b.** 18 and 27

c. 18 and 24   **d.** 12, 18, and 24

e. 15 and 25   **f.** 20, 30, and 40

g. 12 and 15   **h.** 20, 40, and 60

MIXED PRACTICE

Problem set **1.** What is the difference between the product of 12 and 8
(12)   and the sum of 12 and 8?

2. Saturn's average distance from the Sun is one billion,
(12)   four hundred twenty-seven million kilometers. Use digits to write that distance.

3. Which digit in 497,325,186 is in the ten-millions place?
(12)

4. Romulus actually had $427,872, but when Remus asked
(16) him how much money he had, Romulus rounded the
amount to the nearest thousand dollars. How much did
Romulus say he had?

5. The morning temperature was −3°C. By afternoon it
(14) had warmed to 8°C. How many degrees had the
temperature risen?

6. In three games Allen scored 31, 52, and 40 points. What
(18) was the average number of points Allen scored per game?

7. Find the greatest common factor of 12 and 20.
(20)

8. Find the GCF of 9, 15, and 21.
(20)

9. How much money is $\frac{1}{4}$ of $3.24?
(6)

10. 5432 ÷ 10   **11.** $\dfrac{28 + 42}{14}$
(2)   (5)

12. 56,042 + 49,985   **13.** 37,080 ÷ 12
(1)   (2)

14. $6.47 × 10   **15.** 5 × 4 × 3 × 2 × 1
(2)   (5)

Find each missing number. Check your work.

16. $w - 76 = 528$ **17.** $14,009 - w = 9670$
(3)   (3)

18. $6w = 90$ **19.** $q - 365 = 365$
(4)   (3)

20. $365 - p = 365$
(3)

21. Find the missing number in the following sequence:
(10)

_____, 10, 16, 22, 28, …

22. Compare: 50 − 1 ◯ 49 + 1
(9)

23. The first positive odd number is 1. What is the tenth
(10) positive odd number?

24. The perimeter of a square is 100 cm. Describe how to find
(8) the length of each side.

25. Estimate the length of this key to the nearest inch. Then
(17) use a ruler to find the length of the key to the nearest
sixteenth of an inch.

26. A "bit" is $\frac{1}{8}$ of a dollar.
(6)
    (a) How many bits are in a dollar?

    (b) How many bits are in three dollars?

27. In four boxes there are 12, 24, 36, and 48 golf balls
(18) respectively. If the golf balls are rearranged so that there
are the same number of golf balls in each of the four boxes,
how many golf balls will be in each box?

28. Which of the numbers below is a prime number?
(19)
    A. 5   B. 15   C. 25

29. List the whole-number factors of 24.
(19)

30. Ten billion is how much less than one trillion?
(12)

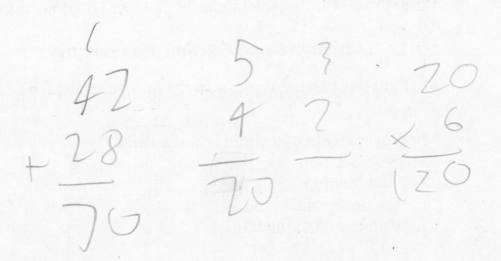

INVESTIGATION 2

Focus on

Investigating Fractions with Manipulatives

In this investigation you will make a set of fraction manipulatives to help you answer questions in this investigation and in future problem sets.

Materials needed:

- Activity Sheets 3–7 (available in *Saxon Math 7/6— Homeschool Tests and Worksheets*)
- scissors
- colored pencils or markers (optional)
- envelope or zip-top plastic bag in which to store fraction pieces

Preparation:

If desired, color each sheet's fraction circles a different color; that is, one color for halves, one color for thirds, and so on. (Color-coding the fraction circles will make sorting the manipulatives easier.) Separate the manipulatives by cutting out the circles and cutting apart the fraction slices along the lines. (*Note:* Cutting the circles of $\frac{1}{8}$ pieces is not necessary.)

Activity: *Using Fraction Manipulatives*

Use your fraction manipulatives to help you with these exercises:

1. What percent of a circle is $\frac{1}{2}$ of a circle?

2. What fraction is half of $\frac{1}{2}$?

3. What fraction is half of $\frac{1}{4}$?

4. Fit three $\frac{1}{4}$ pieces together to form $\frac{3}{4}$ of a circle. Three fourths of a circle is what percent of a circle?

5. Fit four $\frac{1}{8}$ pieces together to form $\frac{4}{8}$ of a circle. Four eighths of a circle is what percent of a circle?

6. Fit three $\frac{1}{6}$ pieces together to form $\frac{3}{6}$ of a circle. Three sixths of a circle is what percent of a circle?

7. Show that $\frac{4}{8}$, $\frac{3}{6}$, and $\frac{2}{4}$ each make one half of a circle. $\left(\text{We say that } \frac{4}{8}, \frac{3}{6}, \text{ and } \frac{2}{4} \text{ all } \textit{reduce} \text{ to } \frac{1}{2}.\right)$

8. The fraction $\frac{2}{8}$ equals which single fraction piece?

9. The fraction $\frac{6}{8}$ equals how many $\frac{1}{4}$'s?

10. The fraction $\frac{2}{6}$ equals which single fraction piece?

11. The fraction $\frac{4}{6}$ equals how many $\frac{1}{3}$'s?

12. The sum $\frac{1}{8} + \frac{1}{8} + \frac{1}{8}$ is $\frac{3}{8}$. If you add $\frac{3}{8}$ and $\frac{2}{8}$, what is the sum?

13. Form a whole circle using six of the $\frac{1}{6}$ pieces. Then remove (subtract) $\frac{1}{6}$. What fraction of the circle is left?

14. Demonstrate subtracting $\frac{1}{3}$ from 1 by forming a circle of $\frac{3}{3}$ and then removing $\frac{1}{3}$. What fraction is left?

15. Use four $\frac{1}{4}$'s to demonstrate the subtraction $1 - \frac{1}{4}$. Then write the answer.

16. Eight $\frac{1}{8}$'s form one circle. If $\frac{3}{8}$ of a circle is removed from one circle $\left(1 - \frac{3}{8}\right)$, then what fraction of the circle remains?

17. What percent of a circle is $\frac{1}{3}$ of a circle?

18. What percent of a circle is $\frac{1}{6}$ of a circle?

Fraction manipulatives can help us compare fractions. Since $\frac{1}{2}$ of a circle is larger than $\frac{1}{3}$ of a circle, we can see that

$$\frac{1}{2} > \frac{1}{3}$$

For problems 19 and 20, use your fraction manipulatives to construct models of the fractions. Use the models to help you write the correct comparison for each problem.

19. Compare: $\frac{2}{3} \bigcirc \frac{3}{4}$

20. Compare: $\frac{2}{3} \bigcirc \frac{3}{8}$

We can also draw pictures to help us compare fractions.

21. Draw two rectangles of the same size. Shade $\frac{1}{3}$ of one rectangle and $\frac{1}{5}$ of the other rectangle. What fraction represents the rectangle that has the larger amount shaded?

22. Draw and shade rectangles to illustrate this comparison:

$$\frac{3}{5} > \frac{3}{10}$$

Problems 23–29 involve **improper fractions.** Improper fractions are fractions that are equal to or greater than 1. In a fraction equal to 1 the numerator equals the denominator (as in $\frac{3}{3}$). In a fraction greater than 1 the numerator is greater than the denominator (as in $\frac{4}{3}$).

23. Show that the improper fraction $\frac{5}{4}$ equals the mixed number $1\frac{1}{4}$ by combining four of the $\frac{1}{4}$ pieces to make a whole circle.

24. The improper fraction $\frac{7}{4}$ equals what mixed number?

25. The improper fraction $\frac{3}{2}$ equals what mixed number?

26. Form $1\frac{1}{2}$ circles using only $\frac{1}{4}$'s. How many $\frac{1}{4}$ pieces are needed to make $1\frac{1}{2}$?

27. How many $\frac{1}{3}$ pieces are needed to make two whole circles?

28. The improper fraction $\frac{4}{3}$ equals what mixed number?

29. Convert $\frac{11}{6}$ to a mixed number.

30. An analog clock can serve as a visual reference for twelfths. At 1 o'clock the hands mark off $\frac{1}{12}$ of a circle; at 2 o'clock the hands mark off $\frac{2}{12}$ of a circle, and so on. How many twelfths are in each of these fractions of a circle? *Hint:* Try holding each fraction piece at arm's length in the direction of the clock (as an artist might extend a thumb toward a subject.)

$$\frac{1}{2}, \frac{1}{4}, \frac{1}{3}, \frac{1}{6}$$

After you have completed the exercises, gather and store your fraction manipulatives in a bag or envelope for later use.

LESSON

21 Divisibility

WARM-UP

NEW CONCEPT

There are ways of discovering whether some numbers are factors of other numbers without actually dividing. For instance, even numbers can be divided by 2. Therefore, 2 is a factor of every even counting number. Since even numbers are "able" to be divided by 2, we say that even numbers are "divisible" by 2. Tests for **divisibility** can help us find the factors of a number. Here we list divisibility tests for the numbers 2, 3, 5, 9, and 10.

Last-Digit Tests

> Inspect the last digit of the number. A number is divisible by …
>
>   2 if the last digit is even.
>
>   5 if the last digit is 0 or 5.
>
>   10 if the last digit is 0.

Sum-of-Digits Tests

> Add the digits of the number and inspect the total. A number is divisible by …
>
>   3 if the sum of the digits is divisible by 3.
>
>   9 if the sum of the digits is divisible by 9.

[†]As a shorthand, we will use commas to separate operations to be performed sequentially from left to right. In this case, 25 × 2 = 50, then 50 − 1 = 49, then 49 ÷ 7 = 7, then 7 + 1 = 8, then 8 ÷ 2 = 4. The answer is 4.

Example 1 Which of these numbers is divisible by 2?

365   1179   1556

Solution To determine whether a number is divisible by 2, we inspect the last digit of the number. If the last digit is an even number, then the number is divisible by 2. The last digits of these three numbers are 5, 9, and 6. Since 5 and 9 are not even numbers, neither 365 nor 1179 is divisible by 2. Since 6 is an even number, 1556 is divisible by 2. It is not necessary to perform the division to answer the question. By inspecting the last digit of each number, we see that the number that is divisible by 2 is **1556.**

Example 2 Which of these numbers is divisible by 3?

365   1179   1556

Solution To determine whether a number is divisible by 3, we add the digits of the number and then inspect the sum. If the sum of the digits is divisible by 3, then the number is also divisible by 3.

The digits of 365 are 3, 6, and 5. The sum of these is 14.

$$3 + 6 + 5 = 14$$

We try to divide 14 by 3 and find that there is a remainder of 2. Since 14 is not divisible by 3, we know that 365 is not divisible by 3 either.

The digits of 1179 are 1, 1, 7, and 9. The sum of these digits is 18.

$$1 + 1 + 7 + 9 = 18$$

We divide 18 by 3 and get no remainder. We see that 18 is divisible by 3, so 1179 is also divisible by 3.

The sum of the digits of 1556 is 17.

$$1 + 5 + 5 + 6 = 17$$

Since 17 is not divisible by 3, the number 1556 is not divisible by 3.

By using the divisibility test for 3, we find that the number that is divisible by 3 is **1179.**

Example 3 Which of the numbers 2, 3, 5, 9, and 10 are factors of 135?

Solution First we will use the last-digit tests. The last digit of 135 is 5, so 135 is divisible by 5 but not by 2 or by 10. Next we use the sum-of-digits tests. The sum of the digits in 135 is 9 (1 + 3 + 5 = 9). Since 9 is divisible by both 3 and 9, we know that 135 is also divisible by 3 and 9. So **3, 5,** and **9** are factors of 135.

LESSON PRACTICE

Practice set

a. Which of these numbers is divisible by 2?

123   234   345

b. Which of these numbers is divisible by 3?

1234   2345   3456

Use the divisibility tests to decide which of the numbers 2, 3, 5, 9, and 10 are factors of the following numbers:

c. 120   **d.** 102

MIXED PRACTICE

Problem set

1. What is the product of the sum of 8 and 5 and the difference of 8 and 5?
(12)

2. In 1787 Delaware became the first state. In 1959 Hawaii became the fiftieth state admitted to the Union. How many years were there between these two events? (Write an equation and solve the problem.)
(13)

3. Mariam figured that the bowling balls on the rack weighed a total of 240 pounds. How many 16-pound bowling balls weigh a total of 240 pounds? (Write an equation and solve the problem.)
(15)

4. An apple pie was cut into four equal slices. One slice was quickly eaten. What fraction of the pie was left?
(6)

5. There are 17 girls at a party with 30 guests. What fraction of the party guests are girls?
(6)

6. Use digits to write the fraction three hundredths.
(6)

7. How much money is $\frac{1}{2}$ of $2.34?
(6)

8. What is the place value of the 7 in 987,654,321?
(12)

9. Describe the rule of the following sequence. Then find
(10) the next term.

$$1, 4, 16, 64, \underline{\hspace{1cm}}, \ldots$$

10. Compare: 64 × 1 ◯ 64 + 1
(9)

11. Which of these numbers is divisible by 9?
(21)

A. 365   B. 1179   C. 1556

12. Estimate the sum of 396, 197, and 203 by rounding each
(16) number to the nearest hundred before adding.

13. What is the greatest common factor (GCF) of 12 and 16?
(20)

14. 100)$\overline{4030}$ **15.** 48,840 ÷ 24
(2)   (2)

16. $\frac{678}{6}$ **17.** $4.75 × 10
(2)   (2)

Find each missing number. Check your work.

18. $10 − w = 87¢   **19.** 463 + 27 + m = 500
(3)   (3)

20. Arrange these numbers in order from least to greatest:
(17)

$$1, \frac{1}{2}, 0, -2, \frac{1}{4}$$

21. What is the average of 12, 16, and 23?
(18)

22. List the whole numbers that are factors of 28.
(19)

23. A regular octagon has eight sides of equal length. What is
(8) the perimeter of a regular octagon with sides 18 cm long?

24. Use an inch ruler to draw a line segment four inches long.
(7) Then use a centimeter ruler to find the length to the
nearest centimeter.

25. (12 × 12) − (11 × 13)
(5)

26. To divide a circle into thirds, John
(Inv. 2) first imagined the face of a clock. From the center of the "clock," he drew one segment up to the 12. Then, starting from the center, John drew two other segments. To which two numbers on the "clock" did John draw the two segments when he divided the circle into thirds?

27. Draw and shade rectangles to illustrate this comparison:
(Inv. 2)

$$\frac{2}{3} < \frac{3}{4}$$

28. A "bit" is $\frac{1}{8}$ of a dollar.
(6)

(a) How many bits are in a dollar?

(b) How many bits are in a half-dollar?

29. What whole numbers are factors of both 20 and 30?
(19)

30. Describe a method for dividing a circle into eight equal
(Inv. 2) parts that involves drawing a plus sign and a times sign. Illustrate the explanation.

LESSON
22

"Equal Groups" Stories with Fractions

WARM-UP

Facts Practice: 100 Subtraction Facts (Test C)

Mental Math: Count by 2's from 2 to 40. Count by 4's from 4 to 40. Count up and down by $\frac{1}{4}$'s between $\frac{1}{4}$ and 12.

a. 4×54

b. 3×56

c. $36 + 29$ *(29 is 30 − 1.)*

d. $359 - 42$

e. $\$10.00 - \$3.50$

f. $\frac{1}{2}$ of 48

g. Start with 100, − 1, ÷ 9, + 1, ÷ 2, − 1, × 5

Problem Solving:

Two dot cubes are tossed. The total number of dots on the two top faces is 6. What is the total of the dots on the bottom faces of the two dot cubes?

NEW CONCEPT

Here we show a collection of six objects. The collection is divided into three equal groups. We see that there are two objects in $\frac{1}{3}$ of the collection. We also see that there are four objects in $\frac{2}{3}$ of the collection.

This collection of twelve objects has been divided into four equal groups. There are three objects in $\frac{1}{4}$ of the collection, so there are nine objects in $\frac{3}{4}$ of the collection.

Example 1 Two thirds of the 12 musicians played guitars. How many of the musicians played guitars?

Solution This is a two-step problem. First we divide the 12 musicians into three equal groups (thirds). Each group contains 4 musicians. Then we count the number of musicians in two of the three groups.

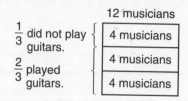

Since there are 4 musicians in each third, the number of musicians in two thirds is 8. We find that **8 musicians** played guitars.

Example 2 Cory has finished $\frac{3}{4}$ of the 28 problems on the assignment. How many problems has Cory finished?

Solution First we divide the 28 problems into four equal groups (fourths). Then we find the number of problems in three of the four groups. Since 28 ÷ 4 is 7, there are 7 problems in each group (in each fourth).

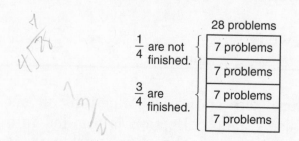

In each group there are 7 problems. So in two groups there are 14 problems, and in three groups there are 21 problems. We see that Cory has finished **21 problems.**

Example 3 How much money is $\frac{3}{5}$ of $3.00?

Solution First we divide $3.00 into five equal groups. Then we find the amount of money in three of the five groups. We divide $3.00 by 5 to find the amount of money in each group.

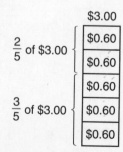

$$\frac{\$0.60}{5\overline{)\$3.00}} \text{ in each group}$$

Now we multiply $0.60 by 3 to find the amount of money in three groups.

$$\begin{array}{r} \$0.60 \\ \times \quad 3 \\ \hline \$1.80 \end{array}$$

We find that $\frac{3}{5}$ of $3.00 is **$1.80.**

Example 4   What number is $\frac{3}{4}$ of 100?

Solution   We divide 100 into four equal groups. Since $100 \div 4$ is 25, there are 25 in each group. We will find the total of three of the parts.

$$3 \times 25 = 75$$

Example 5   (a) What percent of a whole circle is $\frac{1}{5}$ of a circle?

(b) What percent of a whole circle is $\frac{3}{5}$ of a circle?

Solution   A whole circle is 100%. We divide 100% into five equal groups.

(a) One of the five parts $\left(\frac{1}{5}\right)$ is **20%.**

(b) Three of the five parts $\left(\frac{3}{5}\right)$ is $3 \times 20\%$, which equals **60%.**

LESSON PRACTICE

Practice set*   Answer problems **a–f.** Draw a diagram to illustrate each problem.

a. Three fourths of the 12 musicians could play the piano. How many of the musicians could play the piano?

b. How much money is $\frac{2}{3}$ of $4.50?

c. What number is $\frac{4}{5}$ of 60?

      d. What number is $\frac{3}{10}$ of 80?

      e. Five sixths of 24 is what number?

      f. Giovanni answered $\frac{9}{10}$ of the questions correctly. What percent of the questions did Giovanni answer correctly?

MIXED PRACTICE

Problem set

1. When the sum of 15 and 12 is subtracted from the product
⁽¹²⁾ of 15 and 12, what is the difference?

2. There were 13 original states. There are now 50 states.
⁽⁶⁾ What fraction of the states are the original states?

3. A marathon race is 26 miles plus 385 yards long. A mile
^(11, 15) is 1760 yards. Altogether, how many yards long is a marathon? (First use a multiplication pattern to find the number of yards in 26 miles. Then use an addition pattern to include the 385 yards.)

4. If $\frac{2}{3}$ of the 12 jelly beans were eaten, how many were
⁽²²⁾ eaten? Draw a diagram to illustrate the problem.

5. What number is $\frac{3}{4}$ of 16? Draw a diagram to illustrate
⁽²²⁾ the problem.

6. How much money is $\frac{3}{10}$ of $3.50? Draw a diagram to
⁽²²⁾ illustrate the problem.

7. As Shannon rode her bike out of the low desert, the
⁽¹⁴⁾ elevation changed from −100 ft to 600 ft. What was the total elevation change for her ride?

Find each missing number. Check your work.

8. $w - 15 = 8$
⁽³⁾

9. $\dfrac{w}{15} = 345$
⁽⁴⁾

10. 36¢ + $4.78 + $34.09
⁽¹⁾

11. $12.45 ÷ 3
⁽²⁾

12. $35\overline{)1000}$
(2)

13. $\dfrac{7 + 9 + 14}{3}$
(5)

14. Find the product of 36 and 124, and then round the
(16) answer to the nearest hundred.

15. Which digit in 375,426,198,000 is in the ten-millions place?
(12)

16. Find the greatest common factor of 12 and 15.
(20)

17. List the whole numbers that are factors of 30.
(19)

18. The number 100 is divisible by which of these numbers:
(21) 2, 3, 5, 9, 10?

19. Jeb answered $\frac{4}{5}$ of the questions correctly. What percent of
(22) the questions did Jeb answer correctly? Draw a diagram to
illustrate the problem.

20. Compare: $\dfrac{1}{3} \bigcirc \dfrac{1}{2}$
(9)

21. Which of these numbers is not a prime number?
(19)

    A. 19   B. 29   C. 39

22. $(3 + 3) - (3 \times 3)$
(5, 14)

23. Find the number halfway between 27 and 43.
(18)

24. What is the perimeter of the rectangle below?
(8)

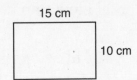

15 cm

10 cm

25. Use an inch ruler to find the length of the line segment
(7) below.

_____

26. An apple pie and a cherry pie were baked in pans of
(Inv. 2) equal size. The apple pie was cut into six equal slices,
and the cherry pie was cut into five equal slices. Which
was larger, a slice of apple pie or a slice of cherry pie?

27. Compare these fractions. Draw and shade rectangles to
(Inv. 2) illustrate the comparison.

$$\frac{2}{4} \bigcirc \frac{3}{5}$$

28. A quarter of a year is $\frac{1}{4}$ of a year. There are 12 months in a
(22) year. How many months are in a quarter of a year? Draw a
diagram to illustrate the problem.

29. A "bit" is one eighth of a dollar.
(6)

(a) How many bits are in a dollar?

(b) How many bits are in a quarter of a dollar?

30. The letters c, p, and t represent three different numbers.
(1) When p is subtracted from c, the answer is t.

$$c - p = t$$

Use these letters to write another subtraction fact and two
addition facts.

LESSON
23 Ratio

WARM-UP

Facts Practice: 64 Multiplication Facts (Test D)

Mental Math: Count up and down by 3's between 3 and 60.
Count up and down by $\frac{1}{4}$'s between $\frac{1}{4}$ and 12.

a. 5 × 62
b. 5 × 36
c. 87 + 9 *(9 is 10 − 1.)*
d. 1200 + 350
e. $20.00 − $15.50
f. $\frac{1}{2}$ of 84
g. 10 × 3, + 2, ÷ 4, + 1, ÷ 3, × 4, ÷ 6

Problem Solving:

Sarah used eight sugar cubes to make a larger cube as shown. The cube she made was two cubes tall, two cubes wide, and two cubes deep. How many cubes will she need to make a cube that has three smaller cubes along each edge?

NEW CONCEPT

A **ratio** is a way to describe a relationship between numbers. If there are 13 boys and 15 girls at the park, then the ratio of boys to girls is 13 to 15. Ratios can be written in several forms. Each of these forms is a way to write the boy-girl ratio:

$$13 \text{ to } 15 \qquad 13{:}15 \qquad \frac{13}{15}$$

Each of these forms is read the same: "Thirteen to fifteen."

In this lesson we will focus on the fraction form of a ratio. When writing a ratio in fraction form, we keep the following points in mind:

1. We write the terms of the ratio in the order we are asked to give them.

2. We reduce ratios in the same manner as we reduce fractions.

3. We leave ratios in fraction form. We do not write ratios as mixed numbers.

Example 1   A team lost 3 games and won 7 games. What was the team's win-loss ratio?

Solution   The question asks for the ratio in the order of wins, then losses. The team's win-loss ratio was 7 to 3, which we write as the fraction $\frac{7}{3}$.

$$\frac{\text{number of games won}}{\text{number of games lost}} = \frac{7}{3}$$

We leave the ratio in fraction form.

Example 2   On a team of 28 players, there are 13 boys. What is the ratio of boys to girls on the team?

Solution   To write the ratio, we need to know the number of girls. If 13 of the 28 players are boys, then 15 of the players are girls. We are asked to write the ratio in "boys to girls" order.

$$\frac{\text{number of boys}}{\text{number of girls}} = \frac{13}{15}$$

LESSON PRACTICE

Practice set   **a.** What is the ratio of dogs to cats in a neighborhood that has 19 cats and 12 dogs?

b. What is the girl-boy ratio in a group of 30 children with 17 boys?

c. If the ratio of cars to trucks in the parking lot is 7 to 2, what is the ratio of trucks to cars in the parking lot?

MIXED PRACTICE

Problem set   **1.** How many millimeters long is a ruler that is 30 cm long?
    (7)

2. Dan has finished $\frac{2}{3}$ of the 27 problems on an assignment.
(22)   How many problems has Dan finished? Draw a diagram to illustrate the problem.

3. William Tell shot at an apple from 100 paces away. If
(15)   each pace was 36 inches, how many inches away was the apple? (Write an equation and solve the problem.)

4. There are 31 days in December. After December 25, what
(6)   fraction of the month remains?

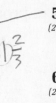

5'10⅔

5. What number is $\frac{3}{5}$ of 25? Draw a diagram to illustrate the problem.
(22)

6. How much money is $\frac{7}{10}$ of $36.00? Draw a diagram to illustrate the problem.
(22)

Use your fraction manipulatives to help answer problems 7–9.

7. What is the sum of $\frac{3}{8}$ and $\frac{4}{8}$?
(Inv. 2)

1'7
1'5

8. The improper fraction $\frac{9}{8}$ equals what mixed number?
(Inv. 2)

9. Two eighths of a circle is what percent of a circle?
(Inv. 2)

10. $3.75 · 16
(2)

11. $\frac{\$3.75}{25}$
(2)

12. What is the place value of the 6 in 36,174,591?
(12)

Bewnand
Size
Comparasion

13. Describe a way to find $\frac{2}{3}$ of a number.
(22)

Find each missing number. Check your work.

14. $0.35n = $35.00
(4)

15. $10.20 − m = $3.46
(3)

16. Compare: $\frac{3}{4}$ ◯ 1
(17)

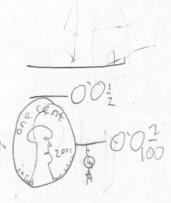

17. The length of a rectangle is 20 inches. The width of the rectangle is half its length. What is the perimeter of the rectangle?
(8)

18. What is the sixth number in the following sequence?
(10)

2, 4, 8, 16, ...

Mhi
bewnand
Size
comparasion

19. Estimate the sum of 3174 and 4790 to the nearest thousand.
(16)

20. Compare: 12 ÷ 6 − 2 ◯ 12 ÷ (6 − 2)
(9)

21. What is the greatest common factor (GCF) of 24 and 32?
(20)

22. What is the sum of the first seven positive odd numbers?
(10)

Use your fraction manipulatives to help answer problems 23–25.

23. (a) How many $\frac{1}{4}$'s are in 1?
(Inv. 2)

   (b) How many $\frac{1}{4}$'s are in $\frac{1}{2}$?

24. One eighth of a circle is what percent of a circle?
(Inv. 2)

25. Write a fraction with a denominator of 8 that is equal to $\frac{1}{2}$.
(Inv. 2)

26. In a group of 23 children, there were 10 boys. What was
(23) the boy-girl ratio?

27. Which prime numbers are greater than 20 but less than 30?
(19)

28. Which of the figures below represents a line?
(7)

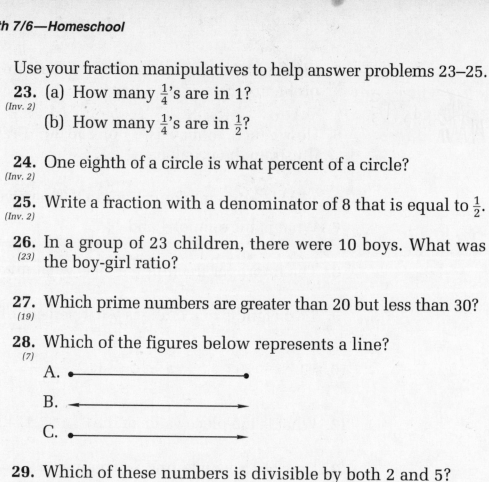

   A.

   B.

   C.

29. Which of these numbers is divisible by both 2 and 5?
(21)

   A. 252   B. 525   C. 250

30. If a team lost 9 games and won 5 games, then what is the
(23) team's win-loss ratio?

LESSON
24
Adding and Subtracting Fractions That Have Common Denominators

WARM-UP

Facts Practice: 100 Subtraction Facts (Test C)

Mental Math: Count by 4's from 4 to 80.
Count up and down by $\frac{1}{4}$'s between $\frac{1}{4}$ and 12.

a. 6×24 **b.** 4×75 **c.** $47 + 39$
d. $1500 - 250$ **e.** $\$20.00 - \$14.50$ **f.** $\frac{1}{2}$ of 68
g. $6 \times 7, -2, \div 5, \times 2, -1, \div 3$

Problem Solving:

Xavier, Yolanda, and Zollie finished first, second, and third in the race, though not necessarily in that order. List all the possible orders of finish. If Xavier was not first, how many possible orders of finish are there?

NEW CONCEPT

Using our fraction manipulatives, we see that when we add $\frac{2}{8}$ to $\frac{3}{8}$ the sum is $\frac{5}{8}$.

$$\frac{3}{8} + \frac{2}{8} = \frac{5}{8}$$

Three eighths plus two eighths equals five eighths.

Likewise, if we subtract $\frac{2}{8}$ from $\frac{5}{8}$, then $\frac{3}{8}$ are left.

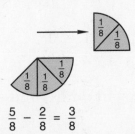

$$\frac{5}{8} - \frac{2}{8} = \frac{3}{8}$$

Five eighths minus two eighths equals three eighths.

Notice that we add the numerators when we add fractions that have the same denominator, and we subtract the numerators when we subtract fractions that have the same denominator. The denominators of the fractions do not change when we add or subtract fractions that have the same denominator.

Example 1 Add: $\frac{1}{4} + \frac{1}{4} + \frac{1}{4}$

Solution The denominators are the same. We add the numerators.

$$\frac{1}{4} + \frac{1}{4} + \frac{1}{4} = \frac{3}{4}$$

Example 2 Add: $\frac{1}{2} + \frac{1}{2}$

Solution One half plus one half is two halves, which is one whole.

$$\frac{1}{2} + \frac{1}{2} = \frac{2}{2} = 1$$

Example 3 Add: $\frac{3}{4} + \frac{3}{4} + \frac{3}{4} + \frac{3}{4}$

Solution The denominators are the same. We add the numerators.

$$\frac{3}{4} + \frac{3}{4} + \frac{3}{4} + \frac{3}{4} = \frac{12}{4} = 3$$

Example 4 Subtract: $\frac{7}{8} - \frac{2}{8}$

Solution The denominators are the same. We subtract the numerators.

$$\frac{7}{8} - \frac{2}{8} = \frac{5}{8}$$

Example 5 Subtract: $\frac{1}{2} - \frac{1}{2}$

Solution If we start with $\frac{1}{2}$ and subtract $\frac{1}{2}$, then what is left is zero.

$$\frac{1}{2} - \frac{1}{2} = \frac{0}{2} = 0$$

LESSON PRACTICE

Practice set Find each sum or difference:

a. $\frac{3}{8} + \frac{4}{8}$

b. $\frac{3}{4} + \frac{1}{4}$

c. $\frac{1}{8} + \frac{1}{8} + \frac{1}{8}$

d. $\frac{4}{8} - \frac{1}{8}$

e. $\frac{3}{4} - \frac{2}{4}$

f. $\frac{1}{4} - \frac{1}{4}$

MIXED PRACTICE

Problem set

1. Martin worked in the yard for five hours and was paid
(11, 15) $6.00 per hour. Then he was paid $5.00 for washing the car. Altogether, how much money did Martin earn? What pattern did you use to find Martin's yard-work earnings? What pattern did you use to find his total earnings?

2. In one bite, Cookie ate $\frac{3}{4}$ of a dozen chocolate chip cookies.
(22) How many cookies did Cookie eat in that bite? Draw a diagram to illustrate the problem.

3. One mile is one thousand, seven hundred sixty yards.
(22) How many yards is $\frac{1}{8}$ of a mile? Draw a diagram to illustrate the problem.

Use your fraction manipulatives to help answer problems 4–8.

4. $\frac{1}{4} + \frac{2}{4}$
(24)

5. $\frac{7}{8} - \frac{4}{8}$
(24)

6. $\frac{1}{2} + \frac{1}{2}$
(24)

7. $\frac{1}{2} - \frac{1}{2}$
(24)

8. What percent of a circle is $\frac{1}{2}$ of a circle plus $\frac{1}{4}$ of a circle?
(Inv. 2)

9. In the bookcase there were 23 nonfiction books and
(23) 41 fiction books. What was the ratio of fiction to nonfiction books in the bookcase?

10. Describe how to find the number halfway between 123
(18) and 321.

11. Paul wanted to fence in a square pasture for Babe, his
(8) blue ox. Each side needed to be 25 miles long. How many
miles of fence did Paul need for the pasture?

12. Which of these numbers is not a prime number?
(19)

    A. 21   B. 31   C. 41

13. $9\overline{)1000}$ **14.** $22{,}422 \div 32$
(2)   (2)

15. $\$350.00 \div 100$ **16.** Compare: $\frac{1}{2} \bigcirc \frac{1}{4}$
(2)   (17)

17. Mr. Johnson rented a moving van and will drive from
(16) Seattle, Washington, to San Francisco, California. On the
way to San Francisco he will go through Portland,
Oregon. The distance from Seattle to Portland is 172
miles, and the distance from Portland to San Francisco is
636 miles. The van rental company charges extra if a van
is driven more than 900 miles. If Mr. Johnson stays with
his planned route, will he be charged extra for the van?
To solve this problem, do you need an exact answer or an
estimate? Explain your thinking.

18. What temperature is shown on the
(10) thermometer at right?

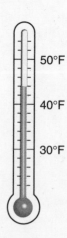

19. $(35 \times 35) - (5 \times 5)$
(5)

20. Round 32,987,145 to the nearest million.
(16)

21. What is the GCF of 21 and 28?
(20)

22. Which of these numbers is divisible by 9?
(21)

    A. 123   B. 234   C. 345

23. Write a fraction equal to 1 that has 4 as the denominator.
(Inv. 2)

Find each missing number. Check your work.

24. $\dfrac{w}{8} = 20$ **25.** $7x = 84$
(4)   (4)

26. $376 + w = 481$ **27.** $m - 286 = 592$
(3)   (3)

Refer to the bar graph shown below to answer problems 28–30.

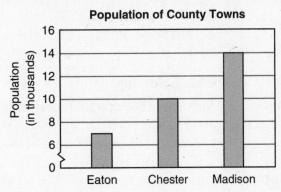

Population of County Towns

28. Which town has about twice the population of Eaton?
(16)

29. About how many more people live in Madison than in
(16) Chester?

30. Copy this graph on your paper, and add a fourth town to
(Inv. 1) your graph: Wilson, population 11,000.

LESSON

25

Writing Division Answers as Mixed Numbers • Multiples

WARM-UP

Facts Practice: 90 Division Facts (Test F)

Mental Math: Count by $\frac{1}{8}$'s from $\frac{1}{8}$ to 2.

   a. 6 × 43
                 b. 3 × 75

   c. 57 + 29 *(29 is 30 − 1.)*
     d. 2650 − 150

   e. $10.00 − $6.25
             f. $\frac{1}{2}$ of 30

   g. 10 × 2, + 1, ÷ 3, + 2, ÷ 3, × 4, ÷ 3

Problem Solving:

Alexis has 6 coins that total exactly $1.00. Name the one coin she **must** have.

NEW CONCEPTS

Writing division answers as mixed numbers
We have been writing division answers with remainders. However, not all questions involving division can be appropriately answered using remainders. Some story problems have answers that are mixed numbers, as we will see in the following example.

Example 1 A 15-inch length of ribbon was cut into four equal lengths. How long was each piece of ribbon?

Solution We divide 15 by 4 and write the answer as a mixed number.

$$
\begin{array}{r}
3\frac{3}{4} \\
4\overline{)15} \\
12 \\
\hline
3
\end{array}
$$

Notice that the remainder is the numerator of the fraction, and the divisor is the denominator of the fraction. We find that the length of each piece of ribbon is $3\frac{3}{4}$ **inches.**

Example 2 A whole circle is 100% of a circle. One third of a circle is what percent of a circle?

Solution   If we divide 100% by 3, we will find the percent equivalent of $\frac{1}{3}$.

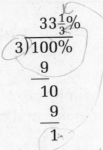

$$
\begin{array}{r}
33\frac{1}{3}\% \\
3\overline{)100\%} \\
\underline{9} \\
10 \\
\underline{9} \\
1
\end{array}
$$

One third of a circle is **$33\frac{1}{3}\%$** of a circle. Notice that our answer matches our fraction manipulative piece for $\frac{1}{3}$.

Example 3   Write $\frac{25}{6}$ as a mixed number.

Solution   The fraction bar in $\frac{25}{6}$ serves as a division symbol. We divide 25 by 6 and write the remainder as the numerator of the fraction.

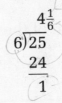

$$
\begin{array}{r}
4\frac{1}{6} \\
6\overline{)25} \\
\underline{24} \\
1
\end{array}
$$

We find that the improper fraction $\frac{25}{6}$ equals the mixed number **$4\frac{1}{6}$**.

Multiples   We find **multiples** of a number by multiplying the number by 1, 2, 3, 4, 5, 6, and so on.

The first six multiples of 2 are 2, 4, 6, 8, 10, and 12.

The first six multiples of 3 are 3, 6, 9, 12, 15, and 18.

The first six multiples of 4 are 4, 8, 12, 16, 20, and 24.

The first six multiples of 5 are 5, 10, 15, 20, 25, and 30.

Example 4   What are the first four multiples of 8?

Solution   Multiplying 8 by 1, 2, 3, and 4 gives the first four multiples: **8, 16, 24,** and **32.**

Example 5   What number is the eighth multiple of 7?

Solution   The eighth multiple of 7 is 8 × 7, which is **56.**

LESSON PRACTICE

Practice set* **a.** A 28-inch long ribbon was cut into eight equal lengths. How long was each piece of ribbon?

b. A whole circle is 100% of a circle. What percent of a circle is $\frac{1}{7}$ of a circle?

c. Divide 467 by 10 and write the quotient as a mixed number.

d. What are the first four multiples of 12?

e. What are the first six multiples of 8?

f. What number is both the third multiple of 8 and the second multiple of 12?

Write each of these improper fractions as a mixed number:

g. $\frac{35}{6}$ **h.** $\frac{49}{10}$ **i.** $\frac{65}{12}$

MIXED PRACTICE

Problem set **1.** What is the difference between the sum of $\frac{1}{2}$ and $\frac{1}{2}$ and the
(Inv. 2) sum of $\frac{1}{3}$ and $\frac{1}{3}$?

2. In three tries Carlos punted the football 35 yards, 30 yards,
(18) and 37 yards. How can Carlos find the average distance of his punts?

3. Earth's average distance from the Sun is one hundred
(12) forty-nine million, six hundred thousand kilometers. Use digits to write that distance.

4. What is the perimeter of the rectangle below?
(8, 24)

$\frac{3}{8}$ in.

$\frac{1}{8}$ in.

5. A 30-inch length of ribbon was cut into 4 equal lengths.
(25) How long was each piece of ribbon? (Write an equation and solve the problem.)

6. Two thirds of the teams finished their games on time. What
(Inv. 2) fraction of the teams did not finish their games on time?

7. Compare: $\frac{1}{2}$ of 12 ◯ $\frac{1}{3}$ of 12
(22)

8. What fraction is half of the fraction that is half of $\frac{1}{2}$?
(Inv. 2)

9. A whole circle is 100% of a circle. What percent of a
(25) circle is $\frac{1}{9}$ of a circle?

10. (a) How many $\frac{1}{6}$'s are in 1?
(Inv. 2)

   (b) How many $\frac{1}{6}$'s are in $\frac{1}{2}$?

11. What fraction of a circle is $33\frac{1}{3}\%$ of a circle?
(25)

12. Divide 365 by 7 and write the answer as a mixed number.
(25)

13. $\frac{2}{3} + \frac{2}{3} + \frac{2}{3}$ **14.** $\frac{6}{6} - \frac{5}{6}$
(24)   (24)

15. 30 × 40 ÷ 60   **16.** $\frac{5}{12} - \frac{5}{12}$
(5)   (24)

17. A team won seven of the twenty games played and lost
(23) the rest. What was the team's win-loss ratio?

18. Cheryl bought 10 pens for 25¢ each. How much did she
(15) pay for all 10 pens? (Write an equation and solve the
   problem.)

19. What is the greatest common factor (GCF) of 24 and 30?
(20)

20. What number is $\frac{1}{100}$ of 100?
(22)

Find each missing number. Check your work.

21. $\frac{5}{8} + m = 1$ **22.** $\frac{144}{n} = 12$
(Inv. 2)   (4)

23. Estimate the sum of 3142, 6328, and 4743 to the nearest
₍₁₆₎ thousand.

24. Two thirds of the 60 children liked hamburgers. How
₍₂₂₎ many of the children liked hamburgers? Draw a diagram
that illustrates the problem.

25. Estimate the length in inches of the line segment below.
₍₁₇₎ Then use an inch ruler to find the length of the line
segment to the nearest sixteenth of an inch.

_____

26. To divide a circle into thirds, Jan imagined the circle was
_(Inv. 2) the face of a clock. Describe how Jan could draw segments
to divide the circle into thirds.

27. Write $\frac{15}{4}$ as a mixed number.
₍₂₅₎

28. Draw and shade rectangles to illustrate and complete this
_(Inv. 2) comparison:

$$\frac{3}{4} \bigcirc \frac{4}{5}$$

29. What are the first four multiples of 25?
₍₂₅₎

30. Which of these numbers is divisible by both 9 and 10?
₍₂₁₎

    A. 910   B. 8910   C. 78,910

LESSON
26

Using Manipulatives to Reduce Fractions • Adding and Subtracting Mixed Numbers

WARM-UP

Facts Practice: 100 Subtraction Facts (Test C)

Mental Math: Count up and down by $\frac{1}{8}$'s between $\frac{1}{8}$ and 2.

   a. 7 × 34   **b.** 4 × 56   **c.** 74 + 19

   d. 475 + 125   **e.** \$5.00 − \$1.75   **f.** $\frac{1}{2}$ of 32

   g. 7 × 5, + 1, ÷ 6, × 3, ÷ 2, + 1, ÷ 5

Problem Solving:

Tad picked up a dot cube. His thumb and forefinger covered opposite faces. He counted the dots on the other four faces. How many dots did he count?

NEW CONCEPTS

Using manipulatives to reduce fractions

You can use fraction manipulatives to model these fractions:

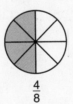   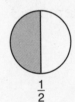

     $\frac{4}{8}$ $\frac{3}{6}$ $\frac{2}{4}$ $\frac{1}{2}$

We see that each picture illustrates half of a circle. The model that uses the fewest pieces is $\frac{1}{2}$. We say that each of the other fractions **reduces** to $\frac{1}{2}$.

We can use our fraction manipulatives to reduce a given fraction by making an equivalent model that uses fewer pieces.

Example 1 Use your fraction manipulatives to reduce $\frac{2}{6}$.

Solution First we use our manipulatives to form $\frac{2}{6}$.

Then we search for a fraction piece equivalent to the $\frac{2}{6}$ model. We find $\frac{1}{3}$.

The models illustrate that $\frac{2}{6}$ reduces to $\frac{1}{3}$.

Adding and subtracting mixed numbers   When adding mixed numbers, we first add the fraction parts, and then we add the whole-number parts. Likewise, when subtracting mixed numbers, we first subtract the fraction parts, and then we subtract the whole-number parts.

Example 2   Two thirds of a circle is what percent of a circle?

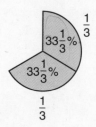

Solution   One third equals $33\frac{1}{3}\%$. So two thirds can be found by adding $33\frac{1}{3}\%$ and $33\frac{1}{3}\%$.

$$\begin{array}{r} 33\frac{1}{3}\% \\ + \ 33\frac{1}{3}\% \\ \hline \mathbf{66\frac{2}{3}\%} \end{array}$$

Example 3   Two sixths of a circle is what percent of a circle?

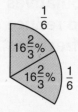

Solution   We add $16\frac{2}{3}\%$ and $16\frac{2}{3}\%$.

$$\begin{array}{r} 16\frac{2}{3}\% \\ + \ 16\frac{2}{3}\% \\ \hline 32\frac{4}{3}\% \end{array}$$

We notice that the fraction part of the answer, $\frac{4}{3}$, is an improper fraction that equals $1\frac{1}{3}$.

So $32\frac{4}{3}\%$ equals $32\% + 1\frac{1}{3}\%$, which is $\mathbf{33\frac{1}{3}\%}$. This makes sense because $\frac{2}{6}$ reduces to $\frac{1}{3}$, which is the same as $33\frac{1}{3}\%$.

Example 4   Rory lives $2\frac{3}{4}$ miles from the lake. He rode his bike from home to the lake and back to home. How far did Rory ride?

Solution This story has an addition pattern.

$$2\tfrac{3}{4}\text{ mi}$$
$$+\ 2\tfrac{3}{4}\text{ mi}$$
$$\overline{4\tfrac{6}{4}\text{ mi}}$$

The fraction part of the answer reduces to $1\tfrac{1}{2}$ $\left(\tfrac{6}{4} = 1\tfrac{2}{4} = 1\tfrac{1}{2}\right)$. So we add $1\tfrac{1}{2}$ to the whole-number part of the answer and find that Rory rode his bike **$5\tfrac{1}{2}$ miles.**

Example 5 Subtract: $5\tfrac{3}{8} - 1\tfrac{1}{8}$

Solution We subtract $\tfrac{1}{8}$ from $\tfrac{3}{8}$, and we subtract 1 from 5. The resulting difference is $4\tfrac{2}{8}$.

$$5\tfrac{3}{8} - 1\tfrac{1}{8} = 4\tfrac{2}{8}$$

We reduce the fraction $\tfrac{2}{8}$ to $\tfrac{1}{4}$ and write the answer as **$4\tfrac{1}{4}$.**

LESSON PRACTICE

Practice set Use your fraction manipulatives to reduce these fractions:

   a. $\dfrac{2}{8}$ **b.** $\dfrac{6}{8}$

Add. Reduce the answer when possible.

   c. $12\tfrac{1}{2}\% + 12\tfrac{1}{2}\%$ **d.** $16\tfrac{2}{3}\% + 66\tfrac{2}{3}\%$

   e. $3\tfrac{3}{4} + 2\tfrac{3}{4}$ **f.** $1\tfrac{1}{8} + 2\tfrac{7}{8}$

   g. $3 + 2\tfrac{2}{3}$ **h.** $\dfrac{3}{4} + 4$

MIXED PRACTICE

Problem set **1.** Maya rode her bike to the park and back. If the trip was
   (26) $3\tfrac{3}{4}$ miles each way, how far did she ride in all?

        2. The young elephant was 36 months old. How many years
   (15)   old was the elephant?

        3. Gwen bought $2\tfrac{1}{2}$ dozen cupcakes for the party. That was
   (6, 15)   enough for how many children to have one cupcake each?

4. There are 100 centimeters in a meter. There are 1000 meters
(15) in a kilometer. How many centimeters are in a kilometer?

5. What is the perimeter of the
(8, 24) equilateral triangle shown?

$\frac{2}{3}$ in.

6. Compare: $\frac{1}{2}$ plus $\frac{1}{2}$ $\bigcirc$ $\frac{1}{2}$ of $\frac{1}{2}$
(Inv. 2)

7. $5\frac{7}{8} + 7\frac{5}{8}$
(26)

8. One eighth of a circle is $12\frac{1}{2}$% of a circle. What percent of
(26) a circle is $\frac{3}{8}$ of a circle?

9. Write a fraction equal to 1 that has a denominator of 12.
(Inv. 2)

10. What is the greatest common factor of 15 and 25?
(20)

11. Describe the rule of the following sequence. Then find
(10) the **seventh** term.

$$8, 16, 24, 32, 40, \ldots$$

12. Write $\frac{14}{5}$ as a mixed number.
(25)

13. Add and simplify: $\frac{2}{5} + \frac{4}{5}$
(26)

Find the missing number. Remember to check your work.

14. $\frac{2}{3} + n = 1$
(Inv. 2)

15. What is the greatest factor of both 12 and 18?
(20)

16. $1 - \frac{3}{4}$ **17.** $3\frac{3}{4} + 3$ **18.** $2\frac{1}{2} - 2\frac{1}{2}$
(Inv. 2)   (26)   (26)

19. Which of the numbers below is divisible by both 2 and 3?
(21)
   A. 4671   B. 3858   C. 6494

20. List the prime numbers between 30 and 40.
(19)

21. Estimate the difference of 5063 and 3987 to the nearest
(16) thousand.

DALLEN IS A DORK!

22. Use your fraction manipulatives to reduce $\frac{6}{8}$.
(26)

23. Find the average of 85, 85, 90, and 100.
(18)

24. How much money is $\frac{3}{5}$ of $30? Draw a diagram to
(22) illustrate the problem.

25. (a) How many millimeters long is the line segment below?
(7, 17)

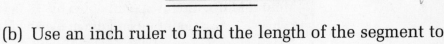

(b) Use an inch ruler to find the length of the segment to
the nearest sixteenth of an inch.

26. Arrange these numbers in order from least to greatest:
(17)

$$\frac{1}{2}, 0, -1, 1$$

Adriana began measuring rainfall when she moved to her
new home. The bar graph below shows the annual rainfall
near Adriana's home during her first three years there. Refer
to this graph to answer problems 27–30.

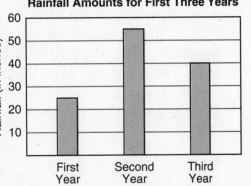

Rainfall Amounts for First Three Years

27. About how many more inches of rain fell during the
(16) second year than during the first year?

28. What was the approximate average annual rainfall during
(18) the first three years?

29. The first year's rainfall was about how many inches below
(18) the average annual rainfall of the first three years?

30. Write a problem with an addition pattern that relates to
(11) the graph. Then answer the problem.

LESSON

27 Measures of a Circle

WARM-UP

Facts Practice: 100 Multiplication Facts (Test E)

Mental Math: Count up and down by 3's between 3 and 60.
Count up and down by 6's between 6 and 60.

a. 7 × 52 **b.** 6 × 33 **c.** 63 + 19
d. 256 + 50 **e.** $10.00 − $7.25 **f.** $\frac{1}{2}$ of 86
g. 8 × 8, − 1, ÷ 7, × 2, + 2, ÷ 2

Problem Solving:

The digits 1 through 9 are used in this subtraction problem. Copy the problem and fill in the missing digits.

$$\begin{array}{r} \overline{}\,\overline{}\,\overline{} \\ -\ 452 \\ \hline 3\,\_\,\_ \end{array}$$

NEW CONCEPT

There are several ways to measure a circle. We can measure the distance around the circle, the distance across the circle, and the distance from the center of the circle to the circle itself. The pictures below identify these measures.

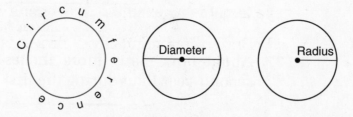

The **circumference** is the distance **around** the circle. This distance is the same as the perimeter of a circle. The **diameter** is the distance **across** a circle through its center. The **radius** is the distance from the center to the circle. The plural of *radius* is **radii.** For any circle, the diameter is twice the length of the radius.

Activity: *Using a Compass*

A **compass** is a tool for drawing a circle. Here we show two types:

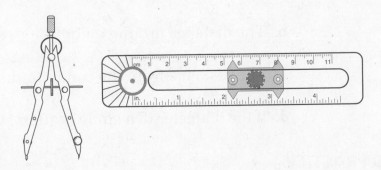

To use a compass, we select a radius and a center point for a circle. Then we rotate the compass about the center point to draw the circle. In this activity you will use a compass and paper to draw circles with given radii.

Materials needed:

- compass and pencil
- plain paper

Draw a circle with each given radius:

a. 2 in.

b. 3 cm

c. $1\frac{3}{4}$ in.

Concentric circles are circles with the same center. A bull's-eye target is an example of concentric circles.

d. Draw three concentric circles with radii of 4 cm, 5 cm, and 6 cm.

Example 1   What is the name for the perimeter of a circle?

Solution   The distance around a circle is its **circumference.**

Example 2   If the radius of a circle is 4 cm, what is its diameter?

Solution   The diameter of a circle is twice its radius—in this case, **8 cm.**

LESSON PRACTICE

Practice set   In problems **a–c**, name the described measure of a circle.

    a. The distance across a circle

    b. The distance around a circle

    c. The distance from the center to the circle

    d. If the diameter of a circle is 10 in., what is its radius?

MIXED PRACTICE

Problem set   **1.** What is the product of the sum of 55 and 45 and the
    ⁽¹²⁾ difference of 55 and 45?

    2. Potatoes are three-fourths water. If a sack of potatoes weighs
    ⁽²²⁾ 20 pounds, how many pounds of water are in the potatoes?
    Draw a diagram to illustrate the problem.

    3. Frankie found three hundred six fleas on his dog. He
    ⁽¹¹⁾ caught two hundred forty-nine of them. How many fleas
    got away? (Write an equation and solve the problem.)

    4. If the diameter of a circle is 5 in., what is the radius of
    ⁽²⁷⁾ the circle?

    5. Which of these numbers is divisible by both 2 and 3?
    ⁽²¹⁾

      A. 122   B. 123   C. 132

    6. Round 1,234,567 to the nearest ten thousand.
    ⁽¹⁶⁾

    7. If ten pounds of apples costs $4.90, what is the price per
    ⁽¹⁵⁾ pound? (Write an equation and solve the problem.)

    8. What is the denominator of $\frac{23}{24}$?
    ⁽⁶⁾

    9. What number is $\frac{3}{5}$ of 65? Draw a diagram to illustrate the
    ⁽²²⁾ problem.

    10. How much money is $\frac{2}{3}$ of $15? Draw a diagram to
    ⁽²²⁾ illustrate the problem.

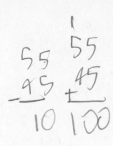

Use your fraction manipulatives to help answer problems 11–18.

11. $\frac{1}{6} + \frac{2}{6} + \frac{3}{6}$
(Inv. 2)

12. $\frac{7}{8} - \frac{3}{8}$
(Inv. 2)

13. $\frac{6}{6} - \frac{5}{6}$
(Inv. 2)

14. $\frac{2}{8} + \frac{5}{8}$
(Inv. 2)

15. (a) How many $\frac{1}{8}$'s are in 1?
(Inv. 2)

    (b) How many $\frac{1}{8}$'s are in $\frac{1}{2}$?

16. Reduce: $\frac{4}{6}$
(26)

17. What fraction is half of $\frac{1}{4}$?
(Inv. 2)

18. What fraction of a circle is 50% of a circle?
(Inv. 2)

19. Divide 2100 by 52 and write the answer with a remainder.
(2)

20. If a 36-inch-long string is made into the shape of a square, how long will each side be?
(8)

21. Convert $\frac{7}{6}$ to a mixed number.
(25)

22. $\frac{432}{18}$
(2)

23. $(55 + 45) \div (55 - 45)$
(5)

24. Which of these numbers is divisible by both 2 and 5?
(21)

    A. 502   B. 205   C. 250

25. Describe a method for determining whether a number is divisible by 9.
(21)

26. Which prime number is not an odd number?
(19)

27. What is the name for the perimeter of a circle?
(27)

28. What is the ratio of even numbers to odd numbers in the square below?
(23)

1	2	3
4	5	6
7	8	9

29. $37\frac{1}{2}\% - 12\frac{1}{2}\%$
(26)

30. $33\frac{1}{3}\% + 16\frac{2}{3}\%$
(26)

LESSON

28   Angles

WARM-UP

Facts Practice: 90 Division Facts (Test F)

Mental Math: Count up and down by 3's between 3 and 60.
Count up and down by 6's between 6 and 60.

a. 8×42 b. 3×85 c. $36 + 49$
d. $1750 - 500$ e. $\$10.00 - \$8.25$ f. $\frac{1}{2}$ of 36
g. $8 \times 4, + 1, \div 3, + 1, \times 2, + 1, \div 5$

Problem Solving:

Grant glued eight wooden blocks together to make a cube. Then he painted all six faces of the cube. Later the cube broke apart into eight blocks. How many faces of each small block were painted?

NEW CONCEPT

In mathematics, a **plane** is a flat surface, such as a tabletop or a sheet of paper. When two lines are drawn in the same plane, they will either cross at one point or they will not cross at all. When lines do not cross but stay the same distance apart, we say that the lines are **parallel.** When lines cross, we say that they **intersect.** When they intersect and make square angles, we call the lines **perpendicular.** If lines intersect at a point but are not perpendicular, then the lines are **oblique.**

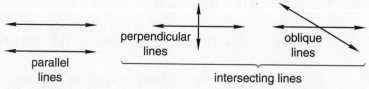

Where lines intersect, **angles** are formed. We show several angles below.

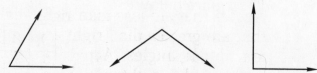

Rays make up the sides of the angles. The rays of an angle originate at a point called the **vertex** of the angle.

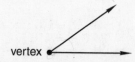

Angles are named in a variety of ways. When there is no chance of confusion, an angle may be named with only one letter: the letter of its vertex. Here is angle *B* (abbreviated ∠*B*):

An angle may also be named with three letters, using a point from one side, the vertex, and a point from the other side. Here is angle *ABC* (∠*ABC*):

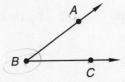

This angle may also be named angle *CBA* (∠*CBA*). However, it may not be named ∠*BAC*, ∠*BCA*, ∠*CAB*, or ∠*ACB*. The vertex must be in the middle. Angles may also be named with a number or letter in the interior of the angle. In the figure below we see ∠1 and ∠2.

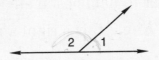

The square angles formed by perpendicular lines, rays, or segments are called **right angles**. We may mark a right angle with a small square.

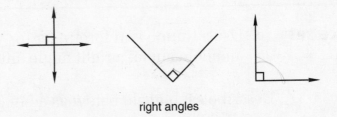

right angles

Angles that are less than right angles are **acute angles**. Angles that are greater than right angles but less than a straight line are **obtuse angles**. A pair of oblique lines forms two acute angles and two obtuse angles.

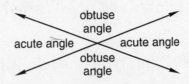

Example 1 (a) Name the acute angle in this figure.

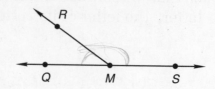

(b) Name the obtuse angle in this figure.

Solution To avoid confusion, we use three letters to name the angles.

(a) The acute angle is ∠*QMR* (or ∠*RMQ*).

(b) The obtuse angle is ∠*RMS* (or ∠*SMR*).

Example 2 In this figure, angle *D* is a right angle.

(a) Which other angle is a right angle?

(b) Which angle is acute?

(c) Which angle is obtuse?

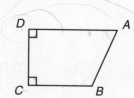

Solution Since there is one angle at each vertex, we may use a single letter to name each angle.

(a) **Angle *C*** is a right angle.

(b) **Angle *A*** is acute.

(c) **Angle *B*** is obtuse.

LESSON PRACTICE

Practice set **a.** Use a thumb and forefinger (or your arms) to approximate an acute angle, a right angle, and an obtuse angle.

Describe each angle below as acute, right, or obtuse.

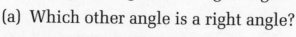

b.   **c.**   **d.**

e. What type of angle is formed by the hands of a clock at 4 o'clock?

f. What type of angle is formed at the corner of a door?

g. Which two angles formed by these oblique lines are acute angles?

h. Draw two parallel line segments.

i. Draw two perpendicular lines.

Refer to the triangle to answer problems **j** and **k**.

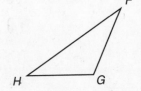

j. Angle *H* is an acute angle. Name another acute angle.

k. Name an obtuse angle.

MIXED PRACTICE

Problem set

1. What is the sum of $\frac{1}{3}$ and $\frac{2}{3}$ and $\frac{3}{3}$?
(24)

2. Two fifths of Robin Hood's one hundred forty men rode with Little John to the castle. How many men rode with Little John? Draw a diagram to illustrate the problem.
(22)

3. Seven hundred sixty-eight peanuts are to be shared equally by the thirty-two children at the party. How many peanuts should each child receive? (Write an equation and solve the problem.)
(15)

4. Columbus arrived in the Americas in 1492. The Declaration of Independence was signed in 1776. How many years passed between Columbus's arrival in the Americas and the signing of the Declaration of Independence? (Write an equation and solve the problem.)
(13)

5. Convert $\frac{23}{3}$ to a mixed number.
(25)

6. $1\frac{2}{3} + 1\frac{2}{3}$ **7.** $3 + 4\frac{2}{3}$ **8.** $3\frac{5}{6} - 1\frac{4}{6}$
(26)   (26)   (26)

9. Use your fraction manipulatives to reduce $\frac{4}{8}$.
(26)

10. How much money is $\frac{2}{3}$ of $24.00? Draw a diagram to illustrate the problem.
(22)

11. (a) What is $\frac{1}{10}$ of 100%?
(22)

    (b) What is $\frac{3}{10}$ of 100%?

12. Twenty-five percent of a circle is what fraction of a circle?
(Inv. 2)

13. What number is 240 less than 250?
(13)

Find each missing number. Remember to check your work.

14. $\frac{1}{4} + m = 1$ **15.** $423 - w = 297$
(Inv. 2)   (3)

16. Refer to the figure below to answer (a) and (b).
(28)

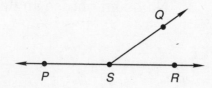

    (a) Name an obtuse angle.

    (b) Name an acute angle.

17. On the last four tests the number of questions Christie
(18) answered correctly was 22, 20, 23, and 23 respectively. She averaged how many correct answers on each test?

18. The three sides of an equilateral triangle are of equal
(8) length. If a 36-inch-long string is formed into the shape of an equilateral triangle, how long will each side of the triangle be?

19. What is the greatest common factor (GCF) of 24, 36, and 60?
(20)

20. $10,010 - 9909$ **21.** $(100 \times 100) - (100 \times 99)$
(1)   (5)

22. If $\frac{1}{10}$ of the team was absent, what percent of the team was
(22) absent? Draw a diagram to illustrate the problem.

23. Divide 5097 by 10 and write the answer as a mixed number.
(25)

24. Three fourths of two dozen eggs is how many eggs? Draw
(22) a diagram to illustrate the problem.

25. (a) Use a ruler to find the length of the line segment
(7, 17)   below to the nearest sixteenth of an inch.

_____

(b) Use a centimeter ruler to find the length of the line
    segment to the nearest centimeter.

26. List the first five multiples of 6 and the first five
(25) multiples of 8. Circle any numbers that are multiples of
      both 6 and 8.

27. Which fraction manipulative covers $\frac{1}{2}$ of $\frac{1}{3}$?
(Inv. 2)

28. There are thirteen stripes on the United States flag. Seven
(23) of the stripes are red, and the rest of the stripes are white.
      What is the ratio of red stripes to white stripes on the
      United States flag?

29. Here we show 24 written as a product of prime numbers:
(19)

$$2 \cdot 2 \cdot 2 \cdot 3$$

Show how prime numbers can be multiplied to equal 27.

30. Which of the numbers below is not divisible by 9?
(21)

A. 234   B. 345   C. 567

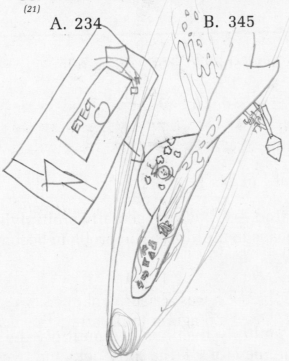

LESSON
29

Multiplying Fractions •
Reducing Fractions by Dividing
by Common Factors

WARM-UP

Facts Practice: 100 Addition Facts (Test B)

Mental Math: Count up and down by $\frac{1}{8}$'s between $\frac{1}{8}$ and 3.

  a. 7×43 **b.** 4×64 **c.** $53 + 39$

  d. $325 + 50$ **e.** $\$20.00 - \$17.25$ **f.** $\frac{1}{2}$ of 70

  g. $4 \times 5, - 6, \div 7, \times 8, + 9, \times 2$

Problem Solving:

Todd flipped a coin three times. It landed heads, heads, then tails. If he flips the coin three more times, what are the possible outcomes?

NEW CONCEPTS

Multiplying fractions

Below we have shaded $\frac{1}{2}$ of $\frac{1}{2}$ of a circle.

We see that $\frac{1}{2}$ of $\frac{1}{2}$ is $\frac{1}{4}$.

When we find $\frac{1}{2}$ of $\frac{1}{2}$, we are actually multiplying. The "of" in "$\frac{1}{2}$ of $\frac{1}{2}$" means to multiply. The problem becomes

$$\frac{1}{2} \times \frac{1}{2} = \frac{1}{4}$$

When we multiply fractions, we multiply the numerators to find the numerator of the product, and we multiply the denominators to find the denominator of the product.

Example 1   What fraction is $\frac{1}{2}$ of $\frac{3}{4}$?

Solution The word *of* in the question means to multiply. We multiply $\frac{1}{2}$ and $\frac{3}{4}$ to find $\frac{1}{2}$ of $\frac{3}{4}$.

$$\frac{1}{2} \times \frac{3}{4} = \frac{3}{8} \quad \begin{matrix} \leftarrow (1 \times 3 = 3) \\ \leftarrow (2 \times 4 = 8) \end{matrix}$$

We find that $\frac{1}{2}$ of $\frac{3}{4}$ is $\frac{3}{8}$. You can illustrate this with your fraction manipulatives.

Example 2 Multiply: $\frac{3}{4} \times \frac{2}{3}$

Solution By performing this multiplication, we will find $\frac{3}{4}$ of $\frac{2}{3}$. We multiply the numerators to find the numerator of the product, and we multiply the denominators to find the denominator of the product.

$$\frac{3}{4} \times \frac{2}{3} = \frac{6}{12}$$

The fraction $\frac{6}{12}$ can be reduced to $\frac{1}{2}$, as we can see in this figure:

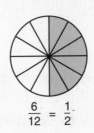

$$\frac{6}{12} = \frac{1}{2}$$

A whole number can be written as a fraction by writing the whole number as the numerator of the fraction and 1 as the denominator of the fraction. Thus, the whole number 2 can be written as the fraction $\frac{2}{1}$. Writing whole numbers as fractions is helpful when multiplying whole numbers by fractions.

Example 3 Multiply: $4 \times \frac{2}{3}$

Solution We write 4 as $\frac{4}{1}$ and multiply.

$$\frac{4}{1} \times \frac{2}{3} = \frac{8}{3}$$

Then we convert the improper fraction $\frac{8}{3}$ to a mixed number.

$$\frac{8}{3} = 2\frac{2}{3}$$

Example 4   Three pennies are placed side by side as shown below. The diameter of one penny is $\frac{3}{4}$ inch. How long is the row of pennies?

$\leftarrow\!\!-\frac{3}{4}\text{ in.}-\!\!\rightarrow$

Solution   We can find the answer by adding or by multiplying. We will show both ways.

Adding: $\frac{3}{4}$ in. $+$ $\frac{3}{4}$ in. $+$ $\frac{3}{4}$ in. $=$ $\frac{9}{4}$ in. $=$ $2\frac{1}{4}$ in.

Multiplying: $\frac{3}{1}$ $\times$ $\frac{3}{4}$ in. $=$ $\frac{9}{4}$ in. $=$ $2\frac{1}{4}$ in.

We find that the row of pennies is **$2\frac{1}{4}$ inches** long.

Reducing fractions by dividing by common factors   We can reduce fractions by dividing the numerator and the denominator by a factor of both numbers. To reduce $\frac{6}{12}$, we will divide both the numerator and the denominator by 6.

$$\frac{6 \div 6}{12 \div 6} = \frac{1}{2}$$

We divided both the numerator and the denominator by 6 because 6 is the largest factor (the GCF) of 6 and 12. If we had divided by 2 instead of by 6, we would not have completely reduced the fraction.

$$\frac{6 \div 2}{12 \div 2} = \frac{3}{6}$$

The fraction $\frac{3}{6}$ can be reduced by dividing the numerator and the denominator by 3.

$$\frac{3 \div 3}{6 \div 3} = \frac{1}{2}$$

It takes two or more steps to reduce fractions if we do not divide by the greatest common factor in the first step.

Example 5   Reduce: $\frac{8}{12}$

Solution   We will show two methods.

Method 1: Divide both numerator and denominator by 2.

$$\frac{8 \div 2}{12 \div 2} = \frac{4}{6}$$

Again divide both numerator and denominator by 2.

$$\frac{4 \div 2}{6 \div 2} = \frac{2}{3}$$

Method 2: Divide both numerator and denominator by 4.

$$\frac{8 \div 4}{12 \div 4} = \frac{2}{3}$$

Either way, we find that $\frac{8}{12}$ reduces to $\frac{2}{3}$. Since the greatest common factor of 8 and 12 is 4, we reduced $\frac{8}{12}$ in one step in Method 2 by dividing the numerator and denominator by 4.

Example 6 Multiply: $2 \times \frac{5}{12}$

Solution We write 2 as $\frac{2}{1}$ and multiply.

$$\frac{2}{1} \times \frac{5}{12} = \frac{10}{12}$$

We can reduce $\frac{10}{12}$ because both 10 and 12 are divisible by 2.

$$\frac{10 \div 2}{12 \div 2} = \frac{5}{6}$$

Example 7 There were 8 baseball bats and 12 baseball gloves in the bag. What was the ratio of bats to gloves in the bag?

Solution We reduce ratios the same way we reduce fractions. The ratio 8 to 12 reduces to $\frac{2}{3}$.

$$\frac{\text{number of bats}}{\text{number of gloves}} = \frac{8}{12} = \frac{2}{3}$$

LESSON PRACTICE

Practice set* Multiply; then reduce if possible.

    a. $\frac{1}{2}$ of $\frac{4}{5}$ **b.** $\frac{1}{4}$ of $\frac{2}{3}$ **c.** $\frac{2}{3} \times \frac{3}{4}$

Multiply; then convert each answer from an improper fraction to a whole number or to a mixed number.

    d. $\frac{5}{6} \times \frac{6}{5}$ **e.** $5 \times \frac{2}{3}$ **f.** $2 \times \frac{4}{3}$

Reduce each fraction:

g. $\dfrac{9}{12}$ h. $\dfrac{6}{10}$ i. $\dfrac{18}{24}$

j. In a group of 30 children, there were 20 girls. What was the ratio of boys to girls?

MIXED PRACTICE

Problem set

1. The African elephant can weigh eight tons. A ton is two thousand pounds. How many pounds can an African elephant weigh?
(15)

2. Sixteen jelly beans weigh one ounce. How many jelly beans weigh one pound (1 pound = 16 ounces)?
(15)

3. If the product of $\frac{1}{2}$ and $\frac{1}{2}$ is subtracted from the sum of $\frac{1}{2}$ and $\frac{1}{2}$, what is the difference?
(Inv. 2, 29)

4. A team won 6 games and lost 8 games. What was the team's win-loss ratio?
(29)

5. Reduce: $\dfrac{16}{24}$
(29)

6. $\dfrac{1}{8} + \dfrac{3}{8}$
(24, 29)

7. $\dfrac{1}{2} \times \dfrac{2}{3}$
(29)

8. $\dfrac{7}{12} - \dfrac{3}{12}$
(24, 29)

9. The Nobel Prize is a famous international award that recognizes important work in physics, chemistry, medicine, economics, peacemaking, and literature. Ninety-six Nobel Prizes in Literature were awarded from 1901 to 1999. One eighth of the prizes were given to Americans. How many Nobel Prizes in Literature were awarded to Americans from 1901 to 1999?
(22)

10. Find the next three numbers in the sequence below:
(10)

1, 4, 7, 10, _____, _____, _____, ...

11. When five months have passed, what fraction of the year remains?
(Inv. 2)

12. $3.60 × 100
(2)

13. 50,000 ÷ 100
(2)

14. Convert $\frac{18}{4}$ to a mixed number. Remember to reduce the
(25) fraction part of the mixed number.

15. The temperature rose from −8°F to 15°F. This was an
(14) increase of how many degrees?

Find each missing number. Remember to check your work.

16. $m + 496 + 2684 = 3217$
(3)

17. $1000 - n = 857$ **18.** $24x = 480$
(3)   (4)

19. $7 \cdot 11 \cdot 13$
(5)

20. Describe how to estimate the quotient of $4963 \div 39$.
(16)

21. Compare: $\frac{2}{3} \times \frac{3}{2} \bigcirc 1$
(29)

22. The perimeter of the rectangle
(8) shown is 60 mm. The width of
the rectangle is 10 mm. What is
its length?

10 mm

23. $12 - 40$ **24.** $\left(\frac{1}{2} \times \frac{1}{2}\right) - \frac{1}{4}$
(14)   (29)

25. (a) Which angles in the figure at
(28)   right are acute angles?

(b) Which angles are obtuse angles?

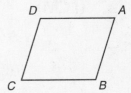

26. What fraction is $\frac{2}{3}$ of $\frac{3}{5}$?
(29)

27. What is the product of $\frac{3}{4}$ and $\frac{4}{3}$?
(29)

28. If the diameter of a bicycle wheel is 24 inches, what is
(27, 29) the ratio of the radius of the wheel to the diameter of
the wheel?

29. What type of an angle is formed by the hands of a clock at
(28) 2 o'clock?

30. What percent of a circle is $\frac{2}{5}$ of a circle?
(22)

LESSON

30

Least Common Multiple (LCM) • Reciprocals

WARM-UP

Facts Practice: 100 Multiplication Facts (Test E)

Mental Math: Count up and down by 3's between 3 and 30.
Count up and down by 4's between 4 and 40.

a. 9×32 **b.** 5×42 **c.** $45 + 49$
d. $436 + 99$ **e.** $\$20.00 - \$12.75$ **f.** $\frac{1}{2}$ of 72
g. $7 \times 7, -1, \div 6, \times 3, +1, \times 2, -1$

Problem Solving:

Use the digits 6, 7, and 8 to complete this multiplication problem.

$$\begin{array}{r} 23\_ \\ \times \_ \\ \hline 166\_ \end{array}$$

NEW CONCEPTS

Least common multiple (LCM)

A number that is a multiple of two or more numbers is called a *common multiple* of those numbers. Here we show some multiples of 2 and 3. We have circled the common multiples.

Multiples of 2: 2, 4, ⑥, 8, 10, ⑫, 14, 16, ⑱, 20, ...

Multiples of 3: 3, ⑥, 9, ⑫, 15, ⑱, 21, ...

We see that 6, 12, and 18 are common multiples of 2 and 3. Since the number 6 is the least of these common multiples, it is called the **least common multiple**. The term *least common multiple* is abbreviated LCM.

Example 1   What is the least common multiple of 3 and 4?

Solution   We will list some multiples of each number and emphasize the common multiples.

Multiples of 3: 3, 6, 9, ⑫, 15, 18, 21, ㉔, ...

Multiples of 4: 4, 8, ⑫, 16, 20, ㉔, 28, ...

We see that the number 12 and 24 are in both lists. Both 12 and 24 are common multiples of 3 and 4. The least common multiple is **12**. When we list the multiples in order, the first number that is a common multiple is always the least common multiple.

Example 2   What is the LCM of 2 and 4?

Solution   We will list some multiples of 2 and 4.

Multiples of 2: 2, ④, 6, ⑧, 10, ⑫, 14, ⑯, ...

Multiples of 4: ④, ⑧, ⑫, ⑯, ...

The first number that is a common multiple of both 2 and 4 is **4.**

Reciprocals   **Reciprocals** are two numbers whose product is 1. For example, the numbers 2 and $\frac{1}{2}$ are reciprocals because $2 \times \frac{1}{2} = 1$.

$$2 \times \frac{1}{2} = 1$$

reciprocals

We say that 2 is the reciprocal of $\frac{1}{2}$ and that $\frac{1}{2}$ is the reciprocal of 2. Sometimes we want to find the reciprocal of a certain number. One way we will practice finding the reciprocal of a number is by solving equations like this:

$$3 \times \boxed{} = 1$$

The number that goes in the box is $\frac{1}{3}$ because 3 times $\frac{1}{3}$ is 1. One third is the reciprocal of 3.

Reciprocals also answer questions like this:

How many $\frac{1}{4}$'s are in 1?

The answer is the reciprocal of $\frac{1}{4}$, which is 4.

Fractions have two **terms,** the numerator and the denominator. To form the reciprocal of a fraction, we reverse the terms of the fraction.

$$\frac{3}{4} \quad\diagdown\!\!\!\!\diagup\quad \frac{4}{3}$$

The new fraction, $\frac{4}{3}$, is the reciprocal of $\frac{3}{4}$.

If we multiply $\frac{3}{4}$ by $\frac{4}{3}$, we see that the product, $\frac{12}{12}$, equals 1.

$$\frac{3}{4} \times \frac{4}{3} = \frac{12}{12} = 1$$

Example 3   How many $\frac{2}{3}$'s are in 1?

Solution   To find the number of $\frac{2}{3}$'s in 1, we need to find the reciprocal of $\frac{2}{3}$. The easiest way to find the reciprocal of $\frac{2}{3}$ is to reverse the positions of the 2 and the 3. The reciprocal of $\frac{2}{3}$ is $\frac{3}{2}$. (We may convert $\frac{3}{2}$ to $1\frac{1}{2}$, but we usually write reciprocals as fractions rather than as mixed numbers.)

Example 4   What number goes into the box to make the equation true?

$$\frac{5}{6} \times \square = 1$$

Solution   When $\frac{5}{6}$ is multiplied by its reciprocal, the product is 1. So the answer is the reciprocal of $\frac{5}{6}$, which is $\frac{6}{5}$. When we multiply $\frac{5}{6}$ by $\frac{6}{5}$, we get $\frac{30}{30}$.

$$\frac{5}{6} \times \frac{6}{5} = \frac{30}{30}$$

The fraction $\frac{30}{30}$ equals 1.

Example 5   What is the reciprocal of 5?

Solution   Recall that a whole number can be written as a fraction that has a denominator of 1. So 5 can be written as $\frac{5}{1}$. (This means "five wholes.") Reversing the positions of the 5 and the 1 gives us the reciprocal of 5, which is $\frac{1}{5}$. This makes sense because five $\frac{1}{5}$'s make 1, and $\frac{1}{5}$ of 5 is 1.

LESSON PRACTICE

Practice set*   Find the least common multiple of each pair of numbers:

    a. 6 and 8   **b.** 3 and 5   **c.** 5 and 10

Write the reciprocal of each number:

    d. $\frac{6}{1}$ **e.** $\frac{2}{3}$ **f.** $\frac{8}{5}$ **g.** $\frac{1}{3}$

For problems **h–k**, find the number that goes into the box to make the equation true.

    h. $\frac{3}{8} \times \square = 1$ **i.** $\frac{4}{1} \times \square = 1$

    j. $\square \times \frac{1}{6} = 1$ **k.** $\square \times \frac{7}{8} = 1$

    l. How many $\frac{2}{5}$'s are in 1?

    m. How many $\frac{5}{12}$'s are in 1?

MIXED PRACTICE

Problem set   **1.** If the fourth multiple of 3 is subtracted from the third
(12, 25)   multiple of 4, what is the difference?

2. About $\frac{2}{3}$ of a person's body weight is water. Albert weighs
(22)   117 pounds. About how many pounds of Albert's weight
is water? Draw a diagram to illustrate the problem.

3. Cynthia ate 42 pieces of popcorn during the first 15 minutes
(15)   of a movie. If she kept eating at the same rate, how many
pieces of popcorn did she eat during the 2-hour movie?
(Write an equation and solve the problem.)

4. What are the first four multiples of 12?
(25)

5. What is the least common multiple (LCM) of 4 and 6?
(30)

6. There were 12 minutes of commercials during the one-
(23)   hour program. What was the ratio of commercial to
noncommercial time during the one-hour program?

7. $\frac{2}{5} + \frac{2}{5} + \frac{2}{5}$ **8.** $1 - \frac{1}{10}$ **9.** $\frac{11}{12} - \frac{1}{12}$
(24)   (Inv. 2)   (24, 29)

10. $\frac{3}{4} \times \frac{4}{3}$ **11.** $5 \times \frac{3}{4}$ **12.** $\frac{5}{2} \times \frac{5}{3}$
(29)   (29)   (29)

13. The number 24 has how many different whole-number
(19)   factors?

14. $3 + $24 + 6.50 **15.** $5 - 1.50
(1)   (1)

16. Estimate the product: 596×405
(16)

17. Which angle of the triangle at right
(28)   is an obtuse angle?

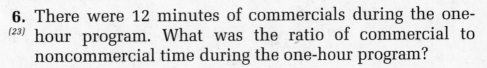

18. Compare: $\frac{2}{3} \times \frac{2}{3} \bigcirc \frac{2}{3} \times 1$
(29)

19. $500{,}000 \div 100$ **20.** $35\overline{)8540}$
(2)   (2)

21. $\dfrac{100\%}{7}$
(25)

22. Reduce: $\dfrac{4}{12}$
(29)

23. What is the average of 375, 632, and 571?
(18)

24. A regular hexagon has six sides of equal length. If a
(8) regular hexagon is made from a 36-inch-long string, what is the length of each side?

25. What is the product of a number and its reciprocal?
(30)

26. How many $\frac{2}{5}$'s are in 1?
(30)

27. What number goes into the box to make the equation true?
(30)

$$\frac{3}{8} \times \boxed{} = 1$$

28. What is the reciprocal of the only even prime number?
(19, 30)

29. Convert $\frac{45}{10}$ to a mixed number. Remember to reduce the
(25) fraction part of the mixed number.

30. Four pennies are placed side by side as shown below.
(29) The diameter of one penny is $\frac{3}{4}$ inch. What is the length of the row of pennies?

$$\vdash\!\!-\tfrac{3}{4}\text{ in.}-\!\!\dashv$$

INVESTIGATION 3

Focus on

Measuring and Drawing Angles with a Protractor

One way to measure angles is with units called **degrees**. A full circle measures 360 degrees, which we abbreviate 360°. A tool to help us measure angles is a **protractor**. To measure an angle, we place the center point of the protractor on the vertex of the angle, and we place one of the zero marks on one ray of the angle. Where the other ray of the angle passes through the scale, we can read the degree measure of the angle.

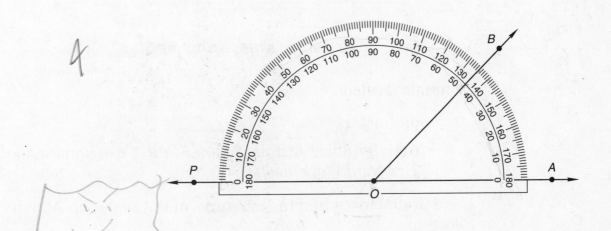

The scale on a protractor has two sets of numbers. One set is for measuring angles starting from the right side, and the other set is for measuring angles starting from the left side. The easiest way to ensure that we are reading from the correct scale is to decide whether the angle we are measuring is acute or obtuse. Looking at ∠AOB, we read the numbers 45° and 135°. Since the angle is less than 90° (acute), it must measure 45°, not 135°. We say that "the measure of angle AOB is 45°," which we may write as follows:

$$m\angle AOB = 45°$$

Practice reading a protractor by finding the measures of these angles:

1. ∠*AOC*   **2.** ∠*AOE*   **3.** ∠*AOF*

4. ∠*AOH*   **5.** ∠*IOH*   **6.** ∠*IOE*

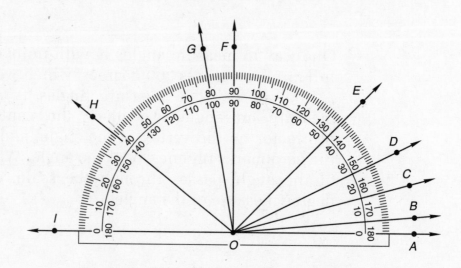

Activity: *Measuring Angles*

Materials needed:

- protractor
- Activity Sheet 8 from *Saxon Math 7/6—Homeschool Tests and Worksheets*

Use a protractor to find the measures of the angles on Activity Sheet 8.

To draw angles with a protractor, follow these steps. Begin by drawing a horizontal ray. The sketch of the ray should be longer than half the diameter of the protractor.

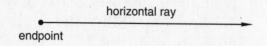

horizontal ray

endpoint

Next, position the protractor so that the center point of the protractor is on the endpoint of the ray and a zero degree mark of the protractor is on the ray.

Then, with the protractor in position, make a dot on the paper at the appropriate degree mark for the angle you intend to draw. Here we show the placement of a dot for drawing a 60° angle:

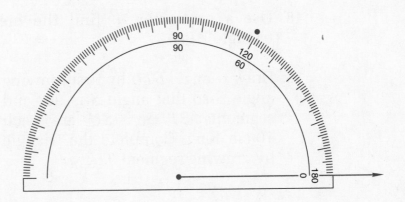

Finally, remove the protractor and draw a ray from the endpoint of the first ray through the dot you made.

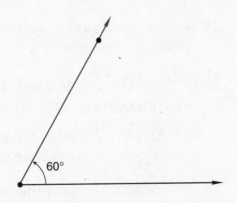

Use your protractor to draw angles with these measures:

7. 30°   **8.** 80°   **9.** 110°

10. 135°   **11.** 45°   **12.** 15°

13. Draw triangle *ABC* by first drawing segment *BC* six inches long. Then draw a 60° angle at vertex *B* and a 60° angle at vertex *C*. Extend the segments so that they intersect at point *A*.

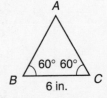

Refer to the triangle you drew in problem 13 to answer problems 14 and 15.

14. Use a ruler to find the lengths of segment *AB* and segment *AC* in triangle *ABC*.

15. Use a protractor to find the measure of angle *A* in triangle *ABC*.

16. Draw triangle *STU* by first drawing angle *S* so that angle *S* is 90° and segments *ST* and *SU* are each 10 cm long. Complete the triangle by drawing segment *TU*.

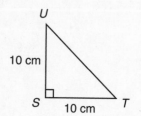

Refer to the triangle you drew in problem 16 to answer problems 17 and 18.

17. Use a protractor to find the measures of angle *T* and angle *U*.

18. Use a centimeter ruler to measure segment *TU* to the nearest centimeter.

Note: Problems intended for additional exposure to the concepts in this investigation are available in the appendix.

LESSON
31   Areas of Rectangles

WARM-UP

Facts Practice: 90 Division Facts (Test F)

Mental Math: Count up and down by $\frac{1}{8}$'s between $\frac{1}{8}$ and 2.

a. 4 × 25

b. 6 × 37

c. 28 + 29 *(29 is 30 – 1.)*

d. $6.25 + $2.50

e. $\frac{1}{3}$ of 63

f. $\frac{600}{10}$

g. 10 × 10, − 20, + 1, ÷ 9, × 2, ÷ 3, × 5, + 2, ÷ 4

Problem Solving:

The bus has 40 seats for passengers. Two passengers can sit in each seat. Sixty passengers got on the bus. What is the largest number of seats that could be empty? What is the largest number of seats that could have just one passenger?

NEW CONCEPT

We have measured and calculated the distance around a shape. This distance is called *perimeter.*

The perimeter of a shape is the distance around it.

We can also measure how much surface is enclosed by the sides of a shape. When we measure the "inside" of a flat shape, we are measuring its **area.**

The area of a shape is the amount of surface enclosed by its sides.

We use a different kind of unit to measure area than we use to measure perimeter. To measure perimeter, we use units of length such as centimeters. Units of area are called **square units.** One example is a square centimeter.

———

This is 1 centimeter.   This is 1 square centimeter.

Other common units of area are square inches, square feet, and square yards. We can think of units of area as floor tiles. The area of a shape is the number of "floor tiles" of a certain size that completely cover the shape. Note that we can use "sq." to abbreviate the word *square*.

Example 1   How many floor tiles, 1 foot on each side, are needed to cover the floor of a room that is 8 feet wide and 12 feet long?

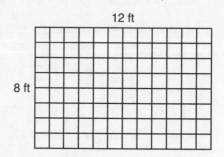

Solution   The surface of the floor is covered with tiles. By answering this question, we are finding the area of the room in square feet. We could count the tiles, but a faster way to find the number of tiles is to multiply. There are 8 rows of tiles with 12 tiles in each row.

$$\begin{array}{r} 12 \text{ tiles in each row} \\ \times \quad 8 \text{ rows} \\ \hline 96 \text{ tiles} \end{array}$$

To cover the floor, **96 tiles** are needed. This means that the area of the room is 96 sq. ft.

Example 2   What is the area of this rectangle?

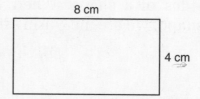

Solution   The diagram shows the length and width of the rectangle in centimeters. Therefore, we will use square centimeters to measure the area of the rectangle. We calculate the number of square-centimeter tiles needed to cover the rectangle by multiplying the length by the width.

$$\text{Length} \times \text{width} = \text{area}$$

$$8 \text{ cm} \times 4 \text{ cm} = \textbf{32 sq. cm}$$

Example 3   The area of a square is 100 square inches.

(a) How long is each side of the square?

(b) What is the perimeter of the square?

Solution   (a) The length and width of a square are equal. So we think, "What number multiplied by itself equals 100?" Since 10 × 10 = 100, we find that each side is **10 inches** long.

(b) Since each of the four sides is 10 inches, the perimeter of the square is **40 inches.**

LESSON PRACTICE

Practice set   Find the number of square units needed to cover the area of these rectangles. For reference, square units have been drawn along the length and width of each rectangle.

a.

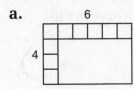

b.

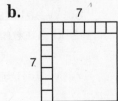

Find the area of these rectangles:

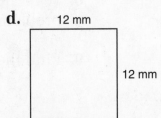

The area of a square is 25 square inches.

e. How long is each side of the square?

f. What is the perimeter of the square?

MIXED PRACTICE

Problem set   **1.** When the third multiple of 4 is divided by the fourth
        (12, 25)   multiple of 3, what is the quotient?

2. The distance the Earth travels around the Sun each year
(12)   is about five hundred eighty million miles. Use digits to write that distance.

3. Convert $\frac{10}{3}$ to a mixed number.
$_{(25)}$

4. How many square stickers with sides 1 centimeter long
$_{(31)}$ would be needed to cover the rectangle below?

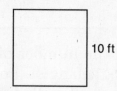

4 cm

2 cm

5. How many square floor tiles with sides 1 foot long would
$_{(31)}$ be needed to cover the square below?

10 ft

6. What is the area of a rectangle 12 inches long and
$_{(31)}$ 8 inches wide?

7. What is the next term in this sequence?
$_{(10)}$

1, 4, 9, 16, 25, 36, …

8. What number is $\frac{2}{3}$ of 24? Draw a diagram to illustrate the
$_{(22)}$ problem.

9. Find the missing number. Remember to check your work.
$_{(3)}$

$$24 + f = 42$$

Write each answer in simplest form:

10. $\frac{1}{8} + \frac{1}{8}$ **11.** $\frac{5}{6} - \frac{1}{6}$
$_{(24)}$ $_{(24)}$

12. $\frac{2}{3} \cdot \frac{1}{2}$ **13.** $\frac{2}{3} \times 5$
$_{(29)}$ $_{(29)}$

14. Estimate the product of 387 and 514.
$_{(16)}$

15. \$20.00 ÷ 10   **16.** (63)47¢   **17.** 4623 ÷ 22
$_{(2)}$ $_{(2)}$ $_{(2)}$

18. What is the reciprocal of the smallest odd prime number?
$_{(19, 30)}$

19. Two thirds of a circle is what percent of a circle?
$_{(26)}$

20. Which of these numbers is closest to 100?
(9)

    A. 90   B. 89   C. 111   D. 109

21. For most of its orbit, Pluto is the farthest planet from the
(12) Sun in our solar system. Pluto's average distance from the Sun is about three billion, six hundred seventy million miles. Use digits to write the average distance between Pluto and the Sun.

22. The diameter of the pizza was 14 inches. What was the
(23, 27) ratio of the radius to the diameter of the pizza?

23. Three of the nine players play outfield. What fraction of
(29) the players play outfield? Write the answer as a reduced fraction.

24. Use an inch ruler to find the length of the line segment
(17) below.

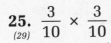

25. $\dfrac{3}{10} \times \dfrac{3}{10}$
(29)

26. How many $\frac{3}{4}$'s are in 1?
(30)

27. Write a fraction equal to 1 with a denominator of 8.
(29)

28. Five sixths of the 24 new drivers scored 80% or higher on
(22) the driving test. How many new drivers scored 80% or higher? Draw a diagram to illustrate the problem.

29. (a) Name an angle in the figure at
(28)   right that measures less than 90°.

    (b) Name an obtuse angle in the figure at right.

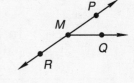

30. Using a ruler, how could you calculate the floor area of a
(31) room in your home?

LESSON

32

Expanded Notation • More on Elapsed Time

WARM-UP

Facts Practice: 30 Fractions to Reduce (Test G)

Mental Math: Count up and down by 25's between 25 and 400.

a. 4×75 b. $380 + 1200$ c. $54 + 19$

d. $\$8.00 - \$1.50$ e. $\frac{1}{2}$ of 240   f. $\frac{600}{100}$

g. $12 \times 3, -1, \div 5, \times 2, +1, \div 3, \times 2$

Problem Solving:

Monifa picked up a dot cube and held it so that she could see the dots on three faces. The total number of dots on the three faces was 6. How many dots were on each of the faces she could see? What was the total number of dots on the faces she could not see?

NEW CONCEPTS

Expanded notation

Recall that in our number system the location of a digit in a number has a value called its *place value*. Consider the value of the 5 in these two numbers:

$$250 \qquad 520$$

In 250 the value of the 5 is 5×10. In 520 the value of the 5 is 5×100.

To find a digit's value within a number, we multiply the digit by the value of the place occupied by the digit. When we write a number in **expanded notation**, we write each nonzero digit times its place value.

Example 1 Write 250 in expanded notation.

Solution The 2 is in the hundreds place, and the 5 is in the tens place. In expanded notation we write

$$(2 \times 100) + (5 \times 10)$$

Since there is a zero in the ones place, we could add a third set of parentheses, (0 × 1). However, since zero times any number equals zero, it is not necessary to include zeros when writing numbers in expanded notation.

Example 2   Write (5 × 1000) + (2 × 100) + (8 × 10) in standard notation.

Solution   Standard notation is our usual way of writing numbers. One way to think about this number is 5000 + 200 + 80. Another way to think about this number is 5 in the thousands place, 2 in the hundreds place, and 8 in the tens place. We may assume a 0 in the ones place. Either way we think about the number, the standard form is **5280.**

More on elapsed time   The hours of the day are divided into two parts: **a.m.** and **p.m.** The 12 "a.m." hours run from midnight (12:00 a.m.) to the moment just before noon (12:00 p.m.). The 12 "p.m." hours run from noon to the moment just before midnight. Recall from Lesson 13 that when we calculate the amount of time between two events, we are calculating elapsed time (the amount of time that has passed). We can use the later-earlier-difference pattern to solve elapsed-time problems about hours and minutes.

Example 3   Jason started the marathon at 7:15 a.m. He finished the race at 10:10 a.m. How long did it take Jason to run the marathon?

Solution   This problem has a subtraction pattern. We find Jason's race time (elapsed time) by subtracting the earlier time from the later time.

Later	10:10 a.m.
– Earlier	– 7:15 a.m.
Difference	

Since we cannot subtract 15 minutes from 10 minutes, we rename one hour as 60 minutes. Those 60 minutes plus 10 minutes equal 70 minutes. (This means 70 minutes after 9:00, which is the same as 10:10.)

$$
\begin{array}{r}
\overset{9\ :70}{\cancel{10:10}} \\
-\ \ 7:15 \\
\hline
2:55
\end{array}
$$

We find that it took Jason **2 hours 55 minutes** to run the marathon.

Example 4   What time is two and a half hours after 10:43 a.m.?

Solution   This is an elapsed time problem, and it has a subtraction pattern. The elapsed time, $2\frac{1}{2}$ hours, is the difference. We write the elapsed time as 2:30. The earlier time is 10:43 a.m.

$$
\begin{array}{cc}
\text{Later} & \text{Later} \\
- \text{ Earlier} & - \text{ 10:43 a.m.} \\
\hline
\text{Difference} & \text{2:30}
\end{array}
$$

We need to find the later time, so we add $2\frac{1}{2}$ hours to 10:43 a.m. We will describe two methods to do this: a mental calculation and a pencil-and-paper calculation. For the mental calculation, we could first count two hours after 10:43 a.m. One hour later is 11:43 a.m. Another hour later is 12:43 p.m. (Note the switch from a.m. to p.m.) From 12:43 p.m., we count 30 minutes (one half hour). To do this, we can count 10 minutes at a time from 12:43 p.m.: 12:53 p.m., 1:03 p.m., 1:13 p.m. We find that $2\frac{1}{2}$ hours after 10:43 a.m. is **1:13 p.m.** (Another mental calculation is to add 3 hours and then subtract 30 minutes.)

To perform a pencil-and-paper calculation, we add 2 hours 30 minutes to 10:43 a.m.

$$
\begin{array}{r}
\text{10:43 a.m.} \\
+ \quad \text{2:30} \\
\hline
\text{12:73 p.m.}
\end{array}
$$

Notice that the time switches from a.m. to p.m. and that the sum, 12:73 p.m., is improper. Seventy-three minutes is more than an hour. We think of 73 minutes as "one hour plus 13 minutes." We add 1 to the number of hours and write 13 as the number of minutes. So $2\frac{1}{2}$ hours after 10:43 a.m. is **1:13 p.m.**

LESSON PRACTICE

Practice set*   Write each of these numbers in expanded notation:

a. 205   b. 1760   c. 8050

Write each of these numbers in standard form:

d. $(6 \times 1000) + (4 \times 100)$ e. $(7 \times 100) + (5 \times 1)$

f. George started the marathon at 7:15 a.m. He finished the race at 11:05 a.m. How long did it take George to run the marathon?

g. What time is $3\frac{1}{2}$ hours after 11:50 p.m.?

MIXED PRACTICE

Problem set

1. When the sum of 24 and 7 is multiplied by the difference
(12) of 18 and 6, what is the product?

2. Davy Crockett was born in Tennessee in 1786 and died at
(13) the Alamo in 1836. How many years did he live? (Write an equation and solve the problem.)

3. A 16-ounce box of a certain cereal costs $2.24. What is
(15) the cost per ounce of the cereal?

4. What time is 3 hours 30 minutes after 6:50 a.m.?
(32)

5. Forty percent equals $\frac{40}{100}$. Reduce $\frac{40}{100}$.
(29)

6. A baseball diamond is the square section formed by the
(31) four bases on a baseball field. On a major league field the distance between home plate and 1st base is 90 feet. What is the area of a baseball diamond?

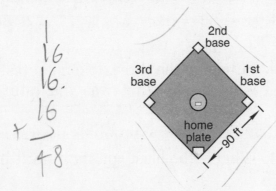

7. What is the perimeter of a baseball diamond, as described
(8) in problem 6?

8. Describe the sequence below. Then find the **eighth** term.
(10)

$$1, 3, 5, 7, \ldots$$

9. Write 7500 in expanded notation.
(32)

10. Which of these numbers is closest to 1000?
(9)

    A. 990   B. 909   C. 1009   D. 1090

11. In three separate bank accounts Sumi has $623, $494,
(18) and $380. What is the average amount of money she has per account?

12. $0.05 × 100   **13.** How many $\frac{2}{5}$'s are in 1?
(2)   (30)

14. How much money is $\frac{3}{4}$ of $24? Draw a diagram to illustrate
(22) the problem.

Write each answer in simplest form:

15. $\frac{3}{5} + \frac{3}{5}$ **16.** $\frac{3}{4} - \frac{1}{4}$ **17.** $\frac{3}{4} \times \frac{1}{3}$
(24)   (24)   (29)

18. $\frac{3}{10} \times \frac{7}{10}$ **19.** $1\frac{2}{3} - 1\frac{1}{3}$
(29)   (26)

20. Three fourths of a circle is what percent of a circle?
(Inv. 2)

Find each missing number. Remember to check your work.

21. $w - 53 = 12$ **22.** $8q = 240$
(3)   (4)

23. Fifteen of the three dozen contestants in the waiting room
(23, 29) were boys. What was the ratio of boys to girls in the waiting room?

24. What is the least common multiple of 4 and 6?
(30)

25. Draw triangle *ABC* so that ∠*C* measures 90°, side *AC*
(Inv. 3) measures 3 in., and side *BC* measures 4 in. Then draw and measure the length of side *AB*.

26. If 24 of the 30 anglers caught a big fish, what fraction of
(29) the anglers caught a big fish?

27. Brad and Sharon began the hike at 6:45 a.m. and finished
(32) at 11:15 a.m. For how long did they hike?

28. Compare: $(3 \times 100) + (5 \times 1) \bigcirc 350$
(32)

29. What fraction is represented by point A on the number
(17) line below?

30. Some grocery stores post the price per ounce of different
(15) cereals to help customers compare costs. How can we
find the cost per ounce of a box of cereal?

LESSON

33

Writing Percents as Fractions, Part 1

WARM-UP

Facts Practice: 30 Fractions to Reduce (Test G)

Mental Math: Count by 7's from 7 to 84.

a. $(4 \times 100) + (4 \times 25)$ b. 7×29
c. $56 + 28$ d. $\$5.50 + \$1.75$
e. Double 120.   f. $\frac{120}{10}$
g. $2 \times 3, +1, \times 8, +4, \div 6, \times 2, +1, \div 3$

Problem Solving:

Grant glued 27 small blocks together to make this cube. Then he painted the 6 faces of the cube. Later the cube broke apart into 27 blocks. How many of the blocks had 3 painted faces? ... 2 painted faces? ... 1 painted face?

NEW CONCEPT

Our fraction manipulatives describe parts of circles as fractions and as percents. The manipulatives show that 50% is equivalent to $\frac{1}{2}$ and that 25% is equivalent to $\frac{1}{4}$. We can find fraction equivalents of other percents by writing a percent as a fraction and then reducing the fraction.

A **percent** is actually a fraction with a denominator of 100. The word *percent* and its symbol, %, mean "per hundred." To write a percent as a fraction, we remove the percent sign and write the number as the numerator and 100 as the denominator.

Example 1   Write 60% as a fraction.

Solution   We remove the percent sign and write 60 over 100.

$$60\% = \frac{60}{100}$$

We can reduce $\frac{60}{100}$ in one step by dividing 60 and 100 by their GCF, which is 20. If we begin by dividing by a number smaller than 20, it will take more than one step to reduce the fraction.

$$\frac{60 \div 20}{100 \div 20} = \frac{3}{5}$$

We find that 60% is equivalent to the fraction $\frac{3}{5}$.

Example 2 Find the reduced fraction that equals 4%.

Solution We remove the percent sign and write 4 over 100.

$$4\% = \frac{4}{100}$$

We reduce the fraction by dividing both the numerator and denominator by 4, which is the GCF of 4 and 100.

$$\frac{4 \div 4}{100 \div 4} = \frac{1}{25}$$

We find that 4% is equivalent to the fraction $\frac{1}{25}$.

LESSON PRACTICE

Practice set* Write each percent as a fraction. Reduce when possible.

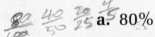

a. 80%   **b.** 5%   **c.** 25% $\frac{25}{100}$ $\frac{1}{4}$

d. 24%   **e.** 23%   **f.** 10%

g. 20%   **h.** 2%   **i.** 75%

MIXED PRACTICE

Problem set **1.** When the product of 10 and 15 is divided by the sum of
$^{(12)}$ 10 and 15, what is the quotient?

2. The Nile River is 6650 kilometers long. The Mississippi
$^{(13)}$ River is 3766 kilometers long. How much longer is the Nile River than the Mississippi? (Write an equation and solve the problem.)

$$6\overline{)6650}$$

Bullschnoz **3.** Some astronomers think that the universe is about fifteen
(12) billion years old. Use digits to write that number of years.

4. Write 3040 in expanded notation.
(32)

5. Write (6 × 100) + (2 × 1) in standard notation.
(32)

6. Write two fractions equal to 1, one with a denominator of
(29) 10 and the other with a denominator of 100.

7. By what number should $\frac{5}{3}$ be multiplied for the product
(30) to be 1?

8. What is the perimeter of this
(8) rectangle?

12 in.

8 in.

9. How many square tiles with sides
(31) 1 inch long would be needed to
cover this rectangle?

10. Which of these numbers is divisible by both 2 and 3?
(21)

    A. 56   B. 75   C. 83   D. 48

11. Estimate the difference of 4968 and 2099.
(16)

12. $4.30 × 100   **13.** $402.00 ÷ 25
(2)   (2)

14. What is $\frac{3}{5}$ of 20?
(29)

Write each answer in simplest form:

15. $\frac{4}{5} + \frac{4}{5}$ **16.** $\frac{5}{8} - \frac{1}{8}$
(24)   (24)

17. $\frac{5}{2} \times \frac{3}{2}$ **18.** $\frac{3}{10} \times \frac{3}{100}$
(29)   (29)

19. Describe the sequence below. Then find the **tenth** term.
(10)

$$2, 4, 6, 8, \ldots$$

Find each missing number. Remember to check your work.

20. $Q - 24 = 23$ **21.** $\frac{1}{2}w = 1$
(3)   (30)

22. Here we show 16 written as a product of prime numbers:
(19)

$$2 \cdot 2 \cdot 2 \cdot 2$$

Write 15 as a product of prime numbers.

23. A meter is about one big step. About how many meters
(7) tall is a door?

24. Five of the 30 skateboards were orange. What fraction of
(29) the skateboards were orange? Write the answer as a
reduced fraction.

25. To what mixed number is the arrow pointing on the
(17) number line below?

26. Write each percent as a reduced fraction:
(33)

(a) 70%   (b) 30%

27. Four fifths of Gina's 20 answers were correct. How many of
(22) Gina's answers were correct? Draw a diagram to illustrate
the problem.

28. By looking at the numerator and denominator of a fraction,
(Inv. 2) how can you tell whether the fraction is greater than or
less than $\frac{1}{2}$?

29. What time is $6\frac{1}{2}$ hours after 8:45 p.m.?
(32)

30. Arrange these fractions in order from least to greatest:
(17)

$$\frac{1}{8}, \frac{1}{4}, \frac{1}{16}, \frac{1}{2}$$

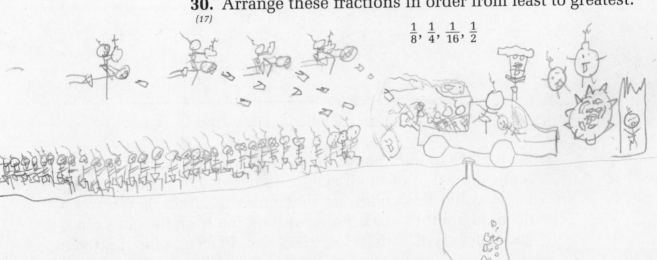

LESSON
34   Decimal Place Value

WARM-UP

Facts Practice: 64 Multiplication Facts (Test D)

Mental Math: Count up and down by $\frac{1}{8}$'s between $\frac{1}{8}$ and 2.

a. $(4 \times 200) + (4 \times 25)$
b. $1480 - 350$
c. $45 + 18$ *(18 is 20 – 2.)*
d. $\$12.00 - \$2.50$
e. Double 250.
f. $\frac{1500}{100}$
g. $3 \times 3, \times 9, - 1, \div 2, + 2, \div 7, \times 2$

Problem Solving:

The deck of 52 cards contained only red and black cards. Kathy selected three cards. Two were red and one was black. If she selects three more cards, what possible color combinations of three cards could she select?

NEW CONCEPT

Since Lesson 12 we have studied place value from the ones place leftward to the hundred trillions place. As we move to the left, each place is ten times as large as the preceding place. If we move to the right instead of to the left, each place is one tenth as large as the preceding place.

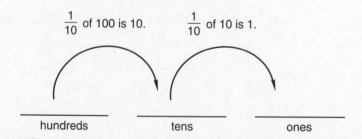

$\frac{1}{10}$ of 100 is 10. $\frac{1}{10}$ of 10 is 1.

hundreds   tens   ones

Each place to the right of the ones place also has a value one tenth the value of the place to its left. Each of these places has a value less than one (but more than zero). We use a **decimal point** to mark the separation between the ones place and places with values less than one. Places to the right of a decimal point are often called **decimal places.** Here we show three decimal places:

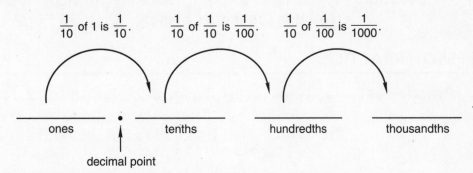

$\frac{1}{10}$ of 1 is $\frac{1}{10}$. $\qquad$ $\frac{1}{10}$ of $\frac{1}{10}$ is $\frac{1}{100}$. $\qquad$ $\frac{1}{10}$ of $\frac{1}{100}$ is $\frac{1}{1000}$.

ones $\qquad$ tenths $\qquad$ hundredths $\qquad$ thousandths

decimal point

Thinking about money is a helpful way to remember decimal place values.

mill

A mill is $\frac{1}{1000}$ of a dollar and $\frac{1}{10}$ of a cent. We do not have a coin for a mill. However, purchasers of gasoline are charged mills at the gas pump. A price of 1.49\frac{9}{10}$ per gallon is one mill less than $1.50 and nine mills more than $1.49.

Of course, decimal place values extend beyond the thousandths place. Moving to the right, the place values get smaller and smaller; each place has one tenth the value of the place to its left. The chart below shows decimal place values from the millions place through the millionths place:

Decimal Place Values

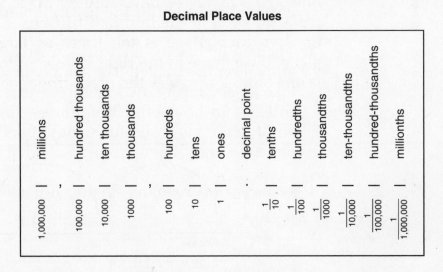

millions	hundred thousands	ten thousands	thousands	hundreds	tens	ones	decimal point	tenths	hundredths	thousandths	ten-thousandths	hundred-thousandths	millionths
1,000,000	100,000	10,000	1000	100	10	1	.	$\frac{1}{10}$	$\frac{1}{100}$	$\frac{1}{1000}$	$\frac{1}{10,000}$	$\frac{1}{100,000}$	$\frac{1}{1,000,000}$

Example 1 Which digit in 123.45 is in the hundredths place?

Solution The *-ths* ending of *hundredths* indicates that the hundredths place is to the right of the decimal point. The first place to the right of the decimal point is the tenths place. The second place is the hundredths place. The digit in the hundredths place is **5.**

Example 2   What is the place value of the 8 in 67.89?

Solution   The 8 is in the first place to the right of the decimal point, which is the **tenths** place.

LESSON PRACTICE

Practice set   **a.** What is the place value of the 5 in 12.345?

b. Which digit in 5.4321 is in the tenths place?

c. In 0.0123, what is the digit in the thousandths place?

d. What is the value of the place held by zero in 50.375?

e. What is the name for one hundredth of a dollar?

f. What is the name for one thousandth of a dollar?

MIXED PRACTICE

Problem set   **1.** Three eighths of the 24 choir members were tenors. How
(22)   many tenors were in the choir? Draw a diagram to illustrate the problem.

2. Mom wants to triple a recipe for cheesecake. If the recipe
(15)   calls for 8 ounces of cream cheese, how many ounces of cream cheese should she use in the mix?

3. The mayfly has the shortest known adult life span of any
(32)   animal on the planet. The mayfly grows underwater in a lake or stream for two or three years, but it lives for as little as an hour after it sprouts wings and becomes an adult. If a mayfly sprouts wings at 8:47 a.m. and lives for one hour and fifteen minutes, at what time does it die?

4. Write each percent as a reduced fraction:
(33)
(a) 60%   (b) 40%

5. Compare: $\dfrac{100}{100} \bigcirc \dfrac{10}{10}$
(29)

6. Write $(6 \times 100) + (5 \times 1)$ in standard notation.
(32)

7. Which digit is in the ones place in $42,876.39?
(34)

8. If the perimeter of a square is 24 inches,
(31)
(a) how long is each side of the square?

(b) what is the area of the square?

9. Draw triangle ABC so that $\angle C$ is a right angle, side AC is
(Inv. 3) $1\frac{1}{2}$ inches, and side BC is 2 inches. Then measure the length of side AB.

10. What is the least common multiple of 6 and 8?
(30)

11. $5.60 ÷ 10
(2)

12. $\frac{9}{10} \cdot \frac{9}{10}$
(29)

13. Estimate the quotient when 898 is divided by 29.
(16)

14. Round 36,847 to the nearest hundred.
(16)

Find each missing number. Remember to check your work.

15. $6d = 144$
(4)

16. $\frac{d}{6} = 144$
(4)

17. Compare: $\frac{5}{2} + \frac{5}{2} \bigcirc 2 \times \frac{5}{2}$
(29)

18. $\frac{3}{8} + \frac{3}{8}$
(24)

19. $\frac{11}{12} - \frac{1}{12}$
(24)

20. $\frac{5}{4} \times \frac{3}{2}$
(29)

21. What is the ratio of the first term to the fifth term of the
(10, 23) sequence below?

$$6, 12, 18, 24, \ldots$$

22. The movie theater sold 86 tickets to the afternoon show
(15) for $5.75 per ticket. What was the total of the ticket sales for the show? (Write an equation and solve the problem.)

23. To what number is the arrow pointing on the number line
(14) below?

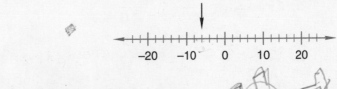

24. $(80 ÷ 40) - (8 ÷ 4)$
(5)

25. Which digit in 2,345.678 is in the thousandths place?
(34)

26. Draw a circle and shade $\frac{2}{3}$ of it.
(Inv. 2)

27. Divide 5225 by 12 and write the quotient as a mixed
(25) number.

28. The first glass contained 12 ounces of water. The second
(18) glass contained 11 ounces of water. The third glass
contained 7 ounces of water. If water was poured from
the first and second glasses into the third glass until each
glass contained the same amount, then how many ounces
of water would be in each glass?

29. The letters r, t, and d represent three different numbers.
(2) The product of r and t is d.

$$rt = d$$

Arrange the letters to form another multiplication fact
and two division facts.

30. Instead of dividing 75 by 5, Sandy mentally doubled both
(2) numbers and divided 150 by 10. Find the quotient
of $75 \div 5$ and the quotient of $150 \div 10$.

LESSON

35 Writing Decimal Numbers as Fractions, Part 1 • Reading and Writing Decimal Numbers

WARM-UP

Facts Practice: 30 Fractions to Reduce (Test G)

Mental Math: Count by 3's from 3 to 60. Count by 7's from 7 to 84.

a. $(4 \times 300) + (4 \times 25)$ b. 8×43

c. $37 + 39$ d. $\$7.50 + \$7.50$

e. $\frac{1}{3}$ of 360   f. $\frac{3600}{10}$

g. $5 \times 5, - 1, \div 3, \times 4, + 1, \div 3, + 1, \div 3$

Problem Solving:

Copy this problem and fill in the missing digits:
$$\begin{array}{r} \_\_\_ \\ +\quad\_\_\ 1 \\ \hline \_\_\_\_ \end{array}$$

NEW CONCEPTS

Writing decimal numbers as fractions, part 1

Decimal numbers are actually fractions. Their denominators come from the sequence 10, 100, 1000, ... The denominator of a decimal fraction is not written. Instead, the denominator is indicated by the number of decimal places.

One decimal place indicates that the denominator is 10.

$$0.3 = \frac{3}{10}$$

Two decimal places indicates that the denominator is 100.

$$0.03 = \frac{3}{100}$$

Three decimal places indicates that the denominator is 1000.

$$0.003 = \frac{3}{1000}$$

Notice that the number of zeros in the denominator equals the number of decimal places in the decimal number.

Example 1 Write 0.23 as a fraction.

Solution The decimal number 0.23 has two decimal places, so the denominator is 100. The numerator is 23.

$$0.23 = \frac{23}{100}$$

Example 2   Write $\frac{9}{10}$ as a decimal number.

Solution   The denominator is 10, so the decimal number has one decimal place. We write the digit 9 in this place.

$$\frac{9}{10} \longrightarrow 0.\underline{} \longrightarrow \mathbf{0.9}$$

Reading and writing decimal numbers   We read numbers to the right of a decimal point the same way we read whole numbers, and then we say the place value of the last digit. We read 0.23 as "twenty-three hundredths" because the last digit is in the hundredths place. To read a mixed decimal number like 20.04, we read the whole number part, say "and," and then read the decimal part.

Example 3   Write 0.023 with words.

Solution   We see 23 and three decimal places. We write **twenty-three thousandths.**

Example 4   Use words to write 20.04.

Solution   The decimal point separates the whole number part of the number from the decimal part of the number. We name the whole number part, write "and," and then name the decimal part.

twenty and four hundredths

Example 5   Write twenty-one hundredths
(a) as a fraction.
(b) as a decimal number.

Solution   The same words name both a fraction form and a decimal form of the number.
(a) The word *hundredths* indicates that the denominator is 100.

$$\frac{\mathbf{21}}{\mathbf{100}}$$

(b) The word *hundredths* indicates that the decimal number has two decimal places.

0.21

Example 6   Write fifteen and two tenths as a decimal number.

Solution   The whole number part is fifteen. The fractional part is two tenths, which we write in decimal form.

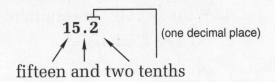

fifteen and two tenths

LESSON PRACTICE

Practice set*   Write each decimal number as a fraction:

   a. 0.1   **b.** 0.21   **c.** 0.321

Write each fraction as a decimal number:

   d. $\frac{3}{10}$ **e.** $\frac{17}{100}$ **f.** $\frac{123}{1000}$

Use words to write each number:

   g. 0.05   **h.** 0.015

   i. 1.2

Write each number first as a fraction, then as a decimal number:

   j. seven tenths

   k. thirty-one hundredths

   l. seven hundred thirty-one thousandths

Write each number as a decimal number:

   m. five and six tenths

   n. eleven and twelve hundredths

   o. one hundred twenty-five thousandths

MIXED PRACTICE

Problem set   **1.** What is the product of three fourths and three fifths?
                 (29)

   2. Bugs planted 360 carrot seeds in his garden. Three fourths
   (22) of them sprouted. How many carrot seeds sprouted? Draw
   a diagram to illustrate the problem.

3. Jan's birthday cake must bake for 2 hours 15 minutes. If it is
(32) put into the oven at 11:45 a.m., at what time will it be done?

4. Write twenty-three hundredths
(35) (a) as a fraction.

(b) as a decimal number.

5. Write 10.01 with words.
(35)

6. Write ten and five tenths as a decimal number.
(35)

7. Write each percent as a reduced fraction:
(33) (a) 25%   (b) 75%

8. Write (5 × 1000) + (6 × 100) + (4 × 10) in standard
(32) notation.

9. Which digit in 1.23 has the same place value as the 5 in
(34) 0.456?

10. What is the area of the rectangle below?
(31)

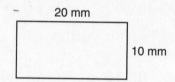

11. In problem 10, what is the perimeter of the rectangle?
(8)

12. There are 100 centimeters in a meter. How many
(7) centimeters are in 10 meters?

13. Arrange these numbers in order from least to greatest:
(34)

$$0.001, \ 0.1, \ 1.0, \ 0.01, \ 0, \ -1$$

14. A meter is about one big step. About how many meters
(7) wide is a door?

15. $\dfrac{3}{5} + \dfrac{2}{5}$ **16.** $\dfrac{5}{8} - \dfrac{5}{8}$ **17.** $\dfrac{2}{3} \times \dfrac{3}{4}$
(24)   (24)   (29)

18. (a) How many $\frac{2}{5}$'s are in 1?
(30)

(b) Use the answer to part (a) to find the number of $\frac{2}{5}$'s in 2.

19. Convert $\frac{20}{6}$ to a mixed number. Remember to reduce the
(25) fraction part of the number.

20. $\frac{100\%}{6}$ **21.** $3\frac{4}{4} - 1\frac{1}{4}$
(25)   (26)

22. Compare: $5 \bigcirc 4\frac{4}{4}$
(29)

23. One sixth of a circle is what percent of a circle?
(Inv. 2)

24. Compare: $3 \times 18 \div 6 \bigcirc 3 \times (18 \div 6)$
(9)

25. To what number is the arrow pointing on the number line
(14) below?

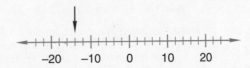

26. Which of these division problems has the greatest quotient?
(2)

   A. $\frac{6}{2}$ B. $\frac{60}{20}$ C. $\frac{12}{4}$ D. $\frac{25}{8}$

27. Write 0.3 and 0.7 as fractions. Then multiply the fractions.
(35) What is the product?

28. Write 21% as a fraction. Then write the fraction as a
(33, 35) decimal number.

29. Instead of solving the division problem $400 \div 50$, Minh
(2) doubled both numbers to form the division $800 \div 100$.
Find both quotients.

30. A 50-inch-long ribbon was cut into four shorter ribbons
(25) of equal length. How long was each of the shorter
ribbons? (Write an equation and solve the problem.)

LESSON

36

Subtracting Fractions and Mixed Numbers from Whole Numbers

WARM-UP

Facts Practice: 90 Division Facts (Test F)

Mental Math: Count up and down by $\frac{1}{8}$'s between $\frac{1}{8}$ and 3.

a. $(4 \times 400) + (4 \times 25)$
b. $2500 + 375$
c. $86 - 39$
d. $\$15.00 - \$2.50$
e. $\frac{1}{2}$ of 320
f. $\frac{4800}{100}$
g. $2 \times 4, \times 5, + 10, \times 2, - 1, \div 9, \times 3, - 1, \div 4$

Problem Solving:

Liam has seven coins that total exactly one dollar. Name three sets of coins that he **could** have.

NEW CONCEPT

Read this "separating" story problem about pies.

> *There were four pies on the shelf. The server sliced one of the pies into sixths and took $2\frac{1}{6}$ pies from the shelf. How many pies were left on the shelf?*

We can illustrate this story with circles. There were four pies on the shelf.

The server sliced one of the pies into sixths. (Then there were $3\frac{6}{6}$ pies, which is another name for 4 pies.)

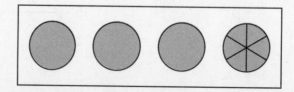

The server took $2\frac{1}{6}$ pies from the shelf.

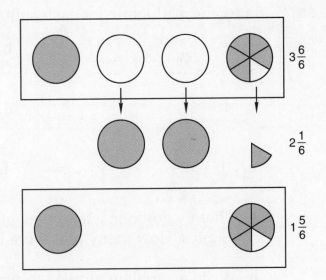

We see $1\frac{5}{6}$ pies left on the shelf.

Now we show the arithmetic for subtracting $2\frac{1}{6}$ from 4.

$$
\begin{array}{r}
4 \text{ pies} \\
- \, 2\frac{1}{6} \text{ pies} \\
\hline
\end{array}
$$

The server sliced one of the pies into sixths, so we change 4 wholes into 3 wholes plus 6 sixths. Then we subtract.

$$
\begin{array}{r}
4 \\
- \, 2\frac{1}{6} \\
\end{array}
\quad \xrightarrow{\;3 + \frac{6}{6}\;} \quad
\begin{array}{r}
3\frac{6}{6} \\
- \, 2\frac{1}{6} \\
\hline
1\frac{5}{6}
\end{array}
$$

Example Subtract: $5 - 1\frac{2}{3}$

Solution To subtract $1\frac{2}{3}$ from 5, we first change 5 to 4 plus $\frac{3}{3}$. Then we subtract.

$$
\begin{array}{r}
5 \\
- \, 1\frac{2}{3} \\
\end{array}
\quad \xrightarrow{\;4 + \frac{3}{3}\;} \quad
\begin{array}{r}
4\frac{3}{3} \\
- \, 1\frac{2}{3} \\
\hline
3\frac{1}{3}
\end{array}
$$

LESSON PRACTICE

Practice set*   Show the arithmetic for each subtraction:

a. $3 - 2\frac{1}{2}$ **b.** $2 - \frac{1}{4}$

c. $4 - 2\frac{1}{4}$ **d.** $3 - \frac{5}{12}$

e. $10 - 2\frac{1}{2}$ **f.** $6 - 1\frac{3}{10}$

g. There were four whole pies on the shelf. The server took $1\frac{5}{6}$ pies. How many pies were left on the shelf?

h. Write a problem similar to problem **g**, and then find the answer.

MIXED PRACTICE

Problem set   **1.** Twenty-five percent of the singers played musical
    (33)   instruments. What fraction of the singers played musical instruments?

2. Grace accidentally sat on her lunch and smashed $\frac{3}{4}$ of her
(36)   sandwich. What fraction of her sandwich was not smashed?

3. A mile is 5280 feet. There are 3 feet in a yard. How many
(15)   yards are in a mile?

4. Which digit in 23.47 has the same place value as the 6
(34)   in 516.9?

5. Write 1.3 with words.
(35)

6. Write the decimal number five hundredths.
(35)

7. Write thirty-one hundredths
(35)   (a) as a fraction.

    (b) as a decimal number.

8. Write $(4 \times 100) + (3 \times 1)$ in standard notation.
(32)

9. Which digit in 4.375 is in the tenths place?
(34)

10. If the area of a square is 9 square inches,
$(8, 31)$

    (a) how long is each side of the square?

    (b) what is the perimeter of the square?

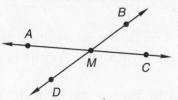

11. Name two obtuse angles in the figure below.
(28)

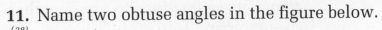

12. $3\frac{1}{4} + 2\frac{1}{4}$ **13.** $3 - 1\frac{1}{4}$ **14.** $3\frac{1}{3} + 2\frac{2}{3}$
(26) (36) (26)

15. What is $\frac{3}{4}$ of 28?   **16.** $\frac{3}{4} \times \frac{4}{6}$
(29) (29)

17. Monte went to the mall with \$24. He spent $\frac{5}{6}$ of his money
(22) in the music store. How much money did he spend in the music store? Draw a diagram to illustrate the problem.

18. What is the average of 42, 57, and 63?
(18)

19. The factors of 6 are 1, 2, 3, and 6. List the factors of 20.
(19)

20. (a) What is the least common multiple of 9 and 6?
$(20, 30)$

    (b) What is the greatest common factor of 9 and 6?

Find each missing number. Remember to check your work.

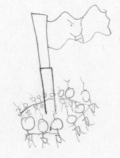

21. $\frac{m}{12} = 6$ **22.** $\frac{12}{n} = 6$
(4) (4)

23. Round 58,742,177 to the nearest million.
(16)

24. Estimate the product of 823 and 680.
(16)

25. How many millimeters long is the line segment shown
(7) below (1 cm = 10 mm)?

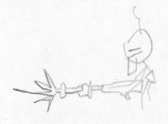

26. Using your fraction manipulatives, you can find that the
(Inv. 2) sum of $\frac{1}{3}$ and $\frac{1}{6}$ is $\frac{1}{2}$.

$$\frac{1}{3} + \frac{1}{6} = \frac{1}{2}$$

Arrange these fractions to form another addition fact and two subtraction facts.

27. Write 0.9 and 0.09 as fractions. Then multiply the
(35) fractions. What is the product?

28. Write $\frac{81}{1000}$ as a decimal number. (*Hint:* Write a zero in the
(35) tenths place.)

29. (a) How many $\frac{3}{4}$'s are in 1?
(30)

(b) Use the answer to part (a) to find the number of $\frac{3}{4}$'s in 3.

30. Freddy drove a stake into the
(23, 27) ground, looped a 12-foot-long rope over it, and walked around the stake to mark off a circle. What was the ratio of the radius to the diameter of the circle?

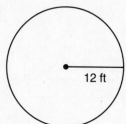

12 ft

THE OLD PROVERB SAYS

He carries the bucktoofian
flag, still dreaming about his
family back home. he must
push on! Can he make it?
will he fall on the field
like so many others? will
his moment of triumpth
be taken in vein? find
out next week!

The
EPIC

Bernard

LESSON
37

Adding and Subtracting Decimal Numbers

WARM-UP

NEW CONCEPT

When we add or subtract numbers using pencil and paper, it is important to align digits that have the same place value. For whole numbers this means lining up the ending digits. When we line up the ending digits (which are in the ones place) we automatically align other digits that have the same place value.

$$\begin{array}{r} 23 \\ 241 \\ + \ 317 \end{array}$$

Lining up the ones place automatically aligns all other digits by their place value.

However, lining up the ending digits of decimal numbers might not properly align all the digits. We use another method for decimal numbers. **We line up decimal numbers for addition or subtraction by lining up the decimal points.** The decimal point in the answer is aligned with the other decimal points. Empty places are treated as zeros.

$$\begin{array}{r} 2.3 \\ 2.41 \\ + \ 31.7 \end{array}$$

Lining up the decimal points automatically aligns digits that have the same place value.

Example 1   Add: 3.4 + 0.26 + 0.3

Solution   Line up the decimal points in the problem and add. The decimal point in the answer is placed in line with the other decimal points.

$$\begin{array}{r} 3.4 \\ 0.26 \\ +\ 0.3 \\ \hline \mathbf{3.96} \end{array}$$

Example 2   Subtract: 4.56 − 2.3

Solution   Line up the decimal points in the problem and subtract.

$$\begin{array}{r} 4.56 \\ -\ 2.3 \\ \hline \mathbf{2.26} \end{array}$$

LESSON PRACTICE

Practice set   Find each sum or difference. Remember to line up the decimal points.

a. 3.46 + 0.2

b. 8.28 − 6.1

c. 0.735 + 0.21

d. 0.543 − 0.21

e. 0.43 + 0.1 + 0.413

f. 0.30 − 0.27

g. 0.6 + 0.7

h. 1.00 − 0.24

i. 0.9 + 0.12

j. 1.23 − 0.4

MIXED PRACTICE

Problem set   **1.** Sixty percent of the people in the theater were girls. What
(33)   fraction of the people in the theater were girls?

2. Penny broke 8 pencils while taking her math test. She
(22)   broke half as many while taking her spelling test. How many pencils did she break in all?

3. What number must be added to three hundred seventy-
(3)   five to get the number one thousand?

4. 3.4 + 0.62 + 0.3
(37)

5. 4.56 − 3.2
(37)

6. $0.37 + $0.23 + $0.48   **7.** $5 − m = 5¢
(1)   (3)

8. What is the next number in this sequence?
(10)

$$1, 10, 100, 1000, \ldots$$

9. Harriet used 100 square floor tiles with sides 1 foot long
(8, 31) to cover the floor of a square room.

(a) What was the length of each side of the room?

(b) What was the perimeter of the room?

10. Which digit is in the ten-millions place in 1,234,567,890?
(12)

11. Three of the numbers shown below are equal. Which
(35) number is not equal to the others?

A. $\frac{1}{10}$ B. 0.1   C. $\frac{10}{100}$ D. 0.01

12. Estimate the product of 29, 42, and 39.
(16)

13. 3210 ÷ 3   **14.** 32,100 ÷ 30
(2)   (2)

15. $10,000 − $345   **16.** $\frac{3}{4} + \frac{3}{4}$
(1)   (24)

17. $3 - 1\frac{3}{5}$ **18.** $\frac{3}{3} - \frac{2}{2}$
(36)   (29)

$\times C \neq C$

19. $1\frac{1}{3} + 2\frac{1}{3} + 3\frac{1}{3}$
(26)

20. Compare: $\frac{1}{4} + \frac{3}{4} \bigcirc \frac{1}{4} \times \frac{3}{4}$
(29)

21. Convert the improper fraction $\frac{100}{7}$ to a mixed number.
(25)

22. What is the average of 90 lb, 84 lb, and 102 lb?
(18)

23. What is the least common multiple of 4 and 5?
(30)

24. The stock's value dropped from $38.50 to $34.00. What
(14) negative number shows the change in value?

25. To what mixed number is the arrow pointing on the
(17) number line below?

26. Write 0.3 and 0.9 as fractions. Then multiply the
(35) fractions. Change the product to a decimal number.

27. Write three different fractions equal to 1. How can you
(Inv. 2) tell whether a fraction is equal to 1?

28. Instead of dividing 6 by $\frac{1}{2}$, Thoreau doubled both
(2) numbers and divided 12 by 1. Do you think both
quotients are the same? Write a one- or two-sentence
reason for your answer.

29. The movie started at 2:50 p.m. and ended at 4:23 p.m.
(32) How long was the movie?

30. Three fifths of the 25 bananas were ripe. How many
(22) bananas were ripe? Draw a diagram to illustrate the
problem.

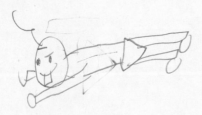

LESSON
38

Adding and Subtracting Decimal Numbers and Whole Numbers • Squares and Square Roots

WARM-UP

Facts Practice: 64 Addition Facts (Test A)

Mental Math: Count by 6's from 6 to 72. Count by 7's from 7 to 84.

a. $(4 \times 600) + (4 \times 25)$
b. $875 - 125$
c. $56 - 19$
d. $\$10.00 - \$6.25$
e. $\frac{1}{2}$ of 150
f. $\frac{\$40.00}{10}$
g. $10 + 10, - 2, \div 3, \times 4, + 1, \times 4, \div 2, + 6, \div 7$

Problem Solving:

Teresa wanted to paint each face of a cube so that the adjacent faces (the faces next to each other) were different colors. She wanted to use fewer than six different colors. What is the least number of colors she could use? Describe how the cube could be painted.

NEW CONCEPTS

Adding and subtracting decimal numbers and whole numbers

Here we show two ways to write three dollars:

$$\$3 \qquad \$3.00$$

We see that we may write dollar amounts with or without a decimal point. We may also write whole numbers with or without a decimal point. In any number written with a decimal point, the decimal point follows the ones place. Here are several ways to write the whole number three:

$$3 \qquad 3. \qquad 3.0 \qquad 3.00$$

As we will see in the following examples, it may be helpful to write a whole number with a decimal point when adding and subtracting with decimal numbers.

Example 1   Add: 12 + 7.5

Solution   When adding decimal numbers, we align decimal points so that we add digits with the same place values. The whole number 12 may be written with a decimal point to the right of the digit 2. We line up the decimal points and add.

$$
\begin{array}{r}
12.0 \\
+\ \ 7.5 \\
\hline
\mathbf{19.5}
\end{array}
$$

Example 2   Subtract: 12.75 − 5.00

Solution   We write the whole number 5 with a decimal point to its right. Then we line up the decimal points and subtract.

$$
\begin{array}{r}
12.75 \\
-\ \ 5.00 \\
\hline
\mathbf{7.75}
\end{array}
$$

Squares and square roots   Recall that we find the area of a square by multiplying the length of a side of the square by itself. For example, the area of a square with sides 5 cm long is 5 cm × 5 cm, which equals 25 sq. cm.

From the model of the square comes the expression "squaring a number." We square a number by multiplying the number by itself.

"Five squared" is 5 × 5, which is 25.

To indicate squaring, we use the **exponent** 2.

$$5^2 = 25$$

"Five squared equals 25."

Notice that the exponent is elevated and written to the right of the 5. An exponent shows how many times the other number, the **base,** is to be used as a factor. In this case, 5 is to be used as a factor twice.

Example 3   (a) What is twelve squared?

(b) Simplify: $3^2 + 4^2$

Solution   (a) "Twelve squared" is 12 × 12, which is **144.**

(b) We apply exponents before adding, subtracting, multiplying, or dividing.

$$3^2 + 4^2 = 9 + 16 = \mathbf{25}$$

We can use an exponent of 2 with a unit of length to indicate square units for measuring area.

$$1 \text{ cm}^2 = 1 \text{ square centimeter}$$

Example 4 What is the area of a square with sides 5 meters long?

Solution We multiply 5 meters by 5 meters. Both the units and the numbers are multiplied.

$$5 \text{ m} \cdot 5 \text{ m} = (5 \cdot 5)(\text{m} \cdot \text{m}) = \textbf{25 m}^2$$

We read 25 m^2 as "twenty-five square meters."

If we know the area of a square, we can find the length of each side. We do this by determining the length whose square equals the area. For example, a square whose area is 49 cm^2 has side lengths of 7 cm because 7 cm $\cdot$ 7 cm equals 49 cm^2.

Determining the length of a side of a square from the area of the square is a model for finding the principal **square root** of a number. Finding the square root of a number is the inverse of squaring a number.

6 squared is 36.

The principal square root of 36 is 6.

The square root symbol looks like this:

We read $\sqrt{100}$ as "the square root of 100." This expression means, "What positive number, when multiplied by itself, has a product of 100?" Since 10 × 10 equals 100, the principal square root of 100 is 10.

$$\sqrt{100} = 10$$

A number is a **perfect square** if it has a square root that is a whole number. Starting with 1, the first four perfect squares are 1, 4, 9, and 16, as illustrated below.

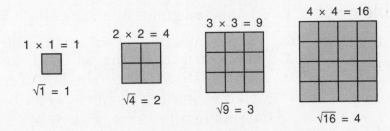

1 × 1 = 1   2 × 2 = 4   3 × 3 = 9   4 × 4 = 16
$\sqrt{1}$ = 1 $\sqrt{4}$ = 2 $\sqrt{9}$ = 3 $\sqrt{16}$ = 4

Example 5   Simplify: $\sqrt{64}$

Solution   The square root of 64 can be thought of in two ways:

1. What is the side length of a square that has an area of 64?

2. What positive number multiplied by itself equals 64?

With either approach, we find that $\sqrt{64}$ equals **8**.

LESSON PRACTICE

Practice set*   Simplify:

a. 4 + 2.1   **b.** 4.3 − 2

c. 3 + 0.4   **d.** 43.2 − 5

e. 0.23 + 4 + 3.7   **f.** 6.3 − 6

g. 12.5 + 10   **h.** 75.25 − 25

i. 9^2 **j.** $\sqrt{81}$

k. $6^2 + 8^2$ **l.** $\sqrt{100} - \sqrt{49}$

m. 15^2 **n.** $\sqrt{144}$

o. 6 ft · 6 ft   **p.** $\sqrt{64 \text{ m}^2}$

q. Starting with 1, the first four perfect squares are as follows:

1, 4, 9, 16

What are the next four perfect squares?

MIXED PRACTICE

Problem set   **1.** What is the greatest factor of both 54 and 45?
      (20)

2. Roberto began saving $3 each week for summer camp,
(15)   which costs $126. How many weeks will it take him to
      save that amount of money? (Write an equation and solve
      the problem.)

3. Gandhi was born in 1869. About how old was he when
(13) he was assassinated in 1948? (Write an equation and solve the problem.)

4. $\sqrt{9}$ + 1.2   **5.** 3.6 + $\sqrt{16}$ **6.** 5.63 − 1.2
(38)   (38)   (37)

7. 5.376 + 0.24   **8.** 4.75 − 0.6   **9.** $\sqrt{16}$ − $\sqrt{9}$
(37)   (37)   (38)

10. Write forty-seven hundredths
(35)

    (a) as a fraction.

    (b) as a decimal number.

11. Write (9 × 1000) + (4 × 10) + (3 × 1) in standard
(32) notation.

12. Which digit is in the hundredths place in $123.45?
(34)

13. The area of a square is 81 square inches.
(8, 38)

    (a) What is the length of each side?

    (b) What is the perimeter of the square?

14. What is the least common multiple of 2, 3, and 4?
(30)

15. $1\frac{2}{3}$ + $2\frac{2}{3}$ **16.** 3^2 − $1\frac{1}{4}$
(26)   (36, 38)

17. What is $\frac{3}{4}$ of $\frac{4}{5}$?   **18.** $\frac{7}{10}$ × $\frac{11}{10}$
(29)   (29)

19. (a) How many $\frac{2}{3}$'s are in 1?
(30)

    (b) Use the answer to part (a) to find the number of $\frac{2}{3}$'s in 2.

20. Six of the nine players got on base. What fraction of the
(29) players got on base?

21. List the factors of 30.
(19)

22. Write each percent as a reduced fraction:
(33)
(a) 35%   (b) 65%

23. Round 186,282 to the nearest thousand.
(16)

24. $\frac{1}{3}m = 1$ **25.** $\frac{22 + 23 + 24}{3}$
(30)   (5)

26. Compare: $24 \div 8 \bigcirc 240 \div 80$
(9)

27. Write 0.7 and 0.21 as fractions. Then multiply the
(35) fractions. Change the product to a decimal number.

28. Peter bought ten carrots for $0.80. What was the cost for
(15) each carrot?

29. Which of these fractions is closest to 1?
(36)
A. $\frac{1}{5}$ B. $\frac{2}{5}$ C. $\frac{3}{5}$ D. $\frac{4}{5}$

30. If you know the perimeter of a square, you can find the
(38) area of the square in two steps. Describe the two steps.

LESSON
39 Multiplying Decimal Numbers

WARM-UP

Facts Practice: 30 Fractions to Reduce (Test G)

Mental Math: Count up and down by $\frac{1}{8}$'s between $\frac{1}{8}$ and 3.

   a. $(4 \times 700) + (4 \times 25)$ **b.** 6×45

   c. $67 - 29$ **d.** \$8.75 + \$0.75

   e. $\frac{1}{2}$ of 350   **f.** $\frac{2500}{100}$

   g. $8 \times 5, \div 2, + 1, \div 7, \times 3, + 1, \div 10, \div 2$

Problem Solving:

Megan has many red socks, white socks, and blue socks in a drawer. In the dark she pulled out two socks that did not match. How many more socks does Megan need to pull from the drawer to be certain to have a matching pair?

NEW CONCEPT

To find the area of a rectangle that is 0.75 meter long and 0.5 meter wide, we multiply 0.75 m by 0.5 m.

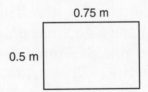

0.75 m

0.5 m

One way to multiply these numbers is to write each decimal number as a proper fraction and then multiply the fractions.

$$0.75 \times 0.5$$

$$\frac{75}{100} \times \frac{5}{10} = \frac{375}{1000}$$

The product $\frac{375}{1000}$ can be written as the decimal number 0.375. We find that the area of the rectangle is 0.375 square meter.

Notice that the product 0.375 has three decimal places and that three is the **total** number of decimal places in the factors, 0.75 (two) and 0.5 (one). When we multiply decimal numbers, the product has the same number of decimal places as there are in all of the factors combined. This fact allows us to multiply decimal numbers as if they were whole numbers. After multiplying, we count the total number of decimal places in the factors. Then we place a decimal point in the product to give it the same number of decimal places as there are in the factors.

$$\left.\begin{array}{r} \text{Three decimal} \\ \text{places in} \\ \text{the factors} \end{array}\right\} \begin{array}{r} 0.7\underline{5} \\ \times \quad 0.\underline{5} \\ \hline 0.\underline{375} \end{array} \quad \begin{array}{l} \text{We do not align decimal} \\ \text{points. We multiply and} \\ \text{then count decimal places.} \end{array}$$

Three decimal places
in the product

Example 1   Multiply: 0.25×0.7

Solution   We set up the problem as though we were multiplying whole numbers, initially ignoring the decimal points. Then we multiply. Next we count the digits to the right of the decimal points in the two factors. There are three, so we place a decimal point in the product three places from the right-hand end. We write .175 as **0.175**.

$$\left.\begin{array}{r} 0.25 \\ \times \quad 0.7 \end{array}\right\} \text{3 places}$$
$$\overline{0.175}$$

Example 2   Simplify: $(2.5)^2$

Solution   We square 2.5 by multiplying 2.5 by 2.5. We set up the problem as if we were multiplying whole numbers. Then we count decimal places in the factors. There are two, so we place a decimal point in the product two places from the right-hand end. We see that $(2.5)^2$ equals **6.25**.

$$\left.\begin{array}{r} 2.5 \\ \times \quad 2.5 \end{array}\right\} \text{2 places}$$
$$\begin{array}{r} 1\ 2\ 5 \\ 5\ 0 \\ \hline 6.2\ 5 \end{array}$$

Example 3 Multiply: 1.6 × 3

Solution We multiply as though we were multiplying whole numbers. Then we count decimal places in the factors. There is only one, so we place a decimal point in the product one place from the right-hand end. The answer is **4.8**.

$$\begin{array}{r} 1.6 \\ \times\ \ 3 \end{array} \Big\} \text{ 1 place}$$
$$\overline{4.8}$$

LESSON PRACTICE

Practice set Simplify:

a. 15 × 0.3

b. 1.5 × 3

c. 1.5 × 0.3

d. 0.15 × 3

e. 1.5 × 1.5

f. 0.15 × 10

g. 0.25 × 0.5

h. 0.025 × 100

i. $(0.8)^2$

j. $(1.2)^2$

MIXED PRACTICE

Problem set

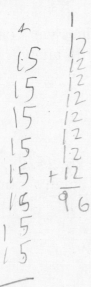

1. Mount Everest, the world's tallest mountain, rises to an
(12) elevation of twenty-nine thousand, thirty-five feet above sea level. Use digits to write that elevation.

2. There are three feet in a yard. How many yards above sea
(7, 25) level is Mount Everest's peak?

3. The sign in front of the old barber shop reads as follows:
(6)

> Shave and a Haircut
> *six bits*

  A bit is $\frac{1}{8}$ of a dollar. How many cents is 6 bits?

4. 0.25 × 0.5
(39)

5. $1.80 × 10
(2)

6. 63 × 0.7
(39)

7. $1.23 + \sqrt{16} + 0.5$
(38)

8. 12.34 − 5.6
(37)

9. $(1.1)^2$
(39)

10. Write ten and three tenths
(35)

(a) as a decimal number.

(b) as a mixed number.

11. Think of two different fractions that are greater than zero
(17) but less than one. Multiply the two fractions to form a third fraction. For your answer to this problem, write the three fractions in order from least to greatest.

12. Write the decimal number one hundred twenty-three
(35) thousandths.

13. Write $(6 \times 100) + (4 \times 10)$ in standard form.
(32)

14. The perimeter of a square is 40 inches. How many square
(38) tiles with sides 1 inch long are needed to cover its area?

15. What is the least common multiple (LCM) of 2, 3, and 6?
(30)

16. Convert $\frac{20}{8}$ to a mixed number. Remember to reduce the
(25) fraction part of the mixed number.

17. $\left(\frac{1}{3} + \frac{2}{3}\right) - 1$ **18.** $\frac{3}{5} \times \frac{2}{3}$ **19.** $\frac{8}{9} \times \frac{9}{8}$
(24)   (29)   (29)

20. A pie was cut into six equal slices. Two slices were eaten.
(36) What fraction of the pie was left? Write the answer as a reduced fraction.

21. What time is $2\frac{1}{2}$ hours before 1 a.m.?
(32)

22. On Hiroshi's last four assignments he had 26, 29, 28, and
(18) 25 correct answers. He averaged how many correct answers on these papers?

23. Estimate the quotient of 7987 divided by 39.
(16)

24. Compare: $365 - 364 \bigcirc 364 - 365$
(14)

25. Which digit in 3.675 has the same place value as the 4
(34) in 14.28?

26. Use an inch ruler to find the length of the segment below
(17) to the nearest sixteenth of an inch.

_____

27. (a) How many $\frac{3}{5}$'s are in 1?
(30)
　　(b) Use the answer to part (a) to find the number of $\frac{3}{5}$'s in 2.

28. Instead of solving the division problem $390 \div 15$, Roosevelt
(2) divided both numbers by 3 to form the division $130 \div 5$.
Then he multiplied both of those numbers by 2 to get
$260 \div 10$. Find all three quotients.

29. Find the area of the rectangle below.
(39)

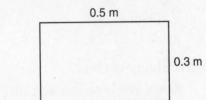

0.5 m

0.3 m

30. In the figure below, what is the ratio of the measure of
(23) $\angle ABC$ to the measure of $\angle CBD$?

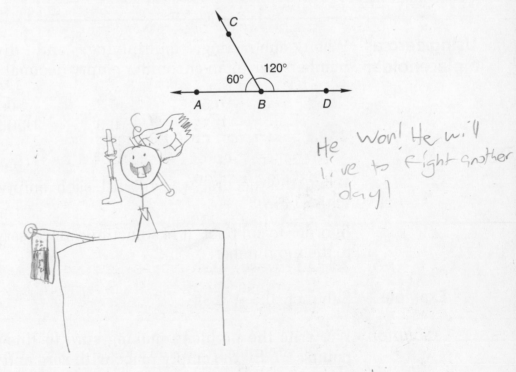

He won! He will
live to fight another
day!

The EPIC bewnawd

Conclusion

LESSON
40

Using Zero as a Placeholder • Circle Graphs

WARM-UP

Facts Practice: 64 Multiplication Facts (Test D)

Mental Math: Count by 8's from 8 to 96.

a. $(4 \times 800) + (4 \times 25)$
b. $1500 + 750$
c. $74 - 39$
d. $\$8.25 - \$1.50$
e. Double 240.
f. $\frac{480}{10}$
g. $4 \times 4, -1, \div 5, \times 6, +2, \times 2, +2, \div 6$

Problem Solving:

Copy this problem and fill in the missing digits:

$$\begin{array}{r} \overline{}\;\overline{}\;\overline{} \\ \times 9 \\ \hline \_\_2 \end{array}$$

NEW CONCEPTS

Using zero as a placeholder

When subtracting, multiplying, and dividing decimal numbers, we often encounter empty decimal places.

$$\begin{array}{r} 0.5\_ \\ -\,0.32 \\ \hline \end{array} \qquad \begin{array}{r} 0.2 \\ \times\,0.3 \\ \hline \_.\_6 \end{array} \qquad \begin{array}{r} \$0.\_4 \\ 3)\overline{\$0.12} \end{array}$$

When this occurs, we will fill each empty decimal place with a zero.

In order to subtract, it is sometimes necessary to attach zeros to the top number.

Example 1 Subtract: $0.5 - 0.32$

Solution We write the problem, making sure to line up the decimal points. We fill the empty place with zero and subtract.

$$\begin{array}{r} 0.5\_ \\ -\,0.32 \\ \hline \end{array} \longrightarrow \begin{array}{r} 0.\overset{4}{\cancel{5}}\overset{1}{0} \\ -\,0.32 \\ \hline \mathbf{0.18} \end{array}$$

Example 2   Subtract: $3 - 0.4$

Solution   We place a decimal point on the back of the whole number and line up the decimal points. We fill the empty place with zero and subtract.

$$
\begin{array}{r} 3.\_ \\ -\ 0.4 \end{array} \longrightarrow \begin{array}{r} \overset{2}{\cancel{3}}.\overset{1}{0} \\ -\ 0.4 \\ \hline \textbf{2.6} \end{array}
$$

When multiplying decimal numbers, we may need to insert one or more zeros between the multiplication answer and the decimal point to hold the other digits in their proper places.

Example 3   Multiply: 0.2×0.3

Solution   We multiply and count two places from the right. We fill the empty place with zero, and we write a zero in the ones place.

$$
\begin{array}{r} 0.2 \\ \times\ 0.3 \\ \hline -.6 \end{array} \longrightarrow \begin{array}{r} 0.2 \\ \times\ 0.3 \\ \hline \textbf{0.06} \end{array}
$$

Example 4   Use digits to write the decimal number twelve thousandths.

Solution   The word *thousandths* tells us that there are three places to the right of the decimal point.

$$. \_\ \_\ \_$$

We fit the two digits of twelve into the last two places.

$$._\underline{1}\,\underline{2}$$

Then we fill the empty place with zero.

0.012

Circle graphs   Circle graphs, which are sometimes called **pie graphs** or **pie charts**, display quantitative information in fractions of a circle. The next example uses a circle graph to display information about neighborhood pets.

Example 5   Brett collected information from people in his neighborhood about their pets. He displayed the information about the number of pets in a circle graph. Use the graph to answer the following questions:

(a) How many pets are represented in the graph?

(b) What fraction of the pets are birds?

(c) What percent of the pets are dogs?

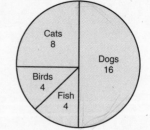

LIDSKIN
DORPSHNOD
CRETATIOUS
PARAKEETS

Solution   (a) We add the number of dogs, cats, birds, and fish. The total is **32.**

(b) Birds are 4 of the 32 pets. The fraction $\frac{4}{32}$ reduces to $\frac{1}{8}$. (The bird portion of the circle is $\frac{1}{8}$ of the whole circle.)

(c) Dogs are 16 of the 32 pets, which means that $\frac{1}{2}$ of the pets are dogs. From our fraction manipulatives we know that $\frac{1}{2}$ equals **50%.** Circle graphs often express portions in percent form. Instead of showing the number of each kind of animal, the graph could have labeled each portion with a percent.

LESSON PRACTICE

Practice set*   Simplify:

a. 0.2×0.3 b. $4.6 - 0.46$

c. 0.1×0.01 d. $0.4 - 0.32$

e. 0.12×0.4 f. $1 - 0.98$

g. $(0.3)^2$ h. $(0.12)^2$

i. Write the decimal number ten and eleven thousandths.

j. In the circle graph in example 5, what percent of the pets are cats?

MIXED PRACTICE

Problem set   **1.** In the circle graph in example 5, what percent of the pets
(40)   are birds? (Use your fraction manipulatives to help you answer the question.)

2. The U.S. Constitution was ratified in 1788. In 1920 the
(13)   19th Amendment to the Constitution was ratified, guaranteeing women the right to vote. How many years after the Constitution was ratified were women guaranteed the right to vote? (Write an equation and solve the problem.)

3. White Rabbit is three-and-a-half-hours late for a very
(32)   important date. If the time is 2:00 p.m., what was the time of his date?

4. $\sqrt{9} - 0.3$ **5.** $1.2 - 0.12$ **6.** $1 - 0.1$
(38, 40)   (40)   (40)

7. 0.12×0.2 **8.** $(0.1)^2$ **9.** 4.8×0.23
(40)   (40)   (39)

10. Write one and two hundredths as a decimal number.
(40)

11. Write $(6 \times 10{,}000) + (8 \times 100)$ in standard form.
(32)

12. A square room has a perimeter of 32 feet. How many
(8, 31)   square floor tiles with sides 1 foot long are needed to cover the floor of the room?

13. What is the least common multiple (LCM) of 2, 4, and 8?
(30)

14. $6\frac{2}{3} + 4\frac{2}{3}$ **15.** $5 - 3\frac{3}{8}$
(26)   (36)

16. $\frac{5}{8} \times \frac{2}{3}$ **17.** $2\frac{5}{6} + 5\frac{2}{6}$
(29)   (26)

18. Compare: $\frac{1}{2} \times \frac{2}{2} \bigcirc \frac{1}{2} \times \frac{3}{3}$
(29)

19. $1000 - w = 567$
(3)

20. Nine whole numbers are factors of 100. Two of the factors
(19) are 1 and 100. List the other seven factors.

21. $9^2 + \sqrt{9}$
(38)

22. Round $4167 to the nearest hundred dollars.
(16)

The circle graph below displays Maggie's grades on recent
assignments. Use the graph to answer problems 23–26.

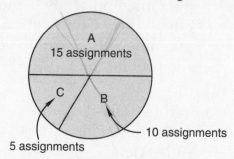

23. How many assignments are displayed?
(40)

24. On what fraction of the assignments did Maggie receive C's?
(40)

25. On what percent of the assignments did Maggie receive A's?
(40)

26. Write a ratio problem that relates to the circle graph. Then
(23, 40) solve the problem.

27. Arrange the numbers in this multiplication fact to form
(40) another multiplication fact and two division facts.

$$0.2 \times 0.3 = 0.06$$

28. Instead of solving the division problem $240 \div 15$,
(2) Elianna divided both numbers by 3 to form the
division $80 \div 5$. Then she doubled both numbers to
get $160 \div 10$. Find all three quotients.

29. Forty percent of the 25 balloons are red. The rest of the
(23, 33) balloons are blue. Write 40% as a reduced fraction. Then
find the ratio of blue balloons to red balloons.

30. What mixed number is represented by point *A* on the
(17) number line below?

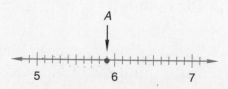

INVESTIGATION 4

Focus on

Data Collection and Surveys

Statistics is the science of gathering and organizing data (information) in such a way that we can draw conclusions from the data. In this investigation we will focus on ways in which statistical data can be collected.

Data can be either **quantitative** or **qualitative** in nature. Quantitative data comes in numbers: the population of a city, the number of pairs of shoes someone owns, or the number of hours per week someone watches television. Qualitative data comes in categories: the month in which someone is born or a person's favorite flavor of ice cream.

In problems 1–5 below, decide whether the data obtained is qualitative or quantitative.

1. Hans buys 50 bags of peanuts and counts the number of peanuts in each bag.

2. For one hour Carlos notes the color of each car that drives past his house.

3. Sharon rides a bus home after work. For two weeks she measures the time the bus trip takes.

4. Brigit asks each person at the party, "Among New Year's, Thanksgiving, and Independence Day, which is your favorite holiday?"

5. Marcello asks each player on his little league team, "Which major league baseball team is your favorite? Which team do you like the least?"

In problems 4 and 5 above, the responses to the survey questions are limited. Remember from Investigation 1 that these kinds of surveys are called **closed-option surveys.** Many closed-option surveys allow only two responses. Surveys that do not limit the choices are called **open-option surveys.**

6. Create both an open-option and a closed-option survey question about television viewing. Then create a question that lists certain possible answers but also allows other choices.

Surveys can be designed to gather data about a certain group of people. This "target group" is called a **population.** For example, if a record company wants to know about teenagers' music preferences, it would not include senior citizens in the population it studies. Often it is not realistic to poll an entire population, so a small part of the population is surveyed. We call this small part a **sample.** Surveyors must carefully select their samples, because different samples will provide different data. It is important that the sample for a population be a **representative sample.** That is, the characteristics of the sample should be similar to those of the entire population.

In problems 7 and 8 below, explain why each sample is not representative. How would you expect the sample's responses to differ from those of the general population?

7. To determine public opinion in the city of Dallas about a proposed leash law for dogs, Sally interviews shoppers in several Dallas pet stores.

8. Tamika wants to know the movie preferences of students at her university. Since she is in the chamber orchestra, she chooses to survey the orchestra members.

Often the results of a survey depend on the way its questions are worded or who is asking the questions. These factors can introduce **bias** into a survey. When a survey is biased, the people surveyed might be influenced to give certain answers over other possible answers.

For problems 9 and 10, identify the bias in the survey that is described. Is a "yes" answer *more likely* or *less likely* because of the bias?

9. The researcher asked the group of adults, "If you were lost in an unfamiliar town, would you be sensible and ask for directions?"

10. Mrs. Wang is having a birthday party for her daughter, Waverly. Mrs. Wang made a lemon cake for the party. She asked each of the children in attendance, "Would you have preferred chocolate cake to my lemon cake?"

Extensions   *Note:* In Investigation 5, you will have the opportunity to display the data collected in extensions **a–c** below.

a. One might guess that adults prefer different seasons than children. Interview exactly 10 children (under the age of 15) and 10 adults (over the age of 20). Ask the question "Which of the following seasons of the year do you like most: fall, winter, spring, or summer?"

b. One might guess that adults drink different things for breakfast than children. Interview exactly 8 children (under 15) and 8 adults (over 20). Ask the question "Which of the following beverages do you most often drink at breakfast: coffee, juice, milk, or something else?" If the choice is "something else," record the person's preferred breakfast drink.

c. Choose a nearby fast-food restaurant. For ten consecutive days at 7 p.m., count the customers in the restaurant. If a different time is better for you, choose that time (but use the same time each day).

d. Record the number of hours you sleep each night for the next 12 nights. Round the times to the nearest hour.

e. With a friend, construct a six-question, true-false quiz on a topic that interests you both (for example, music, animals, or geography). Have your families take the quiz; then record the number of questions that each family member answers correctly.

LESSON

41 Finding a Percent of a Number

WARM-UP

Facts Practice: 30 Fractions to Reduce (Test G)

Mental Math: Count up and down by $\frac{1}{8}$'s between $\frac{1}{8}$ and 2.

a. 4×250 **b.** $625 + 50$ **c.** $47 + 8$

d. $\$3.50 + \$1.75$ **e.** $\frac{1}{2}$ of 700   **f.** $\frac{600}{10}$

g. $5 \times 3, + 1, \div 2, + 1, \div 3, \times 8, \div 2$

Problem Solving:

Tennis is played on a rectangular surface that usually has two different courts drawn on it. Two players compete on a singles court, and four players (two per team) compete on a doubles court. The singles court is 78 feet long and 27 feet wide. The doubles court has the same length as the singles court, but it is 9 feet wider.

a. What is the perimeter of a singles tennis court?

b. How wide is a doubles tennis court?

c. What is the perimeter of a doubles court?

NEW CONCEPT

To describe part of a group, we often use a fraction or a percent. Here are a couple of examples:

Three fourths of the members voted for Imelda.

Tim answered 80% of the questions correctly.

We also use percents to describe financial situations.

Music CDs were on sale for 30% off the regular price.

The sales-tax rate is 7%.

The bank pays 5% interest on savings accounts.

When we are asked to find a certain percent of a number, we usually change the percent to either a fraction or a decimal before performing the calculation. Recall that a percent is really a fraction whose denominator is 100.

25% means $\frac{25}{100}$, which reduces to $\frac{1}{4}$.

5% means $\frac{5}{100}$, which reduces to $\frac{1}{20}$.

A percent is also easily changed to a decimal number. Study the following changes from percent to fraction to decimal:

$$25\% \longrightarrow \frac{25}{100} \longrightarrow 0.25 \qquad 5\% \longrightarrow \frac{5}{100} \longrightarrow 0.05$$

We see that the same nonzero digits are in both the decimal and percent forms of a number. In the decimal form, however, the decimal point is shifted two places to the left.

Example 1 Write 15% in decimal form.

Solution Fifteen percent means $\frac{15}{100}$, which can be written **0.15.**

Example 2 Write 75% as a reduced fraction.

Solution We write 75% as $\frac{75}{100}$ and reduce.

$$\frac{75}{100} = \frac{3}{4}$$

Example 3 What number is 75% of 20?

Solution We can translate this problem into an equation, changing the percent into either a fraction or a decimal. We use a letter for "what number," an equal sign for "is," and a multiplication sign for "of."

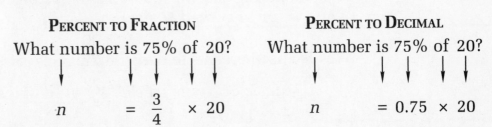

PERCENT TO FRACTION
What number is 75% of 20?

$$n = \frac{3}{4} \times 20$$

PERCENT TO DECIMAL
What number is 75% of 20?

$$n = 0.75 \times 20$$

We show both the fraction form and the decimal form. Often, one form is easier to calculate than the other form.

$$\frac{3}{4} \times 20 = 15 \qquad 0.75 \times 20 = 15.00$$

We find that 75% of 20 is **15.**

Example 4 Jamaal correctly answered 80% of the 25 questions. How many questions did he answer correctly?

Solution We want to find 80% of 25. We can change 80% to a fraction $\left(\frac{80}{100} = \frac{4}{5}\right)$ or to a decimal number (80% = 0.80).

<table>
<tr><td>PERCENT TO A FRACTION</td><td>PERCENT TO A DECIMAL</td></tr>
<tr><td>80% of 25</td><td>80% of 25</td></tr>
<tr><td>$\frac{4}{5} \times 25$</td><td>0.80 × 25</td></tr>
</table>

Then we calculate.

$$\frac{4}{5} \times 25 = 20 \qquad 0.80 \times 25 = 20.00$$

We find that Jamaal correctly answered **20 questions.**

Example 5 The sales-tax rate was 6%. Find the tax on a $12.00 purchase. Then find the total price including tax.

Solution We can change 6% to a fraction $\left(\frac{6}{100} = \frac{3}{50}\right)$ or to a decimal number (6% = 0.06). It seems easier for us to multiply $12.00 by 0.06 than by $\frac{3}{50}$, so we will use the decimal form.

$$6\% \text{ of } \$12.00$$

$$0.06 \times \$12.00 = \$0.72$$

So the tax on the $12.00 purchase was **$0.72.** To find the total price including tax, we add $0.72 to $12.00.

$$\begin{array}{r} \$12.00 \\ +\quad 0.72 \\ \hline \mathbf{\$12.72} \end{array}$$

LESSON PRACTICE

Practice set* Write each percent in problems **a–f** as a reduced fraction:

a. 50%   **b.** 10%   **c.** 25%

d. 75%   **e.** 20%   **f.** 1%

Write each percent in problems **g–l** as a decimal number:

g. 65%   **h.** 7%   **i.** 30%

j. 8%   **k.** 60%   **l.** 1%

m. Mentally find 10% of 350. Describe how to perform the mental calculation.

n. Mentally find 25% of 48. Describe how to perform the mental calculation.

o. How much money is 8% of $15.00?

p. The sales-tax rate is $9\frac{1}{2}$%. Estimate the tax on a $9.98 purchase. Explain how you arrived at your answer.

q. Erika correctly answered 80% of the 30 questions. How many questions did she answer correctly?

MIXED PRACTICE

Problem set

1. Michelle correctly answered 80% of the 20 questions on the test. How many questions did she answer correctly?
(41)

2. Dana ordered items from the menu totaling $8.50. If the sales-tax rate is 8%, how much should be added to the bill for sales tax?
(41)

3. The ten-acre farm was on a square piece of land 220 yards on each side. A fence surrounded the land. How many yards of fencing surrounded the farm?
(8)

4. Describe how to find 20% of 30.
(41)

5. The dinner cost $9.18. Jeb paid with a $20 bill. How much money should he get back? (Write an equation and solve the problem.)
(11)

6. Two hundred eighty-eight chairs were arranged in 16 equal rows. How many chairs were in each row? (Write an equation and solve the problem.)
(15)

7. Dagmar's bowling scores for three games were 126, 102,
(18) and 141. What was her average score for the three games?

8. What is the area of this rectangle?
(31, 39)

2.5 m

2 m

9. Arrange these numbers in order from least to greatest:
(14, 17)

$$\frac{3}{2}, 0, -1, \frac{2}{3}$$

10. List the first eight prime numbers.
(19)

11. By which of these numbers is 600 not divisible?
(21)

A. 2   B. 3   C. 5   D. 9

12. The fans were depressed, for their team had won only 15
(23) of the first 60 games. What was the team's win-loss ratio
after 60 games?

13. To loosen her shoulder, Mary swung her arm around in a
(27) big circle. If her arm was 28 inches long, what was the
diameter of the circle?

14. The map shows three streets in
(28) town.

IVY

VINE   MAIN

(a) Name a street parallel to Vine.

(b) Name a street perpendicular
to Vine.

15. Rob remembers that an acute angle is "a cute little
(28, Inv. 3) angle." Which of the following could be the measure of
an acute angle?

A. 0°   B. 45°   C. 90°   D. 135°

16. $(2.5)^2$ **17.** $\sqrt{81}$
(38, 39)   (38)

18. Write 40% as a reduced fraction.
(41)

19. Write 9% as a decimal number. Then find 9% of $10.
(41)

20. What is the reciprocal of $\frac{2}{3}$?
(30)

Find each missing number. Remember to check your work.

21. $7m = 3500$
(4)

22. $\$6.25 + w = \$10.00$
(3)

23. $\frac{2}{3}n = 1$
(30)

24. $x - 37 = 76$
(3)

25. $6.25 + (4 - 2.5)$
(37)

26. $3\frac{3}{4} + 2\frac{3}{4}$
(26)

27. $\frac{4}{4} - \frac{3}{3}$
(Inv. 2)

28. $\frac{5}{6} \times \frac{3}{5}$
(29)

29. What is $\frac{3}{4}$ of 48?
(29)

30. Fran estimated 9% of $21.90 by first rounding 9% to 10%
(41) and rounding $21.90 to $20. She then mentally calculated 10% of $20 and got the answer $2. Use Fran's method to estimate 9% of $32.17. Describe the steps.

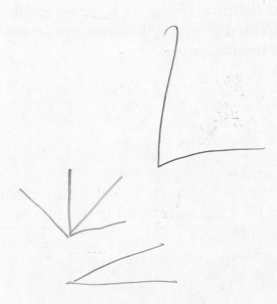

LESSON

42

Renaming Fractions by Multiplying by 1

WARM-UP

Facts Practice: 30 Fractions to Reduce (Test G)

Mental Math: Count by 12's from 12 to 96.

a. 4×125 b. $825 + 50$ c. $67 + 8$

d. $\$6.75 + \$2.50$ e. $\frac{1}{2}$ of 1000   f. $\frac{580}{10}$

g. $3 \times 4, - 2, \times 5, - 2, \div 6, + 1, \div 3$

Problem Solving:

The average of two numbers is 25. If one of the numbers is 19, what is the other number?

NEW CONCEPT

With our fraction manipulatives we have seen that the same fraction can be named many different ways. Here we show six ways to name the fraction $\frac{1}{2}$:

$\frac{1}{2}$ $\frac{2}{4}$ $\frac{3}{6}$ $\frac{4}{8}$ $\frac{5}{10}$ $\frac{6}{12}$

In this lesson we will practice renaming fractions by multiplying them by a fraction equal to 1. Here we show six ways to name 1 as a fraction:

$\frac{1}{1}$ $\frac{2}{2}$ $\frac{3}{3}$ $\frac{4}{4}$ $\frac{5}{5}$ $\frac{6}{6}$

We know that when we multiply a number by 1, the product equals the number multiplied. So if we multiply $\frac{1}{2}$ by 1, the answer is $\frac{1}{2}$.

$$\frac{1}{2} \times 1 = \frac{1}{2}$$

However, if we multiply $\frac{1}{2}$ by fractions equal to 1, we find different names for $\frac{1}{2}$. Here we show $\frac{1}{2}$ multiplied by $\frac{2}{2}$, $\frac{3}{3}$, and $\frac{4}{4}$:

$$\frac{1}{2} \times \frac{2}{2}\mathbf{1} = \frac{2}{4} \qquad \frac{1}{2} \times \frac{3}{3}\mathbf{1} = \frac{3}{6} \qquad \frac{1}{2} \times \frac{4}{4}\mathbf{1} = \frac{4}{8}$$

The fractions $\frac{2}{4}$, $\frac{3}{6}$, and $\frac{4}{8}$ are all equivalent to $\frac{1}{2}$.

Example 1 Write a fraction equal to $\frac{1}{2}$ that has a denominator of 20.

$$\frac{1}{2} = \frac{?}{20}$$

Solution To rename a fraction, we multiply the fraction by a fraction equal to 1. The denominator of $\frac{1}{2}$ is 2. We want to make an equivalent fraction with a denominator of 20.

$$\frac{1}{2} = \frac{?}{20}$$

Since we need to multiply the denominator by 10, we multiply $\frac{1}{2}$ by $\frac{10}{10}$.

$$\frac{1}{2} \times \frac{10}{10} = \frac{10}{20}$$

Example 2 Write $\frac{1}{2}$ and $\frac{1}{3}$ as fractions with denominators of 6. Then add the renamed fractions.

Solution We multiply each fraction by a fraction equal to 1 to form fractions that have denominators of 6.

$$\frac{1}{2} = \frac{?}{6} \qquad \frac{1}{3} = \frac{?}{6}$$

We multiply $\frac{1}{2}$ by $\frac{3}{3}$, and we multiply $\frac{1}{3}$ by $\frac{2}{2}$.

$$\frac{1}{2} \times \frac{3}{3} = \frac{3}{6} \qquad \frac{1}{3} \times \frac{2}{2} = \frac{2}{6}$$

The renamed fractions are $\frac{3}{6}$ and $\frac{2}{6}$. We are told to add these fractions.

$$\frac{3}{6} + \frac{2}{6} = \frac{5}{6}$$

LESSON PRACTICE

Practice set   In problems **a–d**, multiply by a fraction equal to 1 to complete each equivalent fraction.

a. $\dfrac{1}{3} = \dfrac{?}{12}$ **b.** $\dfrac{2}{3} = \dfrac{?}{6}$

c. $\dfrac{3}{4} = \dfrac{?}{8}$ **d.** $\dfrac{3}{4} = \dfrac{?}{12}$

e. Write $\frac{2}{3}$ and $\frac{1}{4}$ as fractions with denominators of 12. Then add the renamed fractions.

f. Write $\frac{1}{6}$ as a fraction with 12 as the denominator. Subtract the renamed fraction from $\frac{5}{12}$. Reduce the subtraction answer.

MIXED PRACTICE

Problem set   **1.** Write $\frac{1}{2}$ and $\frac{2}{3}$ as fractions with denominators of 6. Then add
   $^{(42)}$ the renamed fractions. Write the answer as a mixed number.

2. According to some estimates, our own galaxy, the Milky
   $^{(12)}$ Way, contains about two hundred billion stars. Use digits to write that number of stars.

3. The rectangular park is 120 yards long and 40 yards wide.
   $^{(31)}$ How many square yards is its area?

4. What number is 40% of 30?
   $^{(41)}$

In problems 5 and 6, multiply $\frac{1}{2}$ by a fraction equal to 1 to complete each equivalent fraction.

5. $\dfrac{1}{2} = \dfrac{?}{8}$ **6.** $\dfrac{1}{2} = \dfrac{?}{10}$
   $^{(42)}$ $^{(42)}$

7. $4.32 + 0.6 + \sqrt{81}$ **8.** $6.3 - 0.54$
   $^{(37, 38)}$ $^{(37)}$

9. $(0.15)^2$
   $^{(38, 40)}$

10. What is the reciprocal of $\frac{6}{7}$?
   $^{(30)}$

11. Which digit in 12,345 has the same place value as the 6
   $^{(34)}$ in 67.89?

12. What is the least common multiple of 3, 4, and 6?
   $^{(30)}$

13. $5\frac{3}{5} + 4\frac{4}{5}$ **14.** $\sqrt{36} - 4\frac{2}{3}$ **15.** $\frac{8}{3} \times \frac{1}{2}$
(26)   (36, 38)   (29)

16. $\frac{6}{5} \times 3$ **17.** $1 - \frac{1}{4}$ **18.** $\frac{10}{10} - \frac{5}{5}$
(29)   (36)   (Inv. 2)

19. Form three different fractions that are equal to $\frac{1}{3}$ by
(42) multiplying $\frac{1}{3}$ by three different fraction names for 1.

20. The prime numbers that multiply to form 35 are 5 and 7.
(19) Which prime numbers can be multiplied to form 34?

21. In three games Alma's scores were 12,143; 9870; and 14,261.
(18) Describe how to estimate her average score per game.

22. Estimate the quotient of $\frac{8176}{41}$. Describe how you
(16) performed the estimate.

23. How many doughnuts are in $\frac{2}{3}$ of a dozen? Draw a
(22) diagram to illustrate the problem.

24. Write $\frac{3}{4}$ with a denominator of 8. Subtract the renamed
(42) fraction from $\frac{7}{8}$.

25. What is the perimeter of this
(8) rectangle?

0.4 m

0.2 m

26. What is the area of this rectangle?
(31)

27. The regular price r minus the discount d equals the sale
(1) price s.

$$r - d = s$$

Arrange these letters to form another subtraction fact and
two addition facts.

28. Below we show the same division problem written three
(2) different ways. Identify which number is the divisor,
which is the dividend, and which is the quotient.

$$\frac{20}{4} = 5 \qquad 4\overline{)20} \qquad 20 \div 4 = 5$$

29. What time is $2\frac{1}{2}$ hours after 11:45 a.m.?
(32)

30. (a) How many $\frac{5}{6}$'s are in 1?
(30)

    (b) Use the answer to part (a) to find the number of $\frac{5}{6}$'s in 3.

LESSON
43

Equivalent Division Problems • Missing-Number Problems with Fractions and Decimals

WARM-UP

Facts Practice: 100 Subtraction Facts (Test C)

Mental Math: Count by 6's to 72. Count by 8's to 96.

a. 4 × 225	**b.** 720 − 200	**c.** 37 + 28
d. $200 − $175	**e.** $\frac{1}{2}$ of 1200	**f.** $\frac{\$70.00}{10}$
g. 8 × 4, − 2, × 2, + 3, ÷ 7, × 2, ÷ 3		

Problem Solving:

You can roll six different numbers with one toss of a number cube (1–6). You can roll eleven different numbers with one toss of two number cubes (2–12). How many different numbers can you roll with one toss of three number cubes?

NEW CONCEPTS

Equivalent division problems

The following two division problems have the same quotient. We call them **equivalent division problems.** Which problem seems easier to perform mentally?

$$\text{(a) } 700 \div 14$$

$$\text{(b) } 350 \div 7$$

We can change problem (a) to problem (b) by dividing both 700 and 14 by 2.

$$700 \div 14$$
↓
Divide both 700 and 14 by 2.
↓
$$350 \div 7$$

By dividing both the dividend and divisor by the same number (in this case, 2), we formed an equivalent division problem that was easier to divide mentally. This process simply reduces the terms of the division as we would reduce a fraction.

$$\frac{700}{14} = \frac{350}{7} \quad \begin{array}{l}(700 \div 2 = 350) \\ (14 \div 2 = 7)\end{array}$$

We may also form equivalent division problems by multiplying the dividend and divisor by the same number. Consider the following equivalent problems:

$$\text{(c)} \quad 7\frac{1}{2} \div \frac{1}{2}$$

$$\text{(d)} \quad 15 \div 1$$

We changed problem (c) to problem (d) by doubling both $7\frac{1}{2}$ and $\frac{1}{2}$; that is, by multiplying both numbers by 2.

$$7\frac{1}{2} \div \frac{1}{2}$$

$$\downarrow$$

Multiply both $7\frac{1}{2}$ and $\frac{1}{2}$ by 2.

$$\downarrow$$

$$15 \div 1$$

This process forms an equivalent division problem in the same way we would form an equivalent fraction.

$$\frac{7\frac{1}{2}}{\frac{1}{2}} \cdot \frac{2}{2} = \frac{15}{1}$$

Example 1 Form an equivalent division problem for the division below. Then calculate the quotient.

$$1200 \div 16$$

Solution Instead of dividing 1200 by the two-digit number 16, we can divide both the dividend and the divisor by 2 to form the equivalent division of $600 \div 8$. We then calculate.

$$
16\overline{)1200} \quad \xrightarrow[\text{numbers by 2.}]{\text{Divide both}} \quad
\begin{array}{r}
75 \\
8\overline{)600} \\
\underline{56} \\
40 \\
\underline{40} \\
0
\end{array}
$$

Both quotients are 75, but dividing by 8 is easier than dividing by 16.

Notice that several equivalent division problems can be formed from the original problem $1200 \div 16$:

$$1200 \div 16 \longrightarrow 600 \div 8 \longrightarrow 300 \div 4 \longrightarrow 150 \div 2$$

All of these problems have the same quotient.

Example 2   Form an equivalent division problem for the division problem below. Then calculate the quotient.

$$7\frac{1}{2} \div 2\frac{1}{2}$$

Solution   Instead of performing the division with these mixed numbers, we will double both numbers to form a whole-number division problem.

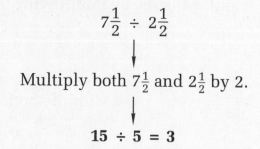

$$7\frac{1}{2} \div 2\frac{1}{2}$$

Multiply both $7\frac{1}{2}$ and $2\frac{1}{2}$ by 2.

$$15 \div 5 = 3$$

Missing-number problems with fractions and decimals   Since Lessons 3 and 4 we have practiced finding missing numbers in whole-number arithmetic problems. Beginning with this lesson we will find missing numbers in fraction and decimal problems. If you are unsure how to find the solution to a problem, try making up a similar, easier problem to help you determine how to find the answer.

Example 3   Solve: $d - 5 = 3.2$

Solution   This problem is similar to the subtraction problem $d - 5 = 3$. We remember that we find the first number of a subtraction problem by adding the other two numbers. So we have the following:

$$\begin{array}{r} 5 \\ + \ 3.2 \\ \hline 8.2 \end{array}$$

We check our work by replacing the letter with the solution and testing the result.

$$d - 5 = 3.2$$
$$8.2 - 5 = 3.2$$
$$3.2 = 3.2$$

Example 4   Solve: $f + \frac{1}{5} = \frac{4}{5}$

Solution This problem is similar to $f + 1 = 4$. We can find a missing addend by subtracting the known addend from the sum.

$$\frac{4}{5} - \frac{1}{5} = \frac{3}{5}$$

$$f = \frac{3}{5}$$

We check the solution by substituting it into the original equation.

$$f + \frac{1}{5} = \frac{4}{5}$$

$$\frac{3}{5} + \frac{1}{5} = \frac{4}{5}$$

$$\frac{4}{5} = \frac{4}{5}$$

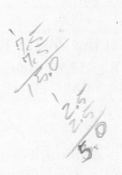

Example 5 Solve: $\frac{3}{5}n = 1$

Solution In this problem two numbers are multiplied, and the product is 1. This can only happen when the two factors are reciprocals. So we want to find the reciprocal of the known factor, $\frac{3}{5}$. Switching the terms of $\frac{3}{5}$ gives us the fraction $\frac{5}{3}$. We check our answer by substituting $\frac{5}{3}$ into the original equation.

$$\frac{3}{5} \times \frac{5}{3} = \frac{15}{15} = 1 \quad \text{check}$$

$$n = \frac{5}{3}$$

LESSON PRACTICE

Practice set **a.** Form an equivalent division problem for $5 \div \frac{1}{3}$ by multiplying both the dividend and divisor by 3. Then find the quotient.

b. Form an equivalent division problem for $266 \div 14$ that has a one-digit divisor. Then find the quotient.

Solve:

c. $5 - d = 3.2$

d. $f - \dfrac{1}{5} = \dfrac{4}{5}$

$m + \dfrac{6}{5} = 4$

e. $m + 1\dfrac{1}{5} = 4$

f. $\dfrac{3}{8}w = 1$

MIXED PRACTICE

Problem set

1. The bike cost $120. The sales-tax rate was 8%. What was
(41) the total cost of the bike including sales tax?

2. If one hundred fifty knights could sit at the Round Table
(11) and only one hundred twenty-eight knights were seated, how many empty places were at the table? (Write an equation and solve the problem.)

3. During the 1996 Summer Olympics in Atlanta, Georgia,
(32, 37) the American athlete Michael Johnson set an Olympic and world record in the men's 200-meter run. He finished the race in 19.32 seconds, breaking the previous Olympic record of 19.75 seconds. By how much did Michael Johnson break the previous Olympic record?

In problems 4 and 5, multiply by a fraction equal to 1 to complete each equivalent fraction.

4. $\dfrac{2}{3} = \dfrac{?}{6}$
(42)

5. $\dfrac{1}{2} = \dfrac{?}{6}$
(42)

Find each missing number. Remember to check your work.

6. $\dfrac{2}{3}n = 1$
(43)

7. $6 - w = 1\dfrac{4}{5}$
(43)

8. $m - 4\dfrac{1}{4} = 6\dfrac{3}{4}$
(43)

9. $c - 2.45 = 3$
(43)

10. $12 - d = 1.43$
(43)

11. $\dfrac{5}{8} \times \dfrac{1}{5}$
(29)

12. $\dfrac{3}{4} \times 5$
(29)

13. $3\dfrac{7}{8} - 1\dfrac{3}{8}$
(26)

14. Which of these numbers is not a prime number?
(19)

    A. 23   B. 33   C. 43

15. Compare: $\frac{2}{2}$ ◯ $\frac{2}{2} \times \frac{2}{2}$
(29)

16. In football a loss of yardage is often expressed as a
(14) negative number. If a quarterback is sacked for a 5-yard loss, the yardage change on the play can be shown as –5. How would a 12-yard loss be shown using a negative number?

17. Write the decimal number for nine and twelve hundredths.
(35)

18. Round 67,492,384 to the nearest million.
(16)

19. 0.37×10^2 **20.** $0.6 \times 0.4 \times 0.2$
(38, 39)   (40)

21. The perimeter of a square room is 80 feet. The area of the
(38) room is how many square feet?

22. Divide 100 by 16 and write the answer as a mixed number.
(25) Reduce the fraction part of the mixed number.

23. (a) Instead of dividing 100 by 16, Sandy divided the
(43)   dividend and divisor by 4. What new division problem did Sandy make? What is the quotient?

     (b) Form an equivalent division problem for $4\frac{1}{2} \div \frac{1}{2}$ by doubling both the dividend and divisor. Then find the quotient.

24. What is the least common multiple (LCM) of 4, 6, and 8?
(30)

25. What are the next three numbers in this sequence?
(17)

$$\frac{1}{16}, \frac{1}{8}, \frac{3}{16}, \frac{1}{4}, \frac{5}{16}, \frac{3}{8}, \frac{7}{16}, \underline{\quad\quad}, \underline{\quad\quad}, \underline{\quad\quad}, \ldots$$

26. Find the length of the segment below to the nearest eighth
(17) of an inch.

27. What mixed number is indicated on the number line below?
(17)

28. Write $\frac{1}{2}$ and $\frac{1}{5}$ as fractions with denominators of 10. Then
(42) add the renamed fractions.

29. Forty percent of the 20 seats on the bus were occupied.
(22, 33) Write 40% as a reduced fraction. Then find the number of seats that were occupied. Draw a diagram to illustrate the problem.

30. Describe each angle in the figure as
(Inv. 3) acute, right, or obtuse.

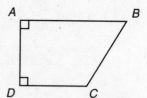

(a) angle *A*

(b) angle *B*

(c) angle *C*

(d) angle *D*

LESSON
44

Simplifying Decimal Numbers •
Comparing Decimal Numbers

WARM-UP

Facts Practice: 30 Fractions to Reduce (Test G)

Mental Math: Count up and down by $\frac{1}{8}$'s between $\frac{1}{8}$ and 3.

a. 4×325 b. $426 + 35$ c. $28 + 57$

d. $\$8.50 + \$2.75$ e. $\frac{1}{2}$ of 1400   f. $\frac{\$15.00}{100}$

g. $6 \times 8, - 3, \div 5, + 1, \times 6, + 3, \div 9$

Problem Solving:

Jeana folded a square piece of paper in half from top to bottom. Then she folded the folded paper in half from left to right so that the four corners were together at the lower right. Then she cut off the lower right corners as shown. Draw a picture of what the paper will look like when it is unfolded. What shape is this?

cut

4 corners

NEW CONCEPTS

Simplifying decimal numbers

Perform these two subtractions with a calculator. Which calculator answer differs from the printed answer?

$$
\begin{array}{r}
425 \\
- 125 \\
\hline
300
\end{array}
\qquad
\begin{array}{r}
4.25 \\
- 1.25 \\
\hline
3.00
\end{array}
$$

Calculators automatically simplify a decimal number with zeros at the end by removing the extra zeros. However, some calculators leave a decimal point at the end of a whole number. So "3.00" might simplify to "3." on a calculator. Using pencil and paper, we usually remove the decimal point from a whole number, so we would write "3" only.

Example 1 Multiply 0.25 by 0.04 and simplify the product.

Solution We multiply.

$$
\begin{array}{r}
0.25 \\
\times\ 0.04 \\
\hline
0.0100
\end{array}
$$

If we perform this multiplication on a calculator, the answer 0.01 is displayed. The calculator simplifies the answer by removing zeros at the end of the decimal number.

0.0100 simplifies to **0.01**

In this book decimal answers are printed in simplified form unless otherwise stated.

Comparing decimal numbers Zeros at the end of a decimal number do not affect the value of the decimal number. Each of these decimal numbers has the same value because the 3 is in the tenths place:

0.3   0.30   0.300

Although 0.3 is the simplified form, sometimes it is useful to attach extra zeros to a decimal number. For instance, comparing decimal numbers can be easier if the numbers being compared have the same number of decimal places.

Example 2 Compare: 0.3 ◯ 0.303

Solution When comparing decimal numbers, it is important to pay close attention to place values. Writing both numbers with the same number of decimal places can make comparing easier. We will attach two zeros to 0.3 so that it has the same number of decimal places as 0.303.

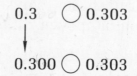

We see that 300 thousandths is less than 303 thousandths. We write our answer like this:

0.3 < 0.303

Example 3 Arrange these numbers in order from least to greatest:

0.3   0.042   0.24   0.235

Solution We write each number with three decimal places.

0.300   0.042   0.240   0.235

Then we arrange the numbers in order, omitting ending zeros.

0.042   0.235   0.24   0.3

LESSON PRACTICE

Practice set Write these numbers in simplified form:

      a. 0.0500   **b.** 50.00

      c. 1.250   **d.** 4.000

Compare:

      e. 0.2 ◯ 0.15   **f.** 12.5 ◯ 1.25

      g. 0.012 ◯ 0.12   **h.** 0.31 ◯ 0.039

      i. 0.4 ◯ 0.40

      j. Write these numbers in order from least to greatest:

                0.120   0.125   0.015   0.2

MIXED PRACTICE

Problem set **1.** What is the sum of the third multiple of four and the
         (12, 25) third multiple of five?

      2. One mile is 5280 feet. How many feet is five miles?
      (15)

The summit of Mt. Everest is 29,035 feet above sea level. The summit of Mt. Whitney is 14,495 feet above sea level. Use this information to answer problems 3 and 4.

      3. Mt. Everest is how many feet taller than Mt. Whitney?
      (13)

      4. The summit of Mt. Everest is how many feet higher than
      (13) 5 miles above sea level? (Refer to problem 2.)

Find each missing number. Remember to check your work.

      5. $5\frac{1}{3} - w = 4$ **6.** $m - 6\frac{4}{5} = 1\frac{3}{5}$
      (43)   (43)

      7. $6.74 + 0.285 + f = 11.025$
      (43)

      8. $0.4 - d = 0.33$
      (43)

9. Wearing shoes, Fiona stands $67\frac{3}{4}$ inches tall. If the heels
(26) of her shoes are $1\frac{1}{4}$ inches thick, then how tall does Fiona
stand without shoes?

10. A $36 dress is on sale for 10% off the regular price.
(41) Mentally calculate 10% of $36. Describe the method you
used to arrive at your answer.

11. Write thirty-two thousandths as a decimal number.
(35)

12. What number is $\frac{1}{6}$ of 24,042?
(29)

13. Compare:
(44)
(a) 0.25 $\bigcirc$ 0.125   (b) 25% $\bigcirc$ 12.5%

14. Write the standard numeral for $(6 \times 100) + (4 \times 1)$.
(32)

15. Form an equivalent division problem for $8\frac{1}{2} \div \frac{1}{2}$ by
(43) doubling both the dividend and divisor. Then find the
quotient.

16. (a) How many $\frac{5}{8}$'s are in 1?
(30)

(b) Use the answer to part (a) to find the number of $\frac{5}{8}$'s in 3.

17. What is the least common multiple of 2, 3, 4, and 6?
(30)

18. $(1.3)^2$ **19.** $\frac{3}{4} = \frac{?}{12}$ **20.** $\frac{2}{3} = \frac{?}{12}$
(39)   (42)   (42)

21. Find the average of 26, 37, 42, and 43.
(18)

22. Round 364,857 to the nearest thousand.
(16)

23. Twelve of the 30 players on the team were girls. What
(23, 29) was the ratio of boys to girls on the team?

24. (a) List the factors of 100.
(19)

(b) Which of the factors of 100 are prime numbers?

25. Write 9% as a fraction. Then write the fraction as a decimal
(33, 35) number.

26. Write $\frac{3}{4}$ and $\frac{2}{3}$ as fractions with denominators of 12. Then
(42) add the renamed fractions.

27. Which percent best describes the
(Inv. 2) shaded portion of this rectangle? Why?

A. 80% B. 40%

C. 60% D. 20%

28. Shelby started working at 10:30 a.m. and finished working
(32) at 2:15 p.m. How long did Shelby work?

29. Which of these numbers is closest to 1?
(44)

A. 0.1   B. 0.8   C. 1.1   D. 1.2

30. What mixed number corresponds to point X on the number
(17) line below?

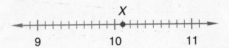

LESSON
45

Dividing a Decimal Number by a Whole Number

WARM-UP

Facts Practice: 90 Division Facts (Test F)

Mental Math: Count up and down by 25's between 25 and 300.

a. 4 × 425   b. 375 + 500   c. 77 + 18

d. $12.00 − $1.25   e. $\frac{1}{2}$ of 1500   f. $\frac{\$40.00}{10}$

g. 4 × 8, − 2, ÷ 3, + 2, ÷ 3, × 5, + 1, ÷ 3

Problem Solving:

Sheldon used six blocks to build this three-step shape. How many blocks would he need to build a six-step shape?

NEW CONCEPT

Dividing a decimal number by a whole number is similar to dividing dollars and cents by a whole number.

$$\frac{\$0.45}{5)\$2.25} \qquad \frac{0.45}{5)2.25}$$

Notice that the decimal point in the quotient is directly above the decimal point in the dividend.

Example 1   Divide: $3)\overline{4.2}$

Solution   The decimal point in the quotient is directly above the decimal point in the dividend.

$$\begin{array}{r} 1.4 \\ 3\overline{)4.2} \\ \underline{3} \\ 1\,2 \\ \underline{1\,2} \\ 0 \end{array}$$

Example 2   Divide: $3)\overline{0.24}$

Solution   The decimal point in the quotient is directly above the decimal point in the dividend. We fill the empty place with zero.

$$\begin{array}{r} 0.08 \\ 3\overline{)0.24} \\ \underline{24} \\ 0 \end{array}$$

Decimal division answers are not written with remainders. Instead, we attach zeros to the end of the dividend and continue dividing.

Example 3 Divide: $5\overline{)0.6}$

Solution The decimal point in the quotient is directly above the decimal point in the dividend. To complete the division, we attach a zero to 0.6, making the equivalent decimal number 0.60. Then we continue dividing.

$$\begin{array}{r} 0.12 \\ 5\overline{)0.60} \\ \underline{5} \\ 10 \\ \underline{10} \\ 0 \end{array}$$

LESSON PRACTICE

Practice set* **a.** The distance from Margaret's house to the store and back is 3.6 miles. How far does Margaret live from the store?

b. The perimeter of a square is 6.4 meters. How long is each side of the square?

Divide:

c. $\dfrac{4.5}{3}$ **d.** $0.6 \div 4$ **e.** $2\overline{)0.14}$

f. $0.4 \div 5$ **g.** $4\overline{)0.3}$ **h.** $\dfrac{0.012}{6}$

i. $10\overline{)1.4}$ **j.** $\dfrac{0.7}{5}$ **k.** $0.1 \div 4$

MIXED PRACTICE

Problem set **1.** By what fraction must $\frac{5}{3}$ be multiplied to get a product of 1?
(30)

2. How many $20 bills equal one thousand dollars?
(15)

3. Cindy made $\frac{2}{3}$ of her 24 shots at the basket. Each basket
(29) was worth 2 points. How many points did she score?

4. $3\overline{)4.5}$ **5.** $8\overline{)0.24}$ **6.** $5\overline{)0.8}$
(45)   (45)   (45)

7. What is the least common multiple (LCM) of 2, 4, 6, and 8?
(30)

Find each missing number. Remember to check your work.

8. $\sqrt{36} - m = 2\frac{3}{10}$
(43)

9. $g - 2\frac{2}{5} = 5\frac{4}{5}$
(43)

10. $m - 1.56 = 1.44$
(43)

11. $3^2 - n = 5.39$
(38, 43)

12. $4\frac{3}{8} - 2\frac{1}{8}$
(26)

13. $\frac{8}{3} \cdot \frac{5}{2}$
(29)

14. Estimate the product of 694 and 412.
(16)

15. $0.7 \times 0.6 \times 0.5$
(39)

16. 0.46×0.17
(40)

17. Brenda's car traveled 177.6 miles on 8 gallons of gas. Her
(15, 45) car traveled an average of how many miles per gallon? Use a multiplication pattern. (Write an equation and solve the problem.)

18. What number is $\frac{3}{8}$ of 6?
(29)

19. A shirt regularly priced at $40 is on sale for 25% off.
(41) Mentally calculate 25% of $40. Describe the method you used to arrive at your answer.

20. Write a fraction equal to $\frac{5}{6}$ that has 12 as the denominator.
(42) Then subtract $\frac{7}{12}$ from the fraction. Reduce the answer.

21. The area of a square is 36 ft².
(38)

(a) How long is each side of the square?

(b) What is the perimeter of the square?

22. Write 27% as a fraction. Then write the fraction as a
(33, 35) decimal number.

23. Use a ruler to find the length of
(17) this rectangle to the nearest eighth of an inch.

24. Seventy-five percent of the 20 answers were correct.
$_{(22,\,33)}$ Write 75% as a reduced fraction. Then find the number of
answers that were correct. Draw a diagram to illustrate
this fractional-parts problem.

25. The product of $\frac{1}{2}$ and $\frac{2}{3}$ is $\frac{1}{3}$.
$_{(29)}$

$$\frac{1}{2} \times \frac{2}{3} = \frac{1}{3}$$

Arrange these fractions to form another multiplication
fact and two division facts.

26. Which percent best describes the
$_{(Inv.\,2)}$ shaded portion of this circle? Why?

A. 80%   B. 60%

C. 40%   D. 20%

27. Write nine hundredths
$_{(35)}$ (a) as a fraction.

(b) as a decimal number.

28. Form an equivalent division problem for $5 \div \frac{1}{3}$ by
$_{(43)}$ multiplying both the dividend and divisor by 3. Then
find the quotient.

29. The average number of birds in each of three trees was
$_{(18)}$ 24. Altogether, how many birds were in the three trees?

30. Coach O'Rourke has a measuring
$_{(27)}$ wheel that records the distance the
wheel is rolled along the ground.
The circumference of the wheel is
one yard. If the wheel is pushed half
a mile, how many times will the
wheel go around (1 mi = 5280 ft)?

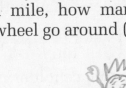

L E S S O N

46

Writing Decimal Numbers in Expanded Notation • Mentally Multiplying Decimal Numbers by 10 and by 100

WARM-UP

Facts Practice: 30 Fractions to Reduce (Test G)

Mental Math: Count by 12's from 12 to 108.

a. 4×525 b. $567 - 120$ c. $38 + 17$
d. $\$5.75 + \$2.50$ e. $\frac{1}{2}$ of 950   f. $\frac{2000}{100}$
g. $9 \times 7, + 1, \div 8, \times 3, + 1, \times 2, - 1, \div 7$

Problem Solving:

Copy this problem and fill in the missing digits:

$$\begin{array}{r} - - - \\ + \underline{ - } \\ \underline{- - - \, 8} \end{array}$$

NEW CONCEPTS

Writing decimal numbers in expanded notation

We may use expanded notation to write decimal numbers just as we have used expanded notation to write whole numbers. The values of some decimal places are shown in this table:

Decimal Place Values

1	$\frac{1}{10}$	$\frac{1}{100}$	$\frac{1}{1000}$
ones	tenths	hundredths	thousandths

We write 4.025 in expanded notation this way:

$$(4 \times 1) + \left(2 \times \frac{1}{100}\right) + \left(5 \times \frac{1}{1000}\right)$$

The zero that serves as a placeholder is usually not included in expanded notation.

Example 1 Write 5.06 in expanded notation.

Solution The 5 is in the ones place, and the 6 is in the hundredths place.

$$(5 \times 1) + \left(6 \times \frac{1}{100}\right)$$

Example 2 Write $\left(4 \times \frac{1}{10}\right) + \left(5 \times \frac{1}{1000}\right)$ as a decimal number.

Solution We write the decimal number with a 4 in the tenths place and a 5 in the thousandths place. No digits in the ones place or the hundredths place are indicated, so we write zeros in those places.

0.405

Mentally multiplying decimal numbers by 10 and by 100

When we multiply whole numbers by 10 or by 100, we can find the product mentally by attaching zeros to the whole number we are multiplying.

$$24 \times 10 = 240$$

$$24 \times 100 = 2400$$

It may seem that we are just attaching zeros, but we are actually shifting the digits to the left. When we multiply 24 by 10, the digits shift one place to the left. When we multiply 24 by 100, the digits shift two places to the left. In each product zeros hold the 2 and the 4 in their proper places.

1000's	100's	10's	1's	
		2	4	24
	2	4	0	24 × 10 (one-place shift)
2	4	0	0	24 × 100 (two-place shift)

When we multiply a decimal number by 10, the digits shift one place to the left. When we multiply a decimal number by 100, the digits shift two places to the left. Here we show the products when 0.24 is multiplied by 10 and by 100.

10's	1's	.	$\frac{1}{10}$'s	$\frac{1}{100}$'s	
	0	.	2	4	0.24
	2	.	4		0.24 × 10 (one-place shift)
2	4	.			0.24 × 100 (two-place shift)

Although it is the digits that are shifting one or two places to the left, we get the same effect by shifting the decimal point one or two places to the right.

$$0.24 \times 10 = 2.4 \qquad 0.24 \times 100 = 24. = 24$$

one-place shift   two-place shift

Example 3   Multiply: 3.75 × 10

Solution   Since we are multiplying by 10, the product will have the same digits as 3.75, but the digits will be shifted one place. The product will be ten times as large, so we mentally shift the decimal point one place to the right.

$$3.75 \times 10 = \textbf{37.5} \text{ (one-place shift)}$$

We do not need to attach any zeros, because the decimal point serves to hold the digits in their proper places.

Example 4   Multiply: 3.75 × 100

Solution   When multiplying by 100, we mentally shift the decimal point two places to the right.

$$3.75 \times 100 = 375. = \textbf{375} \text{ (two-place shift)}$$

We do not need to attach zeros. Since there are no decimal places, we may leave off the decimal point.

Example 5   Multiply: $\dfrac{1.2}{0.4} \times \dfrac{10}{10}$

Solution   Multiplying both 1.2 and 0.4 by 10 shifts each decimal point one place.

$$\frac{1.2}{0.4} \times \frac{10}{10} = \frac{12}{4}$$

The expression $\frac{12}{4}$ means "12 divided by 4."

$$\frac{12}{4} = \textbf{3}$$

LESSON PRACTICE

Practice set*   Write these numbers in expanded notation:

a. 2.05

b. 20.5

c. 0.205

Write these numbers in decimal form:

d. $(7 \times 10) + \left(8 \times \dfrac{1}{10}\right)$

e. $\left(6 \times \dfrac{1}{10}\right) + \left(4 \times \dfrac{1}{100}\right)$

Mentally calculate each product:

f. 0.35×10 **g.** 0.35×100

h. 2.5×10 **i.** 2.5×100

j. 0.125×10 **k.** 0.125×100

For the following statements, answer "true" or "false":

l. If 0.04 is multiplied by 10, the product is a whole number.

m. If 0.04 is multiplied by 100, the product is a whole number.

Multiply as shown. Then complete the division.

n. $\dfrac{1.5}{0.5} \times \dfrac{10}{10}$ **o.** $\dfrac{2.5}{0.05} \times \dfrac{100}{100}$

MIXED PRACTICE

Problem set

1. When a fraction with a numerator of 30 and a denominator of 8 is converted to a mixed number and reduced, what is the result?
(25)

2. Normal body temperature is 98.6° on the Fahrenheit scale. A person with a temperature of 100.2°F would have a temperature how many degrees above normal? (Write an equation and solve the problem.)
(13)

3. Four and twenty blackbirds is how many dozen?
(15)

4. Write $(5 \times 10) + \left(6 \times \dfrac{1}{10}\right) + \left(7 \times \dfrac{1}{1000}\right)$ in decimal form.
(46)

5. Twenty-one percent of the earth's atmosphere is oxygen. Write 21% as a fraction. Then write the fraction as a decimal number.
(33, 35)

6. Twenty-one percent is slightly more than 20%. Twenty
(33) percent is equivalent to what reduced fraction?

7. 5$\overline{)6.35}$
(45)

8. 4$\overline{)0.5}$
(45)

9. 8$\overline{)1.0}$
(45)

Find each missing number:

10. $x + 3\frac{5}{8} = 9$
(43)

11. $y - 16\frac{1}{4} = 4\frac{3}{4}$
(43)

12. $1 - q = 0.235$
(43)

13. $26.9 + 12 + w = 49.25$
(43)

14. Fifty percent of the area of this rectangle is shaded. What
(31, 41) is the area of the shaded region?

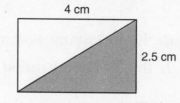

4 cm

2.5 cm

15. What is the ratio of the value of a dime to the value of a
(23) quarter?

16. 3.7×0.25
(39)

17. $\frac{3}{4} = \frac{?}{12}$
(42)

18. What is the least common multiple of 3, 4, and 8?
(30)

19. Compare:
(40, 44)

(a) $\frac{1}{10} \bigcirc 0.1$

(b) $0.1 \bigcirc (0.1)^2$

20. Which digit is in the thousandths place in 1,234.5678?
(34)

21. Estimate the quotient when 3967 is divided by 48.
(16)

22. The area of a square is 100 cm². What is its perimeter?
(38)

23. John carried the football twice in the game. One play
(14) gained 6 yards. The other play lost 8 yards. Use a negative
number to show John's total yardage for the game.

24. $\frac{1}{2} \cdot \frac{4}{5}$
(29)

25. $\left(\frac{3}{4}\right)\left(\frac{5}{3}\right)$
(29)

26. Laquesha bought a 24-inch-diameter wheel for her bicycle
(27) and measured it carefully. Arrange these measures in
order from least to greatest:

circumference, radius, diameter

27. The chef's salad cost $6.95. The sales-tax rate was 8%.
(41) What was the total cost including tax? Describe how to
use estimation to check whether your answer is
reasonable.

28. Use a ruler to find the width of
(17) this rectangle to the nearest eighth
of an inch.

29. (a) How many $\frac{3}{8}$'s are in 1?
(30)
  (b) Use the answer to part (a) to find the number of $\frac{3}{8}$'s in 3.

30. Rename $\frac{1}{2}$ and $\frac{1}{3}$ so that the denominators of the renamed
(42) fractions are 6. Then add the renamed fractions.

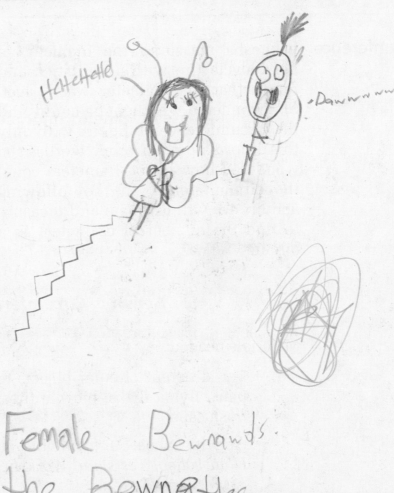

Female Bewrants.
the Bewnettes

LESSON
47   Circumference • Pi (π)

WARM-UP

Facts Practice: 100 Multiplication Facts (Test E)

Mental Math: Count by 9's from 9 to 108.

   a. 4 × 925   **b.** 3 × 87   **c.** 56 − 19

   d. $9.00 − $1.25   **e.** $\frac{1}{2}$ of $12.50   **f.** $\frac{\$25.00}{10}$

   g. 6 × 8, + 2, × 2, − 10, ÷ 9, + 5, ÷ 3, + 1, ÷ 6

Problem Solving:

The average of two numbers is 44. If one of the numbers is 34, what is the other number?

NEW CONCEPTS

Circumference   Laquesha measured the diameter of her bicycle tire with a yardstick and found that the diameter was 2 feet. She wondered whether she could find the circumference of the tire with only this information. In other words, she wondered how many diameters equal the circumference. In the following activity we will estimate and measure to find the number of diameters in a circumference.

Activity: *Circumference*

Materials needed:

- 2–4 different circular objects (e.g., dinner plates, pie pans, flying disks, bicycle tires, can tops or bottoms, trash cans)

- Activity Sheet 9 (available in *Saxon Math 7/6— Homeschool Tests and Worksheets*)

- string or masking tape

- scissors

- cloth tape measure

- calculator

This activity has two parts. In the first part you will cut a length of string as long as the diameter of each object you will measure. (A length of masking tape may be used in place of string.) Then you will wrap the string around the object and estimate the number of diameters needed to reach all the way around. To do this, first mark a starting point on the object. Wrap the string around the object, and mark the point where the string ends. Repeat this process until you reach the starting point, counting the whole lengths of string and estimating any fractional part. Do this for each object you selected.

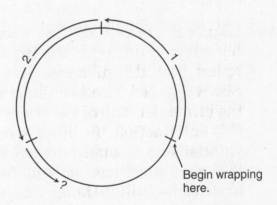

Begin wrapping here.

In the second part of the activity, you will measure the circumference and diameter of the circular objects and record the measurements on a recording sheet. If you have a metric tape measure, record the measurements to the nearest centimeter. If you have a customary tape measure, record your answers to the nearest quarter inch in decimal form $\left(\frac{1}{4} = 0.25; \frac{1}{2} = 0.5; \frac{3}{4} = 0.75\right)$. Using a calculator, divide the circumference of each circle by its diameter to determine the number of diameters in the circumference. Round each quotient to the nearest hundredth.

Record your results on Activity Sheet 9, as shown below.

Part 1: Estimates

Object	Approximate Number of Diameters in the Circumference
plate	$3\frac{1}{5}$
trash can	$3\frac{1}{4}$

Part 2: Measures

Object	Circumference	Diameter	Circumference Diameter
plate	78 cm	25 cm	3.12
trash can	122 cm	38 cm	3.21

Pi (π) If we know the radius or diameter of a circle, we can calculate the approximate circumference of the circle. In the previous activity we found that for any given circle there are a little more than three diameters in the circumference. Some people use 3 as a very rough approximation of the number of diameters in a circumference. The actual number of diameters in a circumference is closer to $3\frac{1}{7}$, which is approximately 3.14. The exact number of diameters in a circumference cannot be expressed as a fraction or as a decimal number, so mathematicians use the Greek letter π **(pi)** to stand for this number.

To find the circumference of a circle, we multiply the diameter of the circle by π. This relationship is shown in the formula below, where C stands for the circumference and d stands for the diameter.

$$C = \pi d$$

Since a diameter is equal to two radii ($2r$), we may replace d in the formula with $2r$. We usually arrange the factors this way:

$$C = 2\pi r$$

Unless otherwise noted, we will use 3.14 as an approximation for π. We may use a "wavy" equal sign to indicate that two numbers are approximately equal, as shown below.

$$\pi \approx 3.14$$

To use a formula such as $C = \pi d$, we **substitute** the measures or numbers we are given in place of the variables in the formula.

Example   Sidney drew a circle with a 2-inch radius. What is the circumference of the circle?

Solution   The radius of the circle is 2 inches, so the diameter is 4 inches. We multiply 4 inches by π (3.14) to find the circumference.

$$C = \pi d$$

$$C = (3.14)(4 \text{ in.})$$

$$C = 12.56 \text{ in.}$$

The circumference of the circle is about **12.56 inches.** This answer is reasonable because π is a little more than 3, and 3 times 4 inches is 12 inches.

LESSON PRACTICE

Practice set   **a.** In this lesson two formulas for the circumference of a circle are shown, $C = \pi d$ and $C = 2\pi r$. Explain why these two formulas are equivalent.

Find the circumference of each of these circles. (Use 3.14 for π.)

b.
2 in.

c.
3   3 cm

d. The diameter of a penny is about $\frac{3}{4}$ of an inch (0.75 inch). Find the circumference of a penny. Round your answer to two decimal places.

e. Roll a penny through one rotation on a piece of paper. Mark the start and the end of the roll. How far did the penny roll in one rotation? Measure the distance to the nearest eighth of an inch.

f. The radius of the great wheel was $14\frac{7}{8}$ ft. Which of these numbers is the best rough estimate of the wheel's circumference? Explain how you decided on your answer.

A. 15 ft   B. 60 ft   C. 90 ft   D. 120 ft

g. Use the formula $C = 2\pi r$ to find the circumference of a circle with a radius of 5 inches. (Use 3.14 for π.)

MIXED PRACTICE

Problem set

1. The first positive odd number is 1. The second is 3. What
(10) is the tenth positive odd number?

2. Giant tidal waves called *tsunamis* can travel 500 miles
(15) per hour. How long would it take a tsunami traveling at that speed to cross 3000 miles of ocean? (Write an equation and solve the problem.)

3. José bought Carmen one dozen red roses, two for each
(15) month he had known her. How long had he known her?

4. If $A = bh$, what is A when $b = 8$ and $h = 4$?
(47)

5. The commutative property of multiplication allows us to
(2) rearrange factors without changing the product. So $3 \cdot 5 \cdot 2$ may be arranged $2 \cdot 3 \cdot 5$. Use the commutative property of multiplication to rearrange these prime factors in order from least to greatest:

$$3 \cdot 7 \cdot 2 \cdot 5 \cdot 2 \cdot 3 \cdot 3 \cdot 5$$

6. If $s = \frac{1}{2}$, what number does $4s$ equal?
(29)

7. Write 6.25 in expanded notation.
(46)

8. Write 99% as a fraction. Then write the fraction as a decimal number.
(33, 35)

9. $12\overline{)0.18}$
(45)

10. $10\overline{)12.30}$
(45)

Find each missing number:

11. $w \div \sqrt{36} = 6^2$
(4, 38)

12. $5y = 1.25$
(43)

13. $n + 5\frac{11}{12} = 10$
(43)

14. $m - 6\frac{2}{5} = 3\frac{3}{5}$
(43)

15. $8\frac{3}{4} + 5\frac{3}{4}$
(26)

16. $\frac{5}{3} \times \frac{5}{4}$
(29)

17. $\frac{3}{4} = \frac{?}{20}$
(42)

18. $\frac{3}{5} = \frac{?}{20}$
(42)

19. Bob's scores on his first five tests were 18, 20, 18, 20, and 20. His average score is closest to which of these numbers?
(18)

A. 17   B. 18   C. 19   D. 20

20. Robert's bicycle tires are 20 inches in diameter. What is the circumference of a 20-inch circle? (Use 3.14 for π.)
(47)

21. Which factors of 20 are also factors of 30?
(19)

22. Mentally calculate the product of 6.25 and 10. Describe how you performed the mental calculation.
(46)

23. Multiply as shown. Then complete the division.
(46)

$$\frac{1.25}{0.5} \cdot \frac{10}{10}$$

24. Shelly answered 90% of the 40 questions correctly. What number is 90% of 40?
(41)

Refer to the chart shown below to answer problems 25 and 26.

Planet	Number of Earth Days to Orbit Sun
Mercury	88
Venus	225
Earth	365
Mars	687

25. Mars takes how many more days than Earth to orbit the
(13)   Sun?

26. In the time it takes Mars to orbit the Sun once, Venus
(15)   orbits the Sun about how many times?

27. Use an inch ruler to find the length
(17)   and width of this rectangle.

28. Calculate the perimeter of the
(8)   rectangle in problem 27.

29. Rename $\frac{2}{5}$ so that the denominator of the renamed fraction
(42)   is 10. Then subtract the renamed fraction from $\frac{9}{10}$. Reduce
       the answer.

30. When we mentally multiply 15 by 10, we can simply
(46)   attach a zero to 15 to make the product 150. When we
       multiply 1.5 by 10, why can't we attach a zero to make
       the product 1.50?

LESSON
48 Subtracting Mixed Numbers with Regrouping, Part 1

WARM-UP

Facts Practice: 30 Fractions to Reduce (Test G)

Mental Math: Count up and down by $\frac{1}{8}$'s between $\frac{1}{8}$ and 3.

a. 8×25	**b.** $630 - 50$	**c.** $62 + 19$
d. $\$4.50 + 75¢$	**e.** $\frac{1}{2}$ of $\$15.00$	**f.** $\frac{\$25.00}{100}$
g. $4 \times 7, -1, \div 3, \times 4, \div 6, \times 3, \div 2$		

Problem Solving:

Brianna scored 85 on each of her first two tests. What score does she need on her third test to have an average score of 90 on her first three tests?

NEW CONCEPT

Here is another "separating" story about pies:

> *There were $4\frac{1}{6}$ pies on the restaurant shelf. The server sliced one of the whole pies into sixths. Then the server removed $1\frac{2}{6}$ pies. How many pies were left on the shelf?*

We may illustrate this story with circles. There were $4\frac{1}{6}$ pies on the shelf.

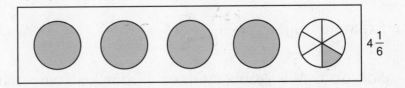

$4\frac{1}{6}$

The server sliced one of the whole pies into sixths. This makes $3\frac{7}{6}$ pies, which equals $4\frac{1}{6}$ pies.

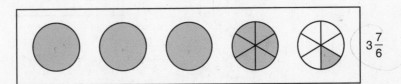

$3\frac{7}{6}$

The server removed $1\frac{2}{6}$ pies. So $2\frac{5}{6}$ pies were left on the shelf.

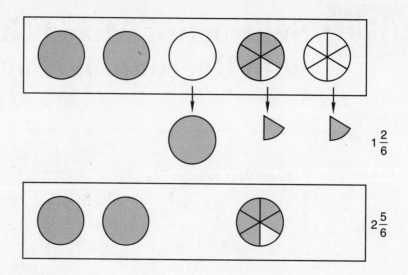

Now we show the arithmetic for subtracting $1\frac{2}{6}$ from $4\frac{1}{6}$.

$$4\frac{1}{6} \text{ pies}$$
$$-\ 1\frac{2}{6} \text{ pies}$$

We cannot subtract $\frac{2}{6}$ from $\frac{1}{6}$, so we rename $4\frac{1}{6}$. Just as the server sliced one of the pies into sixths, so we change one of the four wholes into $\frac{6}{6}$. This makes three whole pies plus $\frac{6}{6}$ plus $\frac{1}{6}$, which is $3\frac{7}{6}$. Now we can subtract.

$$4\frac{1}{6} \xrightarrow{\ 3\ +\ \frac{6}{6}\ +\ \frac{1}{6}\ } 3\frac{7}{6} \text{ pies}$$
$$-\ 1\frac{2}{6} \qquad\qquad\qquad -\ 1\frac{2}{6} \text{ pies}$$
$$\overline{\qquad\qquad 2\frac{5}{6} \text{ pies}}$$

Example   Subtract: $5\frac{1}{3} - 2\frac{2}{3}$

Solution   We cannot subtract $\frac{2}{3}$ from $\frac{1}{3}$, so we rename $5\frac{1}{3}$. We change one of the five wholes into $\frac{3}{3}$. Then we combine 4 and $\frac{3}{3} + \frac{1}{3}$ to get $4\frac{4}{3}$. Now we can subtract.

$$5\frac{1}{3} \xrightarrow{\ 4\ +\ \frac{3}{3}\ +\ \frac{1}{3}\ } 4\frac{4}{3}$$
$$-\ 2\frac{2}{3} \qquad\qquad\qquad -\ 2\frac{2}{3}$$
$$\overline{\qquad\qquad\quad 2\frac{2}{3}}$$

LESSON PRACTICE

Practice set*   Subtract:

a. $4\frac{1}{3}$
$-1\frac{2}{3}$

b. $3\frac{2}{5}$
$-2\frac{3}{5}$

c. $5\frac{2}{4}$
$-1\frac{3}{4}$

d. $5\frac{1}{8}$
$-2\frac{4}{8}$

e. $7\frac{3}{12}$
$-4\frac{10}{12}$

f. $6\frac{1}{4}$
$-2\frac{3}{4}$

MIXED PRACTICE

Problem set

1. The average of two numbers is 10. What is the sum of the two numbers?
(18)

2. What is the cost of 10.0 gallons of gasoline priced at $1.449 per gallon?
(46)

3. The movie started at 11:45 a.m. and ended at 1:20 p.m. The movie was how many hours and minutes long?
(32)

4. Three of the numbers shown below are equal. Which number is not equal to the others?
(44)

A. $\frac{1}{2}$ B. 0.2   C. 0.5   D. $\frac{10}{20}$

5. Arrange these numbers in order from least to greatest:
(44)

1.02, 0.102, 0.12, 1.20, 0, −1

6. $0.1 + 0.2 + 0.3 + 0.4$ **7.** $(8)(0.125)$
(37)   (39)

8. Juan was hiking to a waterfall 3 miles away. After hiking 2.1 miles, how many more miles did he have to hike to reach the waterfall? (Write an equation and solve the problem.)
(11)

9. Estimate the sum of 4967, 8142, and 6890.
(16)

10. $8\overline{)0.144}$ **11.** $6\overline{)0.9}$ **12.** $4\overline{)0.9}$
(45)   (45)   (45)

13. What is the cost of 100 pens priced at 39¢ each?
(46)

14. Write $(5 \times 10) + \left(6 \times \frac{1}{10}\right) + \left(4 \times \frac{1}{100}\right)$ in standard form.
(46)

15. What is the least common multiple of 6 and 8?
(30)

Find each missing number:

16. $w - 7\frac{7}{12} = 5\frac{5}{12}$

17. $12 - m = 5\frac{2}{3}$

18. $n + 2\frac{3}{4} = 5\frac{1}{4}$

19. $x + 3.21 = 4$

20. What fraction is $\frac{2}{3}$ of $\frac{3}{4}$?

21. Sam carried the football three times during the game. He had gains of 3 yards and 5 yards and a loss of 12 yards. Use a negative number to show Sam's overall yardage for the game.

22. If a spool for thread is 2 cm in diameter, then one wind of thread is about how many centimeters long? (Use 3.14 for π.) Describe how to mentally check whether your answer is reasonable.

23. If a rectangle is 12 inches long and 8 inches wide, what is the ratio of its length to its width?

24. The perimeter of this square is 4 feet. What is its perimeter in inches?

25. The area of this square is one square foot. What is its area in square inches?

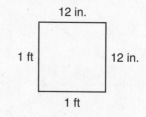

26. If $d = rt$, and if $r = 60$ and $t = 4$, what does d equal?

27. Seventy-five percent of the 32 chairs in the room were occupied. Write 75% as a reduced fraction. Then find the number of chairs that were occupied.

28. Rename $\frac{1}{3}$ and $\frac{1}{4}$ as fractions with denominators of 12. Then add the renamed fractions.

29. Multiply as shown. Then simplify the answer.

$$\frac{3.5}{0.7} \cdot \frac{10}{10}$$

30. There were $3\frac{1}{6}$ pies on the shelf. Explain how the server can take $1\frac{5}{6}$ pies from the shelf.

LESSON
49 Dividing by a Decimal Number

WARM-UP

Facts Practice: 72 Multiplication and Division Facts (Test H)

Mental Math: Count by 12's from 12 to 120.

a. $(8 \times 100) + (8 \times 25)$ b. $290 + 50$

c. $58 - 19$ d. $\$5.00 - \$3.25$

e. $\frac{1}{2}$ of $30.00   f. $\frac{4000}{100}$

g. $5 \times 10, \div 2, + 5, \div 2, + 5, \div 2, \div 2$

Problem Solving:

The children ran around the block, starting and finishing at point A. Instead of running all the way around the block, Nimah took what she called her "shortcut" behind a building, shown by the dotted line in the diagram. How many meters did Nimah save with her "shortcut"?

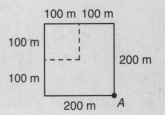

NEW CONCEPT

When the divisor of a division problem is a decimal number, we change the problem so that the divisor is a whole number.

$$\frac{1.24}{0.4} \longleftarrow \quad 0.4\overline{)1.24}$$

The divisor is a decimal number. We change the problem before we divide.

One way to change a division problem is to multiply the divisor and the dividend by 10. Notice in the whole-number division problem below that multiplying both numbers by 10 does not change the quotient.

$$4\overline{)8}^{\,2} \longrightarrow 40\overline{)80}^{\,2}$$

Multiplying 4 and 8 by 10 does not change the quotient.

The quotient is not changed, because we formed an equivalent division problem just as we form equivalent fractions—by multiplying by a form of 1.

$$\frac{8}{4} \times \frac{10}{10} = \frac{80}{40}$$

We can also use this method to change a division by a decimal number to a division by a whole number.

If we multiply the divisor and dividend in $\frac{1.24}{0.4}$ by 10, the new problem has a whole-number divisor.

$$\text{decimal} \longrightarrow \frac{1.24}{0.4} \times \frac{10}{10} = \frac{12.4}{4} \longleftarrow \text{whole-number divisor}$$

We divide 12.4 by 4 to find the quotient.

$$\begin{array}{r} 3.1 \\ 4\overline{)12.4} \end{array}$$

Example 1 Divide: $\dfrac{1.24}{0.04}$

Solution The divisor, 0.04, is a decimal number with two decimal places. To make the divisor a whole number, we will multiply $\frac{1.24}{0.04}$ by $\frac{100}{100}$, which shifts each decimal point two places to the right.

$$\frac{1.24}{0.04} \times \frac{100}{100} = \frac{124}{4}$$

This forms an equivalent division problem in which the divisor is a whole number. Now we perform the division.

$$\begin{array}{r} 31 \\ 4\overline{)124} \\ \underline{12} \\ 04 \\ \underline{4} \\ 0 \end{array}$$

Example 2 Divide: $0.6\overline{)1.44}$

Solution The divisor, 0.6, has one decimal place. If we multiply the divisor and dividend by 10, we will shift the decimal point one place to the right in both numbers.

$$06.\overline{)14.4}$$

This makes a new problem with a whole-number divisor, which we solve below.

$$
\begin{array}{r}
2.4 \\
6\overline{)14.4} \\
\underline{12} \\
2\,4 \\
\underline{2\,4} \\
0
\end{array}
$$

Some people use the phrase "over, over, and up" to keep track of the decimal points when dividing by decimal numbers.

$$
\begin{array}{c}
\text{up} \\
\cdot \\
06.\overline{)14.4} \\
\text{over\ \ over}
\end{array}
$$

LESSON PRACTICE

Practice set*

a. We would multiply the divisor and dividend of $\frac{1.44}{1.2}$ by what number to make the divisor a whole number?

b. We would multiply the divisor and dividend of $0.12\overline{)0.144}$ by what number to make the divisor a whole number?

Change each problem so that the divisor is a whole number. Then divide.

c. $\dfrac{0.24}{0.4}$ **d.** $\dfrac{9}{0.3}$

e. $0.05\overline{)2.5}$ **f.** $0.3\overline{)12}$

g. $0.24 \div 0.8$ **h.** $0.3 \div 0.03$

i. $0.05\overline{)0.4}$ **j.** $0.2 \div 0.4$

MIXED PRACTICE

Problem set

1. When the product of 0.2 and 0.3 is subtracted from the sum of 0.2 and 0.3, what is the difference?
(12, 39)

2. Four fifths of a dollar is how many cents? Draw a diagram to illustrate the problem.
(22)

3. The rectangular, 99-piece "Nano" jigsaw puzzle is only 2.6 inches long and 2.2 inches wide. What is the area of the puzzle in square inches?
(31, 39)

4. Find the perimeter of the puzzle described in problem 3.
(38)

5. Compare:
(44)
    (a) 0.31 ◯ 0.301   (b) 31% ◯ 30.1%

6. 0.67 + 2 + 1.33   **7.** 12(0.25)
(38)   (39)

8. $0.07\overline{)3.5}$ **9.** $0.5\overline{)12}$
(49)   (49)

10. $8\overline{)0.14}$ **11.** (0.012)(1.5)
(45)   (39)

Find each missing number:

12. $n - 6\frac{1}{8} = 4\frac{3}{8}$ **13.** $\frac{4}{5} = \frac{x}{100}$
(43)   (42)

14. $5 - m = 1.37$ **15.** $m + 7\frac{1}{4} = 15$
(43)   (43)

16. Write the decimal number one and twelve thousandths.
(35)

17. $5\frac{7}{10} + 4\frac{9}{10}$ **18.** $\frac{5}{2} \cdot \frac{5}{3}$
(26)   (29)

19. How much money is 40% of $25.00?
(41)

20. There are 24 hours in a day. Jim sleeps 8 hours each night.
(29)
    (a) Eight hours is what fraction of a day?

    (b) What fraction of a day does Jim sleep?

    (c) What fraction of a day does Jim not sleep?

21. List the factors that 12 and 18 have in common (that is, the numbers that are factors of both 12 and 18).
(19)

22. What is the average of 1.2, 1.3, and 1.7?
(18, 45)

23. Jan estimated that 49% of $19.58 is $10. She rounded 49% to 50% and rounded $19.58 to $20. Then she mentally calculated 50% of $20. Use Jan's method to estimate 51% of $49.78. Describe how to perform the estimate.
(41)

24. (a) How many $\frac{3}{4}$'s are in 1?
(30)
    (b) Use the answer to part (a) to find the number of $\frac{3}{4}$'s in 4.

25. Refer to the number line shown below to answer parts
(17) (a)–(c).

(a) Which point is halfway between 1 and 2?

(b) Which point is closer to 1 than 2?

(c) Which point is closer to 2 than 1?

26. Multiply and divide as indicated: $\dfrac{2 \cdot 3 \cdot 2 \cdot 5 \cdot 7}{2 \cdot 5 \cdot 7}$
(5)

27. We can find the number of quarters in three dollars by
(49) dividing $3.00 by $0.25. Show this division using the
pencil-and-paper method taught in this lesson.

28. Use a ruler to find the length of each
(8) side of this square to the nearest
eighth of an inch. Then calculate
the perimeter of the square.

29. A paper-towel tube is about 4 cm in diameter. The
(47) circumference of a paper-towel tube is about how many
centimeters? (Use 3.14 for π.)

30. Sam was given the following division problem:
(49)

$$\frac{2.5}{0.5}$$

Instead of multiplying the numerator and denominator by
10, he accidentally multiplied by 100, as shown below.

$$\frac{2.5}{0.5} \times \frac{100}{100} = \frac{250}{50}$$

Then he divided 250 by 50 and found that the quotient
was 5. Did Sam find the correct answer to $2.5 \div 0.5$?
Why or why not?

LESSON
50

Decimal Number Line (Tenths) • Dividing by a Fraction

WARM-UP

Facts Practice: 30 Fractions to Reduce (Test G)

Mental Math: Count by 7's from 7 to 84.

a. $(8 \times 200) + (8 \times 25)$
b. $565 - 250$
c. $58 + 27$
d. $\$1.45 + 99¢$
e. $\frac{1}{2}$ of $\$25.00$
f. $\frac{5000}{10}$
g. $8 \times 9, + 3, \div 3, - 1, \div 3, + 1, \div 3, \div 3$

Problem Solving:

Susan began building stair-step patterns with blocks. She used one block for a one-step pattern, three blocks for a two-step pattern, and six blocks for a three-step pattern. She wrote the information in a table. Copy the table and complete it through a ten-step pattern.

Blocks Needed to Make Pattern

Steps	Blocks
1	1
2	3
3	6
4	
5	
6	
7	
8	
9	
10	

NEW CONCEPTS

Decimal number line (tenths)

We can locate different kinds of numbers on the number line. We have learned to locate whole numbers, negative numbers, and fractions on the number line. We can also locate decimal numbers on the number line.

On the number line above, the distance between consecutive whole numbers has been divided into ten equal lengths. Each length is $\frac{1}{10}$ of the distance between consecutive whole numbers. The arrow is pointing to a mark three spaces beyond the 1, so it is pointing to $1\frac{3}{10}$. We can rename $\frac{3}{10}$ as the decimal 0.3, so we can say that the arrow is pointing to the mark representing 1.3. When a unit has been divided into ten spaces, we normally use the decimal form instead of the fractional form to name the number represented by the mark.

Example 1 What decimal number is represented by point y on this number line?

Solution The distance from 7 to 8 has been divided into ten smaller segments. Point y is four segments to the right of the whole number 7. So point y represents $7\frac{4}{10}$. We write $7\frac{4}{10}$ as the decimal number **7.4.**

Dividing by a fraction

The following question can be answered by dividing by a decimal number or by dividing by a fraction:

How many quarters are in three dollars?

If we think of a quarter as $\frac{1}{4}$ of a dollar, we have this division problem:

$$3 \div \frac{1}{4}$$

We solve this problem in two steps. First we answer the question, "How many quarters are in one dollar?" The answer is the <u>reciprocal</u> of $\frac{1}{4}$, which is $\frac{4}{1}$, which equals 4.

$$1 \div \frac{1}{4} = \frac{4}{1} = 4$$

For the second step, we use the answer to the question above to find the number of quarters in three dollars. There are four quarters in one dollar, and there are three times as many quarters in three dollars. We multiply 3 by 4 and find that there are 12 quarters in three dollars.

number of dollars ─┐   ┌─ number of quarters in one dollar

$$3 \times 4 = 12$$

└─ number of quarters in three dollars

We will review the steps we took to solve the problem.

Original problem: How many quarters are in $3? $3 \div \frac{1}{4}$

Step 1: Find the number of quarters in $1. $1 \div \frac{1}{4} = 4$

Step 2: Use the number of quarters in $1 to find the number in $3. $3 \times 4 = 12$

Example 2   The diameter of a penny is $\frac{3}{4}$ of an inch. How many pennies are needed to make a row of pennies 6 inches long?

Solution   In effect, this problem asks, "How many $\frac{3}{4}$-inch segments are in 6 inches?" We can write the question this way:

$$6 \div \frac{3}{4}$$

We will take two steps. First we will find the number of pennies (the number of $\frac{3}{4}$-inch segments) in 1 inch. The number of $\frac{3}{4}$'s in 1 is the reciprocal of $\frac{3}{4}$, which is $\frac{4}{3}$.

$$1 \div \frac{3}{4} = \frac{4}{3}$$

We will not convert $\frac{4}{3}$ to the mixed number $1\frac{1}{3}$. Instead, we will use $\frac{4}{3}$ in the second step of the solution. Since there are $\frac{4}{3}$ pennies in 1 inch, there are six times as many in 6 inches. So we multiply 6 by $\frac{4}{3}$.

$$6 \times \frac{4}{3} = \frac{24}{3} = 8$$

We find there are **8 pennies** in a 6-inch row. We will review the steps of the solution.

Original problem: How many $\frac{3}{4}$'s are in 6? $6 \div \frac{3}{4}$

Step 1: Find the number of $\frac{3}{4}$'s in 1. $1 \div \frac{3}{4} = \frac{4}{3}$

Step 2: Use the number of $\frac{3}{4}$'s in 1 to find the number of $\frac{3}{4}$'s in 6. Then simplify the answer. $6 \times \frac{4}{3} = \frac{24}{3}$

$$= 8$$

LESSON PRACTICE

Practice set   To which decimal number is each arrow pointing?

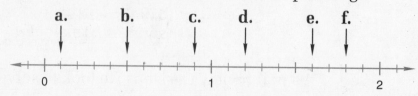

g. Write and solve a division problem to find the number of quarters in four dollars. Use $\frac{1}{4}$ instead of 0.25 for a quarter. Follow this pattern:

Original problem

Step 1

Step 2

h. Write and solve a fraction division problem for this question:

Pads of writing paper were stacked 12 inches high on a shelf. The thickness of each pad was $\frac{3}{8}$ of an inch. How many pads were in a 12-inch stack?

MIXED PRACTICE

Problem set

1. The first three positive odd numbers are 1, 3, and 5. Their
(10) sum is 9. The first five positive odd numbers are 1, 3, 5, 7, and 9. Their sum is 25. What is the sum of the first ten positive odd numbers?

2. Jack keeps his music CDs stacked
(50) in plastic boxes $\frac{3}{8}$ inch thick. Use the method taught in this lesson to find the number of boxes in a stack 6 inches tall.

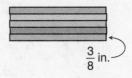

$\frac{3}{8}$ in.

3. The boxing match ended after two minutes of the twelfth
(15) round. Each of the first eleven rounds lasted three minutes. For how many minutes did the contenders box?

4. Compare:
(44)
(a) 3.4 ◯ 3.389
(b) 0.60 ◯ 0.600

Find each missing number:

5. $7.25 + 2 + w = \sqrt{100}$
(43)

6. $6w = 0.144$
(43)

7. $w + \dfrac{5}{12} = 1^2$
(43)

8. $6\dfrac{1}{8} - x = 1\dfrac{7}{8}$
(43)

9. The book cost $20.00. The sales-tax rate was 7%. What
(41) was the total cost of the book including sales tax?

10. $1 - 0.97$
(38)

11. $0.12\overline{)7.2}$
(49)

12. $0.4\overline{)7}$
(49)

13. $6\overline{)0.138}$
(45)

14. $(3.75)(2.4)$
(39)

15. $\dfrac{3}{4} = \dfrac{?}{24}$
(42)

16. Which digit in 4.637 is in the same place as the 2 in 85.21?
(34)

17. One hundred centimeters equals
(31) one meter. How many square
centimeters equal one square meter?

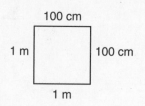

18. What is the least common multiple of 6 and 9?
(30)

19. $6\frac{5}{8} + 4\frac{5}{8}$ **20.** $\frac{8}{3} \cdot \frac{3}{1}$ **21.** $\frac{2}{3} \cdot \frac{3}{4}$
(26)   (29)   (29)

22. The diameter of a soup can is about 7 cm. The label
(47) wraps around the can. About how many centimeters long
must the label be to go all the way around the soup can?
(Use $\frac{22}{7}$ for π.) Describe how to mentally check whether
your answer is reasonable.

23. Find the average of 2.4, 6.3, and 5.7.
(18)

24. What factors do 18 and 24 have in common?
(19)

25. What decimal number corresponds to point A on this
(50) number line?

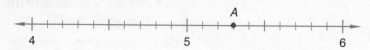

26. $\frac{2 \cdot 3 \cdot 5 \cdot 7}{2 \cdot 5}$ **27.** 0.375×100
(5)   (46)

28. Rename $\frac{1}{3}$ as a fraction with 6 as the denominator. Then
(42) subtract the renamed fraction from $\frac{5}{6}$. Reduce the answer.

29. Points x, y, and z are three points on this number line. Refer
(50) to the number line to answer the questions that follow.

(a) Which point is halfway between 6 and 7?

(b) Which point corresponds to $6\frac{7}{10}$?

(c) Of the points x, y, and z, which point corresponds to
the number that is closest to 6?

30. Which of these numbers is divisible by both 2 and 5?
(21)
A. 552   B. 255   C. 250   D. 525

INVESTIGATION 5

Focus on

Displaying Data

In Investigation 4 we looked at collecting data. In this two-part investigation we will look at various ways to display data. (You might want to use two days to complete this investigation, one day for each part.)

Part 1: *Displaying Qualitative Data*

We have already displayed data in Investigation 1, where we used bar graphs to summarize frequency tables. Now we will use bar graphs to display qualitative data.

In Investigation 4 Brigit asked each guest at the party, "Among New Year's, Thanksgiving, and Independence Day, which is your favorite holiday?" Her survey of 30 guests produced the following results: 10 chose New Year's, 12 chose Thanksgiving, and 8 chose Independence Day. A horizontal bar graph for the data might look like this:

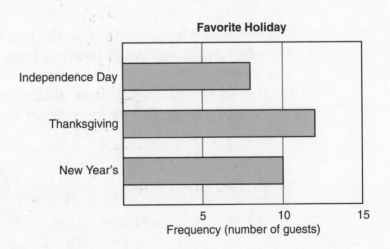

A second way to display this qualitative data is with a **pictograph.** In a pictograph, pictured objects represent the data being counted. Each object represents one or more units of data. In the pictograph below, for example, each object represents two guests from Brigit's data. Choosing objects that remind you of the categories they represent can be fun!

Favorite Holiday

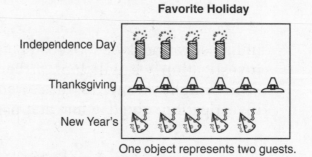

One object represents two guests.

A third possibility is to display the data in a **circle graph**. In a circle graph each category corresponds to a **sector** of the circle. Think of a circle as a pie; a sector is simply a slice of the pie. We use circle graphs when we are interested in the fraction of the group represented by each category and not so interested in the particular number of units in each category. Another name for a circle graph is a **pie chart**.

We use a compass and a protractor to help us create a circle graph. We use the compass to draw the circle and the protractor to divide the circle into sectors of the correct size. To divide the circle, we need to determine how many degrees to assign to each category. One way to do this is to find the number of degrees to assign to each person's response. A whole circle measures 360°. There were 30 guests surveyed. We divide 360° by 30 and find that each response is assigned 12°.

$$360° \div 30 = 12°$$

Since 10 guests chose New Year's, that sector gets 120° of the circle.

$$10 \times 12° = 120°$$

Twelve guests chose Thanksgiving, so that sector gets 144° of the circle.

$$12 \times 12° = 144°$$

Eight guests chose Independence Day, so that sector gets 96° of the circle.

$$8 \times 12° = 96°$$

We position the center of the protractor over the center of the circle and draw a 120° sector for New Year's and a 144° sector for Thanksgiving next to the New Year's sector. The remaining sector of the circle represents Independence Day. How do you know it must measure 96°?

Favorite Holiday

Katie asked 20 of her friends how they get to the park. She found that 9 walk, 4 ride a bicycle, 4 ride a skateboard, and 3 ride in a car.

1. Display Katie's data in a vertical bar graph.

2. Display Katie's data in a horizontal bar graph.

3. Display Katie's data in a pictograph in which one object represents one friend.

4. Display Katie's data in a pictograph in which one object represents two friends. Note that the friends who walk must be represented by $4\frac{1}{2}$ objects, and the friends who ride in a car must be represented by $1\frac{1}{2}$ objects.

5. Display Katie's data in a circle graph. How many degrees represents each friend?

Part 2: *Displaying Quantitative Data*

We now turn to quantitative data. Quantitative data consists of individual measurements or numbers called **data points.** When there are many possible values for data points, we can group them in intervals and display the data in a histogram as we did in Investigation 1. When we group data in intervals, however, the individual data points disappear. In order to display the individual data points, we can use a **line plot.** To do this, we first draw a number line. Then, for each data point, we place an X over its corresponding number on the number line.

Suppose you take 18 tests with 20 possible points each. Your scores, listed in increasing order, are

5, 8, 8, 10, 10, 11, 12, 12, 12, 12, 13, 13, 14, 16, 17, 17, 18, 19

We represent these data in the line plot below.

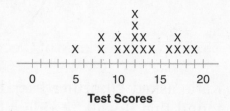

When describing numerical data, we often use terms such as **mean, median, mode,** and **range.** Here we define each of these terms:

  Mean: the average of the numbers

  Median: the middle number when the data are arranged in numerical order

  Mode: the most frequently occurring number

  Range: the difference between the greatest and least of the numbers

To find the mean of the test scores above, we add the 18 scores and divide the sum by 18. The mean is about 12.6. To find the median, we look for the middle score. If the number of scores were an odd number, we would simply select the middle score. But since the number of scores is an even number, 18, we use the average of the ninth and tenth scores for the median. We find that the ninth and tenth scores are both 12, so the median score is 12. From the line plot we can easily see that the most common score is 12. So the mode of the test scores is 12. We also see that the scores range from 5 to 19, so the range is 19 − 5, which is 14.

6. The heights (in inches) of 20 twelve-year-olds are listed below.

Dont drink the yellow water

$$60, 52, 49, 51, 47, 53, 62, 60, 57, 56,$$

$$58, 56, 63, 58, 53, 50, 48, 60, 62, 53$$

Write the heights in increasing order, and display them in a line plot. Begin your number line at 45 inches.

7. What is the median of the heights listed in problem 6?

8. The distribution of the heights in problem 6 is **bimodal** because there are two modes. What are the two modes?

9. What is the range of the heights of the twelve-year-olds? (In this case, the range is the difference in height between the shortest twelve-year-old and the tallest twelve-year-old.)

10. Quantitative data can be displayed in **stem-and-leaf plots.** The beginning of a stem-and-leaf plot for the data in problem 6 is shown below. The "stems" are the tens digits of the data points. The "leaves" for each stem are all the ones digits in the data points that begin with that stem. We have plotted the data points for these heights: 47, 48, 49, 50, 51, 52, 53. Copy this plot. Then insert the rest of the heights from problem 6.

Stem	Leaf
4	7 8 9
5	0 1 2 3 3

5 | 2 represents a height of 52 inches.

Extensions

a. Organize data from your *Saxon Math 7/6* tests. Display your scores on a line plot. Find the median, mode, and range of the scores.

b. Create a stem-and-leaf plot of your math test scores.

Extensions **c–e** apply only if you conducted the extension surveys in Investigation 4.

c. If you surveyed children and adults with respect to their favorite seasons, display the data in horizontal bar graphs and pictographs. For each type of graph, display the data for adults and children separately. Do adults and children seem to prefer the same season?

d. If you surveyed children and adults with respect to their favorite breakfast beverages, display the data in vertical bar graphs and circle graphs. For each type of graph, display the data for adults and children separately. Do preferences seem to change as people get older? In what ways?

e. If you observed the number of people in a fast-food restaurant each day for 10 consecutive days, display your data on a line plot.

Note: Problems intended for additional exposure to the concepts in this investigation are available in the appendix.

Peanot butter and toejam Siadwitch

LESSON
51 Rounding Decimal Numbers

WARM-UP

Facts Practice: 72 Multiplication and Division Facts (Test H)

Mental Math: Count up and down by $\frac{1}{8}$'s between $\frac{1}{8}$ and 3.

  a. 8×125 **b.** 4×68 **c.** $64 - 29$
  d. $\$4.64 + 99¢$ **e.** $\frac{1}{2}$ of $\$150.00$ **f.** $\frac{\$100.00}{100}$
  g. $8 \times 8, -4, \div 2, +2, \div 4, +2, \div 5, \times 10$

Problem Solving:

Copy this problem and fill in the missing digits:

$$\begin{array}{r} 9\_ \\ 9\overline{)\ \_\,9\,\_} \\ \underline{=\,=} \\ -\,- \\ \underline{=\,=} \\ 0 \end{array}$$

NEW CONCEPT

It is often necessary or helpful to round decimal numbers. For instance, money amounts are usually rounded to two places after the decimal point because we do not have a coin smaller than one hundredth of a dollar.

Example 1 Dan wanted to buy a tape for $6.89. The sales-tax rate was 8%. Dan calculated the sales tax. He knew that 8% equaled the fraction $\frac{8}{100}$ and the decimal 0.08. To figure the amount of tax, he multiplied the price ($6.89) by the sales-tax rate (0.08).

$$\begin{array}{r} \$6.89 \\ \times\ \ 0.08 \\ \hline \$0.5512 \end{array}$$

How much tax would Dan pay if he purchased the tape?

Solution Sales tax is rounded to the nearest cent, which is two places to the right of the decimal point. We mark the places that will be included in the answer.

$$\underline{\$0.55}\,12$$

Next we consider the possible answers. We see that $0.5512 is a little more than $0.55 but less than $0.56. We decide whether $0.5512 is closer to $0.55 or $0.56 by looking at the next digit (in this case, the digit in the third decimal place). If the next digit is 5 or more, we round up to $0.56. If it is less than 5, we round down to $0.55. Since the next digit is 1, we round $0.5512 down. If Dan buys the tape, he will need to pay **$0.55** sales tax.

Example 2   Sheila pulled into the gas station and filled the car's tank with 10.381 gallons of gasoline. Round the amount of gasoline she purchased to the nearest tenth of a gallon.

Solution   The tenths place is one place to the right of the decimal point. We mark the places that will be included in the answer.

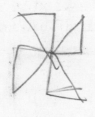

$$\underline{10.3}81$$

Next we consider the possible answers. The number we are rounding is more than 10.3 but less than 10.4. We decide that 10.381 is closer to 10.4 because the digit in the next place is 8, and we round up when the next digit is 5 or more. Sheila bought about **10.4 gallons** of gasoline.

Example 3   Estimate the product of 6.85 and 4.2 by rounding the numbers to the nearest whole number before multiplying.

Solution   We mark the whole-number places.

$$\underline{6}85 \qquad \underline{4}2$$

We see that 6.85 is more than 6 but less than 7. The next digit is 8, so we round 6.85 up to 7. The number 4.2 is more than 4 but less than 5. The next digit is 2, so we round 4.2 down to 4. We multiply the rounded numbers.

$$7 \cdot 4 = 28$$

We estimate that the product of 6.85 and 4.2 is about **28.**

LESSON PRACTICE

Practice set*   Round to the nearest cent:

    a. $6.6666   **b.** $0.4625   **c.** $0.08333

Round to the nearest tenth:

    d. 0.12   **e.** 12.345   **f.** 2.375

Round to the nearest whole number:

g. 16.75 **h.** 4.875 **i.** 73.29

j. If the sales-tax rate is 6%, then how much sales tax is there on a $3.79 purchase? (Round the <u>answer</u> to the nearest cent.)

MIXED PRACTICE

Problem set

1. When the third multiple of 8 is subtracted from the fourth
(12, 25) multiple of 6, what is the difference?

2. From Mona's home to the mall is 3.5 miles. How far does
(37) Mona travel riding from home to the mall and back home?

3. Napoleon I was born in 1769. How old was he when he was
(13) crowned emperor of France in 1804? (Write an equation and solve the problem.)

4. Shelly purchased a music CD for $12.89. The sales-tax
(41, 51) rate was 8%.

(a) What was the tax on the purchase?

(b) What was the total price including tax?

5. Malcomb used a compass to draw a circle with a radius of
(47, 51) 3 inches.

(a) Find the diameter of the circle.

(b) Find the circumference of the circle. Round the answer to the nearest inch. (Use 3.14 for π.)

6. Explain how to round 12.75 to the nearest whole number.
(51)

7. 0.125 + 0.25 + 0.375   **8.** 0.399 + w = 0.4
(37)   (43)

9. $\dfrac{4}{0.25}$ **10.** $4\overline{)0.5}$
(49)   (45)

11. 3.25 ÷ $\sqrt{100}$ **12.** $3\dfrac{5}{12} - 1\dfrac{7}{12}$
(45)   (48)

13. $\dfrac{5}{8} = \dfrac{?}{24}$ **14.** $5^2 - 17\dfrac{3}{4}$
(42)   (48)

15. (0.19)(0.21)
(39)

16. Write 0.01 as a fraction.
(35)

17. Write $(6 \times 10) + \left(7 \times \frac{1}{100}\right)$ as a decimal number.
(46)

18. The area of a square is 64 cm². What is the perimeter of the square?
(38)

19. What is the least common multiple of 2, 3, and 4?
(30)

20. $5\frac{3}{10} + 6\frac{9}{10}$
(26)

21. $\frac{10}{3} \times \frac{1}{2}$
(29)

22. A collection of paperback books was stacked 12 inches high. Each book in the stack was $\frac{3}{4}$ inch thick. Use the method described in Lesson 50 to find the number of books in the stack.
(50)

23. Estimate the quotient when 4876 is divided by 98.
(16)

24. What factors do 16 and 24 have in common?
(19)

25. Estimate the product of 11.8 and 3.89 by rounding the factors to the nearest whole number before multiplying. Explain how you arrived at your answer.
(51)

26. Find the average of the decimal numbers that correspond to points x and y on this number line.
(50)

27. $\dfrac{2 \cdot 2 \cdot 3 \cdot 3 \cdot 5}{2 \cdot 2 \cdot 3 \cdot 5}$
(5)

28. Mentally calculate the total price of ten pounds of bananas at $0.79 per pound. Describe how you performed the mental calculation.
(46)

29. Rename $\frac{2}{3}$ and $\frac{3}{4}$ as fractions with 12 as the denominator. Then add the renamed fractions. Write the sum as a mixed number.
(42)

30. Jason's first nine test scores are shown below. Find the median and mode of the scores.
(Inv. 5)

85, 80, 90, 75, 85, 100, 90, 80, 90

LESSON
52 Mentally Dividing Decimal Numbers by 10 and by 100

WARM-UP

Facts Practice: 30 Fractions to Reduce (Test G)

Mental Math: Count by 12's from 12 to 132.

a. 4 × 250   b. 368 − 150   c. 250 + 99
d. $15.00 + $7.50   e. $\frac{1}{2}$ of 5 f. 20 × 40
g. 5 × 10, + 4, ÷ 6, × 8, + 3, ÷ 3

Problem Solving:

Alycia averaged 88% on her first three tests. What score does she need on her fourth test to have a four-test average of 90%?

NEW CONCEPT

When we divide a decimal number by 10 or by 100, the quotient has the same digits as the dividend. However, the position of the digits is shifted. Here we show 12.5 divided by 10 and by 100:

$$10\overline{)12.50} \quad \begin{array}{c} 1.25 \end{array} \qquad 100\overline{)12.500} \quad \begin{array}{c} .125 \end{array}$$

When we divide by 10, the digits shift one place to the right. When we divide by 100, the digits shift two places to the right. Although it is the digits that are shifting places, we produce the shift by moving the decimal point. When we divide by 10, the decimal point moves one place to the left. When we divide by 100, the decimal point moves two places to the left.

Example 1 Divide: 37.5 ÷ 10

Solution Since we are dividing by 10, the answer will be less than 37.5. We mentally shift the decimal point one place to the left.

$$37.5 ÷ 10 = \textbf{3.75}$$

Example 2   Divide: 3.75 ÷ 100

Solution   Since we are dividing by 100, we mentally shift the decimal point two places to the left. This creates an empty place between the decimal point and the 3, which we fill with a zero. We also write a zero in the ones place.

$$3.75 \div 100 = \mathbf{0.0375}$$

LESSON PRACTICE

Practice set*   Mentally calculate each quotient. Write each answer as a decimal number.

(handwritten: 25 × 100 / 2500)

a. 2.5 ÷ 10   **b.** 2.5 ÷ 100

(handwritten: 25 × 10)

c. 87.5 ÷ 10   **d.** 87.5 ÷ 100

(handwritten: 250)

e. 0.5 ÷ 10   **f.** 0.5 ÷ 100

g. 25 ÷ 10   **h.** 25 ÷ 100

MIXED PRACTICE

Problem set   **1.** What is the product of one half and two thirds?
(29)

2. A piano has 88 keys. Fifty-two of the keys are white. How
(13) many more white keys are there than black keys?

3. Near the island of Puerto Rico, the Atlantic Ocean reaches
(12) its greatest depth of twenty-eight thousand, two hundred thirty-two feet. Use digits to write that number of feet.

4. Mentally calculate each answer. Describe how you
(46, 52) performed each mental calculation.

(a) 3.75 × 10   (b) 3.75 ÷ 10

5. At the summer camp there are 320 campers and
(23) 16 counselors. What is the camper-counselor ratio at the summer camp?

Simplify:

6. $2 \cdot 2 \cdot 2 \cdot 2 \cdot 2$
(5)

7. $(4)(0.125)$
(39)

8. $\dfrac{150}{12}$
(29)

9. $\dfrac{(1 + 0.2)}{(1 - 0.2)}$
(49)

10. $\dfrac{5}{2} \times \dfrac{4}{1}$
(29)

Find each missing number:

11. $5\dfrac{1}{3} - m = 1\dfrac{2}{3}$
(43)

12. $m - 5\dfrac{1}{3} = 1\dfrac{2}{3}$
(43)

13. $\$10 - w = \$0.10$
(43)

14. At a 6% sales-tax rate, what is the tax on an $8.59 purchase?
(41, 51) Round the answer to the nearest cent.

15. The diameter of a tire on the car was 24 inches. Find the
(47, 51) circumference of the tire to the nearest inch. (Use 3.14 for π.)

16. Arrange these numbers in order from least to greatest:
(44)

$$1.02, \ 1.2, \ 0.21, \ 0.201$$

17. What is the missing number in this sequence?
(10)

$$1, \ 2, \ 4, \ 7, \ 11, \ \underline{\hspace{1cm}}, \ 22, \ \ldots$$

18. The perimeter of a square room is 80 feet. How many
(38) floor tiles 1 foot square would be needed to cover the area
of the room?

19. One foot is 12 inches. What fraction of a foot is 3 inches?
(22) Draw a diagram to illustrate the problem.

20. How many cents is $\dfrac{2}{5}$ of a dollar? Draw a diagram to
(22) illustrate the problem.

21. The diameter of a penny is $\dfrac{3}{4}$ inch. How many pennies are
(50) needed to form a row 12 inches long? (To answer this
question, write and solve a fraction division problem
using the method shown in Lesson 50.)

22. What is the least common multiple of 2, 4, and 6?
(30)

23. (a) $\dfrac{4}{4} - \dfrac{2}{2}$ (b) $\sqrt{4} - 2^2$
(29, 38)

24. A meter is about one big step. About how many meters
(7) above the floor is the top of a window in your home?

25. To what decimal number is the arrow pointing on the
(50) number line below?

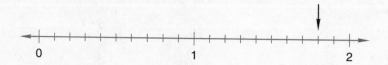

26. $\dfrac{2 \cdot 2 \cdot 2 \cdot 3 \cdot 3}{2 \cdot 3 \cdot 2}$
(5)

27. Rename $\frac{1}{2}$ and $\frac{2}{3}$ as fractions with denominators of 6. Then
(42) add the renamed fractions. Write the sum as a mixed
number.

28. Seventy-eight percent of the earth's atmosphere is nitrogen.
(33, 35) Write 78% as an unreduced fraction. Then write the
fraction as a decimal number.

29. Round 78% to the nearest ten percent. Then write the
(16, 33) answer as a reduced fraction. Approximately what fraction
of the earth's atmosphere is nitrogen?

30. Draw a square with a perimeter of 4 inches. Then shade
(8) 50% of the square.

LESSON
53 Decimals Chart •
Simplifying Fractions

WARM-UP

Facts Practice: 64 Multiplication Facts (Test D)

Mental Math: Count by 9's from 9 to 108.

a. 8×225 b. $256 + 34$ c. $250 - 99$

d. $\$25.00 - \$12.50$ e. Double $2\frac{1}{2}$.   f. $\frac{800}{20}$

g. $10 \times 10, -20, +1, \div 9, \times 5, -1, \div 4$

Problem Solving:

Here is part of a multiplication table. What number is missing?

36	42	48
42	?	56
48	56	64

NEW CONCEPTS

Decimals chart

For many lessons we have been developing our decimal arithmetic skills. We find that arithmetic with decimal numbers is similar to arithmetic with whole numbers. However, in decimal number arithmetic, we need to keep track of the decimal point. The chart below summarizes the rules for arithmetic with decimal numbers by providing keywords to help you keep track of the decimal point.

Across the top of the chart are the four operation signs (+, −, ×, ÷). Below each sign is the rule or memory cue to follow when performing that operation. (There are two kinds of division problems, each with a different cue.)

Decimal Arithmetic Reminders

Operation	+ or −	×	÷ by whole (W)	÷ by decimal (D)
Memory cue	line up $\pm\ \cdot$	×; then count $\times\ \cdot$	up $W)\overline{\cdot}$	over, over, up $D)\overline{\cdot}$

You may need to …
• Place a decimal point to the right of a whole number.
• Fill empty places with zeros.

The bottom of the chart contains two reminders that apply to all of the operations.

Simplifying fractions   We simplify fractions in two ways. We reduce fractions to lowest terms, and we convert improper fractions to mixed numbers. Sometimes a fraction can be reduced and converted to a mixed number.

Example   Simplify: $\frac{4}{6} + \frac{5}{6}$

Solution   By adding the fractions $\frac{4}{6}$ and $\frac{5}{6}$, we get the improper fraction $\frac{9}{6}$. We can simplify this fraction. We may reduce first and then convert the fraction to a mixed number, or we may convert first and then reduce. We show both methods below.

$$\begin{array}{r} \frac{4}{6} \\ + \frac{5}{6} \\ \hline \frac{9}{6} \end{array}$$

REDUCE FIRST

1. Reduce: $\frac{9}{6} = \frac{3}{2}$

2. Convert: $\frac{3}{2} = 1\frac{1}{2}$

CONVERT FIRST

1. Convert: $\frac{9}{6} = 1\frac{3}{6}$

2. Reduce: $1\frac{3}{6} = 1\frac{1}{2}$

LESSON PRACTICE

Practice set   **a.** Discuss how the rules in the decimals chart apply to each of these problems:

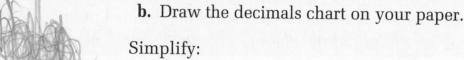

$$5 - 4.2 \qquad 0.4 \times 0.2 \qquad 0.12 \div 3 \qquad 5 \div 0.4$$

b. Draw the decimals chart on your paper.

Simplify:

c. $\frac{10}{12} + \frac{5}{12}$ **d.** $\frac{9}{10} + \frac{6}{10}$ **e.** $\frac{8}{12} + \frac{7}{12}$

MIXED PRACTICE

Problem set   **1.** The decimals chart in this lesson shows that we line up
(53)   the decimal points when we add or subtract decimal numbers. Why do we do that?

2. The turkey must cook for 4 hours 45 minutes. At what time
(32)   must it be put into the oven in order to be done by 3:00 p.m.?

3. Billy won the contest by eating $\frac{1}{4}$ of a berry pie in 7 seconds.
(50)   At this rate, how long would it take Billy to eat a whole berry pie?

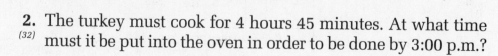

4. In four games the basketball team scored 47, 52, 63, and
(Inv. 5) 66 points. What is the mean of these scores? What is the range of these scores?

Find each missing number:

5. 0.375x = 37.5
(43)

6. $\frac{m}{10}$ = 1.25
(52)

7. Write 1% as a fraction. Then write the fraction as a decimal
(33, 35) number.

8. 3.6 + 4 + 0.39
(38)

9. $\frac{36}{0.12}$
(49)

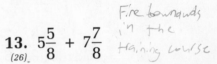

Fire townards
in the
training course

10. $\frac{0.15}{4}$
(45)

11. $6\frac{1}{4} - 3\frac{3}{4}$
(48)

12. $\frac{2}{3} \times \frac{3}{5}$
(29)

13. $5\frac{5}{8} + 7\frac{7}{8}$
(26)

14. Which digit in 3456 has the same place value as the 2
(34) in 28.7?

15. The items ~~Fred~~ ordered for lunch totaled $5.20. The
(41, 51) sales-tax rate was 8%.

(a) Find the sales tax to the nearest cent.

(b) Find the total price for the lunch including sales tax.

16. Which number is closest to 1?
(50)

A. 1.2   B. 0.9   C. 0.1   D. $\frac{1}{2}$

17. The entire group held hands and formed a big circle. If
(47) the circle was 40 feet across the center, then it was how many feet around? Round the answer to the nearest foot. (Use 3.14 for π.)

18. What is the perimeter of this square?
(8)

$\frac{3}{8}$ in.

19. A yard is 36 inches. What fraction of a yard is 3 inches?
(29)

20. (a) List the factors of 11.
₍₁₉₎

(b) What is the name for a whole number that has exactly two factors?

21. Four squared is how much greater than the square root of 4?
_(13, 38)

22. What is the smallest number that is a multiple of both
₍₃₀₎ 6 and 9?

23. The product of $\frac{2}{3}$ and $\frac{3}{2}$ is 1.
₍₃₀₎

$$\frac{2}{3} \cdot \frac{3}{2} = 1$$

Use these numbers to form another multiplication fact and two division facts.

24. $\dfrac{2 \cdot 3 \cdot 2 \cdot 5 \cdot 2 \cdot 5}{2 \cdot 5 \cdot 2 \cdot 5}$ **25.** $\dfrac{5}{6} = \dfrac{?}{24}$
₍₅₎   ₍₄₂₎

26. Copy this rectangle on your paper,
₍₂₉₎ and shade two thirds of it.

27. Thirty percent of the 350 college students enrolled in
_(22, 33) summer school. Find the number of students who enrolled in summer school. Draw a diagram to illustrate the problem.

28. Rename $\frac{1}{4}$ and $\frac{1}{6}$ as fractions with denominators of 12. Then
₍₄₂₎ add the renamed fractions.

29. The number that corresponds to point A is how much less
₍₅₀₎ than the number that corresponds to point B?

30. The bookstore's stock of Huckleberry Finn books fills a
₍₅₀₎ shelf that is 24 inches long. Each book is $\frac{3}{4}$ inch thick. How many books are in the bookstore's stock? (To answer this question, write and solve a fraction division problem using the method shown in Lesson 50.)

LESSON

54 Reducing by Grouping Factors Equal to 1 • Dividing Fractions

WARM-UP

Facts Practice: 30 Fractions to Reduce (Test G)

Mental Math: Count up and down by $\frac{1}{8}$'s between $\frac{1}{8}$ and 3.

a. 6 × 250   b. 736 − 400   c. 375 + 99

d. $8.75 + $5.00   e. $\frac{1}{2}$ of 9   f. 30 × 30

g. 8 × 8, − 1, ÷ 9, × 7, + 1, ÷ 5, × 10

Problem Solving:

Ned walked from his home (*H*) to Natalie's house (*N*) following the path from *H* to *I* to *J* to *K* to *L* to *M* to *N*. Then he walked home from *N* to *C* to *H*. Compare the distance of Ned's walk to Natalie's house to the distance of his walk home.

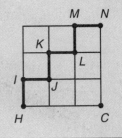

NEW CONCEPTS

Reducing by grouping factors equal to 1

The factors in the problem below are arranged in order from least to greatest. Notice that some factors appear in both the dividend and the divisor.

$$\frac{2 \cdot 2 \cdot 3 \cdot 5}{2 \cdot 2 \cdot 3}$$

Since 2 ÷ 2 equals 1 and 3 ÷ 3 equals 1, we will mark the combinations of factors equal to 1 in this problem.

$$\frac{2 \cdot 2 \cdot 3 \cdot 5}{2 \cdot 2 \cdot 3}$$

Looking at the factors this way, the problem becomes 1 · 1 · 1 · 5, which is 5.

Example 1   Reduce this fraction: $\dfrac{2 \cdot 2 \cdot 2 \cdot 5}{2 \cdot 2 \cdot 3 \cdot 5}$

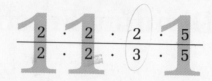

Solution   We will mark combinations of factors equal to 1.

$$\frac{\overset{1}{2} \cdot \overset{1}{2} \cdot 2 \cdot \overset{1}{5}}{\underset{1}{2} \cdot \underset{1}{2} \cdot 3 \cdot \underset{1}{5}}$$

By grouping factors equal to 1, the problem becomes $1 \cdot 1 \cdot 1 \cdot \frac{2}{3}$, which is $\frac{2}{3}$.

Dividing fractions   When we divide 10 by 5, we are answering the question "How many 5's are in 10?" When we divide $\frac{3}{4}$ by $\frac{1}{2}$, we are answering the same type of question. In this case the question is "How many $\frac{1}{2}$'s are in $\frac{3}{4}$?" While it is easy to see how many 5's are in 10, it is not as easy to see how many $\frac{1}{2}$'s are in $\frac{3}{4}$. We remember from Lesson 50 that when the divisor is a fraction, we take two steps to find the answer. We first find how many of the divisors are in 1. This is the reciprocal of the divisor. Then we use the reciprocal to answer the original division problem by multiplying.

Example 2   How many $\frac{1}{2}$'s are in $\frac{3}{4}$? $\left(\frac{3}{4} \div \frac{1}{2} \right)$

Solution   Before we show the two-step process, we will solve the problem with our fraction manipulatives. The question can be stated this way:

How many $\boxed{\frac{1}{2}}$'s are needed to make $\boxed{\frac{1}{4}\frac{1}{4}\frac{1}{4}}$?

We see that the answer is more than one but less than two. If we take one $\boxed{\frac{1}{2}}$ and cut another $\boxed{\frac{1}{2}}$ into two equal parts ($\boxed{\frac{1}{2}}$), then we can fit the first $\boxed{\frac{1}{2}}$ and one of the smaller parts ($\triangledown$) together to make three fourths.

We see that we need $1\frac{1}{2}$ of the $\boxed{\frac{1}{2}}$ pieces to make $\boxed{\frac{1}{4}\frac{1}{4}\frac{1}{4}}$.

Now we will use arithmetic to show that $\frac{3}{4} \div \frac{1}{2}$ equals $1\frac{1}{2}$. The original problem asks, "How many $\frac{1}{2}$'s are in $\frac{3}{4}$?"

$$\frac{3}{4} \div \frac{1}{2}$$

The first step is to find the number of $\frac{1}{2}$'s in 1.

$$1 \div \frac{1}{2} = 2$$

The number of $\frac{1}{2}$'s in 1 is 2, which is the reciprocal of $\frac{1}{2}$. So the number of $\frac{1}{2}$'s in $\frac{3}{4}$ should be $\frac{3}{4}$ of 2. We find $\frac{3}{4}$ of 2 by multiplying.

$$\frac{3}{4} \times 2 = \frac{6}{4}, \text{ which equals } 1\frac{1}{2}$$

We simplified $\frac{6}{4}$ by reducing $\frac{6}{4}$ to $\frac{3}{2}$ and then converting $\frac{3}{2}$ to $1\frac{1}{2}$. We will review the steps we took to solve the problem.

Original problem:

How many $\frac{1}{2}$'s are in $\frac{3}{4}$? $\frac{3}{4} \div \frac{1}{2}$

Step 1: Find the number of $\frac{1}{2}$'s in 1. $1 \div \frac{1}{2} = 2$

Step 2: Use the number of $\frac{1}{2}$'s in 1 to find the number of $\frac{1}{2}$'s in $\frac{3}{4}$. Then simplify the answer.

$$\frac{3}{4} \times 2 = \frac{6}{4} = 1\frac{1}{2}$$

Example 3 How many $\frac{3}{4}$'s are in $\frac{1}{2}$? $\left(\frac{1}{2} \div \frac{3}{4}\right)$

Solution Using our fraction manipulatives, the question can be stated this way:

What fraction of 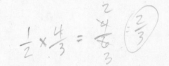 is needed to make $\boxed{\frac{1}{2}}$?

The answer is less than 1. We need to cut off part of 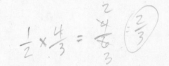 to make $\boxed{\frac{1}{2}}$. If we cut off one of the three parts of three fourths (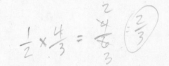), we see that two of the three parts equal $\boxed{\frac{1}{2}}$. So $\frac{2}{3}$ of $\frac{3}{4}$ is needed to make $\frac{1}{2}$.

Now we will show the arithmetic. The original problem asks, "How many $\frac{3}{4}$'s are in $\frac{1}{2}$?"

$$\frac{1}{2} \div \frac{3}{4}$$

First we find the number of $\frac{3}{4}$'s in 1. The number is the reciprocal of $\frac{3}{4}$.

King Nang (王师师) (twinkie)

$$1 \div \frac{3}{4} = \frac{4}{3}$$

The number of $\frac{3}{4}$'s in 1 is $\frac{4}{3}$. So the number of $\frac{3}{4}$'s in $\frac{1}{2}$ is $\frac{1}{2}$ of $\frac{4}{3}$. We find $\frac{1}{2}$ of $\frac{4}{3}$ by multiplying.

$$\frac{1}{2} \times \frac{4}{3} = \frac{4}{6}, \text{ which equals } \frac{2}{3}$$

Again, we will review the steps we took to solve the problem.

Original problem:

How many $\frac{3}{4}$'s are in $\frac{1}{2}$? $\frac{1}{2} \div \frac{3}{4}$

Step 1: Find the number of $\frac{3}{4}$'s in 1. $1 \div \frac{3}{4} = \frac{4}{3}$

Step 2: Use the number of $\frac{3}{4}$'s in 1 to find the number of $\frac{3}{4}$'s in $\frac{1}{2}$. Then simplify the answer. $\frac{1}{2} \times \frac{4}{3} = \frac{4}{6} = \frac{2}{3}$

LESSON PRACTICE

Practice set   Reduce:

a. $\dfrac{2 \cdot 2 \cdot 3 \cdot 5}{2 \cdot 2 \cdot 5}$

b. $\dfrac{2 \cdot 2 \cdot 3 \cdot 3 \cdot 5}{2 \cdot 2 \cdot 3 \cdot 5 \cdot 5}$

c. How many $\frac{3}{8}$'s are in $\frac{1}{2}$? $\left(\frac{1}{2} \div \frac{3}{8} \right)$

d. How many $\frac{1}{2}$'s are in $\frac{3}{8}$? $\left(\frac{3}{8} \div \frac{1}{2} \right)$

MIXED PRACTICE

Problem set   **1.** Draw the decimals chart from Lesson 53.
(53)

2. If 0.4 is the dividend and 4 is the divisor, what is the
(45) quotient?

3. In 1900 the U.S. population was 76,212,168. In 1950 the
(16) population was 151,325,798. Estimate the increase in
population between 1900 and 1950 to the nearest million.

4. Marjani was $59\frac{3}{4}$ inches tall when she turned 11 and $61\frac{1}{4}$
(48) inches tall when she turned 12. How many inches did
Marjani grow during the year? (Write an equation and solve
the problem.)

5. $1000 - (100 - 1)$ **6.** $\dfrac{1000}{24}$
(5)   (25)

7. What number is halfway between 37 and 143?
(18)

Find each missing number:

8. $\$3 - n = 24¢$ **9.** $m + 3\frac{4}{5} = 6\frac{2}{5}$
(3)   (43)

10. $4.2 \div 10^2$ **11.** $(1.2 \div 0.12)(1.2)$
(52)   (53)

12. $\left(\dfrac{4}{3}\right)^2$ **13.** $\sqrt{9} + \sqrt{16}$
(29, 38)   (38)

14. Which digit is in the hundred-thousands place in
(12) 123,456,789?

15. The television cost $289.90. The sales-tax rate was 8%. How
(41, 51) much was the sales tax on the television?

16. Use rulers to compare:
(9)

two centimeters $\bigcirc$ one inch

17. Isadora found the sixth term of this sequence by doubling
(10) six and then subtracting one. She found the seventh term
by doubling seven and subtracting one. What is the twelfth
term of this sequence?

$$1, 3, 5, 7, 9, \ldots$$

18. How many square feet of tile would be needed to cover
(31) the area of a room 14 feet long and 12 feet wide?

19. Nine of the 30 competitors passed the qualifying round.
₍₂₃₎

    (a) What fraction of the competitors passed?

    (b) What was the ratio of competitors who passed to competitors who did not pass?

20. (a) $\dfrac{5}{6} = \dfrac{?}{24}$ (b) $\dfrac{5}{8} = \dfrac{?}{24}$
₍₄₂₎

21. What is the least common multiple of 3, 4, and 6?
₍₃₀₎

22. How many $\frac{1}{2}$'s are in $\frac{2}{3}$? $\left(\frac{2}{3} \div \frac{1}{2}\right)$
₍₅₄₎

23. Eighty percent of the 30 questions were correct. Write 80%
_(22, 33) as a reduced fraction. Then find the number of questions that were correct. Draw a diagram to illustrate the problem.

24. One inch equals 2.54 centimeters. A line 100 inches long is
₍₄₆₎ how many centimeters long?

Reduce:

25. $\dfrac{2 \cdot 3 \cdot 5 \cdot 3 \cdot 2}{2 \cdot 3 \cdot 2 \cdot 5}$ **26.** $\dfrac{2 \cdot 3 \cdot 3 \cdot 5}{2 \cdot 2 \cdot 2 \cdot 3 \cdot 5}$
₍₅₄₎   ₍₅₄₎

27. Rename $\frac{2}{3}$ and $\frac{1}{2}$ as fractions with denominators of 6. Then
₍₄₂₎ add the renamed fractions, and convert the answer to a mixed number.

28. The diameter of the steering wheel in Diane's car is 15
₍₄₇₎ inches. How far does her hand move in one full turn of the steering wheel if she does not let go of the wheel? Round the answer to the nearest inch. (Use 3.14 for π.) Describe how to tell whether your answer is reasonable.

29. Draw a rectangle that is $1\frac{1}{2}$ inches long and 1 inch wide.
_(8, 31)

    (a) What is the perimeter of the rectangle?

    (b) What is the area of the rectangle?

30. Instead of dividing $2\frac{1}{2}$ by $\frac{1}{2}$, Sandra formed an equivalent
₍₄₃₎ division problem with whole numbers by doubling the dividend and the divisor. What equivalent problem did she form, and what is the quotient?

LESSON
55 | Common Denominators, Part 1

WARM-UP

Facts Practice: 28 Improper Fractions to Simplify (Test I)

Mental Math: Count by 7's from 7 to 84.

a. 8 × 325   **b.** 329 + 50   **c.** 375 − 99
d. $12.50 − $5.00   **e.** Double $3\frac{1}{2}$.   **f.** $\frac{600}{20}$
g. 8 × 5, + 2, ÷ 6, × 7, + 7, ÷ 8, × 4, ÷ 7

Problem Solving:

Half of a gallon is a half gallon. Half of a half gallon is a quart. Half of a quart is a pint. Half of a pint is a cup. How many cups of water equals a gallon of water?

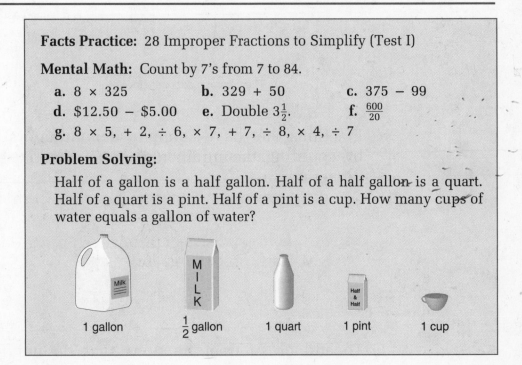

1 gallon $\frac{1}{2}$ gallon   1 quart   1 pint   1 cup

NEW CONCEPT

When the denominators of two or more fractions are equal, we say that the fractions have **common denominators**. The fractions $\frac{3}{5}$ and $\frac{2}{5}$ have common denominators.

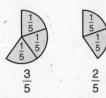

$\frac{3}{5}$ $\frac{2}{5}$

The common denominator is 5.

The fractions $\frac{3}{4}$ and $\frac{1}{2}$ do not have common denominators because the denominators 4 and 2 are not equal.

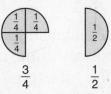

$\frac{3}{4}$ $\frac{1}{2}$

These fractions do not have common denominators.

Fractions that do not have common denominators can be renamed to form fractions that do have common denominators. Since $\frac{2}{4}$ equals $\frac{1}{2}$, we can rename $\frac{1}{2}$ as $\frac{2}{4}$. The fractions $\frac{3}{4}$ and $\frac{2}{4}$ have common denominators.

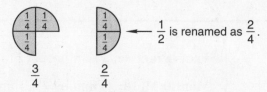

The common denominator is 4.

Fractions that have common denominators can be added by counting the number of parts; that is, by adding the numerators.

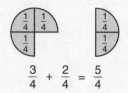

$$\frac{3}{4} \ + \ \frac{2}{4} \ = \ \frac{5}{4}$$

To add or subtract fractions that do not have common denominators, we rename one or more of them to form fractions that do have common denominators. Then we add or subtract. Recall that to rename a fraction, we multiply it by a fraction equal to 1. Here we rename $\frac{1}{2}$ by multiplying it by $\frac{2}{2}$. This forms the equivalent fraction $\frac{2}{4}$, which can be added to $\frac{3}{4}$.

Rename $\frac{1}{2}$.

$$\frac{1}{2} \times \frac{2}{2} = \frac{2}{4}$$

$$+ \ \frac{3}{4} \quad\ \ = \frac{3}{4}$$

Then add.

$$\frac{5}{4} = 1\frac{1}{4}$$

Simplify your answer, if possible.

To find the common denominator of two fractions, we find a common multiple of the denominators. The least common multiple of the denominators is the **least common denominator** of the fractions.

Example   Subtract: $\frac{1}{2} - \frac{1}{6}$

Solution   The denominators are 2 and 6. The least common multiple of 2 and 6 is 6. So 6 is the least common denominator of the two fractions. We do not need to rename $\frac{1}{6}$. We change halves to sixths by multiplying by $\frac{3}{3}$ and then we subtract.

Rename $\frac{1}{2}$.

$$\frac{1}{2} \times \frac{3}{3} = \frac{3}{6}$$
$$-\frac{1}{6} \quad\;\; = \frac{1}{6}$$

Subtract $\frac{1}{6}$ from $\frac{3}{6}$.

$$\frac{2}{6} = \frac{1}{3}$$

Reduce.

LESSON PRACTICE

Practice set   Find each sum or difference:

a. $\frac{1}{2}$ $+ \frac{3}{8}$ b. $\frac{3}{8}$ $+ \frac{1}{4}$ c. $\frac{3}{4}$ $+ \frac{1}{8}$

d. $\frac{1}{2}$ $- \frac{1}{4}$ e. $\frac{5}{8}$ $- \frac{1}{4}$ f. $\frac{3}{4}$ $- \frac{3}{8}$

MIXED PRACTICE

Problem set

1. *(53)* In the decimals chart, the memory cue for dividing by a whole number is "up." What does that mean?

2. *(50)* How many $\frac{3}{8}$-inch-thick CD cases will fit on a 12-inch-long shelf? (To answer the question, write and solve a fraction division problem using the method shown in Lesson 50.)

3. *(15)* The average pumpkin weighs 6 pounds. The prize-winning pumpkin weighs 324 pounds. The prize-winning pumpkin weighs as much as how many average pumpkins?

4. *(55)* $\frac{1}{8} + \frac{1}{2}$ 5. *(55)* $\frac{1}{2} - \frac{1}{8}$ 6. *(55)* $\frac{2}{3} - \frac{1}{6}$

7. 6.28 + 4 + 0.13
(38)

8. 81 ÷ 0.9
(49)

9. 0.2 ÷ 10
(52)

10. (0.17)(100)
(46)

11. $\frac{3}{4} + 3\frac{1}{4}$
(26)

12. $\frac{5}{6} \cdot \frac{2}{3}$
(29)

13. $\frac{5}{8} = \frac{?}{24}$
(42)

14. Write the following in standard notation:
(32)

$$(6 \times 10{,}000) + (4 \times 100) + (2 \times 10)$$

15. Multiply 0.14 by 0.8 and round the product to the nearest
(39, 51) hundredth.

16. Compare: $\frac{2}{3} \bigcirc \frac{2}{3} \times \frac{2}{2}$
(29)

17. If the diameter of a truck tire is 5 feet, then what is the
(47) distance around the tire? (Use 3.14 for π.)

18. A 20-foot rope was used to make a square. How many
(38) square feet of area were enclosed by the rope?

19. What fraction of a dollar is six dimes?
(29)

20. What is the least common multiple (LCM) of 3 and 4?
(30)

21. (a) List the factors of 23.
(19)

(b) What is the name for a whole number that has exactly
two factors?

22. By what fraction should $\frac{2}{5}$ be multiplied to form the
(30) product 1?

23. Compare: $3^2 + 4^2 \bigcirc 5^2$
(9, 38)

24. How many $\frac{2}{5}$'s are in $\frac{1}{2}$? $\left(\frac{1}{2} \div \frac{2}{5}\right)$
(54)

25. How many 12's are in 1212?
(15)

26. The window was 48 inches wide and 36 inches tall. What
(23) was the ratio of the height to the width of the window?

27. What fraction of this group of circles
(29) is shaded?

28. Reduce: $\dfrac{2 \cdot 3 \cdot 2 \cdot 5 \cdot 3 \cdot 7}{2 \cdot 2 \cdot 3 \cdot 5 \cdot 5 \cdot 5}$
(54)

29. The performance began at 7:45 p.m. and concluded at
(32) 10:25 p.m. How long was the performance in hours and
minutes?

30. Triangle *ABC* has three acute angles.
(28)
(a) Which triangle has one right angle?

(b) Which triangle has one obtuse angle?

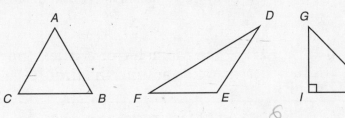

LESSON

56   Common Denominators, Part 2

WARM-UP

Facts Practice: 30 Fractions to Reduce (Test G)

Mental Math: Count up and down by 25's between 25 and 300.

    a. 8 × 250   **b.** 462 − 350   **c.** 150 + 49

    d. $3.75 + $4.50   **e.** $\frac{1}{2}$ of 15   **f.** 30 × 40

    g. 10 × 8, + 1, ÷ 9, × 3, + 1, ÷ 4, × 6, − 2, ÷ 5

Problem Solving:

    Copy this problem and fill in the missing digits:

$$\begin{array}{r} \_56 \\ + \underline{\_\_} \\ \underline{\_000} \end{array}$$

NEW CONCEPT

In Lesson 55 we added and subtracted fractions that did not have common denominators. We renamed one of the fractions in order to add or subtract. In this lesson we will rename both fractions before we add or subtract. To see why this is sometimes necessary, consider the problem $\frac{1}{2} + \frac{1}{3}$. We cannot add the fractions by simply counting the number of parts, because the parts are not the same size (that is, the denominators are different).

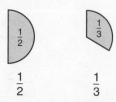

These fractions do not have common denominators.

Renaming $\frac{1}{2}$ as $\frac{2}{4}$ does not help us either. In the fractions $\frac{2}{4}$ and $\frac{1}{3}$, the parts are still of different sizes.

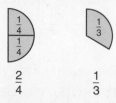

These fractions do not have common denominators.

We must rename both fractions in order to get parts that are the same size. The least common multiple of the denominators can be used as the common denominator of the renamed fractions. The least common multiple of 2 and 3 (the denominators of $\frac{1}{2}$ and $\frac{1}{3}$) is 6.

$$\frac{1}{2} = \frac{3}{6} \qquad \frac{1}{3} = \frac{2}{6}$$

The common denominator is 6.

Example 1   Add: $\frac{1}{2} + \frac{1}{3}$

Solution   The denominators are 2 and 3. The least common multiple of 2 and 3 is 6. We rename each fraction so that 6 is the common denominator. Then we add.

Rename $\frac{1}{2}$ and $\frac{1}{3}$.

$$\frac{1}{2} \times \frac{3}{3} = \frac{3}{6}$$
$$+ \ \frac{1}{3} \times \frac{2}{2} = \frac{2}{6}$$

Then add.

$$\frac{5}{6}$$

Example 2   Subtract: $\frac{3}{4}$

$-\ \frac{2}{3}$

Solution   The least common multiple of 4 and 3 is 12. We rename both fractions so that their denominators are 12 and then subtract.

Rename $\frac{3}{4}$ and $\frac{2}{3}$.

$$\frac{3}{4} \times \frac{3}{3} = \frac{9}{12}$$
$$-\ \frac{2}{3} \times \frac{4}{4} = \frac{8}{12}$$

Then subtract.

$$\frac{1}{12}$$

Renaming one or more fractions can also help us compare fractions. To compare fractions that have common denominators, we simply compare the numerators.

$$\frac{4}{6} < \frac{5}{6}$$

To compare fractions that do not have common denominators, we can rename one or both fractions so that they do have common denominators.

Green

Example 3　Compare: $\frac{3}{8} \bigcirc \frac{1}{2}$

Solution　We rename $\frac{1}{2}$ so that the denominator is 8.

$$\frac{1}{2} \cdot \frac{4}{4} = \frac{4}{8}$$

We see that $\frac{3}{8}$ is less than $\frac{4}{8}$.

$$\frac{3}{8} < \frac{4}{8}$$

Therefore, $\frac{3}{8}$ is less than $\frac{1}{2}$.

$$\frac{\mathbf{3}}{\mathbf{8}} < \frac{\mathbf{1}}{\mathbf{2}}$$

Example 4　Compare: $\frac{2}{3} \bigcirc \frac{3}{4}$

Solution　The denominators are 3 and 4. We rename both fractions with a common denominator of 12.

$$\frac{2}{3} \cdot \frac{4}{4} = \frac{8}{12} \qquad \frac{3}{4} \cdot \frac{3}{3} = \frac{9}{12}$$

We see that $\frac{8}{12}$ is less than $\frac{9}{12}$.

$$\frac{8}{12} < \frac{9}{12}$$

Therefore, $\frac{2}{3}$ is less than $\frac{3}{4}$.

$$\frac{\mathbf{2}}{\mathbf{3}} < \frac{\mathbf{3}}{\mathbf{4}}$$

LESSON PRACTICE

Practice set　Find each sum or difference:

a.　$\frac{2}{3}$
　　$-\frac{1}{2}$

b.　$\frac{1}{4}$
　　$+\frac{2}{5}$

c.　$\frac{3}{4}$
　　$-\frac{1}{3}$

d.　$\frac{2}{3}$
　　$+\frac{1}{4}$

e.　$\frac{1}{3}$
　　$-\frac{1}{4}$

Before comparing, write each pair of fractions with common denominators:

f. $\frac{2}{3} \bigcirc \frac{1}{2}$

g. $\frac{4}{6} \bigcirc \frac{3}{4}$

h. $\frac{2}{3} \bigcirc \frac{3}{5}$

MIXED PRACTICE

Problem set

1. Add $\frac{1}{4}$ and $\frac{1}{3}$. Use 12 as the common denominator.
(56)

2. Subtract $\frac{1}{3}$ from $\frac{1}{2}$. Use 6 as the common denominator.
(56)

3. Of the 88 keys on a piano, 52 are white.
(29)

   (a) What fraction of a piano's keys are white?

   (b) What is the ratio of black keys to white keys?

4. If $4\frac{1}{2}$ apples are needed to make an apple pie, how many
(29) apples would be needed to make two apple pies?

5. Subtract $\frac{1}{4}$ from $\frac{2}{3}$. Use 12 as the common denominator.
(56)

6. Add $\frac{1}{3}$ and $\frac{1}{6}$. Reduce your answer.
(55)

7. Subtract $\frac{1}{2}$ from $\frac{5}{6}$. Reduce your answer.
(55)

8. $3 + $1.75 + 65¢   9. $(0.625)(0.4)$
(1)   (39)

10. $6^2 \div 0.08$ 11. $3\frac{1}{8} - 1\frac{7}{8}$ 12. $\frac{5}{8} \cdot \frac{2}{3}$
(49)   (48)   (29)

13. Harpo answered 40% of the 100 questions correctly.
(22, 33) Write 40% as a reduced fraction. How many questions
did Harpo answer correctly?

14. Write the following as a decimal number:
(46)

$$(8 \times 10) + \left(6 \times \frac{1}{10}\right) + \left(5 \times \frac{1}{100}\right)$$

15. Estimate the sum of 3627 and 4187 to the nearest hundred.
(16)

16. Molly measured the diameter of her bike tire and found
(47) that it was 2 feet across. She estimated that for every turn
of the tire, the bike traveled about 6 feet. Was Molly's
estimate reasonable? Why or why not?

17. What is the mean of 1.2, 1.3, 1.4, and 1.5?
(Inv. 5)

18. The perimeter of a square is 36 inches. What is its area?
(38)

19. Here we show 24 written as a product of prime numbers:
₍₁₉₎

$$2 \cdot 2 \cdot 2 \cdot 3$$

Show how 30 can be written as a product of prime numbers.

Find each missing number:

20. $\frac{2}{3}w = 0$ **21.** $\frac{2}{3}m = 1$ **22.** $\frac{2}{3} - n = 0$
₍₄₃₎ ₍₃₀₎ ₍₄₃₎

Refer to this bar graph to answer problems 23–26.

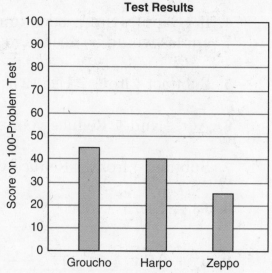

23. How many problems did Harpo miss on the test?
_(Inv. 5)

24. How many more problems did Groucho have correct on the test than Zeppo?
_(Inv. 5)

25. What fraction of the problems on the test did Zeppo answer correctly?
₍₂₉₎

26. Write a percent question that relates to the bar graph, and then answer the question.
₍₄₀₎

27. Reduce: $\dfrac{2 \cdot 3 \cdot 5}{2 \cdot 3 \cdot 5 \cdot 7}$
₍₅₄₎

28. How many $\frac{2}{3}$'s are in $\frac{1}{2}$? $\left(\frac{1}{2} \div \frac{2}{3}\right)$
₍₅₄₎

29. Draw three rectangles that are 2 cm long and 1 cm wide. On each rectangle show a different way to divide the rectangle in half. Then shade half of each rectangle.
₍₇₎

30. Compare: $\frac{2}{3} \bigcirc \frac{5}{6}$
₍₅₆₎

LESSON
57   Adding and Subtracting Fractions: Three Steps

WARM-UP

Facts Practice: 72 Multiplication and Division Facts (Test H)

Mental Math: Count by 12's from 12 to 144.

a. 8×425 b. $465 + 250$ c. $150 - 49$

d. $\$9.75 - \$3.50$ e. Double $4\frac{1}{2}$.   f. $\frac{600}{30}$

g. $2 \times 2, \times 2, \times 2, \times 2, \times 2, \div 8, \div 8$

Problem Solving:

Jermaine was thinking of two different positive odd numbers whose average was 15. Each number contained two digits. Find two pairs of numbers of which Jermaine could have been thinking.

NEW CONCEPT

We follow three steps to solve fraction problems:

Step 1: Put the problem into the correct **shape** or form if it is not already. (When adding or subtracting fractions, the correct form is with common denominators.)

Step 2: Perform the **operation** indicated. (Add, subtract, multiply, or divide.)

Step 3: **Simplify** the answer if possible. (Reduce the fraction or write an improper fraction as a mixed number.)

Example 1   Add: $\frac{1}{2} + \frac{2}{3}$

Solution   We follow the steps described above.

Step 1: Shape: write the fractions with common denominators.

Step 2: Operate: add the renamed fractions.

Step 3: Simplify: convert the improper fraction to a mixed number.

1. Shape

$\frac{1}{2} \times \frac{3}{3} = \frac{3}{6}$

$+ \frac{2}{3} \times \frac{2}{2} = \frac{4}{6}$ 2. Operate

$\frac{7}{6} = 1\frac{1}{6}$

3. Simplify

Example 2   Subtract: $\frac{5}{6} - \frac{1}{3}$

Solution   **Step 1:** Shape: write the fractions with common denominators.
Step 2: Operate: subtract the renamed fractions.
Step 3: Simplify: reduce the fraction.

1. Shape

$\frac{5}{6} = \frac{5}{6}$

$- \frac{1}{3} \times \frac{2}{2} = \frac{2}{6}$ 2. Operate

$\frac{3}{6} = \frac{1}{2}$

3. Simplify

LESSON PRACTICE

Practice set*   Find each sum or difference:

a. $\frac{1}{2} + \frac{1}{6}$ b. $\frac{2}{3} + \frac{3}{4}$ c. $\frac{1}{5} + \frac{3}{10}$

d. $\frac{5}{6} - \frac{1}{2}$ e. $\frac{7}{10} - \frac{1}{2}$ f. $\frac{5}{12} - \frac{1}{6}$

MIXED PRACTICE

Problem set   **1.** What is the difference between the sum of $\frac{1}{2}$ and $\frac{1}{2}$ and the
(12, 29)   product of $\frac{1}{2}$ and $\frac{1}{2}$?

2. Thomas Jefferson was born in 1743. How old was he
(13)   when he was elected president of the United States in 1800? (Write an equation and solve the problem.)

3. Subtract $\frac{3}{4}$ from $\frac{5}{6}$. Use 12 as the common denominator.
(56)

4. $\frac{1}{2} + \frac{2}{3}$
(57)

5. $\frac{1}{2} + \frac{1}{6}$
(57)

6. $\frac{5}{6} + \frac{2}{3}$
(57)

7. How many $\frac{3}{5}$'s are in $\frac{3}{4}$? $\left(\frac{3}{4} \div \frac{3}{5}\right)$
(54)

8. $\$32.50 \div 10$
(52)

9. $\sqrt{4} - (1^2 - 0.2)$
(5, 38)

10. $6 \div 0.12$
(49)

11. 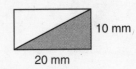 $5\frac{3}{8}$ $5\frac{5}{8}$ $\frac{4}{3}$
(48)   (29)

13. Fifty percent of this rectangle is
(31, 33) shaded. Write 50% as a reduced
fraction. What is the area of the
shaded part of the rectangle?

10 mm

20 mm

14. What is the place value of the 7 in 3.567?
(34)

15. Divide 0.5 by 4 and round the quotient to the nearest tenth.
(45, 51)

16. Arrange these numbers in order from least to greatest:
(44)

$$0.3, \ 3.0, \ 0.03$$

17. In this sequence the first term is 2, the second term is
(10) 4, and the third term is 6. What is the twentieth term of
the sequence?

$$2, \ 4, \ 6, \ 8, \ \ldots$$

18. In a deck of 52 cards there are four aces. What is the ratio
(23) of aces to all cards in a deck of cards?

19. (a) Calculate the perimeter of the
(31, 57)   rectangle shown.

$\frac{1}{4}$ in.

$\frac{1}{2}$ in.

(b) Multiply the length of the
rectangle by its width to find
the area of the rectangle in
square inches.

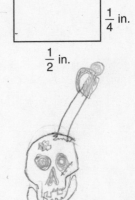

20. What number is $\frac{5}{8}$ of 80?
(29)

$\begin{array}{r} 10 \\ \times\ 20 \\ \hline 200 \end{array}$

21. List the factors of 29.
(19)

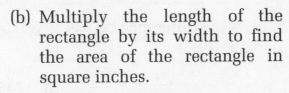

22. What is the least common multiple of 12 and 18?
(30)

23. Compare: $\frac{5}{8} \bigcirc \frac{7}{10}$
(56)

24. What temperature is shown on this
(10) thermometer?

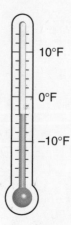

25. If the temperature shown on this
(14) thermometer rose to 12°F, then how many degrees would the temperature have risen?

26. Reduce: $\dfrac{2 \cdot 2 \cdot 3 \cdot 3 \cdot 5 \cdot 7}{2 \cdot 2 \cdot 5 \cdot 5 \cdot 7 \cdot 7}$
(54)

27. What fraction of the group of
(29) circles is shaded?

28. Ling has a 9-inch stack of CDs on the shelf. Each CD is
(50) in a $\frac{3}{8}$-inch-thick plastic case. How many CDs are in the 9-inch stack? (Write and solve a fraction division problem to answer the question.)

29. Subtract $\frac{1}{2}$ from $\frac{4}{5}$. Use 10 as the common denominator.
(55)

30. The diameter of a regulation basketball hoop is 18 inches.
(47) What is the circumference of a regulation basketball hoop? (Use 3.14 for π.)

LESSON
58   Probability and Chance

WARM-UP

Facts Practice: 28 Improper Fractions to Simplify (Test I)

Mental Math: Count up and down by $\frac{1}{8}$'s between $\frac{1}{8}$ and 2.

  a. 2 × 75   **b.** 315 − 150   **c.** 250 + 199
  d. $7.50 + $12.50   **e.** $\frac{1}{2}$ of 25   **f.** 20 × 50
  g. 10 × 10, − 1, ÷ 11, × 8, + 3, ÷ 3, ÷ 5

Problem Solving:

  What are the next three numbers in this sequence?

$$\frac{1}{6}, \frac{1}{3}, \frac{1}{2}, \underline{\quad}, \underline{\quad}, \underline{\quad}, \ldots$$

NEW CONCEPT

We live in a world full of uncertainties.

> *What is the chance of rain on Saturday?*
>
> *What are the odds of winning the big game?*
>
> *What is the probability that I will roll the number needed to land on the winning space?*

The study of **probability** helps us assign numbers to uncertain events and compare the likelihood that various events will occur.

Events that are certain to occur have a probability of one. Events that are certain not to occur have a probability of zero.

> *If I roll a number cube whose faces are numbered 1 through 6, the probability of rolling a 6 or less is one. The probability of rolling a number greater than 6 is zero.*

Events that are uncertain have probabilities that fall anywhere between zero and one. The closer to zero the probability is, the less likely the event is to occur; the closer to one the probability is, the more likely the event is to occur. We typically express probabilities as fractions or as decimals.

In this lesson we will practice assigning probabilities to specific events.

Imagine you are spinning the spinner below. The spinner could land in sector A, in sector B, or in sector C. Since sector B and sector C have the same area, landing in either one is equally likely. Since sector A has the largest area, we can expect the spinner to land in sector A more often than in either sector B or sector C.

The probability of a particular outcome is the fraction of spins we expect to result in that outcome if we spin the spinner many times. Since sector A takes up $\frac{1}{2}$ of the area, the probability that the spinner will land in A is $\frac{1}{2}$, or 0.5. In terms of area, sector B is half the size of sector A $\left(\frac{1}{2} \text{ of } \frac{1}{2}\right)$. Sector B takes up $\frac{1}{4}$ of the area of the spinner. Therefore, the probability that the spinner will land in sector B is $\frac{1}{4}$, or 0.25. This is also the probability that the spinner will land in sector C, since sectors B and C are equal in size. We know that the spinner is certain to land in one of the sectors, so the probabilities of the three outcomes must add up to 1.

$$\frac{1}{2} + \frac{1}{4} + \frac{1}{4} = 1 \qquad \text{or} \qquad 0.5 + 0.25 + 0.25 = 1$$

If we spin the spinner a large number of times, we would expect about $\frac{1}{2}$ of the spins to land in sector A, about $\frac{1}{4}$ of the spins to land in sector B, and about $\frac{1}{4}$ of the spins to land in sector C.

Example 1 Meredith spins the spinner shown above 28 times. About how many times can she expect the spinner to land in sector A? In sector B? In sector C?

Solution The spinner should land in sector A about $\frac{1}{2}$ of 28 times. Instead of multiplying by the fraction $\frac{1}{2}$, we can simply divide by 2.

$$28 \text{ times} \div 2 = 14 \text{ times}$$

The spinner should land in sector B about $\frac{1}{4}$ of 28 times. We divide the total number of spins by 4.

$$28 \text{ times} \div 4 = 7 \text{ times}$$

As we noted before, the probability that the spinner will land in sector B is equal to the probability that it will land in sector C. So Meredith can expect the spinner to land in sector A about **14 times,** in sector B about **7 times,** and in sector C about **7 times.**

It would be very unlikely for the spinner *never* to land in sector A in 28 spins. In 28 spins it also would be very unlikely for the spinner to *always* land in sector A. It would not be unusual, however, if the spinner were to land 12 times in sector A, 10 times in sector B, and 6 times in sector C. It is important to remember that probability indicates expectation; actual results may vary.

In the language of percent, we expect the spinner to land in sector A roughly 50% of the time, and we expect it to land in each of the other sectors roughly 25% of the time. When we express a probability as a percent, it is called a **chance.**

Example 2   In Miami the chance that it will rain on a typical day in September is 60%. Find the probability that it will rain on such a day. What is the probability that it will not rain? (Express each probability in both reduced fraction and decimal forms.)

Solution   To find the probability that it will rain, we convert the chance, 60%, to a fraction and a decimal.

$$60\% = \frac{60}{100} = \frac{6}{10} = \frac{3}{5}$$

$$60\% = \frac{60}{100} = 0.60 = \mathbf{0.6}$$

The probability of rain can be expressed as either $\frac{3}{5}$ or 0.6.

We are certain that it will either rain or not rain. The probabilities of these two possible outcomes must total 1. Subtracting the probability of rain from 1 gives us the probability that it will not rain.

$$1 - \frac{6}{10} = \frac{4}{10} = \frac{2}{5}$$

$$1 - 0.6 = \mathbf{0.4}$$

So the probability that it will not rain can be expressed as either $\frac{2}{5}$ or 0.4.

In some experiments or games, all the outcomes have the same probability. This is true if we flip a coin and observe the upturned side. The probability of the coin landing heads up is $\frac{1}{2}$, and the probability of the coin landing tails up is also $\frac{1}{2}$. Similarly, if we roll a number cube, each number has a probability of $\frac{1}{6}$ of appearing on the cube's upturned side.

An outcome or set of outcomes is called an **event.** To find the probability of an event, we simply add the probabilities of the outcomes that make up the event. If all outcomes of an experiment or game have the *same* probability, then the probability of an event is

$$\frac{\text{number of outcomes in the event}}{\text{number of possible outcomes}}$$

Example 3 A number cube is rolled. Find the probability that the upturned number is greater than 4.

Solution Since a number cube has six faces, there are six possible outcomes for one roll of a number cube. We are asked to find the probability that the upturned number is 5 or 6. So this event consists of two out of the six possible outcomes.

$$\frac{\text{number of outcomes in the event}}{\text{number of possible outcomes}} = \frac{2}{6} = \frac{\mathbf{1}}{\mathbf{3}}$$

Example 4 A bag contains 5 red marbles, 4 yellow marbles, 2 green marbles, and 1 orange marble. Without looking, Brendan draws a marble from the bag and notes its color. What is the probability that the marble drawn is red?

Solution There are 12 possible outcomes in Brendan's experiment (each marble represents one outcome). The event we are considering is drawing a red marble. Since 5 of the possible outcomes are red, we see that the probability that the drawn marble is red is

$$\frac{\text{number of outcomes in the event}}{\text{number of possible outcomes}} = \frac{\mathbf{5}}{\mathbf{12}}$$

Example 5 In the experiment in example 4, what is the probability that the marble Brendan draws is a primary color?

Solution Red and yellow are primary colors. Green and orange are not. Since 5 possible outcomes are red and 4 are yellow, the probability that the drawn marble is a primary color is

$$\frac{5 + 4}{12} = \frac{9}{12} = \frac{\mathbf{3}}{\mathbf{4}}$$

LESSON PRACTICE

Practice set Juan is waiting for the roller coaster at an amusement park. He has been told there is a 40% chance that he will have to wait more than 15 minutes.

  a. Find the probability that Juan's wait will be more than 15 minutes. Write the probability both as a decimal and as a reduced fraction.

  b. Find the probability that Juan's wait will not be more than 15 minutes. Express your answer as a decimal and as a fraction.

  A number cube is rolled. The faces of the cube are numbered 1 through 6.

  c. What is the probability that the number rolled will be odd?

  d. What is the probability that the number rolled will be less than 6?

  Refer to the spinner at right for problems **e–g.**

  e. What is the probability that the spinner will land in the red sector? In the black sector? In either of the white sectors? $\left(\text{Note that } \frac{1}{3} \text{ of } \frac{1}{2} \text{ is } \frac{1}{6}.\right)$

  f. What is the probability that the spinner will not land on white?

  g. If you spin this spinner 30 times, roughly how many times would you expect it to land in each sector?

  h. Roll a number cube 24 times, and make a frequency table for the 6 possible outcomes. Which outcomes occurred more than you expected?

MIXED PRACTICE

Problem set **1.** What is the difference between the sum of $\frac{1}{2}$ and $\frac{1}{3}$ and the
  (12, 55) product of $\frac{1}{2}$ and $\frac{1}{3}$?

  2. The flat of eggs held $2\frac{1}{2}$ dozen eggs. How many eggs are in
  (29) $2\frac{1}{2}$ dozen?

3. In three nights Rumpelstiltskin spun $44,400 worth of
(18) gold thread. What was the average value of the thread he
spun per night?

4. Compare: $\dfrac{5}{8} \bigcirc \dfrac{1}{2}$
(56)

5. Compare: $6^2 + 8^2 \bigcirc 10^2$
(38)

Find each missing number:

6. $m + \dfrac{3}{8} = \dfrac{1}{2}$ **7.** $n - \dfrac{2}{3} = \dfrac{3}{4}$ **8.** $3 - f = \dfrac{5}{6}$
(57)   (57)   (43)

9. $32.50 × 10   **10.** (6.2)(0.48)   **11.** 1.0 ÷ 0.8
(46)   (39)   (49)

12. 120 ÷ 0.5   **13.** $\dfrac{7}{8} \cdot \dfrac{8}{7}$ **14.** $\dfrac{5}{6} \cdot \dfrac{3}{4}$
(49)   (30)   (29)

15. Instead of dividing $7\frac{1}{2}$ by $1\frac{1}{2}$, Julie doubled both numbers
(43) and then divided mentally. What is the division problem
Julie performed mentally, and what is the quotient?

16. Find the total price including 7% tax on a $9.79 purchase.
(41)

17. What number is next in this sequence?
(10)

$$\dots, 0.6, 0.7, 0.8, 0.9, \dots$$

18. The perimeter of this square is 4 cm.
(38) What is the area of the square?

19. How many $\frac{3}{5}$'s are in $\frac{3}{4}$? $\left(\frac{3}{4} \div \frac{3}{5}\right)$
(54)

Find each missing number:
20. $0.32w = 32$ **21.** $x + 3.4 = 5$
(43)   (43)

22. On one roll of a dot cube, what is the probability that the
(58) upturned face will show an even number of dots?

23. Arrange these measurements in order from shortest to
(7)   longest:

$$1 \text{ in.}, \ 3 \text{ cm}, \ 20 \text{ mm}$$

24. Larry correctly answered 45% of the questions.
(29, 33)

    (a) Did Larry correctly answer more than or less than half the questions? How do you know?

    (b) Write 45% as a reduced fraction.

25. Describe how to mentally calculate $\frac{1}{10}$ of $12.50.
(52)

26. Reduce: $\dfrac{2 \cdot 5 \cdot 2 \cdot 3 \cdot 3 \cdot 7}{2 \cdot 2 \cdot 2 \cdot 5 \cdot 5 \cdot 7}$
(54)

27. What is the sum of the decimal numbers represented by points x and y on this number line?
(50)

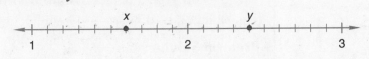

28. Draw a rectangle that is $1\frac{1}{2}$ inches long and $\frac{3}{4}$ inch wide. Then draw a segment that divides the rectangle into two triangles.
(7)

29. What is the perimeter of the rectangle drawn in problem 28?
(8)

30. If $A = lw$, and if $l = 1.5$ and $w = 0.75$, what does A equal?
(47)

LESSON

59 Adding Mixed Numbers

Facts Practice: 30 Fractions to Reduce (Test G)

Mental Math: Count up and down by 3's between 3 and 36.

a. 4×75 b. $279 + 350$ c. $250 - 199$
d. $\$15.00 - \$7.75$ e. Double $\$1.50$. f. $\frac{800}{40}$
g. 4×12, $\div 6$, $\times 8$, $- 4$, $\div 6$, $\times 3$, $\div 2$

Problem Solving:

A dot cube has six faces. Where two faces meet, there is an edge. How many edges does a dot cube have?

NEW CONCEPT

We have been practicing adding mixed numbers since Lesson 26. In this lesson we will rename the fraction parts of the mixed numbers so that the fractions have common denominators. Then we will add.

Example 1 Add: $2\frac{1}{2} + 1\frac{1}{6}$

Solution **Step 1:** Shape: write the fractions with common denominators.

Step 2: Operate: add the renamed fractions and add the whole numbers.

Step 3: Simplify: reduce the fraction.

$$
\begin{array}{r}
\xrightarrow{\text{1. Shape}} \\
2\frac{1}{2} \times \frac{3}{3} = 2\frac{3}{6} \\
+ 1\frac{1}{6} \quad = 1\frac{1}{6} \\
\hline
3\frac{4}{6} = \mathbf{3\frac{2}{3}} \\
\text{3. Simplify}
\end{array}
$$

2. Operate

Example 2 Add: $1\frac{1}{2} + 2\frac{2}{3}$

Solution **Step 1:** Shape: write the fractions with common denominators.

Step 2: Operate: add the renamed fractions and add the whole numbers.

Step 3: Simplify: convert the improper fraction to a mixed number, and combine the mixed number with the whole number.

$$
\begin{array}{l}
\overset{\text{1. Shape}}{\longrightarrow} \\
 1\frac{1}{2} \times \frac{3}{3} = 1\frac{3}{6} \\
+\, 2\frac{2}{3} \times \frac{2}{2} = 2\frac{4}{6} \\
\hline
\phantom{+\, 2\frac{2}{3} \times \frac{2}{2} = } 3\frac{7}{6} = 3 + 1\frac{1}{6} = \mathbf{4\frac{1}{6}}
\end{array}
\quad \text{2. Operate}
$$

3. Simplify

LESSON PRACTICE

Practice set Add:

a. $1\frac{1}{2} + 1\frac{1}{3}$ **b.** $1\frac{1}{2} + 1\frac{2}{3}$ **c.** $5\frac{1}{3} + 2\frac{1}{6}$

d. $3\frac{3}{4} + 1\frac{1}{3}$ **e.** $5\frac{1}{2} + 3\frac{1}{6}$ **f.** $7\frac{1}{2} + 4\frac{5}{8}$

MIXED PRACTICE

Problem set **1.** What is the product of the decimal numbers four tenths
(40)   and four hundredths?

2. Larry looked at the clock. It was 9:45 p.m. His book report
(32)   was due the next morning at 8:30. How many hours and
minutes were there until Larry's book report was due?

3. Pluto's greatest distance from the Sun is seven billion,
(12)   three hundred million kilometers. Use digits to write
that distance.

4. $2\frac{1}{2} + 1\frac{1}{6}$ **5.** $1\frac{1}{2} + 2\frac{2}{3}$
(59)   (59)

6. Compare: $\frac{1}{2} \bigcirc \frac{3}{5}$ **7.** Compare: $\frac{2}{3} \bigcirc \frac{6}{9}$
(56)   (56)

8. $8\frac{1}{5} - 3\frac{4}{5}$
(48)

9. $\frac{3}{4} \cdot \frac{5}{2}$
(29)

10. How many $\frac{1}{2}$'s are in $\frac{2}{5}$? $\left(\frac{2}{5} \div \frac{1}{2}\right)$
(54)

11. (0.875)(40)   **12.** $0.07 \div 4$ **13.** $30 \div d = 0.6$
(39)   (45)   (49)

14. What number is halfway between 0.1 and 0.24?
(50)

15. Round 36,428,591 to the nearest million.
(16)

16. What temperature is 23°F less than 8°F?
(14)

17. Mary wound a garden hose around a circular reel. If the
(47) diameter of the reel was 10 inches, how many inches of
hose was wound on the first full turn of the reel? Round the
answer to the nearest whole inch. (Use 3.14 for π.)

18. How many square inches are needed to cover a square foot?
(38)

19. One centimeter is what fraction of one meter?
(52)

20. Mentally calculate each answer. Describe how you
(46, 52) performed each mental calculation.

(a) 6.25×10 (b) $6.25 \div 10$

21. With one toss of a single number cube, what is the
(58) probability of rolling a number less than three?

22. Compare: $(0.8)^2 \bigcirc 0.8$
(39, 44)

Refer to the line graph below to answer problems 23–25.

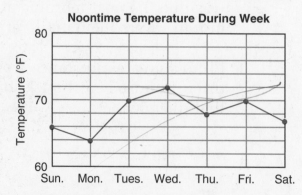

Noontime Temperature During Week

23. What was the range of the noontime temperatures during
(18, Inv. 5) the week?

24. What was the Saturday noontime temperature?
(18)

25. Write a question that relates to this line graph and answer
(18) the question.

26. Nefertiti could pronounce her name in six tenths of a
(50) second. At that rate, how many times could she pronounce her name in 15 seconds? (Write and solve a fraction division problem to answer the question.)

27. One eighth is equivalent to $12\frac{1}{2}\%$. To what percent is
(Inv. 2) three eighths equivalent?

28. Mentally calculate the total cost of 10 gallons of gas priced
(46) at $1.599 per gallon. Describe the process you used.

29. Arrange these three numbers in order from least to greatest:
(56)

$$\frac{3}{4}, \text{ the reciprocal of } \frac{3}{4}, 1$$

30. If $P = 2l + 2w$, and if $l = 4$ and $w = 3$, what does
(47) P equal?

L E S S O N

60   Polygons

WARM-UP

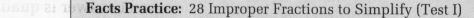

Facts Practice: 28 Improper Fractions to Simplify (Test I)

Mental Math: Count by 7's from 7 to 84.

    a. 2 × 750   **b.** 429 − 250   **c.** 750 + 199

    d. $9.50 + $1.75   **e.** $\frac{1}{2}$ of $5   **f.** 40 × 50

    g. 12 × 3, + 4, × 2, + 20, ÷ 10, × 5, ÷ 2

Problem Solving:

Half of a gallon is a half gallon. Half of a half gallon is a quart. Half of a quart is a pint. Half of a pint is a cup. Into an empty gallon container is poured a half gallon of milk, plus a quart of milk, plus a pint of milk, plus a cup of milk. How much more milk is needed to fill the gallon container?

    1 gallon $\frac{1}{2}$ gallon   1 quart   1 pint   1 cup

NEW CONCEPT

Polygons are closed shapes with straight sides. Polygons are named by the number of sides they have. The chart below names some common polygons.

Polygons

Shape	Number of Sides	Name of Polygon
△	3	triangle
▭	4	quadrilateral
⬠	5	pentagon
⬡	6	hexagon
⯃	8	octagon

Two sides of a polygon meet, or **intersect,** at a vertex (plural: **vertices**). A polygon has the same number of vertices as it has sides.

Example 1   What is the name of a polygon that has four sides?

Solution   The answer is not "square" or "rectangle." Squares and rectangles do have four sides, but not all four-sided polygons are squares or rectangles. The correct answer is quadrilateral. A rectangle is one kind of quadrilateral. A square is a rectangle with sides of equal length.

If all the sides of a polygon have the same length and if all the angles have the same measure, then the polygon is called a regular polygon. A square is a regular quadrilateral, but a rectangle that is longer than it is wide is not a regular quadrilateral.

We often use the word congruent when describing geometric figures. We say that polygons are congruent to each other if they have the same shape and size. We may also refer to segments or angles as congruent. Congruent segments have the same length; congruent angles have the same measure. In a regular polygon all the sides are congruent and all the angles are congruent.

Example 2   A regular octagon has a perimeter of 96 inches. How long is each side?

Solution   An octagon has eight sides. The sides of a regular octagon are the same length. Dividing the perimeter of 96 inches by 8, we find that each side is **12 inches.** (Most of the red stop signs on our roads are regular octagons with sides 12 inches long.)

LESSON PRACTICE

Practice set   **a.** What is the name of this six-sided shape?

b. How many sides does a pentagon have?

c. Can a polygon have 19 sides?

d. What is the name for a corner of a polygon?

e. What is the name for a polygon with four vertices?

MIXED PRACTICE

Problem set

1. What is the cost per ounce of a 42-ounce box of oatmeal priced at $1.26? (Write an equation and solve the problem.)
(15)

2. Amy needs to purchase some gas so that she can mow her lawn. At the station she fills a container with 1.1 gallons of gas. The station charges 1.47\frac{9}{10}$ per gallon. How much will Amy spend on gas? Round your answer to the nearest cent. Describe how to use estimation to check whether your answer is reasonable.
(39, 51)

3. The smallest three-digit whole number is 100. What is the largest three-digit whole number?
(12)

4. $\frac{3}{4} + \frac{5}{8}$
(57)

5. $1\frac{1}{2} + 3\frac{1}{6}$
(59)

6. $\frac{5}{5} \times \left(\frac{4}{4} - \frac{3}{3} \right)$
(29)

7. $\frac{3}{5} \cdot \frac{1}{3}$
(29)

8. How many $\frac{1}{3}$'s are in $\frac{3}{5}$? $\left(\frac{3}{5} \div \frac{1}{3} \right)$
(54)

9. How much money is $\frac{5}{8}$ of $24? Draw a diagram to illustrate the problem.
(22)

10. $(0.65)(0.14)$
(40)

11. $65 \div 0.05$
(49)

12. A quadrilateral has how many sides?
(60)

13. Round the product of 0.24 and 0.26 to the nearest hundredth.
(51)

14. What is the average of 1.3, 2, and 0.81?
(18)

15. What is the sum of the first seven numbers in this sequence?
(10)

$$1, 3, 5, 7, \dots$$

16. How many square feet are needed to cover a square yard?
(38)

17. Ten centimeters is what fraction of one meter?
(52)

Find each missing number:

18. $3x = 1.2 + 1.2 + 1.2$
(43)

19. $\frac{4}{3}y = 1$
(30)

20. $m + 1\frac{3}{5} = 5$
(49)

21. $6\frac{1}{8} - w = 3\frac{5}{8}$
(58)

22. List the prime numbers between 40 and 50.
(19)

23. Ceci cut a lime into thin slices. The largest slice was about 4 cm in diameter. Then she removed the outer peel from the slice. About how long was the outer peel? Round the answer to the nearest centimeter. (Use 3.14 for π.)
(47)

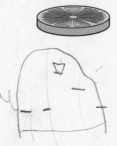

24. (a) To what decimal number is the arrow pointing?
(50, 51)

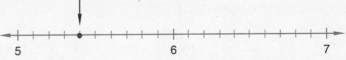

(b) This decimal number rounds to what whole number?

25. The face of this spinner is divided into 8 congruent sectors. If the spinner is spun once, what ratio expresses the probability that it will stop on a 3?
(58)

26. Mary found that the elm tree added about $\frac{3}{8}$ inch to the diameter of its trunk every year. If the diameter of the tree is about 12 inches, then the tree is about how many years old? (Write and solve a fraction division problem to answer the question.)
(50)

27. Duncan's favorite TV show starts at 8 p.m. and ends at 9 p.m. Duncan timed the commercials and found that 12 minutes of commercials aired between 8 p.m. and 9 p.m. Commercials were shown for what fraction of the hour?
(29)

28. Instead of dividing 400 by 16, Chip thought of an equivalent division problem that was easier to perform. Write an equivalent division problem that has a one-digit divisor and find the quotient.
(43)

29. What is the total price of a $6.89 item plus 6% sales tax?
(41)

30. Compare:
(56)

(a) $3\frac{1}{2} \bigcirc \frac{6}{2} + \frac{1}{2}$

(b) $\frac{5}{8} \bigcirc \frac{3}{4}$

INVESTIGATION 6

Focus on

Attributes of Geometric Solids

Polygons are two-dimensional shapes. They have length and width, but they do not have height (depth). The objects we encounter in the world around us are three-dimensional. These objects, called **geometric solids,** have length, width, and height; in other words, they take up space. The table below illustrates some three-dimensional shapes.

Geometric Solids

Shape	Name
	Triangular Prism
	Rectangular Prism
	Cube
	Pyramid
	Cylinder
	Cone
	Sphere

You should learn to recognize, name, and draw each of these shapes.

Name each shape:

1.

2.

3.

4.

5.

6.

Solids can have **faces, edges,** and **vertices.** The illustration below points out a face, an edge, and a vertex of a cube.

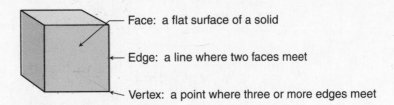

Face: a flat surface of a solid

Edge: a line where two faces meet

Vertex: a point where three or more edges meet

7. A cube has how many faces?

8. A cube has how many edges?

9. A cube has how many vertices?

A pyramid with a square base is shown at right. One face is a square; the others are triangles.

10. How many faces does this pyramid have?

11. How many edges does this pyramid have?

12. How many vertices does this pyramid have?

When solids are drawn, the edges that are hidden from view can be indicated with dashes. To draw a cube, for example, we first draw two squares that overlap as shown. Then we connect the corresponding vertices of the two squares. In both steps we use dashes to represent the edges that are hidden from view. Practice drawing a cube.

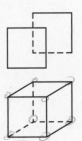

13. Draw a rectangular prism. Begin by drawing two rectangles as shown at right.

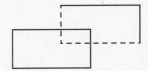

14. Draw a triangular prism. Begin by drawing two triangles as shown at right.

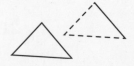

15. Draw a cylinder. Begin by drawing a "flattened circle" as shown at right. This will be the "top" of the cylinder.

If every face of a solid is a polygon, then the solid is called a **polyhedron**. Polyhedrons do not have any curved surfaces. So rectangular prisms and pyramids are polyhedrons, but spheres and cylinders are not.

One way to measure a solid is to find the area of its surfaces. We can find how much surface a polyhedron has by adding the area of its faces. The sum of these areas is called the surface area of the solid.

Each edge of the cube at right is 5 inches long. So each face of the cube is a square with sides 5 inches long. Use this information to answer problems 16 and 17.

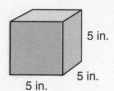

16. What is the area of each face of the cube?

17. What is the total surface area of the cube?

A cereal box has six faces, but not all the faces have the same area. The front and back surfaces have the same area; the top and bottom surfaces have the same area; and the left and right surfaces have the same area. Here we show a cereal box that is 10 inches tall, 7 inches wide, and 2 inches deep.

18. What is the area of the front of the box?

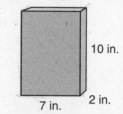

19. What is the area of the top of the box?

20. What is the area of the right panel of the box?

21. Combine the areas of all six panels to find the total surface area of the box.

A container such as a cereal box is constructed out of a flat sheet of cardboard that is printed, cut, folded, and glued to create a colorful three-dimensional container. By cutting apart a cereal box, you can see the six faces of the box at one time. Here we show one way to cut apart a box, but many arrangements are possible.

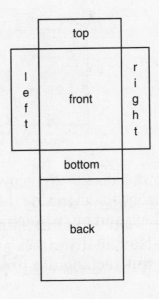

22. Here we show three ways to cut apart a box shaped like a cube. We have also shown an arrangement of six squares that does not fold into a cube. Which pattern below does not form a cube?

A.

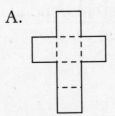

B.

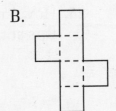

C.

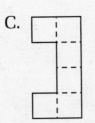

D.

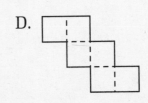

In addition to measuring the surface area of a solid, we can also measure its **volume.** The volume of a solid is the amount of space it occupies. To measure volume, we use units that occupy space, such as cubic centimeters, cubic inches, or cubic feet. Here we show two-dimensional images of a cubic inch and a cubic centimeter:

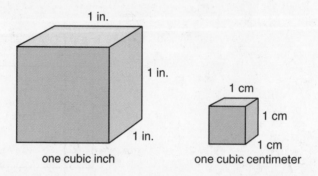

one cubic inch   one cubic centimeter

In problems 23–25 below, we will practice counting the number of cubes to determine the volume of a solid. In a later lesson we will expand our discussion of volume.

23. How many cubes are used to form this rectangular prism?

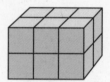

24. How many small cubes are used to form the larger cube at right?

25. How many cubes are used to build this solid?

LESSON

61   Adding Three or More Fractions

Facts Practice: 72 Multiplication and Division Facts (Test H)

Mental Math: Count up and down by $\frac{1}{4}$'s between $\frac{1}{4}$ and 4.

a. 4×750 b. $283 + 250$ c. $750 - 199$
d. $\$8.25 - \$2.50$ e. Double $12\frac{1}{2}$.   f. $\frac{900}{30}$
g. $6 \times 10, \div 3, \times 2, \div 4, \times 5, \div 2, \times 4$

Problem Solving:

Copy this problem and fill in the missing digits:

$$5\overline{)}$$

NEW CONCEPT

To add three or more fractions, we find a common denominator for all the fractions being added. Once we determine a common denominator, we can rename the fractions and add. We usually use the least common denominator, which is the least common multiple of all the denominators.

Example 1 Add: $\frac{1}{2} + \frac{1}{4} + \frac{1}{8}$

Solution First we find a common denominator. The LCM of 2, 4, and 8 is 8, so we rename all fractions as eighths. Then we add and simplify if possible.

$$\frac{1}{2} \times \frac{4}{4} = \frac{4}{8}$$
$$\frac{1}{4} \times \frac{2}{2} = \frac{2}{8}$$
$$+ \frac{1}{8} \times \frac{1}{1} = \frac{1}{8}$$
$$\frac{7}{8}$$

Example 2   Add: $1\frac{1}{2} + 2\frac{1}{3} + 3\frac{1}{6}$

Solution   A common denominator is 6. We rename all fractions. Then we add the whole numbers, and we add the fractions. We simplify the result if possible.

$$1\frac{1}{2} \times \frac{3}{3} = 1\frac{3}{6}$$
$$2\frac{1}{3} \times \frac{2}{2} = 2\frac{2}{6}$$
$$+ \ 3\frac{1}{6} \times \frac{1}{1} = 3\frac{1}{6}$$
$$6\frac{6}{6} = 7$$

LESSON PRACTICE

Practice set   Add:

a. $\dfrac{1}{2} + \dfrac{3}{4} + \dfrac{1}{8}$ **b.** $\dfrac{1}{2} + \dfrac{1}{3} + \dfrac{1}{6}$ **c.** $1\dfrac{1}{2} + 1\dfrac{1}{3} + 1\dfrac{1}{4}$

d. $\dfrac{1}{2} + \dfrac{2}{3} + \dfrac{5}{6}$ **e.** $\dfrac{1}{2} + \dfrac{3}{4} + \dfrac{7}{8}$ **f.** $1\dfrac{1}{4} + 1\dfrac{1}{8} + 1\dfrac{1}{2}$

MIXED PRACTICE

Problem set   **1.** Convert the improper fraction $\frac{20}{6}$ to a mixed number.
(25)   Remember to reduce the fraction part of the mixed number.

2. A fathom is 6 feet. How many feet deep is water that is
(29) $2\frac{1}{2}$ fathoms deep?

3. After 3 days and 7425 guesses, the queen guessed
(18)   Rumpelstiltskin's name. What was the average number of names she guessed per day?

4. $5\dfrac{1}{2} - 1\dfrac{2}{3}$ **5.** $5\dfrac{1}{3} - 2\dfrac{1}{2}$
(57)   (57)

6. $1\dfrac{1}{2} + 2\dfrac{1}{3} + 3\dfrac{1}{4}$ **7.** $3\dfrac{3}{4} + 3\dfrac{1}{3}$
(61)   (59)

8. Compare:
(38, 56)

(a) $\dfrac{2}{3} \bigcirc \dfrac{3}{5}$ (b) $4^2 \bigcirc \sqrt{144}$

9. $\frac{5}{6} \times 6^2$
(29, 38)

10. $\frac{3}{8} \cdot \frac{2}{3}$
(29)

11. How many $\frac{2}{3}$'s are in $\frac{3}{8}$? $\left(\frac{3}{8} \div \frac{2}{3} \right)$
(54)

12. $(4 - 0.4) \div 4$
(53)

13. $4 - (0.4 \div 4)$
(53)

14. Which digit in 49.63 has the same place value as the 7 in 8.7?
(34)

15. Estimate the sum of $642.23 and $861.17 to the nearest hundred dollars. Describe how you arrived at your answer.
(16)

16. Elizabeth used a compass to draw a circle with a radius of 4 cm.
(47)

 (a) What was the diameter of the circle?

 (b) What was the circumference of the circle? (Use 3.14 for π.)

17. What is the next number in this sequence?
(10)

$$..., 100, 10, 1, ...$$

18. The perimeter of a square is 1 foot. How many square inches cover its area?
(38)

19. What is the ratio of the value of a dime to the value of a quarter?
(23)

Find each missing number:

20. $15m = 3 \cdot 10^2$
(4, 38)

21. $\frac{1}{10} = \frac{n}{100}$
(42)

22. By what fraction name for 1 must $\frac{2}{3}$ be multiplied to form a fraction with a denominator of 15?
(42)

23. What time is 5 hours 15 minutes after 9:50 a.m.?
(32)

24. The area of a square is 16 square inches. What is its perimeter?
(38)

25. This figure shows the shape of home plate on a baseball field. What kind of a polygon is shown?
(60)

26. The sales-tax rate was 7%. Dexter bought two items, one
(41) for $4.95 and the other for $2.79. What was the total cost
of the two items including sales tax? Describe how to use
estimation to check whether your answer is reasonable.

27. Rawlings bought a sheet of 100 stamps from the post
(52) office for $37. What was the price of each stamp?

28. Draw a rectangular prism. A rectangular prism has how
(Inv. 6) many

   (a) faces?   (b) edges?   (c) vertices?

Refer to the cube shown below to answer problems 29 and 30.

29. Each face of a cube is a square. What
(Inv. 6) is the area of each face of this cube?

3 cm

30. Find the total surface area of the
(Inv. 6) cube by adding up the area of all of
the faces of the cube.

LESSON
62

Writing Mixed Numbers as Improper Fractions

WARM-UP

Facts Practice: 30 Fractions to Reduce (Test G)

Mental Math: Count by 12's from 12 to 144.

a. 5×40

b. $475 + 1200$

c. 3×84

d. $\$8.50 + \$2.50$

e. $\frac{1}{3}$ of $\$36.00$

f. $\frac{\$25}{10}$

g. $6 \times 8, - 4, \div 4, \times 2, + 2, \div 6, \div 2$

h. Hold your hands one foot apart.

Problem Solving:

The average number of people in each of two rows is 27. If the people are separated into three rows instead of two, what will be the average number of people per row?

NEW CONCEPT

Here is another story about pies. In this story a mixed number is changed to an improper fraction.

There were $3\frac{5}{6}$ pies on the shelf. The restaurant manager asked the server to cut the whole pies into sixths. Altogether, how many slices of pie were there after the server cut the pies?

We illustrate this story with circles. There were $3\frac{5}{6}$ pies on the shelf.

The server cut the whole pies into sixths. Each whole pie then had six slices.

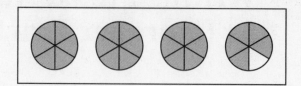

The three whole pies contain 18 slices ($3 \times 6 = 18$). The 5 additional slices from the $\frac{5}{6}$ of a pie bring the total to 23 slices (23 sixths). This story illustrates that $3\frac{5}{6}$ is equivalent to $\frac{23}{6}$.

Now we describe the arithmetic for changing a mixed number such as $3\frac{5}{6}$ to an improper fraction. Recall that a mixed number has a whole-number part and a fraction part.

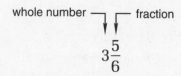

The denominator of the mixed number will also be the denominator of the improper fraction.

$$3\frac{5}{6} = \frac{}{6}$$

same denominator

The denominator indicates the size of the fraction "pieces." In this case the fraction pieces are sixths, so we change the whole number 3 into sixths. We know that one whole is $\frac{6}{6}$, so three wholes is $3 \times \frac{6}{6}$, which is $\frac{18}{6}$. Therefore, we add $\frac{18}{6}$ and $\frac{5}{6}$ to get $\frac{23}{6}$.

$$3\frac{5}{6}$$
$$\frac{6}{6} + \frac{6}{6} + \frac{6}{6} + \frac{5}{6} = \frac{23}{6}$$

$$\frac{18}{6} + \frac{5}{6} = \frac{23}{6}$$

Example 1 Write $2\frac{3}{4}$ as an improper fraction.

Solution The denominator of the fraction part of the mixed number is fourths, so the denominator of the improper fraction will also be fourths.

$$2\frac{3}{4} = \frac{}{4}$$

We change the whole number 2 into fourths. Since 1 equals $\frac{4}{4}$, the whole number 2 equals $2 \times \frac{4}{4}$, which is $\frac{8}{4}$. We add $\frac{8}{4}$ and $\frac{3}{4}$ to get $\frac{11}{4}$.

$$\overset{\lceil 2\frac{3}{4} \rceil}{\frac{8}{4} + \frac{3}{4} = \frac{11}{4}}$$

Example 2 Write $5\frac{2}{3}$ as an improper fraction.

Solution We see that the denominator of the improper fraction will be thirds.

$$5\frac{2}{3} = \frac{}{3}$$

Some people use a quick, mechanical method to find the numerator of the improper fraction. Looking at the mixed number, they multiply the denominator by the whole number and then add the numerator. The result is the numerator of the improper fraction.

$$5\overset{+}{\underset{\times}{\frac{2}{3}}} = \frac{17}{3}$$

Example 3 Write $1\frac{2}{3}$ and $2\frac{2}{5}$ as improper fractions. Then multiply the improper fractions. What is the product?

Solution First we write $1\frac{2}{3}$ and $2\frac{2}{5}$ as improper fractions.

$$\overset{\lceil 1\frac{2}{3} \rceil}{\frac{3}{3} + \frac{2}{3} = \frac{5}{3}} \qquad \overset{\lceil 2\frac{2}{5} \rceil}{\frac{10}{5} + \frac{2}{5} = \frac{12}{5}}$$

Next we multiply $\frac{5}{3}$ by $\frac{12}{5}$.

$$\frac{5}{3} \cdot \frac{12}{5} = \frac{60}{15}$$

The result is an improper fraction, which we simplify.

$$\frac{60}{15} = 4$$

So $1\frac{2}{3} \times 2\frac{2}{5}$ equals **4.**

LESSON PRACTICE

Practice set   Write each mixed number as an improper fraction:

a. $2\frac{4}{5}$ **b.** $3\frac{1}{2}$ **c.** $1\frac{3}{4}$

d. $6\frac{1}{4}$ **e.** $1\frac{5}{6}$ **f.** $3\frac{3}{10}$

g. $2\frac{1}{3}$ **h.** $12\frac{1}{2}$ **i.** $3\frac{1}{6}$

j. Write $1\frac{1}{2}$ and $3\frac{1}{3}$ as improper fractions. Then multiply the improper fractions. What is the product?

MIXED PRACTICE

Problem set   **1.** In music there are whole notes, half notes, quarter notes,
(54)   and eighth notes.

   (a) How many quarter notes equal a whole note?

   (b) How many eighth notes equal a quarter note?

2. Don is 5 feet $2\frac{1}{2}$ inches tall. How many inches tall is that?
(62)

3. Which of these numbers is not a prime number?
(19)   A. 11   B. 21   C. 31   D. 41

4. Write $1\frac{1}{3}$ and $1\frac{1}{2}$ as improper fractions, and multiply the
(62)   improper fractions. What is the product?

5. If the chance of rain is 20%, what is the chance that it
(58)   will not rain?

6. The prices for three pairs of skates were $36.25, $41.50, and
(18) $43.75. What was the average price for a pair of skates?

7. Instead of dividing 15 by $2\frac{1}{2}$, Solomon doubled both
$^{(43)}$ numbers and then divided mentally. What was Solomon's
mental division problem and its quotient?

Find each missing number:

8. $m - 4\frac{3}{8} = 3\frac{1}{4}$
$^{(43, 59)}$

9. $n + \frac{3}{10} = \frac{3}{5}$
$^{(43, 56)}$

10. $6d = 0.456$
$^{(43, 45)}$

11. $0.04w = 1.5$
$^{(43, 49)}$

12. $\frac{1}{2} + \frac{3}{4} + \frac{5}{8}$
$^{(61)}$

13. $\frac{5}{6} - \frac{1}{2}$
$^{(57)}$

14. $\frac{1}{2} \cdot \frac{4}{5}$
$^{(29)}$

15. $\frac{2}{3} \div \frac{1}{2}$
$^{(54)}$

16. $1 - (0.2 - 0.03)$
$^{(40)}$

17. $(0.14)(0.16)$
$^{(39)}$

18. One centimeter equals 10 millimeters. How many
$^{(49)}$ millimeters does 2.5 centimeters equal?

19. List all of the common factors of 18 and 24. Then circle
$^{(19)}$ the greatest common factor.

20. Ten marbles are in a bag. Four of the marbles are red. If
$^{(58)}$ one marble is drawn from the bag, what ratio expresses
the probability that it will be red?

21. If the perimeter of a square is 40 mm, what is the area of
$^{(38)}$ the square?

22. At 6 a.m. the temperature was $-6°$F. At noon the
$^{(14)}$ temperature was 14°F. From 6 a.m. to noon the
temperature rose how many degrees?

23. Lisa used a compass to draw a circle with a radius of
$^{(47)}$ $1\frac{1}{2}$ inches.

(a) What was the diameter of the circle?

(b) What was the circumference of the circle? (Use 3.14
for π.)

The circle graph below shows the favorite sports of 100 people. Refer to the graph to answer problems 24–27.

Favorite Sports of 100 People

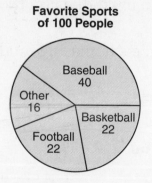

24. How many more people favored baseball than favored
(40)   football?

25. What fraction of the people favored baseball?
(40)

26. Was any sport the favorite sport of the majority of the
(40)   people surveyed? Write one or two sentences to explain
       your answer.

27. Since baseball was the favorite sport of 40 out of 100
(40)   people, it was the favorite sport of 40% of the people
       surveyed. What percent of the people answered that
       football was their favorite sport?

28. What number is 40% of 200?
(41)

29. Here we show 18 written as a product of prime numbers:
(19)

$$2 \cdot 3 \cdot 3$$

Write 20 as a product of prime numbers.

30. Judges awarded Sandra these scores for her performance
(Inv. 5) on the vault:

$$9.1, \, 8.9, \, 9.0, \, 9.2, \, 9.2$$

What is the median score?

LESSON
63
Subtracting Mixed Numbers with Regrouping, Part 2

WARM-UP

Facts Practice: 64 Multiplication Facts (Test D)

Mental Math: Count up and down by $\frac{1}{8}$'s between $\frac{1}{8}$ and 2.

a. 5×140 *(10 × 140 ÷ 2)*   b. $420 - 50$

c. 4×63 d. $\$8.50 - \$2.50$

e. Double $7\frac{1}{2}$.   f. $\frac{\$25}{100}$

g. $5 \times 10, - 20, + 2, \div 4, + 1, \div 3, - 3$

h. Hold your hands one foot apart, then one inch apart.

Problem Solving:

What are the next three numbers in this sequence?

$$\frac{1}{12}, \frac{1}{6}, \frac{1}{4}, \dots$$

NEW CONCEPT

Since Lesson 48 we have practiced subtracting mixed numbers with regrouping. In this lesson we will rename the fractions with common denominators before subtracting.

To subtract $1\frac{1}{2}$ from $3\frac{2}{3}$, we first rewrite the fractions with common denominators. Then we subtract the whole numbers and the fractions. If possible, we simplify.

$$3\frac{2}{3} \times \frac{2}{2} = 3\frac{4}{6}$$
$$- 1\frac{1}{2} \times \frac{3}{3} = 1\frac{3}{6}$$
$$\overline{\qquad\qquad 2\frac{1}{6}}$$

When subtracting, it is sometimes necessary to regroup. We rewrite the fractions with common denominators before regrouping.

$$4\frac{6}{6} + \frac{3}{6} \ 4\frac{9}{6}$$

Example   Subtract: $5\frac{1}{2} - 1\frac{2}{3}$

Solution   We rewrite the fractions with common denominators. Before we can subtract, we must regroup. After we subtract, we simplify if possible.

$$
\begin{aligned}
5\frac{1}{2} \times \frac{3}{3} &= 5\overset{4}{\cancel{3}}\frac{\overset{9}{\cancel{3}}}{6} \\
- \ 1\frac{2}{3} \times \frac{2}{2} &= 1\frac{4}{6} \\
\hline
&\quad\ \ 3\frac{5}{6}
\end{aligned}
$$

LESSON PRACTICE

Practice set*   Subtract:

  a. $5\frac{1}{2} - 3\frac{1}{3}$ **b.** $4\frac{1}{4} - 2\frac{1}{3}$ **c.** $6\frac{1}{2} - 1\frac{3}{4}$

  d. $7\frac{2}{3} - 3\frac{5}{6}$ **e.** $6\frac{1}{6} - 1\frac{1}{2}$ **f.** $4\frac{1}{3} - 1\frac{1}{2}$

  g. $4\frac{5}{6} - 1\frac{1}{3}$ **h.** $6\frac{1}{2} - 3\frac{5}{6}$ **i.** $8\frac{2}{3} - 5\frac{3}{4}$

MIXED PRACTICE

Problem set   **1.** What is the difference between the sum of 0.6 and 0.4
$^{(12,\ 53)}$ and the product of 0.6 and 0.4?

  2. Mt. Whitney, the highest point in California, has an
$^{(14)}$ elevation of 14,494 feet above sea level. From there one can see Death Valley, which contains the lowest point in California, 282 feet **below** sea level. The floor of Death Valley is how many feet below the peak of Mt. Whitney?

  3. The anaconda was 288 inches long. How many feet is
$^{(15)}$ 288 inches?

  4. Write the mixed number $4\frac{2}{3}$ as an improper fraction.
$^{(62)}$

  5. Write $2\frac{1}{2}$ and $1\frac{1}{5}$ as improper fractions. Then multiply the
$^{(62)}$ improper fractions and simplify the product.

  6. What time is $2\frac{1}{2}$ hours after 10:15 a.m.?
$^{(32)}$

7. $(30 \times 15) \div (30 - 15)$
(5)

8. Compare: $\dfrac{5}{8} \bigcirc \dfrac{2}{3}$
(56)

9. $w - 3\dfrac{2}{3} = 1\dfrac{1}{2}$
(43)

10. $\dfrac{6}{8} - \dfrac{3}{4}$
(55)

11. $6\dfrac{1}{4} - 5\dfrac{5}{8}$
(63)

12. $\dfrac{3}{4} \times \dfrac{2}{5}$
(29)

13. $\dfrac{3}{4} \div \dfrac{2}{5}$
(54)

14. $(1 - 0.4)(1 + 0.4)$
(53)

15. How much money is 60% of $45?
(41)

16. $0.4 \div 8$
(45)

17. $8 \div 0.4$
(49)

18. What is the next number in this sequence?
(10)

$$0.2, 0.4, 0.6, 0.8, \ldots$$

19. What is the tenth prime number?
(19)

20. What is the perimeter of this rectangle?
(8, 59)

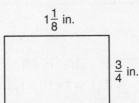

$1\dfrac{1}{8}$ in.

$\dfrac{3}{4}$ in.

21. A triangular prism has how many
(Inv. 6)

    (a) faces?

    (b) edges?

    (c) vertices?

22. Round $678.25 to the nearest ten dollars. Describe how
(16) you decided upon your answer.

23. This rectangle is divided into two
(31) congruent regions. What is the area
of the shaded region?

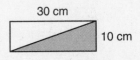

30 cm

10 cm

24. A ton is 2000 pounds. How many pounds is $2\frac{1}{2}$ tons?
(15, 62)

25. Which arrow could be pointing to 0.2 on this number line?
(50)

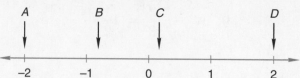

26. The paper cup would not roll straight. One end was 7 cm in diameter, and the other end was 5 cm in diameter. In one roll of the cup,
(47)

Use 3.14 for π.

(a) how far would the larger end roll?

(b) how far would the smaller end roll?

(Round each answer to the nearest centimeter.)

27. Jefferson got a hit 30% of the 240 times he went to bat during the season. Write 30% as a reduced fraction. Then find the number of hits Jefferson got during the season.
(29, 33)

28. Jena has run 11.5 miles of a 26.2-mile race. Find the remaining distance Jena has to run by solving this equation:
(43)

$$11.5 \text{ mi} + d = 26.2 \text{ mi}$$

29. The sales-tax rate was 7%. The two CDs cost $15.49 each. What was the total cost of the two CDs including tax?
(41)

30. Monica is making frosting for a cake. She adds $\frac{1}{2}$ teaspoon of red food coloring and $\frac{3}{4}$ teaspoon of blue food coloring to the frosting to create a dark purple frosting. In all, how much food coloring does Monica add to the frosting?
(57)

LESSON
64   Classifying Quadrilaterals

WARM-UP

Facts Practice: 24 Mixed Numbers to Write as Improper Fractions (Test J)

Mental Math: Count up and down by 25's between 25 and 300.
  a. 5 × 240
  b. 4500 + 450
  c. 7 × 34
  d. $7.50 + $7.50
  e. $\frac{1}{4}$ of $20.00
  f. $\frac{$75}{10}$
  g. 6 × 8, ÷ 2, + 1, ÷ 5, − 1, × 4, ÷ 2
  h. Hold your hands one foot apart, then one yard apart.

Problem Solving:

A dot cube has six faces. Where two faces meet, there is an edge. A "corner" where three edges meet is called a *vertex* (plural: *vertices*). A dot cube has how many vertices?

NEW CONCEPT

We learned in Lesson 60 that quadrilaterals are polygons with four sides. We can classify (sort) quadrilaterals by the characteristics of their sides and angles. The following table describes the various classifications of quadrilaterals:

Classifications of Quadrilaterals

Shape	Characteristic	Name
	No sides parallel	Trapezium
	One pair of parallel sides	Trapezoid
	Two pairs of parallel sides	Parallelogram
	Parallelogram with equal sides	Rhombus
	Parallelogram with right angles	Rectangle
	Rectangle with equal sides	Square

Notice that squares, rectangles, and rhombuses are all parallelograms. Also notice that a square is a special kind of rectangle, which is a special kind of parallelogram, which is a special kind of quadrilateral, which is a special kind of polygon. A square is also a special kind of rhombus.

Example 1   Is the following statement true or false?

All parallelograms are rectangles.

Solution   We are asked to decide whether every parallelogram is a rectangle. Since a rectangle is a special kind of parallelogram, some parallelograms are rectangles. However, some parallelograms are not rectangles. Since not all parallelograms are rectangles, the statement is **false.**

Example 2   Draw a pair of parallel lines. Then draw another pair of parallel lines. These lines should intersect the first pair but not be perpendicular to the first pair. What is the name for the quadrilateral that is formed by the intersecting lines?

Solution   We draw the first pair of parallel lines.

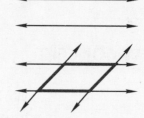

We draw the second pair of lines so that the lines are not perpendicular to the first pair.

At right we have colored the segments that form the quadrilateral. The quadrilateral formed is a **parallelogram.**

LESSON PRACTICE

Practice set   **a.** What is a quadrilateral?

b. Describe the difference between a parallelogram and a trapezoid.

c. Draw a rhombus that is a square.

d. Draw a rhombus that is not a square.

e. True or false: Some rectangles are squares.

f. True or false: All squares are rectangles.

MIXED PRACTICE

Problem set

1. When the sum of 1.3 and 1.2 is divided by the difference
(12, 53) of 1.3 and 1.2, what is the quotient?

2. William Shakespeare was born in 1564 and died in 1616.
(13) How many years did he live?

3. Anna's arrow hit a target 45 yards away. How many feet
(15) did the arrow travel?

4. Why is a square a regular quadrilateral?
(60)

5. A regular hexagon has a perimeter of 36 inches. How long
(60) is each side?

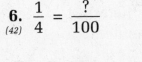

6. $\frac{1}{4} = \frac{?}{100}$
(42)

7. $\frac{8 \times 8}{8 + 8}$
(5)

8. $5\frac{2}{3} + 3\frac{3}{4}$
(59)

9. $\frac{1}{2} + \frac{2}{3} + \frac{1}{4}$
(61)

10. $\frac{9}{10} - \frac{1}{2}$
(57)

11. $6\frac{1}{2} - 2\frac{7}{8}$
(63)

12. Compare: $2 \times 0.4 \bigcirc 2 + 0.4$
(44)

13. 4.8×0.35
(39)

14. $1 \div 0.4$
(49)

15. How many $0.12 pencils can Mr. Jones buy for $4.80?
(15)

16. Round the product of 0.33 and 0.38 to the nearest
(51) hundredth.

17. Multiply the length by the width to
(31) find the area of this rectangle.

$\frac{1}{2}$ in.
$\frac{3}{4}$ in.

18. Is every rectangle a parallelogram?
(64)

19. What is the twelfth prime number?
(19)

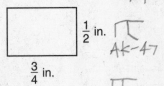
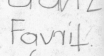

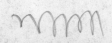

20. The area of a square is 9 cm².
(38)

   (a) How long is each side of the square?

   (b) What is the perimeter of the square?

Refer to the box shown below to answer problems 21 and 22.

21. The top, bottom, and sides of a box
(Inv. 6) are also called the faces of the box.
This box has how many faces?

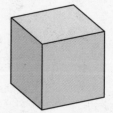

22. If this box is a cube and each edge
(Inv. 6) is 10 inches long, then

   (a) what is the area of each face?

   (b) what is the total surface area of the cube?

23. There are 100 centimeters to a meter. How many
(15) centimeters equal 2.5 meters?

24. Write the mixed numbers $1\frac{1}{2}$ and $2\frac{1}{2}$ as improper
(62) fractions. Then multiply the improper fractions and
simplify the product.

25. The numbers 2, 3, 5, 7, and 11 are prime numbers. The
(19) numbers 4, 6, 8, 9, 10, and 12 are not prime numbers, but
they can be formed by multiplying prime numbers.

$$4 = 2 \cdot 2$$

$$6 = 2 \cdot 3$$

$$8 = 2 \cdot 2 \cdot 2$$

Show how to form 9, 10, and 12 by multiplying prime
numbers.

26. Write 75% as an unreduced fraction. Then write the
(33, 35) fraction as a decimal number.

27. Reduce: $\dfrac{2 \cdot 2 \cdot 2 \cdot 3 \cdot 3}{2 \cdot 2 \cdot 3 \cdot 5 \cdot 5}$
(54)

28. Find the missing distance d in the equation below.
(43)

$$16.6 \text{ mi} + d = 26.2 \text{ mi}$$

Refer to the double-line graph below to answer problems 29 and 30.

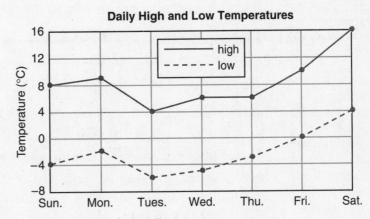

Daily High and Low Temperatures

29. The difference between Tuesday's high and low
(18) temperatures was how many degrees?

30. The difference between the lowest temperature of the
(18) week and the highest temperature of the week was how many degrees?

LESSON

65

Prime Factorization • Division by Primes • Factor Trees

WARM-UP

Facts Practice: 72 Multiplication and Division Facts (Test H)

Mental Math: Count by 9's from 9 to 108.

a. 5×60 b. $586 - 50$ c. 3×65

d. $20.00 - 2.50$ e. Double 75¢.   f. $\frac{\$75}{100}$

g. $9 \times 9, -1, \div 2, +2, \div 6, +3, \div 10$

h. Hold your hands one yard apart, then one meter apart.

Problem Solving:

Nelson bought a pizza and ate half of it. Then his sister ate half of what was left. Then his little brother ate half of what his sister had left. What fraction of the pizza did Nelson's little brother eat? Illustrate the portions of the pizza eaten by each family member.

NEW CONCEPTS

Prime factorization

Every whole number greater than 1 is either a prime number or a **composite number**. A prime number has *only two* factors (1 and itself), while a composite number has *more than two* factors. As we studied in Lesson 19, the numbers 2, 3, 5, and 7 are prime numbers. The numbers 4, 6, 8, and 9 are composite numbers. All composite numbers can be formed by multiplying prime numbers together.

$$4 = 2 \cdot 2$$
$$6 = 2 \cdot 3$$
$$8 = 2 \cdot 2 \cdot 2$$
$$9 = 3 \cdot 3$$

When we write a composite number as a product of its prime factors, we have written the **prime factorization** of the number. The prime factorizations of 4, 6, 8, and 9 are shown above. Notice that if we had written 8 as $2 \cdot 4$ instead of $2 \cdot 2 \cdot 2$, we would not have completed the prime factorization of 8. Since the number 4 is not prime, we would complete the prime factorization by "breaking" 4 into its prime factors of 2 and 2.

In this lesson we will show two methods for factoring a composite number, **division by primes** and **factor trees.** We will use both methods to factor the number 60.

Division by primes To factor a number using division by primes, we write the number in a division box and begin dividing by the smallest prime number that is a factor. The smallest prime number is 2. Since 60 is divisible by 2, we divide 60 by 2 to get 30.

$$\begin{array}{r} 30 \\ 2\overline{)60} \end{array}$$

Since 30 is also divisible by 2, we divide 30 by 2. The quotient is 15. Notice how we "stack" the divisions.

$$\begin{array}{r} 15 \\ 2\overline{)30} \\ 2\overline{)60} \end{array}$$

Although 15 is not divisible by 2, it is divisible by the next-smallest prime number, which is 3. Fifteen divided by 3 produces the quotient 5.

$$\begin{array}{r} 5 \\ 3\overline{)15} \\ 2\overline{)30} \\ 2\overline{)60} \end{array}$$

Five is a prime number. The only prime number that divides 5 is 5.

$$\begin{array}{r} 1 \\ 5\overline{)5} \\ 3\overline{)15} \\ 2\overline{)30} \\ 2\overline{)60} \end{array}$$

By dividing by prime numbers, we have found the prime factorization of 60.

$$60 = 2 \cdot 2 \cdot 3 \cdot 5$$

Factor trees To make a factor tree for 60, we simply think of any two whole numbers whose product is 60. Since 6×10 equals 60, we can use 6 and 10 as the first two "branches" of the factor tree.

The numbers 6 and 10 are not prime numbers, so we continue the process by factoring 6 into 2 · 3 and by factoring 10 into 2 · 5.

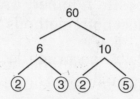

The circled numbers at the ends of the branches are all prime numbers. We have completed the factor tree. We will arrange the factors in order from least to greatest and write the prime factorization of 60.

$$60 = 2 \cdot 2 \cdot 3 \cdot 5$$

Example 1 Use a factor tree to find the prime factorization of 60. Use 4 and 15 as the first branches.

Solution Some composite numbers can be divided into many different factor trees. However, when the factor tree is completed, the same prime numbers appear at the ends of the branches.

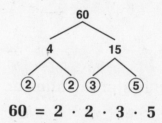

$$\mathbf{60 = 2 \cdot 2 \cdot 3 \cdot 5}$$

Example 2 Use division by primes to find the prime factorization of 36.

Solution We begin by dividing 36 by its smallest prime-number factor, which is 2. We continue dividing by prime numbers until the quotient is 1.[†]

$$
\begin{array}{r}
1 \\
3\overline{)3} \\
3\overline{)9} \\
2\overline{)18} \\
2\overline{)36}
\end{array}
$$

$$\mathbf{36 = 2 \cdot 2 \cdot 3 \cdot 3}$$

[†]Some people prefer to divide only until the quotient is a prime number. When using that procedure, the final quotient is included in the prime factorization of the number.

LESSON PRACTICE

Practice set*   **a.** Which of these numbers are composite numbers?

$$19, \ 20, \ 21, \ 22, \ 23$$

b. Write the prime factorization of each composite number in problem **a.**

c. Use a factor tree to find the prime factorization of 36.

d. Use division by primes to find the prime factorization of 48.

e. Write 125 as a product of prime factors.

f. Write the prime factorization of 10 and of 100. Then look for a pattern to help you find the prime factorization of 1000.

MIXED PRACTICE

Problem set   **1.** The total land area of the world is about fifty-seven million,
   (12)   two hundred eighty thousand square miles. Use digits to write that number of square miles.

2. Kwame photographed an African white rhinoceros that
   (15)   stood $6\frac{1}{2}$ feet tall. How many inches is $6\frac{1}{2}$ feet?

3. Jenny shot 10 free throws and made 6. What fraction of
   (29, 42)   her shots did she make? What percent of her shots did she make?

4. Make a factor tree for 40. Then write the prime
   (65)   factorization of 40.

5. Which of these numbers is a composite number?
   (65)
   A. 21   B. 31   C. 41

6. Write $2\frac{2}{3}$ as an improper fraction. Then multiply the
   (62)   improper fraction by $\frac{3}{8}$. What is the product?

7. Four of the ten marbles in the bag are red. If one marble is
   (58)   drawn from the bag, what ratio expresses the probability that the marble will not be red?

8. $8\frac{1}{2} + 1\frac{1}{3} + 2\frac{1}{6}$
(61)

9. $\frac{1}{12} + \frac{1}{6} + \frac{1}{2}$
(61)

Find each missing number:

10. $15\frac{3}{4} - m = 2\frac{1}{8}$
(43)

11. $\frac{4}{25} = \frac{n}{100}$
(42)

12. $12w = 0.0144$
(45)

13. $\frac{3}{8} \times \frac{1}{3} = y$
(29)

14. Compare: $\frac{1}{2} - \frac{1}{3} \bigcirc \frac{2}{3} - \frac{1}{2}$
(56)

15. $1 - (0.2 + 0.48)$
(38)

16. What is the total cost of two dozen erasers that are priced
(15, 41) at 50¢ each if 8% sales tax is added? Describe a way to perform the calculation mentally.

17. The store manager put $20.00 worth of $0.25 pieces in the
(15) change drawer. How many $0.25 pieces are in $20.00?

18. A pyramid with a square base has
(Inv. 6) how many

(a) faces?

(b) edges?

(c) vertices?

19. Use division by primes to find the prime factorization of 50.
(65)

20. What is the name of a six-sided polygon? How many
(60) vertices does it have?

21. Write $3\frac{4}{7}$ as an improper fraction.
(62)

22. The area of a square is 36 square inches.
(38)

(a) What is the length of each side?

(b) What is the perimeter of the square?

23. Write 16% as a reduced fraction.
(33)

24. How many millimeters long is the line segment below?
(7)

25. A meter is about one big step. About how many meters
(7) long is an automobile?

26. Write the prime factorization of 375 and of 1000.
(65)

27. Reduce: $\dfrac{3 \cdot 5 \cdot 5 \cdot 5}{2 \cdot 2 \cdot 2 \cdot 5 \cdot 5 \cdot 5}$
(54)

28. The radius of the merry-go-round is 15 feet. If the merry-
(47) go-round turns around once, a person riding on the outer
edge will travel how far? Round the answer to the nearest
foot. (Use 3.14 for π.) Describe how to mentally check
whether the answer is reasonable.

29. Eighty percent of the 20 answers were correct. How many
(29, 33) answers were correct?

30. The "rect-" part of *rectangle* means "right." A rectangle is a
(60) "right-angle" shape. Why is every square also a rectangle?

LESSON
66 Multiplying Mixed Numbers

WARM-UP

Facts Practice: 24 Mixed Numbers to Write as Improper Fractions (Test J)

Mental Math: Count up and down by $\frac{1}{8}$'s between $\frac{1}{8}$ and 2.

a. 5 × 160   b. 376 + 99   c. 8 × 23
d. $1.75 + $1.75   e. $\frac{1}{3}$ of $60.00   f. $\frac{\$30}{10}$
g. 8 × 8, − 4, ÷ 2, + 3, ÷ 3, + 1, ÷ 6, ÷ 2
h. Hold your hands one meter apart, then one yard apart.

Problem Solving:

Copy this problem and fill in the missing digits:

$$
\begin{array}{r}
\_\,4\,\_ \\
6\,)\overline{\_\,\_\,6} \\
\underline{\_\,\_} \\
\_\,\_ \\
\underline{\_\,\_} \\
3\,\_ \\
\underline{} \\
0
\end{array}
$$

NEW CONCEPT

Recall from Lesson 57 the three steps to solving an arithmetic problem with fractions.

Step 1: Put the problem into the correct shape (if it is not already).

Step 2: Perform the operation indicated.

Step 3: Simplify the answer if possible.

Remember that putting fractions into the correct shape for adding and subtracting means writing the fractions with common denominators. To multiply or divide fractions, we do not need to use common denominators. However, we must write the fractions in **fraction form.** This means we will write mixed numbers and whole numbers as improper fractions. We write a whole number as an improper fraction by making the whole number the numerator of a fraction with a denominator of 1.

Example 1   Multiply: $2\frac{2}{3} \times \frac{4}{1}$

Solution   First, we write $2\frac{2}{3}$ and 4 in fraction form.

$$\frac{8}{3} \times \frac{4}{1}$$

Second, we multiply the numerators to find the numerator of the product, and we multiply the denominators to find the denominator of the product.

$$\frac{8}{3} \times \frac{4}{1} = \frac{32}{3}$$

Third, we simplify the product by converting the improper fraction to a mixed number.

$$\frac{32}{3} = \mathbf{10\frac{2}{3}}$$

Example 2   Multiply: $2\frac{1}{2} \times 1\frac{1}{3}$

Solution   First, we write the numbers in fraction form.

$$\frac{5}{2} \times \frac{4}{3}$$

Second, we multiply the terms of the fractions.

$$\frac{5}{2} \times \frac{4}{3} = \frac{20}{6}$$

Third, we simplify the product.

$$\frac{20}{6} = 3\frac{2}{6} = \mathbf{3\frac{1}{3}}$$

LESSON PRACTICE

Practice set*   Multiply:

a. $1\frac{1}{2} \times \frac{2}{3}$ b. $1\frac{2}{3} \times \frac{3}{4}$ c. $1\frac{1}{2} \times 1\frac{2}{3}$

d. $1\frac{2}{3} \times 3$ e. $2\frac{1}{2} \times 2\frac{2}{3}$ f. $3 \times 1\frac{3}{4}$

g. $3\frac{1}{3} \times 1\frac{2}{3}$ h. $2\frac{3}{4} \times 2$ i. $2 \times 3\frac{1}{2}$

MIXED PRACTICE

Problem set **1.** Fifty percent of the 60 questions on the test are multiple
(29, 33) choice. Find the number of multiple-choice questions
on the test.

2. Twelve of the 30 racers on the team are boys.
(58)
(a) What is the ratio of boys to girls on the team?

(b) If each racer's name is placed in a hat and one name
is drawn, what is the probability that it will be the
name of a girl?

3. Some railroad rails weigh 155 pounds per yard and are
(15) 33 feet long. How much would a 33-foot-long rail weigh?

4. $1\frac{1}{2} \times 2\frac{2}{3}$
(66)

5. $2\frac{2}{3} \times 2$
(66)

6. The sum of five numbers is 200. What is the average of
(18) the numbers?

7. $\dfrac{100 + 75}{100 - 75}$
(5)

8. $1\frac{1}{5} + 3\frac{1}{2}$
(59)

9. $\dfrac{1}{3} + \dfrac{1}{6} + \dfrac{1}{12}$
(61)

10. $35\frac{1}{4} - 12\frac{1}{2}$
(63)

11. $\dfrac{4}{5} \times \dfrac{1}{2}$
(29)

12. $\dfrac{4}{5} \div \dfrac{1}{2}$
(54)

13. $0.25 \div 5$
(45)

14. $5 \div 0.25$
(49)

15. What is the product of the answers to problems 13 and 14?
(39)

16. Which of the following is equal to $\dfrac{1}{2} + \dfrac{1}{2}$?
(54)

A. $\dfrac{1}{2} - \dfrac{1}{2}$ B. $\left(\dfrac{1}{2}\right)^2$ C. $\dfrac{1}{2} \div \dfrac{1}{2}$

17. Use a factor tree to find the prime factorization of 30.
(65)

18. If three pencils cost a total of 75¢, how much would six
(15) pencils cost?

19. Seven and one half percent is equivalent to the decimal
(41) number 0.075. If the sales-tax rate is $7\frac{1}{2}$%, what is the
sales tax on a $10.00 purchase?

20. One side of a regular pentagon measures 0.8 meter. What
⁽⁶⁰⁾ is the perimeter of the regular pentagon?

21. Twenty minutes is what fraction of an hour?
⁽²⁹⁾

22. The temperature dropped from 12°C to –8°C. This was a
⁽¹⁴⁾ drop of how many degrees?

Refer to the bar graph below to answer problems 23–25.

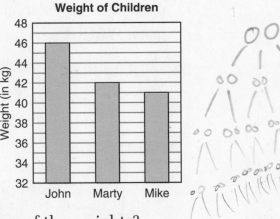

Weight of Children

23. What is the range of the weights?
⁽ᴵⁿᵛ. ⁵⁾

24. What is the mean weight of the three boys?
⁽ᴵⁿᵛ. ⁵⁾

25. Write a story problem with a subtraction pattern that
⁽¹³⁾ relates to the graph, and then answer the problem.

26. Use division by primes to find the prime factorization
⁽⁶⁵⁾ of 400.

27. Simon covered the floor of a square room with 144
⁽³⁸⁾ square floor tiles. How many floor tiles were along each
wall of the room?

28. The weight of a 1-kilogram object is about 2.2 pounds. A
⁽⁴⁶⁾ large man may weigh 100 kilograms. About how many
pounds is that?

29. Reduce: $\dfrac{5 \cdot 5 \cdot 5 \cdot 7}{2 \cdot 2 \cdot 2 \cdot 5 \cdot 5 \cdot 5}$
⁽⁵⁴⁾

30. Which of these polygons is not a regular polygon?
⁽⁶⁰⁾

A.   B.   C.   D.

LESSON
67

Using Prime Factorization to Reduce Fractions

WARM-UP

Facts Practice: 24 Mixed Numbers to Write as Improper Fractions (Test J)

Mental Math: Count down by 2's from 10 to negative 10.

a. 5 × 260   b. 341 − 50   c. 3 × 48
d. $9.25 − 75¢   e. Double $1.25.   f. $\frac{\$30}{100}$
g. 6 × 6, − 1, ÷ 5, × 2, + 1, ÷ 3, ÷ 2
h. Hold your hands one foot apart, then nine inches apart.

Problem Solving:

There are about 520 nine-inch-long noodles in a 1-pound package of spaghetti. Placed end to end, how many **feet** would the noodles in a package of spaghetti reach?

NEW CONCEPT

One way to reduce fractions with large terms is to factor the terms and then reduce the common factors. To reduce $\frac{125}{1000}$, we could begin by writing the prime factorizations of 125 and 1000.

$$\frac{125}{1000} = \frac{5 \cdot 5 \cdot 5}{2 \cdot 2 \cdot 2 \cdot 5 \cdot 5 \cdot 5}$$

We see three pairs of 5's that can be reduced. Each $\frac{5}{5}$ reduces to $\frac{1}{1}$.

$$\frac{\overset{1}{\cancel{5}} \cdot \overset{1}{\cancel{5}} \cdot \overset{1}{\cancel{5}}}{2 \cdot 2 \cdot 2 \cdot \underset{1}{\cancel{5}} \cdot \underset{1}{\cancel{5}} \cdot \underset{1}{\cancel{5}}} = \frac{1}{8}$$

We multiply the remaining factors and find that $\frac{125}{1000}$ reduces to $\frac{1}{8}$.

Example Reduce: $\frac{375}{1000}$

Solution We write the prime factorization of both the numerator and the denominator.

$$\frac{375}{1000} = \frac{3 \cdot 5 \cdot 5 \cdot 5}{2 \cdot 2 \cdot 2 \cdot 5 \cdot 5 \cdot 5}$$

Then we reduce the common factors and multiply the remaining factors.

$$\frac{3 \cdot \overset{1}{\cancel{5}} \cdot \overset{1}{\cancel{5}} \cdot \overset{1}{\cancel{5}}}{2 \cdot 2 \cdot 2 \cdot \underset{1}{\cancel{5}} \cdot \underset{1}{\cancel{5}} \cdot \underset{1}{\cancel{5}}} = \frac{3}{8}$$

We find that $\frac{375}{1000}$ reduces to $\frac{3}{8}$.

LESSON PRACTICE

Practice set Write the prime factorization of both the numerator and the denominator of each fraction. Then reduce each fraction.

a. $\dfrac{875}{1000}$ b. $\dfrac{48}{400}$

c. $\dfrac{125}{500}$ d. $\dfrac{36}{81}$

MIXED PRACTICE

Problem set

1. Allison is making a large collage of a beach scene. She
(66) will use ribbon, magazine clippings, and pieces of cloth to complete the collage. She needs 2 yards of blue ribbon for the ocean, $\frac{1}{2}$ yard of yellow ribbon for the sun, and $\frac{3}{4}$ yard of green ribbon for the grass. Ribbon costs $2 a yard. How much money will Allison need for ribbon?

2. A mile is 5280 feet. A nautical mile is about 6080 feet. A
(13) nautical mile is about how much longer than a mile?

3. Instead of dividing $1.50 by $0.05, Marcus formed an
(43) equivalent division problem by mentally multiplying both the dividend and the divisor by 100. Then he performed the equivalent division problem. What is the equivalent division problem Marcus formed, and what is the quotient?

Find each missing number:

4. $6 \text{ cm} + k = 11 \text{ cm}$
(3)

5. $8g = 9.6$
(43)

6. $\dfrac{7}{10} - w = \dfrac{1}{2}$
(43)

7. $\dfrac{3}{5} = \dfrac{n}{100}$
(42)

8. The combined length of four sticks is 172 inches. What is
(18) the average length of each stick?

9. $100.00 − ($46.75 + $9.68)
(5)

10. (2 × 0.3) − (0.2 × 0.3)
(53)

11. $4\frac{1}{4} - 2\frac{7}{8}$ **12.** $2\frac{2}{3} \times \sqrt{9}$
(63)   (38, 66)

13. $3\frac{1}{3} + 2\frac{3}{4}$ **14.** $1\frac{1}{3} \times 2\frac{1}{4}$
(59)   (66)

15. 1.44 ÷ 60   **16.** $6.00 ÷ $0.15
(45)   (49)

17. Five dollars was divided evenly among 4 people. How
(15) much money did each receive?

18. The area of a regular quadrilateral is 100 square inches.
(60) What is its perimeter?

19. Write the prime factorizations of 625 and of 1000. Then
(67) reduce $\frac{625}{1000}$.

20. What is the area of the rectangle shown below?
(31, 66)

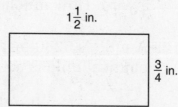

$1\frac{1}{2}$ in.

$\frac{3}{4}$ in.

21. Thirty-six of the 88 piano keys are black. What fraction of
(29) the piano keys are black?

22. Draw a rectangular prism. Begin by drawing two
(Inv. 6) congruent rectangles.

23. $1\frac{1}{2} \times \boxed{} = 1$
(30, 62)

24. There are 1000 meters in a kilometer. How many meters
(15) are in 2.5 kilometers?

25. Which arrow could be pointing to 0.1 on the number line?
(50)

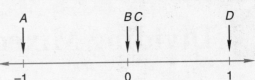

26. If the tip of the minute hand is 6 inches from the center of
(47) the clock, how far does the tip travel in one hour? Round
the answer to the nearest inch. (Use 3.14 for π.)

27. A basketball is an example of which geometric solid?
(Inv. 6)

28. Write 51% as a fraction. Then write the fraction as a
(33, 35) decimal number.

29. What is the probability of rolling a
(19, 58) prime number with one toss of a
dot cube?

30. This quadrilateral has one pair of
(64) parallel sides. What kind of
quadrilateral is it?

when in doubt,
use turtles

LESSON

68 Dividing Mixed Numbers

WARM-UP

Facts Practice: 28 Improper Fractions to Simplify (Test I)

Mental Math: Count down by 5's from 25 to negative 25.

   a. 5 × 80   **b.** 275 + 1500   **c.** 7 × 42

   d. $5.75 + 50¢   **e.** $\frac{1}{4}$ of $48.00   **f.** $\frac{\$120}{10}$

   g. 7 × 8, − 1, ÷ 5, × 2, − 1, ÷ 3, − 8

   h. Hold your hands one meter apart, then 90 cm apart.

Problem Solving:

   Which whole number greater than 90 but less than 100 is a prime number?

NEW CONCEPT

Recall the three steps to solving an arithmetic problem with fractions.

Step 1: Put the problem into the correct shape (if it is not already).

Step 2: Perform the operation indicated.

Step 3: Simplify the answer if possible.

In this lesson we will practice dividing mixed numbers. Recall from Lesson 66 that the correct shape for multiplying and dividing fractions is fraction form. So when dividing, we first write any mixed numbers or whole numbers as improper fractions.

Example 1 Divide: $2\frac{2}{3} \div 4$

Solution We write the numbers as improper fractions.

$$\frac{8}{3} \div \frac{4}{1}$$

To divide, we find the number of 4's in 1. (That is, we find the reciprocal of 4.) Then we use the reciprocal of 4 to find the number of 4's in $\frac{8}{3}$.

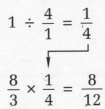

$$1 \div \frac{4}{1} = \frac{1}{4}$$

$$\frac{8}{3} \times \frac{1}{4} = \frac{8}{12}$$

We simplify the answer.

$$\frac{8}{12} = \frac{2}{3}$$

Notice that dividing a number by 4 is equivalent to finding $\frac{1}{4}$ of the number. Instead of dividing $2\frac{2}{3}$ by 4, we could have directly found $\frac{1}{4}$ of $2\frac{2}{3}$.

Example 2 Divide: $2\frac{2}{3} \div 1\frac{1}{2}$

Solution We write the mixed numbers as improper fractions.

$$\frac{8}{3} \div \frac{3}{2}$$

To divide, we find the number of $\frac{3}{2}$'s in 1. $\left(\text{That is, we find the reciprocal of } \frac{3}{2}.\right)$ Then we use the reciprocal of $\frac{3}{2}$ to find the number of $\frac{3}{2}$'s in $\frac{8}{3}$.

$$1 \div \frac{3}{2} = \frac{2}{3}$$

$$\frac{8}{3} \times \frac{2}{3} = \frac{16}{9}$$

We simplify the improper fraction $\frac{16}{9}$ as shown below.

$$\frac{16}{9} = 1\frac{7}{9}$$

LESSON PRACTICE

Practice set* Find each product or quotient:

  a. $1\frac{3}{5} \div 4$ **b.** $\frac{1}{4}$ of $1\frac{3}{5}$

  c. $2\frac{2}{5} \div 3$ **d.** $\frac{1}{3}$ of $2\frac{2}{5}$

e. $1\frac{2}{3} \div 2\frac{1}{2}$ **f.** $2\frac{1}{2} \div 1\frac{2}{3}$

g. $1\frac{1}{2} \div 1\frac{1}{2}$ **h.** $7 \div 1\frac{3}{4}$

MIXED PRACTICE

Problem set

1. What is the difference between the sum of $\frac{1}{2}$ and $\frac{1}{4}$ and the
$^{(12,\ 55)}$ product of $\frac{1}{2}$ and $\frac{1}{4}$?

2. Bill ran a half mile in two minutes fifty-five seconds.
$^{(15)}$ How many seconds is that?

3. The gauge of a railroad—the distance between the two
$^{(15)}$ tracks—is usually 4 feet $8\frac{1}{2}$ inches. How many inches is that?

4. $1\frac{1}{2} \div 2\frac{2}{3}$ **5.** $1\frac{1}{3} \div 4$
$^{(68)}$ $^{(68)}$

6. In six games Yvonne scored a total of 108 points. How
$^{(18)}$ many points per game did she average?

7. Write the prime factorizations of 24 and 200. Then
$^{(67)}$ reduce $\frac{24}{200}$.

Find each missing number:

8. $m - 5\frac{3}{8} = 1\frac{3}{16}$ **9.** $3\frac{3}{5} + 2\frac{7}{10} = n$
$^{(59)}$ $^{(59)}$

10. $25d = 0.375$ **11.** $\frac{3}{4} = \frac{w}{100}$
$^{(45)}$ $^{(42)}$

12. $5\frac{1}{8} - 1\frac{1}{2}$ **13.** $3\frac{1}{3} \times 1\frac{1}{2}$ **14.** $3\frac{1}{3} \div 1\frac{1}{2}$
$^{(63)}$ $^{(66)}$ $^{(68)}$

15. What is the area of a rectangle that is 4 inches long and
$^{(31,\ 66)}$ $1\frac{3}{4}$ inches wide?

16. $(3.2 + 1) - (0.6 \times 7)$ **17.** $12.5 \div 0.4$
$^{(53)}$ $^{(49)}$

18. The product 3.2×10 equals which of the following?
$^{(52)}$

   A. $32 \div 10$ B. $320 \div 10$ C. $0.32 \div 10$

19. Estimate the sum of 6416, 5734, and 4912 to the nearest
(16) thousand.

20. Instead of dividing 800 by 24, Arturo formed an
(43) equivalent division problem by dividing both the
dividend and the divisor by 8. Then he quickly found the
quotient of the equivalent problem. What is the equivalent
problem Arturo formed, and what is the quotient? Write
the quotient as a mixed number.

21. The perimeter of a square is 2.4 meters.
(38)
(a) How long is each side of the square?

(b) What is the area of the square?

22. What is the tax on an $18,000 car if the tax rate is 8%?
(41)

23. If the probability of winning the lottery is 1 chance in a
(58) million, then what is the probability of not winning?

24. Why is a circle not a polygon?
(60)

25. Compare: $\frac{1}{3} \times 4\frac{1}{2} \bigcirc 4\frac{1}{2} \div 3$
(66, 68)

26. Use a ruler to find the length of this line segment to the
(17) nearest eighth of an inch.

27. Which angle in this figure is an obtuse angle?
(28)

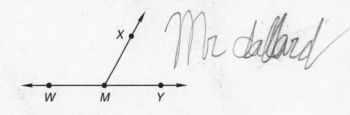

28. Write 3% as a fraction. Then write the fraction as a
(33, 35) decimal number.

29. A shoe box is an example of which geometric solid?
(Inv. 6)

30. Sunrise occurred at 6:20 a.m., and sunset occurred at
(32) 5:45 p.m. How many hours and minutes were there from
sunrise to sunset?

LESSON

69

Lengths of Segments • Complementary and Supplementary Angles

WARM-UP

NEW CONCEPTS

Lengths of segments   Letters are often used to designate points. Recall that we may use two points to identify a line, a ray, or a segment. Below we show a line that passes through points *A* and *B*. This line may be referred to as line *AB* or line *BA*. We may abbreviate line *AB* as $\overleftrightarrow{AB}$.

The ray that begins at point *A* and passes through point *B* is ray *AB*, which may be abbreviated $\overrightarrow{AB}$. The portion of line *AB* between and including points *A* and *B* is segment *AB* (or segment *BA*), which can be abbreviated $\overline{AB}$ (or $\overline{BA}$).

In the figure below we can identify three segments: $\overline{WX}$, $\overline{XY}$, and $\overline{WY}$. The length of $\overline{WX}$ plus the length of $\overline{XY}$ equals the length of $\overline{WY}$.

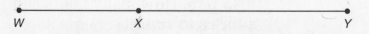

Example 1   In this figure the length of $\overline{LM}$ is 4 cm, and the length of $\overline{LN}$ is 9 cm. What is the length of $\overline{MN}$?

Solution   The length of $\overline{LM}$ plus the length of $\overline{MN}$ equals the length of $\overline{LN}$. With the information in the problem, we can write the equation shown below, where the letter l stands for the unknown length:

$$4 \text{ cm} + l = 9 \text{ cm}$$

Since 4 cm plus 5 cm equals 9 cm, we find that the length of $\overline{MN}$ is **5 cm.**

Complementary and supplementary angles   **Complementary angles** are two angles whose measures total 90°. In the figure on the left below, $\angle PQR$ and $\angle RQS$ are complementary angles. In the figure on the right, $\angle A$ and $\angle B$ are complementary angles.

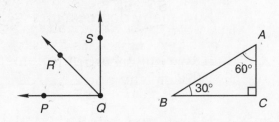

We say that $\angle A$ is the complement of $\angle B$ and that $\angle B$ is the complement of $\angle A$.

Supplementary angles are two angles whose measures total 180°. Below, $\angle 1$ and $\angle 2$ are supplementary, and $\angle A$ and $\angle B$ are supplementary. So $\angle A$ is the supplement of $\angle B$, and $\angle B$ is the supplement of $\angle A$.

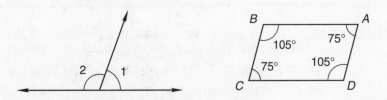

Example 2   In the figure at right, $\angle RWT$ is a right angle.

(a) Which angle is the supplement of $\angle RWS$?

(b) Which angle is the complement of $\angle RWS$?

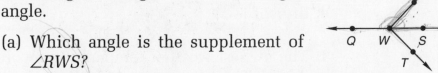

Solution   (a) Supplementary angles total 180°. Angle *QWS* is 180° because it forms a line. So the angle that is the supplement of ∠*RWS* is ∠*QWR* (or ∠*RWQ*).

(b) Complementary angles total 90°. Angle *RWT* is 90° because it is a right angle. So the complement of ∠*RWS* is ∠*SWT* (or ∠*TWS*).

LESSON PRACTICE

Practice set   **a.** In this figure the length of $\overline{AC}$ is 60 mm and the length of $\overline{BC}$ is 26 mm. Find the length of $\overline{AB}$.

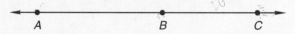

b. The complement of a 60° angle is an angle that measures how many degrees?

c. The supplement of a 60° angle is an angle that measures how many degrees?

d. If two angles are supplementary, can they both be acute? Why or why not?

e. Name two angles in the figure at right that appear to be supplementary.

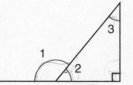

f. Name two angles that appear to be complementary.

MIXED PRACTICE

Problem set   **1.** Draw a pair of parallel lines. Then draw a second pair of
(28, 64)   parallel lines that are perpendicular to the first pair. Trace the quadrilateral that is formed by the intersecting pairs of lines. What kind of quadrilateral did you trace?

2. What is the quotient if the dividend is $\frac{1}{2}$ and the
(54)   divisor is $\frac{1}{8}$?

3. The highest weather temperature recorded was 136°F in
(14)   Africa. The lowest was −127°F in Antarctica. How many degrees difference is there between these temperatures?

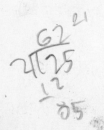

4. A dollar bill is about 6 inches long. Placed end to end,
(15) about how many **feet** would 1000 dollar bills reach?

5. Write the prime factorization of both the numerator and
(67) the denominator of this fraction. Then reduce the fraction.

$$\frac{45}{72}$$

6. In quadrilateral *QRST,* which segment appears to be
(64) parallel to $\overline{RS}$?

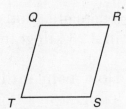

7. In 10 days Donatella saved $27.50. On average, how
(18) much did she save per day?

8. $\dfrac{1 \times 2 \times 3 \times 4 \times 5}{1 + 2 + 3 + 4 + 5}$
(5)

9. $3\frac{1}{2} + 2\frac{3}{4} + 1\frac{5}{8}$
(61)

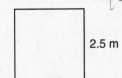

Find each missing number:

10. $m + 1\frac{3}{4} = 5\frac{3}{8}$
(63)

11. $\frac{3}{4} - f = \frac{1}{3}$
(56)

12. $\frac{2}{5}w = 1$
(30)

13. $\frac{8}{25} = \frac{n}{100}$
(42)

14. $1\frac{2}{3} \div 2$
(68)

15. $2\frac{2}{3} \times 1\frac{1}{5}$
(66)

16. $\dfrac{2.4}{0.08}$
(49)

17. (a) What is the perimeter of this
(38)   square?

(b) What is the area of this square?

2.5 m

18. How can you determine whether a counting number is a
(65) composite number?

19. Make a factor tree to find the prime factorization of 250.
(65)

20. A stop sign has the shape of an eight-sided polygon. What
(60) is the name of an eight-sided polygon?

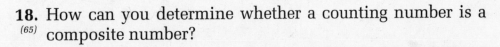

21. There were 15 boys and 12 girls at the birthday party.
(29)

    (a) What fraction of the party was made up of girls?

    (b) What was the ratio of boys to girls at the party?

22. Instead of dividing $4\frac{1}{2}$ by $1\frac{1}{2}$, Carla doubled both numbers
(43) before dividing mentally. What was Carla's mental division problem and its quotient?

23. What is the reciprocal of $2\frac{1}{2}$?
(30, 62)

24. There are 1000 grams in 1 kilogram. How many grams are
(15) in 2.25 kilógrams?

25. How many **millimeters** long is the line below?
(7)

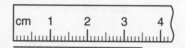

26. The length of $\overline{WX}$ is 53 mm. The length of $\overline{XY}$ is 35 mm.
(69) What is the length of $\overline{WY}$?

27. Draw a cylinder.
(Inv. 6)

28. Arrange these numbers in order from least to greatest:
(50)

$$0.1, \ 1, \ -1, \ 0$$

29. Draw a circle and shade $\frac{1}{4}$ of it. What percent of the circle
(33) is shaded?

30. How many smaller cubes are in the large cube shown
(Inv. 6) below?

LESSON
70   Reducing Fractions Before Multiplying

WARM-UP

Facts Practice: 30 Fractions to Reduce (Test G)

Mental Math: Count up and down by 2's between negative 10 and 10.

a. 5×280 b. $476 + 99$ c. 3×54

d. $\$4.50 + \$1.75$ e. $\frac{1}{3}$ of $\$90.00$ f. $\frac{\$250}{10}$

g. $5 \times 10, \div 2, + 5, \div 2, - 5, \div 10, - 1$

h. Hold your hands one yard apart, then two feet apart.

Problem Solving:

Every even number greater than 2 is not prime but can be written as a product of primes. Chris thinks that every even number greater than 2 can also be written as a sum of primes (e.g., $4 = 2 + 2$, $6 = 3 + 3$, $8 = 5 + 3$). Show how the even numbers from 10 to 20 can be written as sums of prime numbers.

NEW CONCEPT

Before two or more fractions are multiplied, we might be able to reduce the fraction terms, even if the reducing involves different fractions. For example, in the multiplication below we see that the number 3 appears as a numerator and as a denominator in different fractions.

$$\frac{3}{5} \times \frac{2}{3} = \frac{6}{15} \qquad \frac{6}{15} \text{ reduces to } \frac{2}{5}$$

We may reduce the common terms (the 3's) before multiplying. We reduce $\frac{3}{3}$ to $\frac{1}{1}$ by dividing both 3's by 3. Then we multiply the remaining terms.

$$\frac{\overset{1}{\cancel{3}}}{5} \times \frac{2}{\underset{1}{\cancel{3}}} = \frac{2}{5}$$

By reducing before we multiply, we avoid the need to reduce after we multiply. Reducing before multiplying is also known as **canceling**.

Example 1   Simplify: $\frac{5}{6} \times \frac{1}{5}$

Solution   We reduce before we multiply. Since 5 appears as a numerator and as a denominator, we reduce $\frac{5}{5}$ to $\frac{1}{1}$ by dividing both 5's by 5. Then we multiply the remaining terms.

$$\frac{\overset{1}{\cancel{5}}}{6} \times \frac{1}{\underset{1}{\cancel{5}}} = \frac{1}{6}$$

Example 2   Simplify: $1\frac{1}{9} \times 1\frac{1}{5}$

Solution   First we write the numbers in fraction form.

$$\frac{10}{9} \times \frac{6}{5}$$

We mentally pair 10 with 5 and 6 with 9.

We reduce $\frac{10}{5}$ to $\frac{2}{1}$ by dividing both 10 and 5 by 5. We reduce $\frac{6}{9}$ to $\frac{2}{3}$ by dividing both 6 and 9 by 3.

$$\frac{\overset{2}{\cancel{10}}}{\underset{3}{\cancel{9}}} \times \frac{\overset{2}{\cancel{6}}}{\underset{1}{\cancel{5}}} = \frac{4}{3}$$

We multiply the remaining terms. Then we simplify the product.

$$\frac{4}{3} = 1\frac{1}{3}$$

Example 3   Simplify: $\frac{5}{6} \div \frac{5}{2}$

Solution   This is a division problem. We first find the number of $\frac{5}{2}$'s in 1. The answer is the reciprocal of $\frac{5}{2}$. We then use the reciprocal of $\frac{5}{2}$ to find the number of $\frac{5}{2}$'s in $\frac{5}{6}$.

$$1 \div \frac{5}{2} = \frac{2}{5}$$

$$\frac{5}{6} \times \frac{2}{5}$$

Now we have a multiplication problem. We cancel before we multiply.

$$\overset{1}{\underset{3}{\cancel{5}}} \times \overset{1}{\underset{1}{\cancel{2}}} = \frac{1}{3}$$

Note: We may cancel the terms of fractions only when multiplying. A division problem must be rewritten as a multiplication problem before we may cancel the terms of the fractions. We do not cancel the terms of fractions in addition or subtraction problems.

LESSON PRACTICE

Practice set Reduce before multiplying:

$$\textbf{a.}\ \frac{3}{4} \cdot \frac{4}{5} \qquad\qquad \textbf{b.}\ \frac{2}{3} \cdot \frac{3}{4} \qquad\qquad \textbf{c.}\ \frac{8}{9} \cdot \frac{9}{10}$$

Write in fraction form. Then reduce before multiplying.

$$\textbf{d.}\ 2\frac{1}{4} \times 4 \qquad\qquad \textbf{e.}\ 1\frac{1}{2} \times 2\frac{2}{3} \qquad\qquad \textbf{f.}\ 3\frac{1}{3} \times 2\frac{1}{4}$$

Rewrite each division problem as a multiplication problem. Then reduce before multiplying.

$$\textbf{g.}\ \frac{2}{5} \div \frac{2}{3} \qquad\qquad \textbf{h.}\ \frac{8}{9} \div \frac{2}{3} \qquad\qquad \textbf{i.}\ \frac{9}{10} \div 1\frac{1}{5}$$

MIXED PRACTICE

Problem set

1. Alaska was purchased from Russia in 1867 for seven
(12) million, two hundred thousand dollars. Use digits to write that amount.

2. How many eighth notes equal a half note?
(54)

3. Instead of dividing $12\frac{1}{2}$ by $2\frac{1}{2}$, Shannon doubled both
(43) numbers and then divided. Write the division problem Shannon formed, as well as its quotient.

Reduce before multiplying:

$$\textbf{4.}\ \frac{5}{6} \cdot \frac{4}{5} \qquad\qquad \textbf{5.}\ \frac{5}{6} \div \frac{5}{2} \qquad\qquad \textbf{6.}\ \frac{9}{10} \cdot \frac{5}{6}$$
(70) (70) (70)

7. What number is halfway between $\frac{1}{2}$ and 1 on the
(17) number line?

8. $\sqrt{100} + 10^2$
(38)

9. $3\frac{2}{3} + 4\frac{5}{6}$
(59)

10. $7\frac{1}{8} - 2\frac{1}{2}$
(63)

11. $4.37 + 12.8 + 6$
(38)

12. $0.46 \div 5$
(45)

13. $60 \div 0.8$
(49)

14. What is the average of the three numbers marked by the
(18) arrows on this decimal number line? (First estimate
whether the average will be more than 5 or less than 5.)

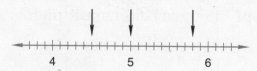

15. The division problem $1.5 \div 0.06$ is equivalent to which
(49) of the following?

    A. $15 \div 6$ B. $150 \div 6$ C. $150 \div 60$

16. There are 1000 milliliters in 1 liter. How many milliliters
(39) are in 3.8 liters?

Find each missing number:

17. $\frac{2}{3} + n = 1$
(43)

18. $\frac{2}{3}m = 1$
(30)

19. $f - \frac{3}{4} = \frac{5}{6}$
(56)

20. A pyramid with a triangular base
(60, Inv. 6) has how many

    (a) faces?

    (b) edges?

    (c) vertices?

Write the numbers in fraction form. Then reduce before
multiplying.

21. $1\frac{2}{3} \times 1\frac{1}{5}$
(70)

22. $\frac{8}{9} \div 2\frac{2}{3}$
(70)

Refer to the line graph below to answer problems 23–25.

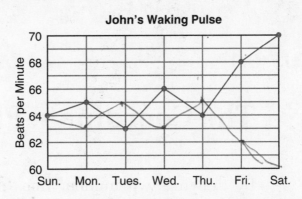

John's Waking Pulse

23. When John woke on Saturday, his pulse was how many
(18) beats per minute more than it was on Tuesday?

24. On Monday John took his pulse for 3 minutes before
(18) marking the graph. How many times did his heart beat in
those 3 minutes?

25. Write a question that relates to the graph and answer the
(18) question.

26. Write the prime factorization of both the numerator and
(67) the denominator of this fraction. Then reduce the fraction.

$$\frac{72}{300}$$

In rectangle *ABCD* the length of $\overline{AB}$ is 2.5 cm, and the length
of $\overline{BC}$ is 1.5 cm. Use this information and the figure to answer
problems 27–30.

27. What is the perimeter of this
(8) rectangle?

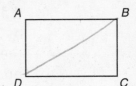

28. What is the area of this rectangle?
(31)

29. Name two segments perpendicular to $\overline{DC}$.
(64)

30. If $\overline{BD}$ were drawn on the figure to divide the rectangle
(31) into two equal parts, what would be the area of each part?

374   *Saxon Math 7/6—Homeschool*

INVESTIGATION 7

Focus on

The Coordinate Plane

By drawing two number lines perpendicular to each other and by extending the unit marks, we can create a grid called a **coordinate plane.**

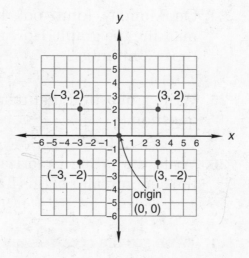

The point at which the number lines intersect is called the **origin.** The horizontal number line is called the **x-axis,** and the vertical number line is called the **y-axis.** We **graph** a point by marking a dot at the location of the point. We can name the location of any point on this coordinate plane with two numbers. The numbers that tell the location of a point are called the **coordinates** of the point.

The coordinates of a point are written as an **ordered pair** of numbers in parentheses; for example, (3, −2). The first number is the *x*-coordinate. It shows the horizontal (↔) direction and distance from the origin. The second number, the *y*-coordinate, shows the vertical (↕) direction and distance from the origin. The sign of the coordinate shows the direction. Positive coordinates are to the right or up, and negative coordinates are to the left or down.

To graph (3, –2), we begin at the origin and move three units to the right along the *x*-axis. From there we move down two units and mark a dot. We may label the point we graphed (3, –2). On the previous coordinate plane, we have graphed three additional points and identified their coordinates. Notice that each pair of coordinates is different and designates a unique point.

Refer to the coordinate plane below to answer problems 1–6.

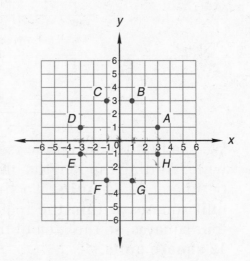

1. What are the coordinates of point *A*?

2. Which point has the coordinates (–1, 3)?

3. What are the coordinates of point *E*?

4. Which point has the coordinates (1, –3)?

5. What are the coordinates of point *D*?

6. Which point has the coordinates (3, –1)?

The coordinate plane is useful in many fields of mathematics, including algebra and geometry.

In the next section of this investigation we will designate points on the plane as vertices of rectangles. Then we will calculate the perimeter and area of each rectangle.

Suppose we are told that the vertices of a rectangle are located at (3, 2), (–1, 2), (–1, –1), and (3, –1). We graph the points and then draw segments between the points to draw the rectangle.

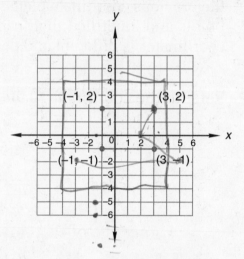

We see that the rectangle is four units long and three units wide. Adding the lengths of the four sides, we find that the perimeter is **14 units.** To find the area, we can count the unit squares within the rectangle. There are three rows of four squares, so the area of the rectangle is 3 × 4, which is **12 square units.**

For problems 7–9, use Activity Sheets 10–12 (available in *Saxon Math 7/6—Homeschool Tests and Worksheets*).

7. The vertices of a rectangle are located at (–2, –1), (2, –1), (2, 3), and (–2, 3).

   (a) Graph the rectangle. What do we call this special type of rectangle?

   (b) What is the perimeter of the rectangle?

   (c) What is the area of the rectangle?

8. The vertices of a rectangle are located at (–4, 2), (0, 2), (0, 0), and (–4, 0).

   (a) Graph the rectangle. Notice that one vertex is located at (0, 0). What is the name for this point on the coordinate plane?

   (b) What is the perimeter of the rectangle?

   (c) What is the area of the rectangle?

9. Three vertices of a rectangle are located at (3, 1), (−2, 1), and (−2, −3).

   (a) Graph the rectangle. What are the coordinates of the fourth vertex?

   (b) What is the perimeter of the rectangle?

   (c) What is the area of the rectangle?

As the following activity illustrates, we can use coordinates to give directions for making a drawing.

Activity: *Drawing on the Coordinate Plane*

Materials needed:

- Activity Sheets 13–15 (available in *Saxon Math 7/6—Homeschool Tests and Worksheets*)

10. Christy made this drawing on a coordinate plane. Then she wrote directions for making the drawing. Follow Christy's directions below to make a similar drawing on your coordinate plane. The coordinates of the vertices are listed in order, as in a "dot-to-dot" drawing.

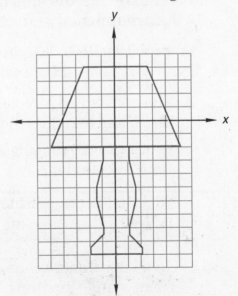

Draw segments to connect the following points in order:

   a. (−1, −2)   **b.** (−1, −3)   **c.** $(-1\frac{1}{2}, -5)$ **d.** $(-1\frac{1}{2}, -6)$

   e. (−1, −8)   **f.** $(-1, -8\frac{1}{2})$ **g.** $(-2, -9\frac{1}{2})$ **h.** (−2, −10)

   i. (2, −10)   **j.** $(2, -9\frac{1}{2})$ **k.** $(1, -8\frac{1}{2})$ **l.** (1, −8)

   m. $(1\frac{1}{2}, -6)$ **n.** $(1\frac{1}{2}, -5)$ **o.** (1, −3)   **p.** (1, −2)

Lift your pencil and restart:

a. $(-2\frac{1}{2}, 4)$ **b.** $(2\frac{1}{2}, 4)$ **c.** $(5, -2)$

d. $(-5, -2)$ **e.** $(-2\frac{1}{2}, 4)$

11. Jenny wrote the following directions for a drawing. Follow her directions to make the drawing on your own paper. Draw segments to connect the following points in order:

a. $(-9, 0)$ **b.** $(6, -1)$ **c.** $(8, 0)$

d. $(7, 1)$ **e.** $(6, \frac{1}{2})$ **f.** $(6, -1)$

g. $(9, -2\frac{1}{2})$ **h.** $(10, -2)$ **i.** $(7, 1)$

j. $(6, 1\frac{1}{2})$ **k.** $(-10\frac{1}{2}, 3)$ **l.** $(-11, 2)$

m. $(-10\frac{1}{2}, 0)$ **n.** $(-10, -1\frac{1}{2})$ **o.** $(9, -2\frac{1}{2})$

p. $(-3, -3\frac{1}{2})$ **q.** $(-7, -8)$ **r.** $(-10, -8)$

s. $(-9, -1\frac{1}{2})$

Lift your pencil and restart:

a. $(-10\frac{1}{2}, 0)$ **b.** $(-11, -\frac{1}{2})$ **c.** $(-12, \frac{1}{2})$

d. $(-11\frac{1}{2}, 1)$ **e.** $(-12, 1\frac{1}{2})$ **f.** $(-11\frac{1}{2}, 2)$

g. $(-12, 2\frac{1}{2})$ **h.** $(-11, 3\frac{1}{2})$ **i.** $(-10\frac{1}{2}, 3)$

j. $(-11\frac{1}{2}, 8)$ **k.** $(-9\frac{1}{2}, 8)$ **l.** $(-7, 3)$

m. $(-6, 2\frac{1}{2})$ **n.** $(-7, 3)$ **o.** $(-6, 5)$

p. $(-4, 5)$ **q.** $(-1, 2)$

12. On a coordinate plane, make a straight-segment drawing. Then write directions for making the drawing by listing the coordinates of the vertices in "dot-to-dot" order.

Note: Problems intended for additional exposure to the concepts in this investigation are available in the appendix.

LESSON
71   Parallelograms

WARM-UP

Facts Practice: 64 Multiplication Facts (Test D)

Mental Math: Count by 12's from 12 to 144.

a. 5 × 480   b. 367 − 99   c. 8 × 43

d. $10.00 − $8.75   e. Double $2.25.   f. $\frac{\$250}{100}$

g. 8 × 9, + 3, ÷ 3, × 2, − 10, ÷ 5, + 3, ÷ 11

h. Hold your hands one yard apart, then one meter apart.

Problem Solving:

Copy this factor tree and fill in the missing numbers:

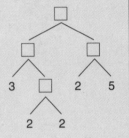

NEW CONCEPT

In this lesson we will learn about various properties of parallelograms. The following example describes some angle properties of parallelograms.

Example 1   In parallelogram *ABCD*, the measure of angle *A* is 60°.

(a) What is the measure of ∠*C*?

(b) What is the measure of ∠*B*?

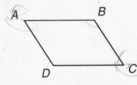

Solution   (a) Angles *A* and *C* are opposite angles in that they are opposite to each other in the parallelogram. The opposite angles of a parallelogram have equal measures. So the measure of angle *C* equals the measure of angle *A*. Thus the measure of ∠*C* is **60°**.

(b) Angles *A* and *B* are adjacent angles in that they share a side. (Side *AB* is a side of ∠*A* and a side of ∠*B*.) The adjacent angles of a parallelogram are supplementary. So ∠*A* and ∠*B* are supplementary, which means their measures total 180°. Since ∠*A* measures 60°, ∠*B* must measure **120°** for their sum to be 180°.

A flexible model of a parallelogram is useful for illustrating some properties of a parallelogram. A model can be constructed of brads and stiff tagboard or cardboard.

Lay two 8-in. strips of tagboard or cardboard over two parallel 10-in. strips as shown. Punch a hole at the center of the overlapping ends. Then fasten the corners with brads to hold the strips together.

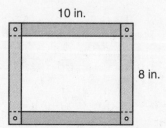

10 in.

8 in.

If we move the sides of the parallelogram back and forth, we see that opposite sides always remain parallel and equal in length. Though the angles change size, opposite angles remain equal and adjacent angles remain supplementary.

With this model we also can observe how the area of a parallelogram changes as the angles change. We hold the model with two hands and slide opposite sides in opposite directions. The maximum area occurs when the angles are 90°. The area reduces to zero as opposite sides come together.

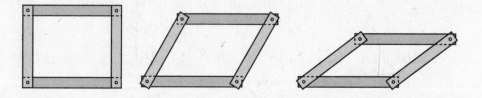

Although the area of a parallelogram changes as the angles change, the perimeter does not change.

The flexible model shows that parallelograms may have sides that are equal in length but areas that are different. To find the area of a parallelogram, we multiply two **perpendicular** measurements. We multiply the **base** by the **height** of the parallelogram.

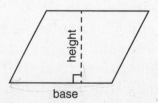

height

base

The base of a parallelogram is the length of one of the sides. The height of a parallelogram is the perpendicular distance from the base to the opposite side. The following activity will illustrate why the area of a parallelogram equals the base times the height.

Activity: *Area of a Parallelogram*

Materials needed:

- graph paper
- ruler
- pencil
- scissors

Tracing over the lines on the graph paper, draw two parallel segments the same number of units long but shifted slightly as shown.

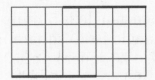

Then draw segments between the endpoints of the pair of parallel segments to complete the parallelogram.

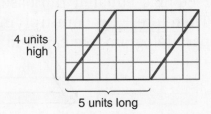

The base of the parallelogram we drew has a length of 5 units. The height of the parallelogram is 4 units. Your parallelogram might be different. How many units long and high is your parallelogram? Can you easily count the number of square units in the area of your parallelogram?

Use scissors to cut out your parallelogram.

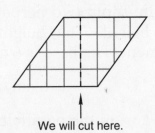

We will cut here.

Then select a line on the graph paper that is perpendicular to opposite sides of the parallelogram and that crosses both of these sides. Cut the parallelogram into two pieces along this line.

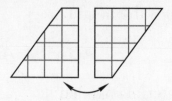

Rearrange the two pieces of the parallelogram to form a rectangle. What is the length and width of the rectangle? How many square units is the area of the rectangle?

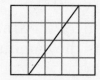

Our rectangle is 5 units long and 4 units wide. The area of the rectangle is 20 square units. So the area of the parallelogram is also 20 square units.

By making a perpendicular cut across the parallelogram and rearranging the pieces, we formed a rectangle having the same area as the parallelogram. The length and width of the rectangle equaled the base and height of the parallelogram. Therefore, by multiplying the base and height of a parallelogram, we can find its area.

Example 2 Find the area of this parallelogram:

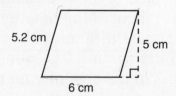

Solution We multiply two perpendicular measurements, the base and the height. The height is often shown as a dashed line segment. The base is 6 cm. The height is 5 cm.

$$6 \text{ cm} \times 5 \text{ cm} = 30 \text{ sq. cm}$$

The area of the parallelogram is **30 sq. cm.**

LESSON PRACTICE

Practice set   Refer to parallelogram *QRST* to answer problems **a–d**.

a. Which angle is opposite ∠Q?

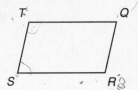

b. Which angle is opposite ∠T?

c. Name two angles that are supplements of ∠T.

d. If the measure of ∠R is 100°, what is the measure of ∠Q?

Calculate the perimeter and area of each parallelogram:

e.

10 m   12 m   8 m

f.
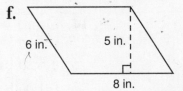
6 in.   5 in.   8 in.

g. A formula for finding the area of a parallelogram is *A* = *bh*. This formula means

$$Area = base \times height$$

The base is the length of one side. The height is the perpendicular distance to the opposite side. Here we show the same parallelogram in two different positions, so the area of the parallelogram is the same in both drawings. What is the height in the figure on the right?

9 cm   6 cm   12 cm   12 cm   *h*   9 cm

MIXED PRACTICE

Problem set   1. What is the least common multiple of 6 and 10?
     (30)

2. The highest point on land is Mt. Everest, whose peak is
   (14) 29,035 feet above sea level. The lowest point on land is the Dead Sea, which dips to 1348 feet below sea level. What is the difference in elevation between these two points?

3. The movie lasted 105 minutes. If the movie started at
   (32) 1:15 p.m., at what time did it end?

In problems 4–7, reduce the fractions, if possible, before multiplying.

4. $\dfrac{2}{3} \cdot \dfrac{3}{8}$
(70)

5. $1\dfrac{1}{4} \cdot 2\dfrac{2}{3}$
(70)

6. $\dfrac{3}{4} \div \dfrac{3}{8}$
(70)

7. $4\dfrac{1}{2} \div 6$
(70)

8. $6 + 3\dfrac{3}{4} + 2\dfrac{1}{2}$
(59)

9. $5 - 3\dfrac{1}{8}$
(63)

10. $5\dfrac{1}{4} - 1\dfrac{7}{8}$
(63)

11. $(3.5)^2$
(39)

12. $15\overline{)\$75.00}$
(2)

13. $(1 + 0.6) \div (1 - 0.6)$
(53)

14. Quan ordered a $4.50 meal at the drive-in. The tax rate
(41)   was $7\frac{1}{2}\%$ (which equals 0.075). He paid for the meal with a $20 bill.

(a) What was the tax on the meal?

(b) What was the total price including tax?

(c) How much money should Quan get back from his payment?

15. What is the name for the point on the coordinate plane
(Inv. 7)   that has the coordinates (0, 0)?

Refer to the coordinate plane below to answer problems 16 and 17.

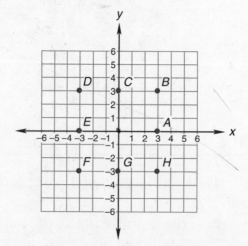

16. Name the points that have the following coordinates:
(Inv. 7)
(a) (–3, 3)   (b) (0, –3)

17. Identify the coordinates of the following points:
(Inv. 7)
(a) *H*   (b) *E*

Find each missing number:

18. $1.2f = 120$
(49)

19. $\dfrac{120}{f} = 1.2$
(49)

20. Write the prime factorization of both the numerator and
(67) the denominator of this fraction. Then reduce the fraction.

$$\frac{64}{224}$$

21. The perimeter of a square is 6.4 meters. What is its area?
(38)

22. What fraction of this circle is not
(Inv. 2) shaded?

23. If the radius of this circle is 1 cm,
(47) what is the circumference of the
circle? (Use 3.14 for π.)

24. A centimeter is about as long as this segment: ———.
(7) About how many centimeters long is your little finger?

25. Water freezes at 32° Fahrenheit.
(10) The temperature shown on the
thermometer is how many degrees
Fahrenheit above the freezing point
of water?

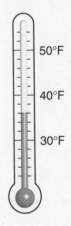

26. Ray watched TV for one hour. He determined that
(29, 33) commercials were shown 20% of that hour. Write 20% as
a reduced fraction. Then find the number of minutes that
commercials were shown during the hour.

27. Name the geometric solid shown
(Inv. 6) at right.

28. This square and regular triangle
(60) share a common side. The perimeter
of the square is 24 cm. What is the
perimeter of the triangle?

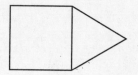

29. (a) What is the perimeter of this
(71) parallelogram?

(b) What is the area of this
parallelogram?

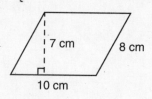

30. In this figure ∠BMD is a right
(69) angle. Name two angles that are

(a) supplementary.

(b) complementary.

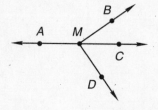

LESSON
72
Fractions Chart •
Multiplying Three Fractions

Facts Practice: 72 Multiplication and Division Facts (Test H)

Mental Math: Count up and down by $\frac{1}{8}$'s between $\frac{1}{8}$ and 2.

   a. 3 × 125   **b.** 275 + 50

   c. 3 × $0.99 *(3 × $1.00 − 3 × 1¢)*   **d.** $20.00 − $9.99

   e. $\frac{1}{3}$ of $6.60   **f.** $2.50 × 10

   g. 2 × 2, × 2, × 2, × 2, − 2, ÷ 2

   h. Hold your hands one foot apart, then six inches apart.

Problem Solving:

Kioko was thinking of two numbers whose average was 24. If one of the numbers was half of 24, what was the other number?

NEW CONCEPTS

Fractions chart

We have learned three steps to take when performing pencil-and-paper arithmetic with fractions and mixed numbers:

Step 1: Write the problem in the correct **shape.**

Step 2: Perform the **operation.**

Step 3: **Simplify** the answer.

The letters S.O.S. can help us remember the steps as "shape," "operate," and "simplify." We summarize the rules we have learned in the following fractions chart. Below the + and − symbols we list the steps we take when adding or subtracting fractions. Below the × and ÷ symbols, we list the steps we take when multiplying or dividing fractions.

Fractions Chart

	+ −	× ÷	
1. Shape	Write fractions with common denominators.	Write numbers in fraction form.	
2. Operate	Add or subtract the numerators.	× cancel. ↰ $\frac{n \times n}{d \times d}$	÷ Find reciprocal of divisor; then ...
3. Simplify	Reduce fractions. Convert improper fractions.		

The "shape" step for addition and subtraction is the same; we write the fractions with common denominators. Likewise, the "shape" step for multiplication and division is the same; we write both numbers in fraction form.

At the "operate" step, however, we separate multiplication and division. When multiplying fractions, we may reduce (cancel) before we multiply. Then we multiply the numerators to find the numerator of the product, and we multiply the denominators to find the denominator of the product. When dividing fractions, we first replace the divisor of the division problem with its reciprocal and change the division problem to a multiplication problem. We cancel terms, if possible, and then multiply.

The "simplify" step is the same for all four operations. We reduce answers when possible and convert answers that are improper fractions to mixed numbers.

Multiplying three fractions

To multiply three or more fractions, we follow the same steps we take when multiplying two fractions:

Step 1: We write the numbers in fraction form.

Step 2: We cancel terms by reducing numerator-denominator pairs that have common factors. Then we multiply the remaining terms.

Step 3: We simplify if possible.

Example Multiply: $\frac{2}{3} \times 1\frac{3}{5} \times \frac{3}{4}$

Solution First we write $1\frac{3}{5}$ as the improper fraction $\frac{8}{5}$. Then we reduce where possible before multiplying. Multiplying the remaining terms, we find the product.

$$\frac{2}{3} \times \frac{8}{5} \times \frac{3}{4} = \frac{4}{5}$$

LESSON PRACTICE

Practice set   **a.** Draw the fractions chart from this lesson.

b. Describe the three steps for adding fractions.

c. Describe the steps for dividing fractions.

Multiply:

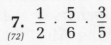

d. $\dfrac{2}{3} \cdot \dfrac{4}{5} \cdot \dfrac{3}{8}$ **e.** $2\dfrac{1}{2} \times 1\dfrac{1}{10} \times 4$

MIXED PRACTICE

Problem set   **1.** What is the average of 4.2, 2.61, and 3.6?
(18)

2. Four tablespoons equals $\dfrac{1}{4}$ cup. How many tablespoons
(54) would equal one full cup?

3. The temperature on the moon ranges from a high of
(14) 134^{d}C to a low of about −170°C. This is a difference of
how many degrees?

4. Four of the 12 marbles in the bag are blue. If one marble
(58) is taken from the bag, what is the probability that the
marble is

(a) blue?   (b) not blue?

5. The diameter of a circle is 1 meter. The circumference is
(7, 47) how many centimeters? (Use 3.14 for π.)

6. What fraction of a dollar is a nickel?
(29)

7. $\dfrac{1}{2} \cdot \dfrac{5}{6} \cdot \dfrac{3}{5}$ **8.** $3 \times 1\dfrac{1}{2} \times 2\dfrac{2}{3}$
(72)   *(72)*

9. $\dfrac{3}{4} \div 2$ **10.** $1\dfrac{1}{2} \div 1\dfrac{2}{3}$
(54)   *(68)*

11. (0.12)(0.24)   **12.** $0.6 \div 0.25$
(39)   *(49)*

Find each missing number:

13. $n - \dfrac{1}{2} = \dfrac{3}{5}$
(56)

14. $1 - w = \dfrac{7}{12}$
(43)

15. $w + 2\dfrac{1}{2} = 3\dfrac{1}{3}$
(59)

16. $1 - w = 0.23$
(43)

17. Write the standard decimal number for the following:
(46)

$$(6 \times 10) + \left(4 \times \dfrac{1}{10}\right) + \left(3 \times \dfrac{1}{100}\right)$$

18. Which of these numbers is closest to 1?
(50)

    A. -1 B. 0.1 C. 10

19. What is the largest prime number that is less than 100?
(19)

20. Which of these figures is not a parallelogram?
(64)

    A.   B.   C.   D.

21. Figure *ABCD* is a rectangle.
(69)

    (a) Name an angle complementary to $\angle DCM$.

    (b) Name an angle supplementary to $\angle AMC$.

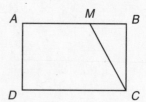

22. A loop of string two feet around is formed to make a square.
(38)

    (a) How many inches long is each side of the square?

    (b) What is the area of the square in square inches?

Refer to this menu and the information that follows to answer problems 23–25.

Menu

Grilled Chicken Sandwich	$3.49	Drinks: Small	$0.89
Taco Salad	$3.29	Medium	$1.09
Pasta Salad	$2.89	Large	$1.29

From this menu the Johnsons ordered two grilled chicken sandwiches, one taco salad, one small drink, and two medium drinks.

23. What was the total price of the Johnsons' order?
(1)

24. If 7% tax is added to the bill, and if the Johnsons pay for
$^{(41)}$ the food with a $20 bill, how much money should they
get back?

25. Make up an order from the menu. Then calculate the bill,
$^{(1)}$ not including tax.

26. If $A = lw$, and if l equals 2.5 and w equals 0.4, what does
$^{(47)}$ A equal?

27. Write the prime factorization of both the numerator and
$^{(66)}$ the denominator of this fraction. Then reduce the
fraction.

$$\frac{72}{120}$$

Refer to the coordinate plane below to answer problems
28 and 29.

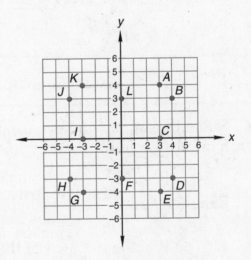

28. Identify the coordinates of the following points:
$^{(Inv. 7)}$

(a) K (b) F

29. Name the points that have the following coordinates:
$^{(Inv. 7)}$

(a) $(3, -4)$ (b) $(-3, 0)$

30. Draw a pair of parallel lines. Then draw a second pair of
$^{(64)}$ parallel lines perpendicular to the first pair of lines and
about the same distance apart. Trace the quadrilateral
that is formed by the intersecting lines. Is the
quadrilateral a rectangle?

LESSON
73

Exponents • Writing Decimal Numbers as Fractions, Part 2

WARM-UP

Facts Practice: 24 Mixed Numbers to Write as Improper Fractions (Test J)

Mental Math: Count up and down by 25's between 25 and 300.

a. 4×112
b. $475 - 150$
c. $4 \times \$0.99$
d. $\$2.99 + \$1.99$
e. Double $3.50.
f. $\$3.50 \div 10$
g. $3 \times 3, \times 3, + 3, \div 3, - 3, \times 3$
h. Hold your hands one yard apart, then one meter apart.

Problem Solving:

Describe the steps from the fractions chart that would be used to find the answer to the following problem:

$$\frac{3}{4} + \frac{2}{3}$$

NEW CONCEPTS

Exponents Since Lesson 38 we have used the exponent 2 to indicate that a number is multiplied by itself.

$$5^2 \text{ means } 5 \cdot 5$$

Exponents indicate repeated multiplication, so

$$5^3 \text{ means } 5 \cdot 5 \cdot 5$$
$$5^4 \text{ means } 5 \cdot 5 \cdot 5 \cdot 5$$

The exponent indicates how many times the base is used as a factor.

We read numbers with exponents as **powers**. Note that when the exponent is 2, we usually say "squared," and when the exponent is 3, we usually say "cubed." The following examples show how we read expressions with exponents:

5^2 "five to the second power" or "five squared"

10^3 "ten to the third power" or "ten cubed"

3^4 "three to the fourth power"

2^5 "two to the fifth power"

Example 1 Compare: $3^4 \bigcirc 4^3$

Solution We find the value of each expression.

3^4 means $3 \cdot 3 \cdot 3 \cdot 3$, which equals 81.

4^3 means $4 \cdot 4 \cdot 4$, which equals 64.

Since 81 is greater than 64, we find that 3^4 is greater than 4^3.

$$3^4 > 4^3$$

Example 2 Write the prime factorization of 1000, using exponents to group factors.

Solution Using a factor tree or division by primes, we find the prime factorization of 1000.

$$1000 = 2 \cdot 2 \cdot 2 \cdot 5 \cdot 5 \cdot 5$$

We group the three 2's and the three 5's with exponents.

$$1000 = 2^3 \cdot 5^3$$

Example 3 Simplify: $100 - 10^2$

Solution We perform operations with exponents before we add, subtract, multiply, or divide. Ten squared is 100. So when we subtract 10^2 from 100, the difference is zero.

$$100 - 10^2$$
$$100 - 100 = 0$$

Writing decimal numbers as fractions, part 2 We will review changing a decimal number to a fraction or mixed number. Recall from Lesson 35 that the number of places after the decimal point indicates the denominator of the decimal fraction (10 or 100 or 1000, etc.). The digits to the right of the decimal point make up the numerator of the fraction.

Example 4 Write 0.5 as a common fraction.

Solution We read 0.5 as "five tenths," which also names the fraction $\frac{5}{10}$. We reduce the fraction.

$$\frac{5}{10} = \frac{1}{2}$$

Example 5   Write 3.75 as a mixed number.

Solution   The whole-number part of 3.75 is 3, and the fraction part is 0.75. Since 0.75 has two decimal places, the denominator is 100.

$$3.75 = 3\frac{75}{100}$$

We reduce the fraction.

$$3\frac{75}{100} = 3\frac{3}{4}$$

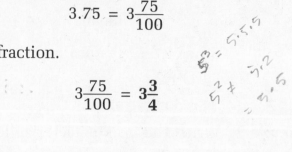

LESSON PRACTICE

Practice set*   Find the value of each expression:

a. 10^4 b. $2^3 + 2^4$ c. $2^2 \cdot 5^2$

d. Write the prime factorization of 72 using exponents.

Write each decimal number as a fraction or mixed number:

e. 12.5   f. 1.25   g. 0.125

h. 0.05   i. 0.24   j. 10.2

MIXED PRACTICE

Problem set   **1.** Tomas's temperature was 102°F. Normal body
      (38)   temperature is 98.6°F. How many degrees above normal
                 was Tomas's temperature? (Write an equation and solve
                 the problem.)

2. Jill has read 42 pages of a 180-page book. How many
(11)   pages are left for her to read? (Write an equation and
        solve the problem.)

3. If Jill wants to finish the book in the next three days, then
(18)   she should read an average of how many pages per day?
        (Write an equation and solve the problem.)

4. Write 2.5 as a reduced mixed number.
(73)

5. Write 0.35 as a reduced fraction.
(73)

6. What is the total cost of a $12.60 item when $7\frac{1}{2}\%$ (0.075)
(41)   sales tax is added?

7. $\frac{3}{4} \times 2 \times 1\frac{1}{3}$
(72)

8. $(100 - 10^2) \div 5^2$
(73)

9. $3 + 2\frac{1}{3} + 1\frac{3}{4}$
(61)

10. $5\frac{1}{6} - 3\frac{1}{2}$
(63)

11. $\frac{3}{4} \div 1\frac{1}{2}$
(68)

12. $7 \div 0.4$
(49)

13. Compare:
(44, 73)

(a) $5^2 \bigcirc 2^5$

(b) $0.3 \bigcirc 0.125$

14. The diameter of a quarter is about 2.4 cm.
(47)

(a) What is the circumference of a quarter? (Use 3.14 for π.)

(b) What is the ratio of the radius of the quarter to the diameter of the quarter?

Find each missing number:

15. $25m = 0.175$
(45)

16. $1.2 + y + 4.25 = 7$
(43)

17. Which digit is in the ten-thousands place in 123,456.78?
(34)

18. Arrange these numbers in order from least to greatest:
(56)

$$1, \frac{1}{2}, \frac{1}{10}, \frac{1}{4}, 0$$

19. Write the prime factorization of 200 using exponents.
(73)

20. The store offered a 20% discount on all tools. The regular price of a hammer was $18.00.
(41)

(a) How much money is 20% of $18.00?

(b) What was the price of the hammer after the discount?

21. The length of $\overline{AB}$ is 16 mm. The length of $\overline{AC}$ is 50 mm. What is the length of $\overline{BC}$?
(69)

22. One half of the area of this square is shaded. What is the area of the shaded region?
(31)

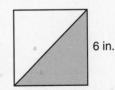

6 in.

23. Is every square a rectangle?
(64)

24. $\dfrac{2^2 + 2^3}{2}$
(73)

25. The fractions chart from Lesson 72 says that the proper
(72) "shape" for multiplying fractions is "fraction form."
What does that mean?

Refer to this coordinate plane to answer problems 26 and 27.

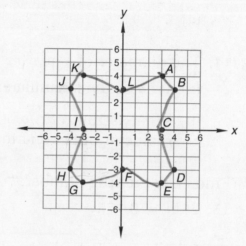

26. Identify the coordinates of the following points:
(Inv. 7)
(a) *H*   (b) *L*

27. Name the points that have the following coordinates:
(Inv. 7)
(a) (−4, 3)   (b) (3, 0)

28. If *s* equals 9, what does s^2 equal?
(73)

29. Draw a cylinder.
(Inv. 6)

30. The measure of ∠*W* in parallelogram
(71) *WXYZ* is 75°.

(a) What is the measure of ∠*X*?

(b) What is the measure of ∠*Y*?

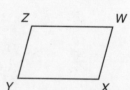

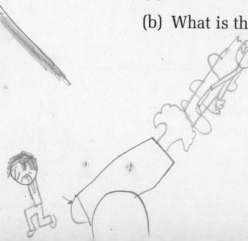

LESSON
74

Writing Fractions as Decimal Numbers

WARM-UP

Facts Practice: 28 Improper Fractions to Simplify (Test I)

Mental Math: Count up and down by 5's between negative 25 and 25.

   a. 3 × 230   **b.** 430 + 270   **c.** 5 × $0.99

   d. $5.00 − $1.98   **e.** $\frac{1}{4}$ of $2.40   **f.** $1.25 × 10

   g. 5 × 5, − 5, × 5, ÷ 2, + 5, ÷ 5

   h. Hold your hands one meter apart, then 100 centimeters apart.

Problem Solving:

In this figure a square and a regular pentagon share a common side. The area of the square is 25 square centimeters. What is the perimeter of the pentagon?

NEW CONCEPT

We learned earlier that a fraction bar indicates division. So the fraction $\frac{1}{2}$ also means "1 divided by 2," which we can write as $2\overline{)1}$. By attaching a decimal point and zero, we can perform the division and write the quotient as a decimal number.

$$\frac{1}{2} \longrightarrow \begin{array}{r} 0.5 \\ 2\overline{)1.0} \\ \underline{1\ 0} \\ 0 \end{array}$$

We find that $\frac{1}{2}$ equals the decimal number 0.5. To convert a fraction to a decimal number, we divide the numerator by the denominator.

Example 1   Convert $\frac{1}{4}$ to a decimal number.

Solution   The fraction $\frac{1}{4}$ means "1 divided by 4," which is $4\overline{)1}$. By attaching a decimal point and zeros, we can complete the division.

$$\begin{array}{r} \mathbf{0.25} \\ 4\overline{)1.00} \\ \underline{8} \\ 20 \\ \underline{20} \\ 0 \end{array}$$

D:

Example 2 Use a calculator to convert $\frac{15}{16}$ to a decimal number.

Solution We begin by clearing the calculator. Then we enter the fraction with these keystrokes.

$$\boxed{1}\ \boxed{5}\ \boxed{\div}\ \boxed{1}\ \boxed{6}\ \boxed{=}$$

After pressing the equal sign, the display shows the decimal equivalent of $\frac{15}{16}$:

0.9375

The answer is reasonable because both $\frac{15}{16}$ and 0.9375 are less than but close to 1.

Example 3 Write $7\frac{2}{5}$ as a decimal number.

Solution The whole number part of $7\frac{2}{5}$ is 7, which we write to the left of the decimal point. We convert $\frac{2}{5}$ to a decimal by dividing 2 by 5.

$$\frac{2}{5} \longrightarrow 5\overline{)2.0}^{\,0.4}$$

Since $\frac{2}{5}$ equals 0.4, the mixed number $7\frac{2}{5}$ equals **7.4.**

LESSON PRACTICE

Practice set* Convert each fraction or mixed number to a decimal number:

    a. $\frac{3}{4}$
    b. $4\frac{1}{5}$
    c. $\frac{1}{8}$

    d. $\frac{7}{20}$
    e. $3\frac{3}{10}$
    f. $\frac{7}{25}$

You may use a calculator to convert these numbers to decimal numbers:

    g. $\frac{11}{16}$
    h. $\frac{31}{32}$
    i. $3\frac{24}{64}$

MIXED PRACTICE

Problem set **1.** What is the difference when five squared is subtracted
(73) from four cubed?

  2. On a certain map, 1 inch represents a distance of 10 miles.
(15) How many miles apart are two towns that are 3 inches apart on the map?

  3. Steve hit the baseball 400 feet. Lesley hit the golf ball 300
(13) yards. How many feet farther did the golf ball travel than the baseball? After converting yards to feet, write an equation and solve the problem.

4. Convert $2\frac{3}{4}$ to a decimal number.
(74)

5. To what decimal number is $\frac{4}{5}$ equal?
(74)

6. Write 0.24 as a reduced fraction.
(73)

7. If $A = bh$, and if b equals 12 and h equals 8, then what
(2) does A equal?

8. Compare: $3^2 \bigcirc 3 + 3$
(73)

9. $\dfrac{1}{2} + \dfrac{2}{3} + \dfrac{1}{6}$
(61)

10. $3\dfrac{1}{4} - 1\dfrac{7}{8}$
(63)

11. $\dfrac{5}{8} \cdot \dfrac{3}{5} \cdot \dfrac{4}{5}$
(72)

12. $3\dfrac{1}{3} \times 3$
(66)

13. $\dfrac{3}{4} \div 1\dfrac{1}{2}$
(68)

14. $(4 + 3.2) - 0.01$
(38)

15. Draw a triangular prism.
(Inv. 6)

16. Nancy bought a dozen golf balls for $10.44. What was the
(15) cost of each golf ball? (Write an equation and solve the
problem.)

17. Estimate the product of 81 and 38.
(16)

18. In four days Jill read 42 pages, 46 pages, 35 pages, and
(18) 57 pages. What was the average number of pages she
read per day?

19. What is the least common multiple of 6, 8, and 12?
(30)

20. $24 + c + 96 = 150$
(3)

21. Write the prime factorization of both the numerator and
(67) the denominator of this fraction. Then reduce the fraction.

$$\frac{40}{96}$$

22. If the perimeter of this square is
(47) 40 centimeters, then

    (a) what is the diameter of the
circle?

    (b) what is the circumference of
the circle? (Use 3.14 for π.)

23. Twenty-four of the three dozen cyclists rode mountain
(29) bikes. What fraction of the cyclists rode mountain bikes?

24. Why are some rectangles not squares?
(64)

25. Which arrow could be pointing to $\frac{3}{4}$?
(17)

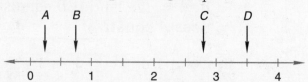

26. In quadrilateral *PQRS,* which
(64) segment appears to be
(a) parallel to $\overline{PQ}$?

(b) perpendicular to $\overline{PQ}$?

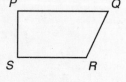

27. The figure at right shows a cube
(Inv. 6) with edges 3 feet long.
(a) What is the area of each face of
the cube?

(b) What is the total surface area of
the cube?

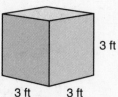

Refer to this coordinate plane to answer problems 28 and 29.

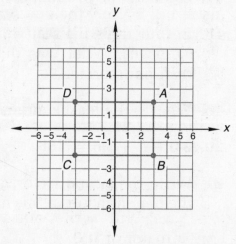

28. Identify the coordinates of the following points:
(Inv. 7)
(a) *C*   (b) origin

29. One pair of parallel segments in rectangle *ABCD* is $\overline{AB}$ and
(64) $\overline{DC}$. Name a second pair of parallel segments.

30. Farmer Ruiz planted corn on 60% of his 300 acres. Find
(41) the number of acres planted with corn.

LESSON
75

Writing Fractions and Decimals as Percents, Part 1

WARM-UP

Facts Practice: Linear Measurement (Test K)

Mental Math: Count by 12's from 12 to 144.

a. 504×6
b. $625 - 250$
c. $3 \times \$1.99$
d. $\$2.50 + \$1.99$
e. Double $\$1.60$.
f. $\$12.50 \div 10$
g. $6 \times 6, -6, \div 6, -5, \times 2, +1$
h. Hold your hands one yard apart, then 18 inches apart.

Problem Solving:

Describe the steps from the fractions chart that would be used to find the answer to the following problem:

$$3\frac{1}{3} \div 2\frac{1}{2}$$

NEW CONCEPT

A percent is actually a fraction with a denominator of 100. Instead of writing the denominator 100, we can use a percent sign (%). So $\frac{25}{100}$ equals 25%.

Example 1 Write $\frac{3}{100}$ as a percent.

Solution A percent is a fraction with a denominator of 100. Instead of writing the denominator, we write a percent sign. We write $\frac{3}{100}$ as **3%**.

Example 2 Write $\frac{3}{10}$ as a percent.

Solution First we will write an equivalent fraction that has a denominator of 100.

$$\frac{3}{10} = \frac{?}{100}$$

We multiply $\frac{3}{10}$ by $\frac{10}{10}$.

$$\frac{3}{10} \cdot \frac{10}{10} = \frac{30}{100}$$

We write the fraction $\frac{30}{100}$ as **30%**.

Example 3   Of the 30 people in the pool, 15 swam laps. What percent of the people swam laps?

Solution   Fifteen of the 30 people swam laps. We write this as a fraction and reduce.

$$\frac{15}{30} = \frac{1}{2}$$

To write $\frac{1}{2}$ as a fraction with a denominator of 100, we multiply $\frac{1}{2}$ by $\frac{50}{50}$.

$$\frac{1}{2} \cdot \frac{50}{50} = \frac{50}{100}$$

The fraction $\frac{50}{100}$ equals **50%**.

Example 4   Write 0.12 as a percent.

Solution   The decimal number 0.12 is twelve hundredths.

$$0.12 = \frac{12}{100}$$

Twelve hundredths is equivalent to **12%**.

Example 5   Write 0.08 as a percent.

Solution   The decimal 0.08 is eight hundredths.

$$0.08 = \frac{8}{100}$$

Eight hundredths is equivalent to **8%**.

Example 6   Write 0.8 as a percent.

Solution   The decimal number 0.8 is eight tenths. If we place a zero in the hundredths place, the decimal is eighty hundredths.

$$0.8 = 0.80 = \frac{80}{100}$$

Eighty hundredths equals **80%**.

Notice that when a decimal number is converted to a percent, the decimal point is shifted two places to the right. In fact, shifting the decimal point two places to the right is a quick and useful way to write decimal numbers as percents.

LESSON PRACTICE

Practice set* Write each fraction as a percent:

a. $\frac{31}{100}$ b. $\frac{1}{100}$ c. $\frac{1}{10}$

d. $\frac{3}{50}$ e. $\frac{7}{25}$ f. $\frac{2}{5}$

g. Twelve of the 30 flowers were lilies. What percent of the flowers were lilies?

h. Jorge correctly answered 18 of the 20 questions on the test. What percent of the questions did he answer correctly?

Write each decimal number as a percent:

i. 0.25   j. 0.3   k. 0.05

l. 1.0   m. 0.7   n. 0.15

MIXED PRACTICE

Problem set

1. What is the reciprocal of two and three fifths?
(30, 62)

2. What time is one hour thirty-five minutes after 2:30 p.m.?
(32)

3. A 1-pound box of candy cost $4.00. What was the cost per ounce (1 pound = 16 ounces)?
(15)

4. Freda bought a sandwich for $4.00 and a drink for 94¢. Her grandson ordered a meal for $6.35. What was the total price of all three items when 8% sales tax was added? Describe how to use estimation to check whether your answer is reasonable.
(41)

5. If the chance of rain is 50%, then what is the chance it will not rain?
(58)

6. Draw a cube.
(Inv. 6)

7. (a) Write $\frac{3}{4}$ as a decimal number.
(74, 75)

    (b) Write the answer to part (a) as a percent.

8. (a) Write $\frac{3}{20}$ as a fraction with a denominator of 100.
(75)

    (b) Write $\frac{3}{20}$ as a percent.

9. Write 12% as a reduced fraction. Then write the fraction
(33, 74) as a decimal number.

Find each missing number:

10. $\dfrac{7}{10} = \dfrac{n}{100}$
(42)

11. $5 - m = 3\frac{1}{8}$
(43)

12. $1 - w = 0.95$
(38, 43)

13. $m + 1\frac{2}{3} = 3\frac{1}{6}$
(59)

14. $\left(\dfrac{1}{2} + \dfrac{1}{3}\right) - \dfrac{1}{6}$
(57)

15. $3\frac{1}{2} \times 1\frac{1}{3} \times 1\frac{1}{2}$
(72)

16. $(0.43)(2.6)$
(39)

17. $0.26 \div 5$
(45)

18. Nathan correctly answered 17 of the 20 questions on the
(75) test. What percent of the questions did Nathan answer
correctly?

19. The diameter of the big tractor tire was about 5 feet. As
(47) the tire rolled one full turn, the tire rolled about how
many feet? Round the answer to the nearest foot. (Use
3.14 for π.)

20. Which digit in 4.87 has the same place value as the 9
(34) in 0.195?

21. Write the prime factorization of both the numerator and
(67) denominator of $\frac{18}{30}$. Then reduce the fraction.

22. What is the greatest common factor of 18 and 30?
(20)

23. If the product of two numbers is 1, then the two numbers
(30) are which of the following?

    A. equal   B. reciprocals C. opposites D. prime

24. Why is every rectangle a quadrilateral?
(64)

25. If b equals 8 and h equals 6, what does $\frac{bh}{2}$ equal?
(47)

26. Find the prime factorization of 400 using a factor tree.
(65, 73) Then write the prime factorization of 400 using exponents.

27. Draw a coordinate plane on graph paper. Then draw a
(Inv. 7) rectangle with vertices located at (3, 1), (3, −1), (−1, 1), and (−1, −1).

28. Refer to the rectangle drawn in problem 27 to answer
(8, 31) parts (a) and (b) below.

(a) What is the perimeter of the rectangle?

(b) What is the area of the rectangle?

29. (a) What is the perimeter of this parallelogram?
(71)

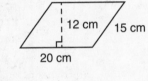

(b) What is the area of this parallelogram?

30. Draw two parallel segments of different lengths. Then
(64) form a quadrilateral by drawing two segments that connect the endpoints of the parallel segments. Is the quadrilateral a rectangle?

LESSON
76

Comparing Fractions by Converting to Decimal Form

WARM-UP

Facts Practice: 30 Fractions to Reduce (Test G)

Mental Math: Count up and down by 2's between negative 12 and 12.

a. 4 × 208

b. 380 + 155

c. 4 × $1.99

d. $10.00 − $4.99

e. $\frac{1}{5}$ of $4.50

f. $0.95 × 100

g. 8 × 8, − 4, ÷ 2, + 2, ÷ 4, × 3, + 1, ÷ 5

h. Hold your hands one foot apart, then 24 inches apart.

Problem Solving:

Celina used division by primes to find the prime factorization of a number. Copy her work and fill in the missing numbers.

$$
\begin{array}{r}
1 \\
7\overline{)\square} \\
5\overline{)\square} \\
3\overline{)\square} \\
2\overline{)\square} \\
2\overline{)\square}
\end{array}
$$

NEW CONCEPT

We have compared fractions by drawing pictures of fractions and by writing fractions with common denominators. Another way to compare fractions is to convert the fractions to decimal form.

Example 1 Compare these fractions. First convert each fraction to decimal form.

$$\frac{3}{5} \bigcirc \frac{5}{8}$$

Solution We convert each fraction to a decimal number by dividing the numerator by the denominator.

$$\frac{3}{5} \longrightarrow 5\overline{)3.0}^{\,0.6} \qquad \frac{5}{8} \longrightarrow 8\overline{)5.000}^{\,0.625}$$

We write both numbers with the same number of decimal places. Then we compare the two numbers.

$$0.600 < 0.625$$

Since 0.6 is less than 0.625, we know that $\frac{3}{5}$ is less than $\frac{5}{8}$.

$$\frac{3}{5} < \frac{5}{8}$$

Example 2 Compare: $\frac{3}{4} \bigcirc 0.7$

Solution First we write the fraction as a decimal.

$$\frac{3}{4} \longrightarrow 4\overline{)3.00} \quad \begin{array}{c} 0.75 \end{array}$$

Then we compare the decimal numbers.

$$0.75 > 0.70$$

Since 0.75 is greater than 0.7, we know that $\frac{3}{4}$ is greater than 0.7.

$$\frac{3}{4} > 0.7$$

LESSON PRACTICE

Practice set Change the fractions to decimals to compare these numbers:

a. $\frac{3}{20} \bigcirc \frac{1}{8}$ b. $\frac{3}{8} \bigcirc \frac{2}{5}$ c. $\frac{15}{25} \bigcirc \frac{3}{5}$

d. $0.7 \bigcirc \frac{4}{5}$ e. $\frac{2}{5} \bigcirc 0.5$ f. $\frac{3}{8} \bigcirc 0.325$

MIXED PRACTICE

Problem set **1.** What is the product of ten squared and two cubed?
(73)

2. What number is halfway between 4.5 and 6.7?
(50)

3. It is said that one year of a dog's life is the same as 7 years
(15) of a human's life. Using that thinking, a dog that is 13 years old is how many "human" years old? (Write an equation and solve the problem.)

4. Compare. First convert each fraction to decimal form.
(76)

$$\frac{2}{5} \bigcirc \frac{1}{4}$$

5. (a) What fraction of this circle is shaded?
(74, 75)

(b) Convert the answer from part (a) to a decimal number.

(c) What percent of this circle is shaded?

6. Convert $2\frac{1}{2}$ to a decimal number.
(74)

7. Write 3.45 as a reduced mixed number.
(73)

8. (a) Write 0.04 as a reduced fraction.
(73, 75)

(b) Write 0.04 as a percent.

9. Instead of dividing 200 by 18, Sam found half of each number and then divided. Show Sam's division problem and write the quotient as a mixed number.
(43)

10. $6\frac{1}{3} + 3\frac{1}{4} + 2\frac{1}{2}$
(61)

11. $\frac{4}{5} = \frac{?}{100}$
(42)

12. $\left(2\frac{1}{2}\right)\left(3\frac{1}{3}\right)\left(1\frac{1}{5}\right)$
(72)

13. $5 \div 2\frac{1}{2}$
(68)

Find each missing number:

14. $6.7 + 0.48 + n = 8$
(43)

15. $12 - d = 4.75$
(43)

16. 0.35×0.45
(39)

17. $4.3 \div 10^2$
(38, 52)

18. Find the median of these numbers:
(Inv. 5)

$$0.3, \ 0.25, \ 0.313, \ 0.2, \ 0.27$$

19. Estimate the sum of 3926 and 5184 to the nearest thousand.
(16)

20. List all the prime numbers between 40 and 50.
(19)

21. Twelve of the 25 animals were herbivores. What percent of the animals were herbivores?
(75)

Refer to the triangle to answer problems 22 and 23.

22. What is the perimeter of this
(8) triangle?

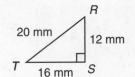

23. Angles T and R are complementary.
(69) If the measure of $\angle R$ is 53°, then
what is the measure of $\angle T$?

24. About how many **millimeters** long is this line segment?
(7)

25. This parallelogram is divided into two congruent triangles.
(71)

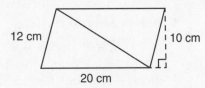

(a) What is the area of the parallelogram?

(b) What is the area of one of the triangles?

26. How many small cubes were used
(Inv. 6) to form this rectangular prism?

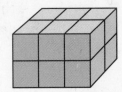

27. Sketch a coordinate plane on graph paper. Graph point A
(Inv. 7) (1, 2), point B (−3, −2), and point C (1, −2). Then draw
segments to connect the three points. What type of polygon
is figure ABC?

28. In the figure drawn in problem 27,
(28) (a) which segment is perpendicular to $\overline{AC}$?

(b) which angle is a right angle?

29. If b equals 12 and h equals 9, what does $\frac{bh}{2}$ equal?
(47)

30. Draw a pair of parallel lines. Draw a third line perpendicular
(64) to the parallel lines. Complete a quadrilateral by drawing a
fourth line that intersects but is not perpendicular to the pair
of parallel lines. Trace the quadrilateral that is formed. Is the
quadrilateral a rectangle?

LESSON

77 Finding Unstated Information in Fraction Problems

WARM-UP

NEW CONCEPT

Often fractional-parts statements contain more information than what is directly stated. Consider this fractional-parts statement:

> *Three fourths of the 28 children at the park are boys.*

This sentence directly states information about the number of boys at the park. It also *indirectly* states information about the number of girls at the park. In this lesson we will practice finding several pieces of information from fractional-parts statements.

Example   Diagram this statement. Then answer the questions that follow.

> *Three fourths of the 28 children at the park are boys.*

(a) Into how many parts is the group of children divided?

(b) How many children are in each part?

(c) How many parts of the group are boys?

(d) How many boys are at the park?

(e) How many parts are girls?

(f) How many girls are at the park?

Solution   We draw a rectangle to represent the group of children. Since the statement uses fourths to describe a part of the group, we divide the rectangle into four parts. Dividing the total number of children by four, we find there are seven children in each part. We identify three of the four parts as boys and one of the four parts as girls. Now we answer the questions.

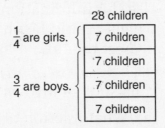

(a) The denominator of the fraction indicates that the group of children is divided into **four parts** for the purpose of this statement. It is important to distinguish between the number of *parts* (as indicated by the denominator) and the number of *categories*. There are two categories of children implied by the statement—boys and girls.

(b) In each of the four parts there are **seven children.**

(c) The numerator of the fraction indicates that **three parts** are boys.

(d) Since three parts are boys and since there are seven children in each part, we find that there are **21 boys** at the park.

(e) Three of the four parts are boys, so only **one part** is girls.

(f) There are seven children in each part. One part is girls, so there are **seven girls.**

LESSON PRACTICE

Practice set*   Diagram this statement. Then answer the questions that follow.

> *Three eighths of the 40 little engines could climb the hill.*

a. Into how many parts was the group divided?

b. How many engines were in each part?

c. How many parts could climb the hill?

d. How many engines could climb the hill?

e. How many parts could not climb the hill?

f. How many engines could not climb the hill?

MIXED PRACTICE

Problem set

1. The weight of an object on the Moon is about $\frac{1}{6}$ of its weight
on Earth. A person weighing 114 pounds on Earth would
weigh about how much on the Moon?
(29)

2. Use the information in problem 1 to calculate what your
weight would be on the Moon. Round your answer to the
nearest pound.
(22)

3. Hala shot 24 arrows and hit the bull's-eye 6 times.
(29, 75)

(a) What fraction of Hala's shots hit the bull's-eye?

(b) What percent of her shots hit the bull's-eye?

4. Diagram this statement. Then answer the questions that
follow.
(77)

> *There are 30 dogs in the show. Three fifths of
> them are male.*

(a) Into how many parts is the group divided?

(b) How many dogs are in each part?

(c) How many males are in the show?

(d) How many females are in the show?

5. (a) In the figure below, what fraction of the group is
shaded?
(74, 75)

(b) Convert the fraction in part (a) to a decimal number.

(c) What percent of the group is shaded?

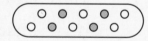

6. Write the decimal number 3.6 as a mixed number.
(73)

Find each missing number:

7. $3.6 + a = 4.15$
(43)

8. $\frac{2}{5}x = 1$
(30)

9. If the chance of rain is 60%, is it more likely to rain or not
to rain? Why?
(58)

10. Three fifths of a circle is what percent of a circle?
(75)

11. A temperature of −3°F is how many degrees below the
(10, 14) freezing temperature of water?

12. Compare:
(73, 76)

(a) $0.35 \bigcirc \dfrac{7}{20}$ (b) $3^2 \bigcirc 2^3$

13. $\dfrac{1}{2} + \dfrac{2}{3}$ **14.** $3\dfrac{1}{5} - 1\dfrac{3}{5}$
(57)   *(63)*

15. $\dfrac{1}{2} + \dfrac{3}{4} + \dfrac{7}{8}$ **16.** $3 \times 1\dfrac{1}{3}$
(61)   *(66)*

17. $3 \div 1\dfrac{1}{3}$ **18.** $1\dfrac{1}{3} \div 3$
(68)   *(68)*

19. What is the perimeter of this
(8) rectangle?

1.5 cm
0.9 cm

20. What is the area of this rectangle?
(31)

21. Write the prime factorization of 1000 using exponents.
(73)

22. Coats were on sale for 40% off. One coat was regularly
(41) priced at $80.

   (a) How much money would be taken off the regular price
      of the coat during the sale?

   (b) What would be the sale price of the coat?

23. Patricia bought a coat that cost $38.80. The sales-tax rate
(41) was 7%.

   (a) What was the tax on the purchase?

   (b) What was the total purchase price including tax?

24. Is every quadrilateral a polygon?
(64)

25. What time is one hour fourteen minutes before noon?
(32)

26. What percent of this rectangle
(Inv. 2) appears to be shaded?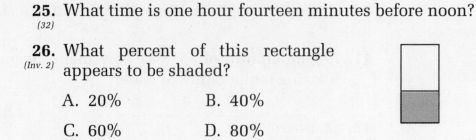

    A. 20%   B. 40%

    C. 60%   D. 80%

27. Sketch a coordinate plane on graph paper. Graph point W
(Inv. 7) (2, 3), point X (1, 0), point Y (–3, 0), and point Z (–2, 3).
Then draw $\overline{WX}$, $\overline{XY}$, $\overline{YZ}$, and $\overline{ZW}$.

28. (a) Which segment in problem 27 is parallel to $\overline{WX}$?
(71)

    (b) Which segment in problem 27 is parallel to $\overline{XY}$?

29. Write the prime factorization of both the numerator and
(67) the denominator of this fraction. Then reduce the
fraction.

$$\frac{210}{350}$$

30. (a) The Moon has the shape of what geometric solid?
(47, Inv. 6)

    (b) The diameter of the Moon is about 2160 miles.
Calculate the approximate circumference of the
Moon using 3.14 for π. Round the answer to the
nearest ten miles.

LESSON
78 Capacity

Facts Practice: 24 Mixed Numbers to Write as Improper Fractions (Test J)

Mental Math: Count up and down by 3's between negative 15 and 15.

a. 4 × 325   b. 1500 + 275   c. 3 × $2.99

d. $20.00 − $2.99   e. $\frac{1}{3}$ of $2.40   f. 1.75 × 100

g. 9 × 11, + 1, ÷ 2, − 1, ÷ 7, − 2, × 5

h. Hold your hands one foot apart, then 18 inches apart.

Problem Solving:

Jim was thinking of a prime number between 75 and 100 that did **not** have a 9 as one of its digits. Of what number was he thinking?

NEW CONCEPT

To measure quantities of liquid in the U.S. Customary System, we use the units gallons (gal), quarts (qt), pints (pt), and ounces (oz). In the metric system we use liters (L) and milliliters (mL). The relationships between units within each system are shown in the following table:

Equivalence Table for Units of Liquid Measure

U.S. Customary System	Metric System
1 gallon = 4 quarts	
1 quart = 2 pints	1 liter = 1000 milliliters
1 pint = 2 cups	
1 cup = 8 ounces	

Common container sizes based on the U.S. Customary System are illustrated below. These containers are named by their **capacity**, that is, by the amount of liquid they can contain. Notice that each container size has half the capacity of the next-largest container. Also, notice that a quart is one "quarter" of a gallon.

1 gallon $\frac{1}{2}$ gallon   1 quart   1 pint   1 cup

Food and beverage containers often have both U.S. Customary and metric capacities printed on the containers. Using the information found on 2-liter soda pop bottles and $\frac{1}{2}$-gallon milk cartons, we find that one liter is a little more than one quart. So the capacity of a 2-liter bottle is a little more than the capacity of a $\frac{1}{2}$-gallon container.

2-liter bottle
$\frac{1}{2}$-gallon container

Example 1   A half gallon of milk is how many pints of milk?

Solution   Two pints equals a quart, and two quarts equals a half gallon. So a half gallon of milk is **4 pints.**

Example 2   Which has the greater capacity, a 12-ounce soda pop can or a 1-pint container?

Solution   A pint equals 16 ounces. **So a 1-pint container has more capacity than a 12-ounce soda pop can.**

LESSON PRACTICE

Practice set   **a.** What fraction of a gallon is a quart?

b. A 2-liter soda pop bottle has a capacity of how many milliliters?

c. A half gallon of orange juice will fill how many 8-ounce cups?

d. The entire contents of a full 2-liter bottle are poured into an empty half-gallon carton. Will the half-gallon container overflow? Why or why not?

MIXED PRACTICE

Problem set   **1.** What is the difference when the product of $\frac{1}{2}$ and $\frac{1}{2}$ is
(12, 55)   subtracted from the sum of $\frac{1}{2}$ and $\frac{1}{2}$?

2. The claws of a Siberian tiger are 10 centimeters long. How
(7)   many millimeters is that?

3. Sue was thinking of a number between 40 and 50 that is a
(25)   multiple of 3 and 4. Of what number was she thinking?

4. Diagram this statement. Then answer the questions that follow.
(77)

Four fifths of the 60 lights were on.

(a) Into how many parts have the 60 lights been divided?

(b) How many lights are in each part?

(c) How many lights were on?

(d) How many lights were off?

5. Which counting number is neither a prime number nor a composite number?
(65)

Find each missing number:

6. $\frac{4}{5}m = 1$
(30)

7. $\frac{4}{5} + w = 1$
(43)

8. $\frac{4}{5} \div x = 1$
(43)

9. $y - \frac{4}{5} = 1$
(43)

10. (a) What fraction of the rectangle below is shaded?
(74, 75)

(b) Write the answer to part (a) as a decimal number.

(c) What percent of the rectangle is shaded?

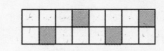

11. Convert the decimal number 1.15 to a mixed number.
(73)

12. Compare:
(73, 76)

(a) $\frac{3}{5} \bigcirc 0.35$

(b) $\sqrt{100} \bigcirc 1^4 + 2^3$

13. $\frac{5}{6} - \frac{1}{2}$
(57)

14. $\frac{3}{4} = \frac{?}{100}$
(42)

15. $\frac{1}{2} + \frac{2}{3} + \frac{5}{6}$
(61)

16. $1\frac{1}{2} \times 2\frac{2}{3}$
(66)

17. $1\frac{1}{2} \div 2\frac{2}{3}$
(68)

18. $2\frac{2}{3} \div 1\frac{1}{2}$
(68)

19. (a) What is the perimeter of this square?
(38)

(b) What is the area of this square?

$\frac{1}{2}$ in.

20. "The opposite sides of a rectangle are parallel." True or
(64) false?

21. What is the average of 3^3 and 5^2?
(73)

22. Round 1.3579 to the hundredths place.
(51)

23. How many inches is $2\frac{1}{2}$ feet?
(66)

24. Which arrow below could be pointing to 0.1?
(50)

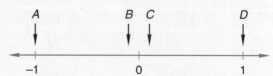

25. Draw a quadrilateral that is not a rectangle.
(64)

26. Find the prime factorization of 900 by using a factor tree.
(65, 73) Then write the prime factorization using exponents.

27. Three vertices of a rectangle have the coordinates (5, 3),
(Inv. 7) (5, −1), and (−1, −1). What are the coordinates of the
fourth vertex of the rectangle?

28. Refer to this table to answer (a) and (b).
(78)

3 teaspoons	=	1 tablespoon
16 tablespoons	=	1 cup
2 cups	=	1 pint
2 pints	=	1 quart
4 quarts	=	1 gallon

(a) A teaspoon of soup is what fraction of a tablespoon
of soup?

(b) How many cups of milk is a gallon of milk?

29. A liter is closest in size to which of the following?
(78)

    A. pint   B. quart   C. $\frac{1}{2}$ gallon   D. gallon

30. In 1881 Clara Barton founded the American Red Cross, an
(78) organization that helps people during emergencies. The
Red Cross organizes "blood drives" in which people can
donate a pint of blood to help hospital patients who will
undergo surgery. How many ounces is a pint?

LESSON
79 Area of a Triangle

WARM-UP

Facts Practice: Linear Measurement (Test K)

Mental Math: Count up and down by 12's between 12 and 144.

a. 307×6 b. $1000 - 420$ c. $4 \times \$2.99$

d. $\$5.75 + \$2.99$ e. Double \$24.   f. 0.125×100

g. $2 \times 2, \times 2, \times 2, - 1, \times 2, + 2, \div 2, \div 2$

h. Hold your hands one inch apart, then one centimeter apart.

Problem Solving:

The perimeter of this rectangle is 48 inches. The width is 8 inches. What is the length?

8 in.

NEW CONCEPT

In this lesson we will demonstrate that the area of a triangle is half the area of a parallelogram with the same base and height.

Activity: *Area of a Triangle*

Materials needed:

- pencil and paper
- ruler
- scissors

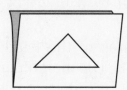

Fold the paper in half, and draw a triangle on the folded paper.

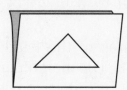

While the paper is folded, use your scissors to cut out the triangle so that you cut out two congruent triangles.

Arrange the two triangles to form a parallelogram. What fraction of the area of the parallelogram is the area of one of the triangles?

We find that whatever the shape of a triangle, its area is half the area of the parallelogram with the same base and height.

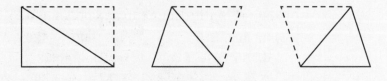

Recall that the area of a parallelogram can be found by multiplying its base by its height ($A = bh$). So the area of a triangle can be determined by finding half of the product of its base and height.

$$\text{Area of a triangle} = \frac{1}{2}bh$$

Since multiplying by $\frac{1}{2}$ and dividing by 2 are equivalent operations, the formula may also be written as

$$A = \frac{bh}{2}$$

In the following examples we will use both formulas stated above. From our calculations of the areas of rectangles and parallelograms, we remember that the base and the height are *perpendicular* measurements.

Example 1 Find the area of the triangle at right.

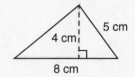

Solution The area of the triangle is half the product of the base and height. The height must be *perpendicular* to the base. The height in this case is 4 cm. Half the product of 8 cm × 4 cm is 16 cm^2.

$$A = \frac{1}{2}(8 \text{ cm})(4 \text{ cm})$$

$$A = \textbf{16 cm}^2$$

Example 2 Find the area of this right triangle:

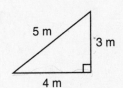

Solution We find the area by multiplying the base by the height and then dividing by 2. All right triangles have two sides that are perpendicular, so we use the perpendicular sides as the base and height.

$$A = \frac{(4 \text{ m})(3 \text{ m})}{2}$$

$$A = 6 \text{ m}^2$$

LESSON PRACTICE

Practice set* Find the area of each triangle:

a.
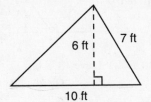

b.
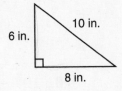

c.
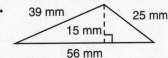

d.
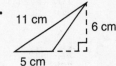

MIXED PRACTICE

Problem set

1. If you know both the perimeter and the length of a rectangle,
 (8) how can you determine the width of the rectangle?

2. A 2-liter bottle contained 2 qt 3.6 oz of beverage. Use this
 (78) information to compare a liter and a quart:

$$1 \text{ liter} \bigcirc 1 \text{ quart}$$

3. Uncle Bill was 38 years old when he started his job. He
 (11) worked for 33 years. How old was he when he retired?

4. Answer "true" or "false" for each statement:
 (64)
   (a) "Every rectangle is a square."

   (b) "Every rectangle is a parallelogram."

5. Ninety percent of the 30 people were right-handed.
_(23, 41)

    (a) What percent of the people were left-handed?

    (b) How many people were right-handed?

    (c) What was the ratio of left-handed to right-handed people?

6. Eighteen of the twenty-four runners finished the race.
_(75, 77)

    (a) What fraction of the runners finished the race?

    (b) What fraction of the runners did not finish the race?

    (c) What percent of the runners did not finish the race?

7. This parallelogram is divided into two congruent triangles.
₍₇₉₎ What is the area of each triangle?

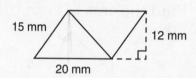

8. $10^3 \div 10^2$
₍₇₃₎

9. $6.42 + 12.7 + 8$
₍₃₈₎

10. $1.2(0.12)$
₍₃₉₎

11. $64 \div 0.08$
₍₄₉₎

12. $3\frac{1}{3} \times \frac{1}{5} \times \frac{3}{4}$
₍₇₂₎

13. $2\frac{1}{2} \div 3$
₍₆₈₎

Find each missing number:

14. $10 - q = 9.87$
₍₄₃₎

15. $24m = 0.288$
₍₄₅₎

16. $n - 2\frac{3}{4} = 3\frac{1}{3}$
₍₆₃₎

17. $w + \frac{1}{4} = \frac{5}{6}$
₍₅₇₎

18. The perimeter of a square is 80 cm. What is its area?
₍₃₈₎

19. Write the decimal number for the following:
₍₄₆₎

$$(9 \times 10) + (6 \times 1) + \left(3 \times \frac{1}{100}\right)$$

20. Bart set the radius on the compass to 10 cm and drew a
₍₄₇₎ circle. What was the circumference of the circle? (Use 3.14 for π.)

21. Which of these numbers is closest to zero?
(50)

    A. −2   B. 0.2   C. 1   D. $\frac{1}{2}$

22. Estimate the product of 6.7 and 7.3 by rounding each
(51) number to the nearest whole number before multiplying. Describe how you arrived at your answer.

23. The expression 2^4 (two to the fourth power) is the prime
(73) factorization of 16. The expression 3^4 is the prime factorization of what number?

24. What number is halfway between 0.2 and 0.3?
(18, 45)

25. To what decimal number is the arrow pointing on the
(50) number line below?

26. Which quadrilateral has only one pair of parallel sides?
(64)

27. The coordinates of the vertices of a quadrilateral are (−5, 5),
(64, Inv. 7) (1, 5), (3, 1), and (−3, 1). What is the name for this kind of quadrilateral?

In the figure below, a square and a regular hexagon share a common side. The area of the square is 100 sq. cm. Use this information to answer problems 28 and 29.

28. (a) What is the length of each side
(38)   of the square?

    (b) What is the perimeter of the square?

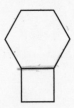

29. (a) What is the length of each side of the hexagon?
(8)

    (b) What is the perimeter of the hexagon?

30. Write the prime factorization of both the numerator and
(67) the denominator of this fraction. Then reduce the fraction.

$$\frac{32}{48}$$

80

Using Scale Factor to Solve Ratio Problems

WARM-UP

Facts Practice: 28 Improper Fractions to Simplify (Test I)

Mental Math: Count up and down by 7's between negative 35 and 35.

a. 4×315 b. $380 + 170$ c. $5 \times \$2.99$
d. $\$10.00 - \$7.99$ e. $\frac{1}{4}$ of $\$4.80$ f. $37.5 \div 100$
g. $5 \times 5, \times 5, - 25, \div 4, \div 5, - 5$
h. Hold your hands one meter apart, then four feet apart.

Problem Solving:

A seven-digit phone number consists of a three-digit prefix followed by four digits. How many different phone numbers are possible for a particular prefix?

NEW CONCEPT

Consider the following ratio problem:

> *The ratio of little boys to little girls in the nursery was 3 to 2. If there were 6 little girls, how many little boys were there?*

We see two uses for numbers in ratio problems. One use is to express a ratio. The other use is to express an actual count. A ratio box can help us sort the two uses by placing the ratio numbers in one column and the actual counts in another column. We write the items being compared along the left side of the rows.

	Ratio	Actual Count
Boys	3	
Girls	2	6

We are told that the ratio of boys to girls was 3 to 2. We place these numbers in the ratio column, assigning 3 to the "boys" row and 2 to the "girls" row. We are given an actual count of 6 girls, which we record in the box. We are asked to find the actual count of boys, so that portion of the ratio box is empty.

Ratio numbers and actual counts are related by a **scale factor**. If we multiply the terms of a ratio by the correct scale factor, we can find the actual count. Recall that a ratio is a reduced form of an actual count. If we can determine the factor by which the actual count was reduced to form the ratio, then we can recreate the actual count.

	Ratio	Actual Count
Boys	3 × scale factor	?
Girls	2 × scale factor	6

In the girls row we see that 2 can be multiplied by 3 to get 6. So 3 is the scale factor in this problem. That means we can multiply each ratio term by 3. Multiplying the ratio term 3 in the boys row by the scale factor 3 gives us an actual count of 9 boys.

Example Sadly, the ratio of flowers to weeds in the garden was 2 to 5. If there were 30 flowers in the garden, how many weeds were there?

Solution We will begin by drawing a ratio box.

	Ratio	Actual Count
Flowers	2	30
Weeds	5	

To determine the scale factor, we study the row that has two numbers. In the "flowers" row we see the ratio number 2 and the actual count 30. If we divide 30 by 2, we find the scale factor, which is 15. Now we will use the scale factor to find the actual count of weeds in the garden.

$$\text{Ratio} \times \text{scale factor} = \text{actual count}$$

$$5 \quad \times \quad 15 \quad = 75$$

There were **75 weeds** in the garden.

LESSON PRACTICE

Practice set   Draw a ratio box and use scale factor to solve each ratio problem:

a. The ratio of boys to girls in the theater was 6 to 5. If there were 60 girls, how many boys were there?

b. The ratio of ants to flies at the picnic was 8 to 3. If there were 24 flies, how many ants were there?

MIXED PRACTICE

Problem set   **1.** What is the mean of 96, 49, 68, and 75? What is the range?
(Inv. 5)

2. The average depth of the ocean beyond the edges of the continents is $2\frac{1}{2}$ miles. How many feet is that (1 mile = 5280 ft)?
(66)

3. The 168 girls who signed up for soccer were divided equally into 12 teams. How many players were on each team? (Write an equation and solve the problem.)
(15)

Parallelogram *ABCD* is divided into two congruent triangles. Segments *BA* and *CD* measure 3 in. Segments *AD* and *BC* measure 5 in. Segment *BD* measures 4 in. Refer to this figure to answer problems 4 and 5.

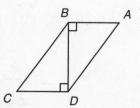

4. (a) What is the perimeter of the parallelogram?
(71)
   (b) What is the area of the parallelogram?

5. (a) What is the perimeter of each triangle?
(79)
   (b) What is the area of each triangle?

6. This quadrilateral has one pair of parallel sides. What is the name of this kind of quadrilateral?
(64)

7. "All squares are rectangles." True or false?
(64)

8. If four fifths of the 30 chorus members were present, then how many chorus members were absent?
(77)

9. The ratio of dogs to cats in the neighborhood was 2 to 5. If
(80) there were 10 dogs, how many cats were there?

10. Write as a percent:
(75)
   (a) $\frac{19}{20}$ (b) 0.6

11. What percent of the perimeter of a square is the length of
(75) one side?

12. Compare: $0.5 \bigcirc \frac{3}{4}$
(76)

13. Write 4.4 as a reduced mixed number.
(73)

14. Write $\frac{1}{8}$ as a decimal number.
(74)

15. $\frac{5}{6} + \frac{1}{2}$ **16.** $\frac{5}{8} - \frac{1}{4}$ **17.** $2\frac{1}{2} \times 1\frac{1}{3} \times \frac{3}{5}$
(57)   (57)   (72)

Find each missing number:

18. $4 - a = 2.6$ **19.** $3n = 1\frac{1}{2}$
(43)   (68)

20. $5x = 0.36$ **21.** $0.9y = 63$
(45)   (49)

22. Round 0.4287 to the hundredths place.
(51)

Refer to the bar graph below to answer problems 23–25.

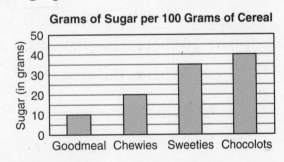

Grams of Sugar per 100 Grams of Cereal

23. Sweeties contains about how many grams of sugar per
(Inv. 5) 100 grams of cereal?

24. Fifty grams of Chocolots would contain about how many
(Inv. 5) grams of sugar?

25. Write a problem about comparing that refers to the bar
(Inv. 5) graph, and then answer the problem.

26. There was one quart of milk in the carton. Oscar poured
(78) one cup of milk on his cereal. How many cups of milk
were left in the carton?

27. Three vertices of a square are (3, 0), (3, 3), and (0, 3).
(Inv. 7)
   (a) What are the coordinates of the fourth vertex of the
   square?

   (b) What is the area of the square?

28. Which of these angles could be the complement of a
(69) 30° angle?

   A.   B.   C.

29. If $A = \frac{1}{2}bh$, and if $b = 6$ and $h = 8$, then what does
(47) A equal?

30. Draw a pair of parallel segments that are the same length.
(64) Form a quadrilateral by drawing two segments between
the endpoints of the parallel segments. Is the quadrilateral
a parallelogram?

Day of the COLLOSAL CHICKEN!

INVESTIGATION 8

Focus on

Geometric Construction of Bisectors

Since Lesson 27 we have used a compass to draw circles of various sizes. We can also use a compass together with a straightedge and a pencil to construct and divide various geometric figures.

Materials needed:

- compass
- ruler
- pencil
- several sheets of unlined paper
- Activity Sheet 16 (available in *Saxon Math 7/6— Homeschool Tests and Worksheets*)

Activity 1: *Perpendicular Bisectors*

The first activity in this investigation is to **bisect** a segment. The word *bisect* means "to cut into two equal parts." We bisect a line segment when we draw a line (or segment) through the midpoint of the line segment. Below, segment *AB* is bisected by line *r* into two parts of equal length.

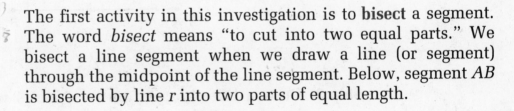

In Section A of Activity Sheet 16, you will see segment *AB*. Follow these directions to bisect the segment:

Step 1: Set your compass so that the distance between the pivot point of the compass and the pencil point is more than half the length of the segment. Then place

the pivot point of the compass on an endpoint of the segment and "swing an arc" on both sides of the segment, as illustrated.

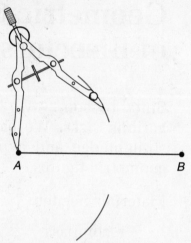

Step 2: Without resetting the radius of the compass, move the pivot point of the compass to the other endpoint of the segment. Swing an arc on both sides of the segment so that the arcs intersect as shown. (It may be necessary to return to the first endpoint to extend the first set of arcs until the arcs intersect on both sides of the segment.)

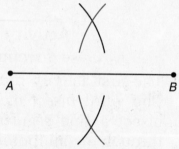

Step 3: Draw a line through the two points where the arcs intersect.

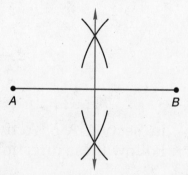

The line bisects the segment and is perpendicular to it. Thus the line is called a **perpendicular bisector** of the segment.

Check your work with a ruler. You should find that the perpendicular bisector has divided segment *AB* into two smaller segments of equal length.

Practice the procedure again by drawing your own line segment on a blank sheet of paper. Position the segment on the page so that there is enough area above and below the segment to draw the arcs you need to bisect the segment. Refer to the directions above if you need to refresh your memory.

Activity 2: *Angle Bisectors*

In Section B of Activity Sheet 16 is an angle. You will bisect the angle by drawing a ray halfway between the two sides of the angle.

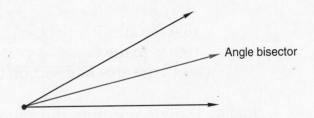

Follow these directions to bisect the angle:

Step 1: Place the pivot point of the compass on the vertex of the angle, and sweep an arc across both sides of the angle.

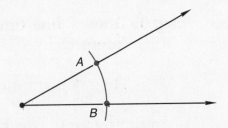

The arc intersects the sides of the angle at two points, which we have labeled *A* and *B*. Point *A* and point *B* are both the same distance from the vertex.

Step 2: Set the compass so that the distance between the pivot point of the compass and the pencil point is more than half the distance from point *A* to point *B*.

Place the pivot point of the compass on point *A*, and sweep an arc as shown.

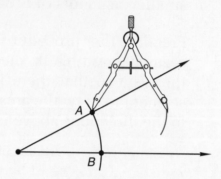

Step 3: Without resetting the radius of the compass, move the pivot point of the compass to point *B* and sweep an arc that intersects the arc drawn in Step 2.

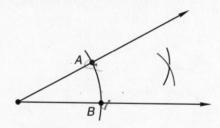

Step 4: Draw a ray from the vertex of the angle through the intersection of the arcs.

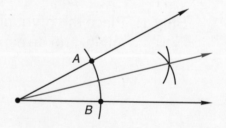

The ray is the **angle bisector** of the angle.

Use a protractor to check your work. You should find that the angle bisector has divided the angle into two smaller angles of equal measure.

Practice the procedure again by drawing an angle on a blank sheet of paper. Make the angle a different size from the one on the activity sheet. Then bisect the angle using the method presented in this activity.

Activity 3: *Constructing Bisectors*

In this activity you will make a worksheet like Activity Sheet 16. On an unlined sheet of paper, draw both a line segment and an angle. The sizes of the segment and angle should be different from the sizes of the segment and angle on Activity Sheet 16.

Using a compass and straightedge, construct the perpendicular bisector of the segment and the angle bisector of the angle on the sheet you made. The arcs you draw in the construction should be visible so that your teacher can check your work.

election 2016: An unknown conspiracy

LESSON

81 Arithmetic with Units of Measure

Facts Practice: Linear Measurement (Test K)

Mental Math: Count up and down by 25's between negative 150 and 150.

a. 311 × 7   b. 2000 − 1250   c. 4 × $9.99

d. $2.50 + $9.99   e. Double $5.50.   f. 0.075 × 100

g. 8 × 8, + 6, ÷ 2, + 1, ÷ 6, × 3, ÷ 2

h. Hold your hands 6 inches apart, then 6 cm apart.

Problem Solving:

Copy this problem and fill in the missing digits:

$$
\begin{array}{r}
\_\,3 \\
\_\,\overline{)\,\_\,\_\,\_} \\
\underline{\_\,\_} \\
\_\,\_ \\
\underline{\_\,\_} \\
2
\end{array}
$$

NEW CONCEPT

Recall that the operations of arithmetic are addition, subtraction, multiplication, and division. In this lesson we will practice adding, subtracting, multiplying, and dividing units of measure.

We may add or subtract measurements that have the same units. If the units are not the same, we first convert one or more measurements so that the units are the same. Then we add or subtract.

Example 1 Add: 2 ft + 12 in.

Solution The units are not the same. Before we add, we either convert 2 feet to 24 inches or we convert 12 inches to 1 foot.

Convert to Inches	Convert to Feet
2 ft + 12 in.	2 ft + 12 in.
24 in. + 12 in. = **36 in.**	2 ft + 1 ft = **3 ft**

Either answer is correct, because 3 feet equals 36 inches.

Notice that in each equation in example 1, the units of the sum are the same as the units of the addends. The units do not change when we add or subtract measurements. However, the units *do* change when we multiply or divide measurements.

When we find the area of a figure, we multiply the lengths. Notice how the units change when we multiply.

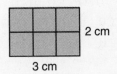

To find the area of this rectangle, we multiply 2 cm by 3 cm. The product has a different unit of measure than the factors.

$$2 \text{ cm} \cdot 3 \text{ cm} = 6 \text{ sq. cm}$$

A centimeter and a square centimeter are two different kinds of units. A centimeter is used to measure length. It can be represented by a line segment.

—————
1 cm

A square centimeter is used to measure area. It can be represented by a square that is 1 centimeter on each side.

1 sq. cm

The unit of the product is a different unit because we multiplied the units of the factors. When we multiply 2 cm by 3 cm, we multiply both the numbers and the units.

$$2 \text{ cm} \cdot 3 \text{ cm} = \underbrace{2 \cdot 3}_{6} \underbrace{\text{cm} \cdot \text{cm}}_{\text{sq. cm}}$$

Instead of writing "sq. cm," we may use exponents to write "cm · cm" as "cm^2." We read cm^2 as "square centimeters."

$$2 \text{ cm} \cdot 3 \text{ cm} = \underbrace{2 \cdot 3}_{6} \underbrace{\text{cm} \cdot \text{cm}}_{\text{cm}^2}$$

Example 2   Multiply: 6 ft · 4 ft

Solution   We multiply the numbers. We also multiply the units.

$$6 \text{ ft} \cdot 4 \text{ ft} = \underbrace{6 \cdot 4}_{24} \underbrace{\frac{\text{ft} \cdot \text{ft}}{}}_{\text{ft}^2}$$

The product is **24 ft²**, which can also be written as "24 sq. ft."

Units also change when we divide measurements. For example, if we know both the area and the length of a rectangle, we can find the width of the rectangle by dividing.

To find the width of this rectangle, we divide 21 cm² by 7 cm.

$$\frac{21 \text{ cm}^2}{7 \text{ cm}} = \frac{\overset{3}{\cancel{21}}}{\underset{1}{\cancel{7}}} \frac{\text{cm} \cdot \text{cm}}{\cancel{\text{cm}}}$$

We divide the numbers and write "cm²" as "cm · cm" in order to reduce the units. The quotient is 3 cm, which is the width of the rectangle.

Example 3   Divide: $\dfrac{25 \text{ mi}^2}{5 \text{ mi}}$

Solution   To divide the units, we write "mi²" as "mi · mi" and reduce.

$$\frac{\overset{5}{\cancel{25}}}{\underset{1}{\cancel{5}}} \frac{\cancel{\text{mi}} \cdot \text{mi}}{\cancel{\text{mi}}}$$

The quotient is **5 mi.**

Sometimes when we divide measurements, the units will not reduce. When units will not reduce, we leave the units in fraction form. For example, if a car travels 300 miles in 6 hours, we can find the average speed of the car by dividing.

$$\frac{300 \text{ mi}}{6 \text{ hr}} = \frac{\overset{50}{\cancel{300}} \text{ mi}}{\underset{1}{\cancel{6}} \text{ hr}}$$

The quotient is $50 \frac{\text{mi}}{\text{hr}}$, which is 50 miles per hour (50 mph).

The word *per* means "for each" and is used in place of the division bar. Notice that speed is a quotient of distance divided by time.

Example 4 Divide: $\frac{300 \text{ mi}}{10 \text{ gal}}$

Solution We divide the numbers. The units do not reduce.

$$\frac{300 \text{ mi}}{10 \text{ gal}} = \frac{\overset{30}{\cancel{300}} \text{ mi}}{\underset{1}{\cancel{10}} \text{ gal}}$$

The quotient is **30** $\frac{\text{mi}}{\text{gal}}$, which is 30 miles per gallon.

LESSON PRACTICE

Practice set Simplify:

a. 2 ft − 12 in. (Write the difference in inches.)

b. 2 ft × 4 ft²   **c.** $\frac{12 \text{ cm}^2}{3 \text{ cm}}$ **d.** $\frac{300 \text{ mi}}{5 \text{ hr}}$

MIXED PRACTICE

Problem set

1. The Jones family had two gallons of milk before
(78) breakfast. The family used two quarts of milk during breakfast. How many quarts of milk did the Jones family have after breakfast?

2. One quart of milk is about 945 milliliters of milk. Use this
(78) information to compare a quart and a liter:

1 quart ◯ 1 liter

3. Carol cut $2\frac{1}{2}$ inches off her hair three times last year. How
(66) much longer would her hair have been at the end of the year if she had not cut it?

4. The plane flew 1200 miles in 3 hours. Divide the distance
(81) by the time to find the average speed of the plane.

5. Write the prime factorization of both the numerator and
(67) the denominator of this fraction. Then reduce the fraction.

$$\frac{54}{135}$$

6. The basketball team scored 60% of its 80 points in the
(29, 33) second half. Write 60% as a reduced fraction. Then find
the number of points the team scored in the second half.

7. What is the area of this
(71) parallelogram?

8. What is the perimeter of this
(71) parallelogram?

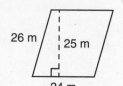

26 m   25 m

24 m

9. "Some rectangles are trapezoids." True or false?
(64)

10. The ratio of red marbles to blue marbles in the bag was
(80) 3 to 4. If 24 marbles were blue, how many were red?

11. Arrange these numbers in order from least to greatest:
(76)

$$\tfrac{1}{2}, \tfrac{1}{5}, 0.4$$

12. (a) What decimal number is equivalent to $\frac{4}{25}$?
(74, 75)

(b) What percent is equivalent to $\frac{4}{25}$?

13. $(10 - 0.1) \times 0.1$ **14.** $(0.4 + 3) \div 2$
(53)   *(53)*

15. $\dfrac{5}{8} + \dfrac{3}{4}$ **16.** $3 - 1\dfrac{1}{8}$ **17.** $4\dfrac{1}{2} - 1\dfrac{3}{4}$
(57)   *(63)*   *(63)*

18. $\dfrac{5}{6} \cdot \dfrac{4}{5} \cdot \dfrac{3}{8}$ **19.** $4\dfrac{1}{2} \times 1\dfrac{1}{3}$ **20.** $3\dfrac{1}{3} \div 1\dfrac{2}{3}$
(72)   *(66)*   *(68)*

The perimeter of this square is two
meters. Refer to this figure to answer
problems 21 and 22.

21. How many centimeters long is
(8) each side of the square (1 meter =
100 centimeters)?

22. (a) What is the diameter of the circle?
(47)

(b) What is the circumference of the circle? (Use 3.14 for π.)

23. If the sales-tax rate is 6%, what is the tax on a $12.80
(41) purchase?

24. What time is two-and-one-half hours after 10:40 a.m.?
(32)

25. Use a ruler to find the length of this line segment to the
(17) nearest sixteenth of an inch.

_____

26. What is the area of a quadrilateral with the vertices (0, 0),
(Inv. 7) (4, 0), (6, 3), and (2, 3)?

27. What is the name of the geometric
(Inv. 6) solid at right?

28. If the area of a square is one square foot, what is the
(38) perimeter?

29. Simplify:
(81) (a) 2 yd + 3 ft   (b) 5 m × 3 m
(Write the sum in yards.)

(c) $\dfrac{36 \text{ ft}^2}{6 \text{ ft}}$ (d) $\dfrac{400 \text{ miles}}{20 \text{ gallons}}$

30. Draw a pair of parallel segments that are not the same
(64) length. Form a quadrilateral by drawing two segments
between the endpoints of the parallel segments. What is
the name of this type of quadrilateral?

LESSON

82   Volume of a Rectangular Prism

WARM-UP

Facts Practice: Liquid Measurement (Test L)

Mental Math: Count by 25's from 25 to 400.

  a. 20 · 20   **b.** 284 − 150   **c.** $1.99 + $2.99

  d. $\frac{1}{3}$ of $7.50   **e.** 2.5 × 10   **f.** $\frac{800}{40}$ *(Reduce: $\frac{80\cancel{0}}{4\cancel{0}}$)*

  g. 10 × 10, − 1, ÷ 3, − 1, ÷ 4, + 1, ÷ 3

  h. Hold your thumb and forefinger one inch apart, then one centimeter apart.

Problem Solving:

After three games Beth's average bowling score was 110. The score for her fourth game was 115, and the score for her fifth game was 120. What was her average score for all five games?

NEW CONCEPT

The **volume** of a shape is the amount of space that the shape occupies. We measure volume by using units that take up space, such as cubic centimeters or cubic inches. The number of cubic units of space that the shape occupies is the volume measurement of that shape. Imagining sugar cubes can help us visualize this idea.

To calculate the volume of a rectangular prism, we can begin by finding the area of the base of the prism. Then we imagine building layers of sugar cubes on the base up to the height of the prism.

Example 1   How many 1-inch sugar cubes are needed to form this rectangular prism? (A 1-inch sugar cube is a sugar cube whose edges are 1 inch long.)

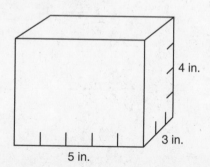

4 in.

3 in.

5 in.

Solution The area of the base is 5 inches times 3 inches, which equals 15 square inches. Thus 15 sugar cubes are needed to make the bottom layer of the prism.

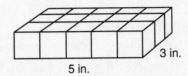

The prism is 4 inches high, so we will have 4 layers. The total number of cubes is 4 times 15, which is **60 sugar cubes.**

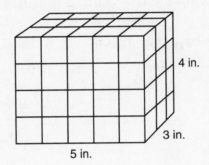

Notice that we multiplied the length by the width to find the area of the base. Then we multiplied the area of the base by the height to find the volume.

We can calculate the volume *V* of a rectangular prism by multiplying the three perpendicular dimensions of the prism: the length *l*, the width *w*, and the height *h*.

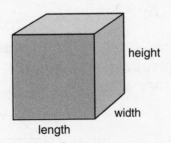

Thus the formula for finding the volume of a rectangular prism is

$$V = lwh$$

Example 2   What is the volume of a cube whose edges are 10 centimeters long?

Solution   The area of the base is 10 cm × 10 cm, or 100 sq. cm. Thus we could set 100 one-centimeter sugar cubes on the bottom layer. There will be 10 layers, so it would take a total of 10 × 100, or 1000 sugar cubes, to fill the cube. Thus the volume is **1000 cu. cm.**

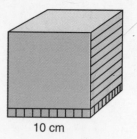

10 cm

Example 3   Find the volume of a rectangular prism that is 4 feet long, 3 feet wide, and 2 feet high.

Solution   For *l, w,* and *h* we substitute 4 ft, 3 ft, and 2 ft. Then we multiply.

$$V = lwh$$

$$V = (4 \text{ ft})(3 \text{ ft})(2 \text{ ft})$$

$$V = \textbf{24 ft}^3$$

Notice that ft^3 means "cubic feet."

LESSON PRACTICE

Practice set
a. How many 1-cm sugar cubes would be needed to build a cube 4 cm on each edge?

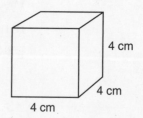

4 cm

4 cm

4 cm

b. What is the volume of a rectangular box that is 5 feet long, 3 feet wide, and 2 feet tall?

c. The interior dimensions of a rectangular box are 10 inches by 6 inches by 4 inches. The box is to be filled with 1-inch cubes. How many cubes can fit on the bottom layer? How many cubes can fit in the box?

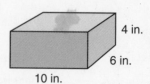

4 in.

6 in.

10 in.

MIXED PRACTICE

Problem set

1. Write the number twenty-one and five hundredths.
(35)

2. Tennis balls are sold in cans containing 3 balls. What would be the total cost of one dozen tennis balls if the price per can was $2.49?
(15)

3. A cubit is about 18 inches. If Ruben was 4 cubits tall, about how many feet tall was he?
(15)

4. (a) Write $\frac{7}{100}$ as a percent.
(75)

    (b) Write $\frac{7}{10}$ as a percent.

5. Write 90% as a reduced fraction. Then write the fraction as a decimal number.
(33, 74)

6. Of the 50 people who took the tour, 23 learned new facts. What percent of the people learned new facts?
(75)

7. Write $\frac{9}{25}$ as a percent.
(75)

8. A box of cereal has the shape of what geometric solid?
(Inv. 6)

Find each missing number:

9. $w - 3\frac{5}{6} = 2\frac{1}{3}$
(63)

10. $3\frac{1}{4} - y = 1\frac{5}{8}$
(63)

11. $6n = 0.12$
(45)

12. $0.12m = 6$
(49)

13. $5n = 10^2$
(38)

14. $1\frac{1}{2}w = 6$
(68)

15. (a) What fraction of this group is shaded?
(75)

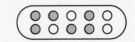

    (b) What percent of this group is shaded?

16. $0.5 + (0.5 \div 0.5) + (0.5 \times 0.5)$
(53)

17. $\frac{1}{2} + \frac{1}{5} + \frac{1}{10}$
(61)

18. $1\frac{4}{5} \times 1\frac{2}{3}$
(66)

19. Which digit in 6.3457 has the same place value as the 8
(34) in 128.90?

20. Estimate the product of 39 and 41.
(16)

21. In a bag there are 12 red marbles and 36 blue marbles.
(23, 58)

(a) What is the ratio of red marbles to blue marbles?

(b) If one marble is taken from the bag, what is the
probability that the marble will be red?

22. What is the area of this
(71) parallelogram?

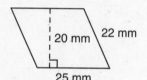

23. What is the perimeter of this
(71) parallelogram?

24. Write the prime factorization of 252 using exponents.
(73)

25. "Some triangles are quadrilaterals." True or false? Why?
(60)

26. A quadrilateral has vertices with the coordinates (−2, −1),
(64, Inv. 7) (1, −1), (3, 3), and (−3, 3). Draw the quadrilateral on graph
paper. The figure is what type of quadrilateral?

27. This cube is constructed of 1-inch
(82) cubes. What is the volume of the
larger cube?

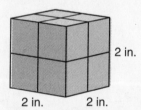

28. Simplify:
(81)

(a) 3 quarts + 2 pints (Write the sum in quarts.)

(b) $\dfrac{49 \text{ m}^2}{7 \text{ m}}$

(c) $\dfrac{400 \text{ miles}}{8 \text{ hours}}$

29. Three of the dozen eggs were cracked. What percent of
(75) the eggs were cracked?

30. A pint of milk weighs about a pound. About how many
(78) pounds does a gallon of milk weigh?

LESSON

83 Proportions

WARM-UP

Facts Practice: 28 Improper Fractions to Simplify (Test I)

Mental Math: Count by $\frac{1}{4}$'s from $\frac{1}{4}$ to 4.

   a. 30 · 30   **b.** 1000 − 125   **c.** 3 × \$3.99
   d. Double $3\frac{1}{2}$.   **e.** 2.5 ÷ 100   **f.** 20 × 34
   g. 9 × 9, − 1, ÷ 2, + 2, ÷ 6, + 2, ÷ 3
   h. Hold your hands one meter apart, then one yard apart.

Problem Solving:

Compare the following quantities. Then describe how you performed the comparison.
$$1\tfrac{7}{8} + 2\tfrac{5}{6} + 3\tfrac{4}{5} \bigcirc 2 + 3 + 4$$

NEW CONCEPT

A **proportion** is a true statement that two ratios are equal. Here is an example of a proportion:

$$\frac{3}{4} = \frac{6}{8}$$

We read this proportion as "Three is to four as six is to eight." Two ratios that are not equivalent are not proportional.

Example 1 Which ratio forms a proportion with $\frac{2}{3}$?

  A. $\frac{2}{4}$ B. $\frac{3}{4}$ C. $\frac{4}{6}$ D. $\frac{3}{2}$

Solution Equivalent ratios form a proportion. The ratio equivalent to $\frac{2}{3}$ is **C.** $\frac{4}{6}$.

Example 2 Write this proportion with digits: Four is to six as six is to nine.

Solution We write "four is to six" as one ratio and "six is to nine" as the equivalent ratio. We are careful to write the numbers in the order stated.

$$\frac{4}{6} = \frac{6}{9}$$

We can use proportions to solve a variety of problems. Proportion problems often involve finding a missing term. The letter a represents a missing term in this proportion:

$$\frac{3}{5} = \frac{6}{a}$$

One way to find a missing term in a proportion is to determine the fractional name for 1 that can be multiplied by one ratio to form the equivalent ratio. The first terms in these ratios are 3 and 6. Since 3 times 2 equals 6, we find that the scale factor is 2. So we multiply $\frac{3}{5}$ by $\frac{2}{2}$ to form the equivalent ratio.

$$\frac{3}{5} \cdot \frac{2}{2} = \frac{6}{10}$$

We find that a represents the number 10.

Example 3   Complete this proportion: Two is to six as what number is to 30?

Solution   We write the terms of the proportion in the stated order, using a letter to represent the unknown number.

$$\frac{2}{6} = \frac{n}{30}$$

We are not given both first terms, but we are given both second terms, 6 and 30. The scale factor is 5, since 6 times 5 equals 30. We multiply $\frac{2}{6}$ by $\frac{5}{5}$ to complete the proportion.

$$\frac{2}{6} \times \frac{5}{5} = \frac{10}{30}$$

The missing term of the proportion is **10**.

$$\frac{2}{6} = \frac{10}{30}$$

LESSON PRACTICE

Practice set   **a.** Which ratio forms a proportion with $\frac{5}{2}$?

A. $\frac{3}{2}$ B. $\frac{4}{10}$ C. $\frac{15}{6}$ D. $\frac{5}{20}$

b. Write this proportion with digits: Six is to eight as nine is to twelve.

c. Write and complete this proportion: Four is to three as twelve is to what number?

d. Write and complete this proportion: Six is to nine as what number is to thirty-six?

MIXED PRACTICE

Problem set

1. What is the product when the sum of 0.2 and 0.2 is
$(12, 53)$ multiplied by the difference of 0.2 and 0.2?

2. Arabian camels travel about 3 times as fast as Bactrian
(66) camels. If Bactrian camels travel at $1\frac{1}{2}$ miles per hour, at how many miles per hour do Arabian camels travel?

3. Mark was paid at a rate of $4 per hour for cleaning up a
(32) neighbor's yard. If he worked from 1:45 p.m. to 4:45 p.m., how much was he paid?

4. Write 55% as a reduced fraction.
(33)

5. (a) Write $\frac{9}{100}$ as a percent.
(75)
  (b) Write $\frac{9}{10}$ as a percent.

6. The entire family was present at the reunion. What
(75) percent of the family was present?

7. A century is 100 years. A decade is 10 years.
$(29, 75)$
  (a) What fraction of a century is a decade?

  (b) What percent of a century is a decade?

8. (a) Write 0.48 as a reduced fraction.
$(73, 75)$
  (b) Write 0.48 as a percent.

9. Write $\frac{7}{8}$ as a decimal number.
(74)

10. $\left(1\frac{1}{3} + 1\frac{1}{6}\right) - 1\frac{2}{3}$ **11.** $1\frac{1}{2} \times 3 \times 1\frac{1}{9}$
(48) (72)

12. $4\frac{2}{3} \div 1\frac{1}{6}$ **13.** $0.1 + (1 - 0.01)$
(68) (38)

14. Write and complete this proportion: Three is to four as
(83) nine is to what number?

15. Which ratio below forms a proportion with $\frac{3}{5}$?
(83)

    A. $\frac{3}{10}$ B. $\frac{6}{15}$ C. $\frac{12}{20}$ D. $\frac{5}{3}$

16. Write the standard numeral for the following:
(32)

$$(8 \times 10{,}000) + (4 \times 100) + (2 \times 10)$$

17. Compare: $2^4 \bigcirc 4^2$
(73)

18. Write the prime factorization of both the numerator and
(67) the denominator of this fraction. Then reduce the
fraction.

$$\frac{24}{32}$$

19. What is the greatest common factor of 24 and 32?
(20)

20. If the diameter of a ceiling fan is 4 ft, then the tip of one
(47) of the blades on the fan moves about how far during one
full turn? Choose the closest answer.

    A. 8 ft   B. 12 ft   C. $12\frac{1}{2}$ ft   D. 13 ft

21. What is the perimeter of this
(8) trapezoid?

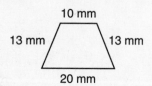

22. A cube has edges 3 cm long. What is the volume of the cube?
(82)

23. (a) What is the area of this
(71, 79)   parallelogram?

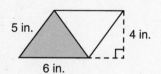

    (b) What is the area of the shaded
        triangle?

24. One fourth of the 120 campers chopped wood. How
(77) many campers did not chop wood?

25. How many millimeters is 2.5 centimeters?
(7)

26. What is the name of this geometric
(Inv. 6) solid?

27. Simplify:
(81)
(a) 3 quarts + 2 pints (Write the sum in pints.)

(b) $\dfrac{64 \text{ cm}^2}{8 \text{ cm}}$ (c) $\dfrac{60 \text{ newspapers}}{3 \text{ stacks}}$

28. DeShawn answered 20 of the 25 questions correctly. What
(75) percent of the questions did DeShawn answer correctly?

29. Draw a triangle that has two perpendicular sides.
(28, 60)

30. The ratio of dimes to nickels in Jason's change box is $\frac{2}{3}$.
(15, 83) Jason has $0.75 in nickels.

(a) How many nickels does Jason have?

(b) How many dimes does Jason have? (*Hint:* Write and
complete a proportion using the ratio given above and
your answer to part (a).)

(c) In all, how much money does Jason have in his
change box?

LESSON

84   Order of Operations, Part 2

WARM-UP

Facts Practice: Linear Measurement (Test K)

Mental Math: Count by 12's from 12 to 144.

    a. 40 · 40   **b.** 980 − 136   **c.** $5.99 + $2.99

    d. $\frac{1}{4}$ of $10.00   **e.** 7.5 × 100   **f.** $\frac{480}{20}$

    g. 8 × 8, − 4, ÷ 3, + 4, ÷ 4, + 2, ÷ 4

    h. Hold your hands one foot apart, then one inch apart.

Problem Solving:

Two shaded triangles in this figure together form a third shaded triangle. Find the area of each of the three shaded triangles.

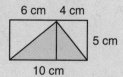

NEW CONCEPT

Recall that the four operations of arithmetic are addition, subtraction, multiplication, and division. When more than one type of operation occurs in the same expression, we perform the operations in the order described below.

ORDER OF OPERATIONS

1. Perform operations within parentheses.

2. Multiply and divide from left to right.

3. Add and subtract from left to right.

Example 1 Simplify: 2 × 8 + 2 × 6

Solution Multiplication and addition occur in this expression. We multiply first.

$$\underbrace{2 \times 8}_{16} + \underbrace{2 \times 6}_{12}$$

Then we add.

$$16 + 12 = \textbf{28}$$

Some calculators are designed to recognize the standard order of operations and some are not. You can test the design of your calculator by using the expression from example 1. Enter these keystrokes:

"Algebraic logic" calculators should display the following after the equal sign is pressed:

$$28.$$

Example 2 Simplify: $0.5 + 0.5 \div 0.5 - 0.5 \times 0.5$

Solution First we multiply and divide from left to right.

$$0.5 + \underbrace{0.5 \div 0.5}_{1} - \underbrace{0.5 \times 0.5}_{0.25}$$
$$0.5 +$$

Then we add and subtract from left to right.

$$0.5 + 1 - 0.25 = \mathbf{1.25}$$

Example 3 Simplify: $2(8 + 6)$

Solution First we perform the operation within the parentheses.

$$2(8 + 6)$$

$$2(14)$$

Then we multiply.

$$2(14) = \mathbf{28}$$

LESSON PRACTICE

Practice set* Simplify:

a. $5 + 5 \times 5 - 5 \div 5$ **b.** $2(10) + 2(6)$

c. $5 + 4 \times 3 \div 2 - 1$ **d.** $32 + 1.8(20)$

e. $3 + (3 \times 3) - (3 \div 3)$

MIXED PRACTICE

Problem set

1. What is the ratio of prime numbers to composite numbers in this list?
(23, 65)

$$2, 3, 4, 5, 6, 7, 8, 9, 10$$

2. Bianca poured four cups of milk from a full half-gallon container. How many cups of milk were left in the container?
(78)

3. $6 + 6 \times 6 - 6 \div 6$
(84)

4. Write 30% as a reduced fraction. Then write the fraction as a decimal number.
(33, 74)

Find the area of each triangle:

5.
(79)

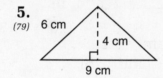
6 cm
4 cm
9 cm

6.
(79)

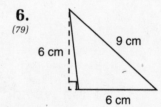
6 cm
9 cm
6 cm

7. (a) Write $\frac{1}{20}$ as a decimal number.
(74, 75)

(b) Write $\frac{1}{20}$ as a percent.

8. "Some parallelograms are rectangles." True or false?
(64)

9. What is the area of this parallelogram?
(71)

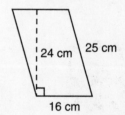
24 cm
25 cm
16 cm

10. What is the perimeter of this parallelogram?
(71)

11. $\left(3\frac{1}{8} + 2\frac{1}{4}\right) - 1\frac{1}{2}$
(48)

12. $\frac{5}{6} \times 2\frac{2}{3} \times 3$
(72)

13. $8\frac{1}{3} \div 100$
(68)

14. $(4 - 3.2) \div 10$
(53)

15. $0.5 \times 0.5 + 0.5 \div 0.5$
(84)

16. $8 \div 0.04$
(49)

17. Which digit is in the hundredths place in 12.345678?
(34)

18. Describe how to round $5\frac{1}{8}$ to the nearest whole number.
(51)

19. Write the prime factorization of 700 using exponents.
(75)

20. $8m = 4^2$
(38)

21. The perimeter of a square is 1 meter. How many
(8) centimeters long is each side?

22. Fong scored 9 of the team's 45 points.
(29, 75)
(a) What fraction of the team's points did Fong score?

(b) What percent of the team's points did she score?

23. What time is 5 hours 30 minutes after 9:30 p.m.?
(32)

24. Write and complete this proportion: Six is to four as
(83) what number is to eight?

25. Figure *ABCD* is a parallelogram. Its
(69) opposite angles ($\angle A$ and $\angle C$, $\angle B$
and $\angle D$) are congruent. Its adjacent
angles (such as $\angle A$ and $\angle B$) are
supplementary. If angle *A* measures
70°, what are the measures of $\angle B$,
$\angle C$, and $\angle D$?

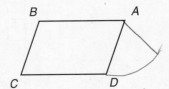

26. If each small cube has a volume of
(82) 1 cm³, what is the volume of this
rectangular prism?

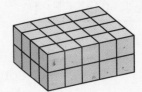

Simplify:

27. 2 ft + 24 in. (Write the sum in inches.)
(81)

28. (a) $\dfrac{100 \text{ cm}^2}{10 \text{ cm}}$ (b) $\dfrac{180 \text{ pages}}{4 \text{ days}}$
(81)

29. A triangle has vertices at the coordinates (4, 4) and (4, 0)
(Inv. 7) and at the origin. Draw the triangle on graph paper.
Notice that inside the triangle are some full squares and
some half squares.

(a) How many full squares are in the triangle?

(b) How many half squares are in the triangle?

30. What is the ratio of men to women in a group of 24 people
(23) with 16 women?

LESSON

85

Using Cross Products to Solve Proportions

WARM-UP

Facts Practice: Liquid Measurement (Test L)

Mental Math: Count up and down by 5's between −25 and 25.

a. 50 · 50   b. 1000 − 625   c. 4 × $3.99
d. Double $1.25.   e. 7.5 ÷ 10   f. 20 × 35
g. 7 × 7, + 1, ÷ 2, − 1, ÷ 2, × 5, ÷ 2
h. Hold your hands one yard apart, then two feet apart.

Problem Solving:

Here is part of a multiplication-facts table, but one number is wrong. What is the wrong number?

48	54
56	63
64	81

NEW CONCEPT

One way to determine whether two fractions are equal is to compare their cross products. If the cross products are equal, then the fractions are equal as well. The cross products of two fractions are found by cross multiplication, as we show below.

$$8 \times 3 = 24 \qquad \frac{3}{4} \times \frac{6}{8} \qquad 4 \times 6 = 24$$

Both cross products are 24. Since the cross products are equal, we can conclude that the fractions are equal.

Equal fractions have equal cross products.

Example 1 Use cross products to determine whether $\frac{3}{5}$ and $\frac{4}{7}$ are equal.

Solution To find the cross products, we multiply the numerator of each fraction by the denominator of the other fraction. We write the cross product above the numerator that is multiplied.

$$21 \diagdown \underset{5}{\overset{3}{}} \times \underset{7}{\overset{4}{}} \diagup 20$$

The cross products are not equal, so **the fractions are not equal.** The greater cross product is above the greater fraction. So $\frac{3}{5}$ is greater than $\frac{4}{7}$.

Cross products do not work by magic. When we find the cross products of two fractions, we are simply renaming the fractions with common denominators. The common denominator is the product of the two denominators and is usually not written. Look again at the two fractions we compared:

$$\frac{3}{5} \qquad \frac{4}{7}$$

The denominators are 5 and 7.

If we multiply $\frac{3}{5}$ by $\frac{7}{7}$ and multiply $\frac{4}{7}$ by $\frac{5}{5}$, we form two fractions that have common denominators.

$$\frac{3}{5} \times \frac{7}{7} = \frac{21}{35} \qquad \frac{4}{7} \times \frac{5}{5} = \frac{20}{35}$$

The numerators of the renamed fractions are 21 and 20, which are the cross products of the fractions. So when we compare cross products, we are actually comparing the numerators of the renamed fractions.

Example 2 Do these two ratios form a proportion?

$$\frac{8}{12}, \frac{12}{18}$$

Solution If the cross products of two ratios are equal, then the ratios are equal and therefore form a proportion. To find the cross products of the ratios above, we multiply 8 by 18 and 12 by 12.

$$144 \diagdown \underset{12}{\overset{8}{}} \times \underset{18}{\overset{12}{}} \diagup 144$$

The cross products are 144 and 144, so **the ratios form a proportion.**

$$\frac{8}{12} = \frac{12}{18}$$

Since equivalent ratios have equal cross products, we can use cross products to find a missing term in a proportion. By cross multiplying, we form an equation. Then we solve the equation to find the missing term of the proportion.

Example 3   Use cross products to complete this proportion: $\frac{6}{9} = \frac{10}{m}$

Solution   The cross products of a proportion are equal. So 6 times m equals 9 times 10, which is 90.

$$\frac{6}{9} = \frac{10}{m}$$

$$6m = 9 \cdot 10$$

We solve this equation:

$$6m = 90$$

$$m = 15$$

The missing term is 15. We complete the proportion.

$$\frac{6}{9} = \frac{10}{15}$$

Example 4   Use cross products to find the missing term in this proportion: Fifteen is to twenty-one as what number is to seventy?

Solution   We write the ratios in the order stated.

$$\frac{15}{21} = \frac{w}{70}$$

The cross products of a proportion are equal.

$$15 \cdot 70 = 21w$$

To find the missing term, we divide 15 · 70 by 21. Notice that we can reduce as follows:

$$\frac{\overset{5}{15} \cdot \overset{10}{70}}{\underset{1}{\overset{}{21}}} = w$$

The missing term is **50.**

LESSON PRACTICE

Practice set* Use cross products to determine whether each pair of ratios forms a proportion:

a. $\dfrac{6}{9}, \dfrac{7}{11}$ b. $\dfrac{6}{8}, \dfrac{9}{12}$

Use cross products to complete each proportion:

c. $\dfrac{6}{10} = \dfrac{9}{x}$ d. $\dfrac{12}{16} = \dfrac{y}{20}$

e. Use cross products to find the missing term in this proportion: 10 is to 15 as 30 is to what number?

MIXED PRACTICE

Problem set

1. Twenty-one of the 25 answers on Scott's test were
(75) correct. What percent of the answers were correct?

2. By the time the blizzard was over, the temperature had
(14) dropped from 17°F to −6°F. This was a drop of how many degrees?

3. The cost to place a collect call was $1.50 for the first
(12) minute plus $1.00 for each additional minute. What was the cost of a 5-minute phone call?

4. The ratio of runners to walkers at the 10K fund-raiser
(80) was 5 to 7. If there were 350 runners, how many walkers were there?

5. The two acute angles in $\triangle ABC$ are
(69) complementary. If the measure of $\angle B$ is 55°, what is the measure of $\angle A$?

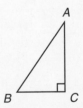

6. Athletic shoes are on sale for 20% off. Toni wants to buy
(41) a pair of running shoes that are regularly priced at $55.

    (a) How much money will be subtracted from the regular price if she buys the shoes on sale?

    (b) What will be the sale price of the shoes?

7. Freddy bought a pair of shoes for a sale price of $39.60.
(41) The sales-tax rate was 8%.

    (a) What was the sales tax on the purchase?

    (b) What was the total price including tax?

8. (a) Write $\frac{1}{25}$ as a decimal number.
(74, 75)

    (b) Write $\frac{1}{25}$ as a percent.

9. Use cross products to determine whether this pair of
(85) ratios forms a proportion:

$$\frac{5}{11} \diagup \frac{6}{13}$$

10. Use cross products to find the missing term in this
(85) proportion: 4 is to 6 as 10 is to what number?

11. $10 \div 2\frac{1}{2}$
(68)

12. $6.5 - (4 - 0.32)$
(38)

13. $(6.25)(1.6)$
(39)

14. $0.06 \div 12$
(45)

Find each missing number:

15. $2\frac{1}{2} + x = 3\frac{1}{4}$
(59)

16. $4\frac{1}{8} - y = 1\frac{1}{2}$
(48)

17. $\dfrac{9}{12} = \dfrac{n}{20}$
(85)

18. Arrange in order from least to greatest:
(76)

$$\tfrac{1}{2}, 0.4, 30\%$$

19. At a camp with 300 campers and 15 counselors, what is
(23) the camper-counselor ratio?

20. If a number cube is rolled once, what is the probability
(58, 65) that it will stop with a composite number on top?

21. One fourth of the 32 marshmallows burned in the fire.
(77) How many did not burn?

22. What is the area of a parallelogram that has vertices with
(Inv. 7, 71) the coordinates (0, 0), (4, 0), (5, 3), and (1, 3)?

23. $2 + (2 \times 2 - (2 \div 2)$
(84)

24. Jim started the 10-kilometer race at 8:22 a.m. He finished the
(32) race at 9:09 a.m. How long did it take him to run the race?

25. Refer to the table below to answer this question: Ten
(15) kilometers is about how many miles? Round the answer
to the nearest mile.

| 1 meter ≈ 1.093 yards |
| 1 kilometer ≈ 0.621 mile |

Note: The symbol ≈ means "is approximately equal to."

26. Lindsey packed boxes that were
(82) 1 foot long, 1 foot wide, and 1 foot
tall into a larger box that was 5 feet
long, 4 feet wide, and 3 feet tall.

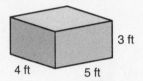

    (a) How many boxes could be
packed on the bottom layer of
the larger box?

    (b) Altogether, how many small boxes could be packed in
the larger box?

27. Simplify:
(81)
    (a) 2 ft + 24 in. (Write the sum in feet.)

    (b) 3 yd · 3 yd

28. A quart is what percent of a gallon?
(75, 78)

This figure shows a square inside a
circle, which is itself inside a larger
square. Refer to this figure to answer
problems 29 and 30.

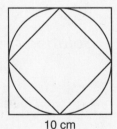

29. The area of the smaller square is
(31) half the area of the larger square.

    (a) What is the area of the larger square?

    (b) What is the area of the smaller square?

30. Based on your answers to the questions in problem 29,
(31) make an educated guess as to the area of the circle.

LESSON

86   **Area of a Circle**

WARM-UP

Facts Practice: 30 Fractions to Reduce (Test G)

Mental Math: Count up and down by 2's between −10 and 10.

    a. 60 · 60   **b.** 850 − 170   **c.** $8.99 + $4.99

    d. $\frac{1}{5}$ of $2.50   **e.** 0.08 × 100   **f.** $\frac{360}{120}$

    g. 6 × 6, − 6, ÷ 2, − 1, ÷ 2, × 8, − 1, ÷ 5

Problem Solving:

Copy this problem and fill in the missing digits. No two digits in the problem may be alike.

$$\begin{array}{r} \overline{}\;\overline{} \\ \times\;\overline{}\;8 \\ \hline \overline{}\;\overline{} \end{array}$$

NEW CONCEPT

We can estimate the area of a circle drawn on a grid by counting the number of square units enclosed by the figure.

Example 1 This circle is drawn on a grid.

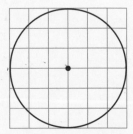

(a) How many units is the radius of the circle?

(b) Estimate the area of the circle.

Solution (a) To find the radius of the circle, we may either find the diameter of the circle and divide by 2, or we may locate the center of the circle and count units to the circle. We find that the radius is **3 units.**

(b) To estimate the area of the circle, we count the square units enclosed by the circle. We show the circle again, this time shading the squares that lie completely or mostly within the circle. We count 24 such squares. We have also marked with dots the squares that have about half their area inside the circle. We count 8 such "half squares." Since $\frac{1}{2}$ of 8 is 4, we add 4 square units to 24 square units to get an estimate of **28 square units** for the area of the circle.

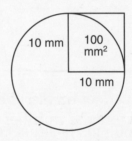

Finding the exact area of a circle involves the number π. To find the area of a circle, we first find the area of a square built on the radius of the circle. The circle below has a radius of 10 mm, so the area of the square is 100 mm². Notice that four of these squares would cover more than the area of the circle. However, the area of three of these squares is less than the area of the circle.

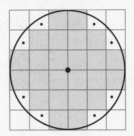

The area of the circle is exactly equal to π times the area of one of these squares. To find the area of this circle, we multiply the area of the square by π. We will continue to use 3.14 for the approximation of π.

$$3.14 \times 100 \text{ mm}^2 = 314 \text{ mm}^2$$

The area of the circle is approximately 314 mm².

Example 2 The radius of a circle is 3 cm. What is the area of the circle? (Use 3.14 for π. Round the answer to the nearest square centimeter.)

Solution We will find the area of a square whose sides equal the radius. Then we multiply that area by 3.14.

Area of square: 3 cm × 3 cm = 9 cm²

Area of circle: (3.14)(9 cm²) = 28.26 cm²

We round 28.26 cm² to the nearest whole number of square centimeters and find that the area of the circle is approximately **28 cm²**.

The area of any circle is π times the area of a square built on a radius of the circle. The following formula uses A for the area of a circle and r for the radius of the circle to relate the area of a circle to its radius:

$$A = \pi r^2$$

LESSON PRACTICE

Practice set* **a.** The radius of this circle is 4 units. Estimate the area of the circle by counting the squares within the circle.

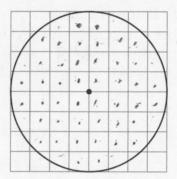

In problems **b–e**, use 3.14 for π.

b. Calculate the area of a circle with a radius of 4 cm.

c. Calculate the area of a circle with a radius of 2 feet.

d. Calculate the area of a circle with a diameter of 2 feet.

e. Calculate the area of a circle with a diameter of 10 inches.

MIXED PRACTICE

Problem set **1.** What is the quotient when the decimal number ten and
(49) six tenths is divided by four hundredths?

2. The time in Los Angeles is 3 hours earlier than the time
(32) in New York. If it is 1:15 p.m. in New York, what time is it in Los Angeles?

3. Geraldine paid with a $10 bill for 1 dozen photographs that
(12) cost 75¢ each. How much should she get back in change?

4. $32 + 1.8(50)$
(84)

5. If each block has a volume of one
(82) cubic inch, what is the volume of
this tower?

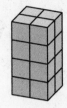

6. The ratio of hardbacks to paperbacks in the bookstore
(80) was 5 to 2. If there were 600 hardbacks, how many
paperbacks were there?

7. Nate missed three of the 20 questions on the test. What
(75) percent of the questions did he miss?

8. The credit card company charges 1.5% (0.015) interest on
(41) the unpaid balance each month. If Mr. Jones has an
unpaid balance of $2000, how much interest does he
need to pay this month?

9. (a) Write $\frac{4}{5}$ as a decimal number.
(74, 75)

(b) Write $\frac{4}{5}$ as a percent.

10. Serena is stuck on a multiple-choice question that has
(58) four choices. She has no idea what the correct answer is,
so she just guesses. What is the probability that her guess
is correct?

11. $5\frac{1}{2} + 3\frac{7}{8}$ **12.** $3\frac{1}{4} - \frac{5}{8}$ **13.** $\left(4\frac{1}{2}\right)\left(\frac{2}{3}\right)$
(59)   (63)   (66)

14. $12\frac{1}{2} \div 100$ **15.** $5 \div 1\frac{1}{2}$ **16.** $\frac{5}{6}$ of $30
(68)   (68)   (29)

Find each missing number:

17. $4.72 + 12 + n = 50.4$ **18.** $\$10 - m = \$9.87$
(43)   (3)

19. $3n = 0.48$ **20.** $\frac{w}{8} = \frac{25}{20}$
(45)   (85)

21. What are the next three terms in this sequence of perfect
(38) squares?

1, 4, 9, 16, 25, 36, 49, 64, 81, 100, _____, _____, _____, ...

22. This parallelogram is divided into
(71, 79) two congruent triangles.

(a) What is the area of the parallelogram?

(b) What is the area of each triangle?

23. Sydney drew a circle with a radius
(86) of 10 cm. What was the area of the circle? (Use 3.14 for π.)

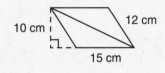

Refer to the line graph below to answer problems 24–27.

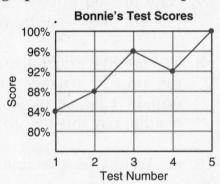

24. What is the range of Bonnie's test scores?
(Inv. 5)

25. What is the median of Bonnie's five test scores?
(Inv. 5)

26. What is the mean of Bonnie's five test scores?
(Inv. 5)

27. Write a question that relates to the line graph and answer
(18) the question.

28. The opposite angles of a
(69, 71) parallelogram are congruent. The adjacent angles are supplementary. If $\angle X$ measures 110°, then what are the measures of $\angle Y$ and $\angle Z$?

29. The diameter of each wheel on the lawn mower is 10
(47) inches. How far must the lawn mower be pushed in order for each wheel to complete one full turn? Round the answer to the nearest inch. (Use 3.14 for π.)

30. The coordinates of three vertices of a parallelogram are
(Inv. 7) (−3, 3), (2, 3), and (4, −1). What are the coordinates of the fourth vertex?

LESSON
87   Finding Missing Factors

Facts Practice: 64 Multiplication Facts (Test D)

Mental Math: Count up and down by $\frac{1}{8}$'s between $\frac{1}{8}$ and 2.

   a. 70 · 70   **b.** 1000 − 375   **c.** 5 × $4.99

   d. Double $0.85.   **e.** 62.5 ÷ 100   **f.** 20 × 45

   g. 5 × 5, − 5, × 5, ÷ 2, − 1, ÷ 7, × 3, − 1, ÷ 2

   h. Hold your hands 50 cm apart, then 25 cm apart.

Problem Solving:

Samantha averaged 85 points on her first three tests. She averaged 95 on her next two tests. What was Samantha's average score for her first five tests?

NEW CONCEPT

Since Lesson 4 we have practiced solving missing factor problems. In this lesson we will solve problems in which the unknown factor is a mixed number or a decimal number. Remember that we can find an unknown factor by dividing the product by the known factor.

Example 1   Solve: $5n = 21$

Solution   To find an unknown factor, we divide the product by the known factor.

$$\begin{array}{r} 4\frac{1}{5} \\ 5\overline{)21} \\ \underline{20} \\ 1 \end{array}$$

$$n = 4\frac{1}{5}$$

Note: We will write the answer as a mixed number unless there are decimal numbers in the problem.

Example 2   Solve: $0.6m = 0.048$

Solution   Again, we find the unknown factor by dividing the product by the known factor. Since there are decimal numbers in the problem, we write our answer as a decimal number.

$$\begin{array}{r} 0.08 \\ 0\,6.\overline{)0\,0.48} \\ \underline{48} \\ 0 \end{array}$$

$$m = 0.08$$

Example 3   Solve: $45 = 4x$

Solution   This problem might seem "backward" because the multiplication is on the right-hand side. However, an equal sign is not directional. It simply states that the quantities on either side of the sign are equal. In this case, the product is 45 and the known factor is 4. We divide 45 by 4 to find the unknown factor.

$$\begin{array}{r} 11\frac{1}{4} \\ 4\overline{)45} \\ \underline{4} \\ 05 \\ \underline{4} \\ 1 \end{array}$$

$$x = 11\frac{1}{4}$$

LESSON PRACTICE

Practice set   Solve:

a. $6w = 21$ **b.** $50 = 3f$ **c.** $5n = 36$

d. $0.3t = 0.24$ **e.** $8m = 3.2$ **f.** $0.8 = 0.5x$

MIXED PRACTICE

Problem set   **1.** If the divisor is 12 and the quotient is 24, what is the dividend?
   (4)

2. The brachiosaurus, one of the largest dinosaurs, weighed only $\frac{1}{4}$ as much as a blue whale. A blue whale can weigh 140 tons. How much could a brachiosaurus have weighed?
   (29)

3. Fourteen of the 32 cookies are still warm. What is the ratio of warm cookies to cooled cookies?
   (23)

Find each missing number:

4. $0.3m = 0.27$ **5.** $31 = 5n$
(87)   (87)

6. $3n = 6^2$ **7.** $4n = 0.35$
(38)   (87)

8. Write 0.25 as a fraction and add it to $3\frac{1}{4}$. What is the sum?
   (73)

9. Write $\frac{3}{5}$ as a decimal and add it to 6.5. What is the sum?
   (74)

10. Write $\frac{1}{50}$ as a decimal number and as a percent.
(74, 75)

11. $12\frac{1}{5} - 3\frac{4}{5}$ **12.** $6\frac{2}{3} \times 1\frac{1}{5}$ **13.** $11\frac{1}{9} \div 100$
(63)   (66)   (68)

14. $4.75 + 12.6 + 10$ **15.** $35 - (0.35 \times 100)$
(38)   (84)

16. Solve this proportion: $\dfrac{12}{m} = \dfrac{18}{9}$
(85)

17. Write the decimal numeral twelve and five hundredths.
(35)

18. Find the volume of this rectangular
(82) prism.

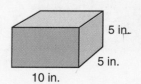

19. If *a* equals 15, then what number does $2a - 5$ equal?
(47)

20. What is the area of this
(71) parallelogram?

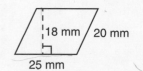

21. What is the perimeter of this
(8) parallelogram?

22. "All rectangles are parallelograms." True or false?
(64)

23. Charles spent $\frac{1}{10}$ of his 100 shillings. How many shillings
(77) does he still have?

24. The temperature rose from −18°F to 19°F. How many
(14) degrees did the temperature increase?

25. How many **centimeters** long is the line below?
(7)

26. Johann poured 500 mL of soda pop from a full 2-liter
(78) container. How many milliliters of soda pop were left in
the container?

27. Name this geometric solid.
(Inv. 6)

28. Simplify:
(81)
(a) 2 meters + 100 centimeters (Write the answer in
meters.)

(b) 2 m · 4 m

29. $4 + 4 \times 4 - 4 \div 4$
(84)

30. What is the perimeter of a rectangle with vertices at (−4, −4),
(Inv. 7) (−4, 4), (4, 4), and (4, −4)?

htewl pplnZ yhzf ta koo wtsnnop

LESSON

88

Using Proportions to Solve Ratio Problems

WARM-UP

Facts Practice: Liquid Measurement (Test L)

Mental Math: Count up and down by 25's between –150 and 150.

a. $80 \cdot 80$ **b.** $720 - 150$ **c.** $\$1.98 + \$1.98$

d. $\frac{1}{10}$ of $\$5.00$ **e.** 0.15×100 **f.** $\frac{750}{250}$

g. $4 \times 4, - 1, \times 2, + 3, \div 3, - 1, \times 10, - 1, \div 9$

Problem Solving:

Silvia was thinking of a number less than 90 that she says when counting by sixes and when counting by fives, but not when counting by fours. Of what number was she thinking?

NEW CONCEPT

Proportions can be used to solve many types of word problems. In this lesson we will use proportions to solve ratio word problems such as those in the following examples.

Example 1 The ratio of salamanders to frogs was 5 to 7. If there were 20 salamanders, how many frogs were there?

Solution In this problem there are two kinds of numbers: ratio numbers and actual-count numbers. The ratio numbers are 5 and 7. The number 20 is an actual count of the salamanders. We will arrange these numbers in two columns and two rows to form a ratio box.

	Ratio	Actual Count
Salamanders	5	20
Frogs	7	f

We were not given the actual count of frogs, so we use the letter f to stand for the actual number of frogs.

Instead of using scale factors in this lesson, we will practice using proportions. We use the positions of numbers in the ratio box to write a proportion. By solving the proportion, we find the actual number of frogs.

	Ratio	Actual Count
Salamanders	5	20
Frogs	7	f

$$\longrightarrow \frac{5}{7} = \frac{20}{f}$$

We can solve the proportion in two ways. We can multiply $\frac{5}{7}$ by $\frac{4}{4}$, or we can use cross products. Here we show the solution using cross products:

$$\frac{5}{7} = \frac{20}{f}$$

$$5f = 7 \cdot 20$$

$$f = \frac{7 \cdot 20}{5}$$

$$f = 28$$

We find that there were **28 frogs.**

Example 2 The ratio of humpback whales to killer whales was 2 to 7. If there were 42 killer whales, how many humpbacks were there?

Solution We begin by drawing a ratio box for this problem. (Ratio boxes are useful because they help us to write proportions correctly.)

	Ratio	Actual Count
Humpback	2	h
Killer	7	42

$$\longrightarrow \frac{2}{7} = \frac{h}{42}$$

We see that we can solve this proportion by multiplying $\frac{2}{7}$ by $\frac{6}{6}$. We can also solve this proportion using cross products.

$$\frac{2}{7} = \frac{h}{42}$$

$$2 \cdot 42 = 7h$$

$$\frac{2 \cdot 42}{7} = h$$

$$12 = h$$

We find that there were **12 humpback whales.**

LESSON PRACTICE

Practice set For each problem, draw a ratio box. Then solve the problem using proportions.

a. There were more damsels than knights in the court. In fact, the ratio of damsels to knights was 5 to 4. If there were 60 knights, how many damsels were there?

b. At the party the boy-girl ratio was 5 to 3. If there were 30 boys, how many girls were there?

MIXED PRACTICE

Problem set

1. Mason scored 12 of the team's 20 points. What percent of
(75) the team's points did Mason score?

2. If Pinocchio's nose grows $\frac{1}{4}$ inch per lie, then how many
(50) lies has he told if his nose has grown 4 inches?

3. Eamon wants to buy a new baseball glove that costs $50.
(87) He has $14 and he earns $6 per hour cleaning yards. How many hours must he work to have enough money to buy the glove?

4. Find the area of this triangle.
(79)

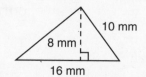

5. Draw a ratio box for this problem. Then solve the
(88) problem using a proportion.

The ratio of skaters to bikers was 3 to 5. If there were 15 bikers, how many skaters were there?

6. What is the volume of this
(82) rectangular prism?

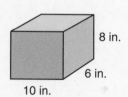

Find each missing factor:

7. $10w = 25$
(87)

8. $20 = 9m$
(87)

9. (a) What is the perimeter of this
(8, 79)   triangle?

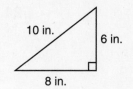

10 in.   6 in.

8 in.

(b) What is the area of this triangle?

10. Write 5% as a
(33, 75)

(a) decimal number.

(b) fraction.

11. Write $\frac{2}{5}$ as a decimal, and multiply it by 2.5. What is the
(74)   product?

12. Compare: $\frac{2}{3} + \frac{3}{2} \bigcirc \frac{2}{3} \cdot \frac{3}{2}$
(29, 56)

13. $\frac{1}{3} \times \frac{100}{1}$ **14.** $6 \div 1\frac{1}{2}$
(70)   (68)

15. $12 \div 0.25$ **16.** 0.025×100
(49)   (46)

17. If the tax rate is 7%, what is the tax on a $24.90 purchase?
(41)

18. The prime factorization of what number is $2^2 \cdot 3^2 \cdot 5^2$?
(73)

19. Which of these is a composite number?
(65)

     A. 61   B. 71   C. 81   D. 101

20. Round the decimal number one and twenty-three
(51)   hundredths to the nearest tenth.

21. Albert baked 5 dozen cookies and gave away $\frac{7}{12}$ of them.
(77)   How many cookies were left?

22. $6 \times 3 - 6 \div 3$
(84)

23. How many milliliters is 4 liters?
(78)

24. Draw a line segment $2\frac{1}{4}$ inches long. Label the endpoints
(69) A and C. Then make a dot at the midpoint of $\overline{AC}$ (the point
halfway between points *A* and *C*), and label the dot *B*.
What are the lengths of $\overline{AB}$ and $\overline{BC}$?

25. On graph paper draw a rectangle with vertices at (–2, –2),
(Inv. 7) (4, –2), (4, 2), and (–2, 2). What is the area of the
rectangle?

26. What is the ratio of the length to the width of the rectangle
(23) in problem 25?

27. Describe how to calculate the area of a triangle.
(79)

28. In the figure at right, angles *ADB*
(69) and *BDC* are supplementary.

(a) What is m∠*ADB*?

(b) What is the ratio of m∠*BDC* to
m∠*ADB*? Write the answer as a
reduced fraction.

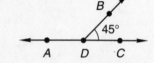

29. Nathan drew a circle with a radius
(86) of 10 cm. Then he drew a square
around the circle.

(a) What was the area of the square?

(b) What was the area of the circle?
(Use 3.14 for *π*.)

30. Solve this proportion: $\dfrac{6}{8} = \dfrac{w}{100}$
(85)

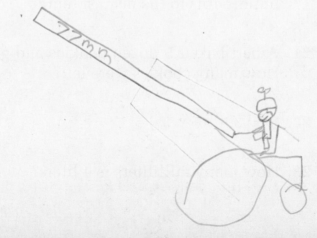

LESSON

89   Estimating Square Roots

WARM-UP

Facts Practice: Linear Measurement (Test K)

Mental Math: Count up and down by 3's between −15 and 15.

a. 90 · 90
b. 1000 − 405
c. 6 × $7.99
d. Double $27.00.
e. 87.5 ÷ 100
f. 20 × 36
g. 3 × 3, + 2, × 5, − 5, × 2, ÷ 10, + 5, ÷ 5
h. About how many meters tall is a classroom door?

Problem Solving:

The perimeter of this rectangle is 1 m. What is its length?

21 cm

NEW CONCEPT

We have practiced finding square roots of perfect squares from 1 to 100. In this lesson we will find the square roots of perfect squares greater than 100. We will also use a guess-and-check method to estimate the square roots of numbers that are not perfect squares. As we practice, our guesses will improve and we will begin to see clues to help us estimate.

Example 1   Simplify: $\sqrt{400}$

Solution   We need to find a number that, when multiplied by itself, has a product of 400.

$$\Box \times \Box = 400$$

We know that $\sqrt{400}$ is more than 10, because 10 × 10 equals 100. We also know that $\sqrt{400}$ is much less than 100, because 100 × 100 equals 10,000. Since $\sqrt{4}$ equals 2, the 4 in $\sqrt{400}$ hints that we should try 20.

$$20 \times 20 = 400$$

We find that $\sqrt{400}$ equals **20.**

Example 2   Simplify: $\sqrt{625}$

Solution   In example 1 we found that $\sqrt{400}$ equals 20. Since $\sqrt{625}$ is greater than $\sqrt{400}$, we know that $\sqrt{625}$ is greater than 20. We find that $\sqrt{625}$ is less than 30, because 30×30 equals 900. Since the last digit is 5, perhaps $\sqrt{625}$ is 25. We multiply to find out.

$$
\begin{array}{r}
25 \\
\times\ 25 \\
\hline
125 \\
50\ \ \\
\hline
625 \\
\end{array}
$$

We find that $\sqrt{625}$ equals **25**.

We have practiced finding the square roots of numbers that are perfect squares. Now we will practice estimating the square root of numbers that are not perfect squares.

Example 3   Between which two consecutive whole numbers is $\sqrt{20}$?

Solution   Notice that we are not asked to find the square root of 20. To find the whole numbers on either side of $\sqrt{20}$, we can first think of the perfect squares that are on either side of 20. Here we show the first few perfect squares, starting with 1.

$$1, 4, 9, 16, 25, 36, 49$$

We see that 20 is between the perfect squares 16 and 25. So $\sqrt{20}$ is between $\sqrt{16}$ and $\sqrt{25}$.

$$\sqrt{16},\ \sqrt{20},\ \sqrt{25}$$

Since $\sqrt{16}$ is 4 and $\sqrt{25}$ is 5, we see that $\sqrt{20}$ is between **4** and **5**.

Using the reasoning in example 3, we know there must be some number between 4 and 5 that is the square root of 20. We try 4.5.

$$4.5 \times 4.5 = 20.25$$

We see that 4.5 is too large, so we try 4.4.

$$4.4 \times 4.4 = 19.36$$

We see that 4.4 is too small. So $\sqrt{20}$ is greater than 4.4 but less than 4.5. (It is closer to 4.5.) If we continued this process, we would never find a decimal number or fraction that exactly equals $\sqrt{20}$. This is because $\sqrt{20}$ belongs to a number family called the **irrational numbers.**

Irrational numbers cannot be expressed exactly as a ratio (that is, as a fraction or decimal). We can only use fractions or decimals to express the *approximate* value of an irrational number.

$$\sqrt{20} \approx 4.5$$

Recall from Lesson 47 that the wavy equal sign means "is approximately equal to." The square root of 20 is approximately equal to 4.5.

Example 4 Use a calculator to approximate the value of $\sqrt{20}$ to two decimal places.

Solution We clear the calculator and then enter $\boxed{\sqrt{}}$ $\boxed{2}$ $\boxed{0}$ $\boxed{=}$ (or $\boxed{2}$ $\boxed{0}$ $\boxed{\sqrt{}}$).[†] The display will show 4.472135955. The actual value of $\sqrt{20}$ contains an infinite number of decimal places. The display approximates $\sqrt{20}$ to nine or so decimal places (depending on the model). We are asked to show two decimal places, so we round the displayed number to **4.47.**

LESSON PRACTICE

Practice set* Find each square root:

a. $\sqrt{169}$ b. $\sqrt{484}$ c. $\sqrt{961}$

Each of these square roots is between which two consecutive whole numbers? Find the answer without using a calculator.

d. $\sqrt{2}$ e. $\sqrt{15}$ f. $\sqrt{40}$

g. $\sqrt{60}$ h. $\sqrt{70}$ i. $\sqrt{80}$

Use a calculator to approximate each square root to two decimal places:

j. $\sqrt{3}$ k. $\sqrt{10}$ l. $\sqrt{50}$

[†]The order of keystrokes depends on the model of calculator. See the instructions for your calculator if the keystroke sequences described in this lesson do not work for you.

MIXED PRACTICE

Problem set   **1.** What is the difference when the product of $\frac{1}{2}$ and $\frac{1}{2}$ is
(12, 72)   subtracted from the sum of $\frac{1}{4}$ and $\frac{1}{4}$?

2. A dairy cow can give 4 gallons of milk per day. How many
(78)   cups of milk is that (1 gallon = 4 quarts; 1 quart = 4 cups)?

3. The recipe called for $\frac{3}{4}$ cup of sugar. If the recipe is
(29)   doubled, how much sugar should be used?

4. Draw a ratio box for this problem. Then solve the problem
(88)   using a proportion.

> *The recipe called for sugar and flour in the ratio
> of 2 to 9. If the chef used 18 pounds of flour, how
> many pounds of sugar were needed?*

5. Which of these numbers is greater than 6 but less than 7?
(89)
   A. $\sqrt{6.5}$ B. $\sqrt{67}$ C. $\sqrt{45}$ D. $\sqrt{76}$

6. Express the missing factor as a mixed number:
(87)
$$7n = 30$$

7. Amanda used a compass to draw a
(47)   circle with a radius of 4 inches.

   (a) What is the diameter of the
       circle?

   (b) What is the circumference of
       the circle?

Use 3.14 for π.

8. In problem 7 what is the area of the circle Amanda drew?
(86)

9. What is the area of the triangle at
(79)   right?

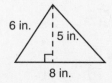

10. (a) What is the area of this
(71)   parallelogram?

   (b) What is the perimeter of this
       parallelogram?

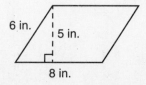

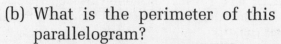

11. Write 0.5 as a fraction and subtract it from $3\frac{1}{4}$. What is the
(63, 73) difference?

12. Write $\frac{3}{4}$ as a decimal, and multiply it by 0.6. What is the
(74) product?

13. $2 \times 15 + 2 \times 12$ **14.** $\sqrt{900}$
(84)   (89)

15. $\$6 \div 8$ **16.** $1\frac{3}{5} \times 10 \times \frac{1}{4}$
(2)   (66)

17. $37\frac{1}{2} \div 100$ **18.** $3 \div 7\frac{1}{2}$
(68)   (68)

19. What is the place value of the 7 in 987,654.321?
(34)

20. Write the decimal number five hundred ten and five
(35) hundredths.

21. $30 + 60 + m = 180$
(3)

22. Half of the guests are girls. Half of the girls have brown
(72) hair. Half of the brown-haired girls wear their hair long. Of
the 32 guests, how many are girls with long, brown hair?

Refer to the pictograph below to answer problems 23–25.

Books Read This Year

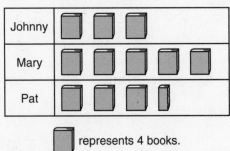

represents 4 books.

23. How many books has Johnny read?
(Inv. 5)

24. Mary has read how many more books than Pat?
(Inv. 5)

25. Write a question that relates to this graph and answer the
(Inv. 5) question.

26. Solve this proportion: $\dfrac{12}{8} = \dfrac{21}{m}$
(85)

27. The face of this spinner is divided
(58) into 12 congruent regions. If the
spinner is spun once, what is the
probability that it will stop on a 3?

28. If two angles are complementary, and if one angle is
(28, 69) acute, then the other angle is what kind of angle?

A. acute   B. right   C. obtuse

29. Simplify:
(81)
(a) 100 cm + 100 cm (Write the answer in meters.)

(b) $\dfrac{(5 \text{ in.})(8 \text{ in.})}{2}$

30. If each small block has a volume of
(82) 1 cubic inch, then what is the
volume of this cube?

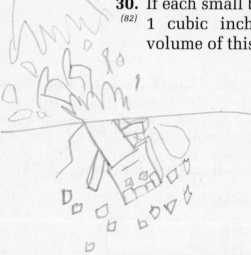

LESSON
90   Measuring Turns

Facts Practice: 24 Mixed Numbers to Write as Improper Fractions (Test J)

Mental Math: Count up and down by $\frac{1}{2}$'s between −3 and 3.

a. $\sqrt{100}$ b. $781 - 35$ c. $\$1.98 + \$2.98$

d. $\frac{1}{3}$ of $24.00   e. 0.375×100 f. $\frac{1200}{300}$

g. $2 \times 2, \times 2, \times 2, - 1, \times 2, + 2, \div 4, \div 4$

h. About how many feet wide is a classroom door?

Problem Solving:

One state used a license plate that included one letter followed by five digits. How many different license plates could be made that started with the letter A?

NEW CONCEPT

Turns can be measured in degrees. A full turn is a 360° turn. If you turn 360°, you will end up facing the same direction you were facing before you turned. A half turn is half of 360°, which is 180°. If you turn 180°, you will end up facing opposite the direction you were facing before you turned.

If you are facing north and turn 90°, you will end up facing either east or west, depending on the direction in which you turned. To avoid confusion, we often specify the direction of a turn as well as the measure of the turn. Sometimes the direction is described as being to the left or to the right. Other times it is described as clockwise or counterclockwise.

Example 1   Leila was traveling north. At the light she turned 90° to the left and traveled one block to the next intersection. At the intersection she turned 90° to the left. What direction was Leila then traveling?

Solution   A picture may help us answer the question. Leila was traveling north when she turned 90° to the left. (The dashes indicate the direction in which Leila would have continued had she not turned.) After that first turn Leila was traveling west. When she turned 90° to the left a second time, she began traveling to the **south.** Notice that the two turns in the same direction (left) total 180°. So we would expect that after the two turns Leila was heading in the direction opposite to her starting direction.

Example 2   Andy and Barney were both facing north. Andy made a quarter turn (90°) clockwise to face east, while Barney turned counterclockwise to face east. How many degrees did Barney turn?

Solution   We will draw the two turns. Andy made a quarter turn clockwise. We see that Barney made a three-quarter turn counterclockwise. We can calculate the number of degrees in three quarters of a turn by finding $\frac{3}{4}$ of 360°.

Andy

Barney

$$\frac{3}{4} \times 360° = 270°$$

Another way to find the number of degrees is to recognize that each quarter turn is 90°. So three quarters is three times 90°.

$$3 \times 90° = 270°$$

Barney turned **270°** counterclockwise.

Example 3   As Elizabeth ran each lap around the park, she made six turns to the left (and no turns to the right). What was the average number of degrees of each turn?

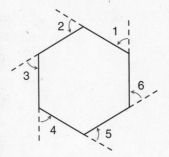

Solution We are not given the measure of any of the turns, but we do know that Elizabeth made six turns to the left to get completely around the park. That is, after six turns she once again faced the same direction she faced before the first turn. So after six turns she had turned a total of 360°. (Notice that turns are measured by the **exterior angle** and not by the **interior angle**.) We find the average number of degrees in each turn by dividing 360° by 6.

$$360° \div 6 = 60°$$

Each of Elizabeth's turns averaged **60°**.

LESSON PRACTICE

Practice set **a.** Tammy was heading south on her bike. When she reached Sycamore, she turned 90° to the right. Then at Highland she turned 90° to the right, and at Elkins she turned 90° to the right again. Assuming each street was straight, in which direction was Tammy heading on Elkins?

b. Rob made one full turn counterclockwise. Mary made two full turns clockwise. How many degrees did Mary turn?

c. David ran three laps around the park. On each lap he made five turns to the left and no turns to the right. What was the average number of degrees in each of David's turns? Draw a picture of the problem.

MIXED PRACTICE

Problem set **1.** What is the mean of 4.2, 4.8, and 5.1?
(Inv. 5)

2. The movie is 120 minutes long. If it begins at 7:15 p.m.,
(32) when will it be over?

Fifteen of the 25 children are sleeping. Use this information to answer problems 3 and 4.

3. What percent of the children are sleeping?
(75)

4. What is the ratio of sleeping to awake children?
(23)

5. This triangular prism has how many more edges than vertices?
(Inv. 6)

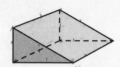

6. The carpenter cut a 12-inch diameter circle from a sheet of plywood.
(86)

(a) What was the radius of the circle?

(b) What was the area of the circle? (Use 3.14 for π.)

7. Describe the appearance of a trapezoid.
(64)

8. Arrange these numbers in order from least to greatest:
(14, 17)

$$1, -2, 0, -4, \tfrac{1}{2}$$

9. Express the missing factor as a mixed number:
(87)

$$25n = 70$$

10. What is the area of this triangle?
(79)

11. What is the perimeter of this triangle?
(8)

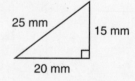

12. Write 6.25 as a mixed number. Then subtract $\frac{5}{8}$ from the mixed number. What is the difference?
(63, 73)

13. Ali was facing north. Then he turned to his left 180°. What direction was he facing after he turned?
(90)

14. Write 28% as a reduced fraction.
(41)

15. $\dfrac{n}{12} = \dfrac{20}{30}$
(85)

16. $0.625 \div 10$
(52)

17. $\dfrac{25}{0.8}$
(49)

18. $3\dfrac{3}{8} + 3\dfrac{3}{4}$
(59)

19. $5\frac{1}{8} - 1\frac{7}{8}$
(48)

20. $6\frac{2}{3} \times \frac{3}{10} \times 4$
(72)

21. One third of the two dozen knights were on horseback.
(77) How many knights were not on horseback?

22. Weights totaling 38 ounces were placed on the left side of
(18) this scale, while weights totaling 26 ounces were placed on
the right side of the scale. How many ounces of weights
should be moved from the left side to the right side to
balance the scale? (*Hint:* Find the average of the weights on
the two sides of the scale.)

23. The cube at right is made up of
(82) smaller cubes that each have a
volume of 1 cubic centimeter. What
is the volume of the larger cube?

24. Round forty-eight hundredths to the nearest tenth.
(51)

25. $\sqrt{144} - \sqrt{121}$
(89)

26. The ratio of dogs to cats in the neighborhood is 6 to 5.
(23) What is the ratio of cats to dogs?

27. $10 + 10 \times 10 - 10 \div 10$
(84)

28. The Thompsons drink a gallon of milk every two days.
(78) There are four people in the Thompson family. Each person
drinks an average of how many pints of milk per day?

29. Simplify:
(81)
(a) 10 cm + 100 mm (Write the answer in millimeters.)

(b) 300 books ÷ 30 shelves

30. On graph paper draw a segment from point A (−3, −1) to
(Inv. 7) point B (5, −1). What are the coordinates of the point on $\overline{AB}$
that is halfway between points A and B?

INVESTIGATION 9

Focus on

Experimental Probability

In Lesson 58 we determined probabilities for the outcomes of experiments without actually performing the experiments. For example, in the case of rolling a number cube we assumed that all six outcomes were equally likely and that each outcome therefore had a probability of $\frac{1}{6}$. In the spinner example we assumed that the likelihood of the spinner landing in a particular sector is proportional to the area of the sector. Thus, if the area of sector A is twice the area of sector B, the probability of the spinner landing in sector A is twice the probability of the spinner landing in sector B.

Probability that is calculated by performing "mental experiments" (as we have been doing since Lesson 58) is called **theoretical probability.** Probabilities associated with many real-world situations, though, cannot be determined by theory. Instead, we must perform the experiment repeatedly or collect data from a sample experiment. Probability determined in this way is called **experimental probability.**

A survey is one type of probability experiment. Suppose a pizza company is going to sell individual pizzas at a football game. Three types of pizzas will be offered: cheese, sausage, and mushroom. The company wants to know how much of each type of pizza to prepare, so it surveys a representative sample of 500 customers. The company finds that 175 of these customers would order cheese pizzas, 225 would order sausage pizzas, and 100 would order mushroom pizzas. To estimate the probability that a particular pizza will be ordered, the company uses **relative frequency.** This means they divide the frequency (the number in each category) by the total (in this case, 500).

	Frequency	Relative Frequency
Cheese	175	$\frac{175}{500} = 0.35$
Sausage	225	$\frac{225}{500} = 0.45$
Mushroom	100	$\frac{100}{500} = 0.20$

Notice that the sum of the three relative frequencies is 1. This means that the entire sample is represented. We can change the relative frequencies from decimals to percents.

$$0.35 \longrightarrow 35\% \qquad 0.45 \longrightarrow 45\% \qquad 0.20 \longrightarrow 20\%$$

Recall from Lesson 58 that we use the term *chance* to describe a probability expressed as a percent. So the company makes these estimates about any given sale: The chance that a cheese pizza will be ordered is **35%**. The chance that a sausage pizza will be ordered is **45%**. The chance that a mushroom pizza will be ordered is **20%**. The company plans to make 3000 pizzas for the football game, so about 20% of the 3000 pizzas should be mushroom. How many pizzas is that?

Now we will apply these ideas to another survey. Suppose a small town has only four markets: Bob's Market, The Corner Grocery, Express Grocery, and Fine Foods. A representative sample of 80 adults was surveyed. Each person chose his or her favorite market: 30 chose Bob's Market, 12 chose Corner Grocery, 14 chose Express Grocery, and 24 chose Fine Foods.

1. Present the data in a relative frequency table similar to the one for pizza.

2. Estimate the probability that in this town an adult's favorite market is Express Grocery. Write your answer as a decimal.

3. Estimate the probability that in this town an adult's favorite market is Bob's Market. Write your answer as a fraction in reduced form.

4. Estimate the chance that in this town an adult's favorite market is Fine Foods. Write your answer as a percent.

5. Suppose the town has 4000 adult residents. The Corner Grocery is the favorite market of about how many adults in the town?

A survey is just one way of conducting a probability experiment. In the following activity we will perform an experiment that involves drawing two marbles out of a bag. By performing the experiment repeatedly and recording the results, we gather information that helps us determine the probability of various outcomes.

Activity: *Probability Experiment*

Materials needed:

- 6 marbles (4 red and 2 white)

- small, opaque bag from which to draw the marbles

- pencil and paper

Note: If marbles are unavailable, different-colored plastic chips, craft sticks, or slips of paper can be used. (It is important that differences in the objects cannot be determined by touch.)

The purpose of this experiment is to determine the probability that two marbles drawn from the bag at the same time will be red. We will create a relative frequency table to answer the question. You will work with your teacher or a partner to complete the experiment.

6. To estimate the probability, put 4 red marbles and 2 white marbles in a bag. Shake the bag; then remove two marbles at the same time. Record the result by marking a tally in a table like the one below. Replace the marbles and repeat this process until you have performed the experiment exactly 25 times.

Outcome	Tally
Both red	
Both white	
One of each	

7. Use your tally table to make a relative frequency table. (Divide each row's tally by 25 and express the quotient as a decimal.)

While you make your frequency table, your teacher or partner should begin drawing two marbles at a time and recording the results in a separate tally table. He or she should replace the marbles and repeat the process until the experiment has been performed exactly 25 times.

8. Using the data in your table, estimate the probability that both marbles drawn will be red. Write your answer as a reduced fraction and as a decimal.

If, for example, you drew two red marbles 11 times out of 25 draws, your best estimate of the probability of drawing two red marbles would be

$$\frac{11}{25} = \frac{44}{100} = 0.44$$

But this is only an estimate. The more times you draw, the more likely it is that the estimate will be close to the theoretical probability. It is better to repeat the experiment 50 times than 25 times. Thus, if you combine your results with your teacher's or partner's results, you will likely produce a better estimate. To combine results, add each of your tallies to the corresponding tally in your teacher's or partner's table; then calculate each new frequency.

Extensions

a. Ask 25 people the following question: "What is your favorite activity among the following: surfing the Internet, watching television, playing sports, talking on the telephone, playing video games, or reading?" Record each response. Create a relative frequency table of your results.

b. Roll two number cubes 100 times. Each time, observe the sum of the upturned faces, and fill out a relative frequency table like the one below.

Sum	2	3	4	5	6	7	8	9	10	11	12
Frequency											
Relative frequency											

Estimate the probability that the sum of a roll will be 8. Estimate the probability that the sum will be at least 10. Estimate the probability that the sum will be odd.

Die,
heank,

LESSON
91 Geometric Formulas

WARM-UP

Facts Practice: 72 Multiplication and Division Facts (Test H)

Mental Math: Count by $\frac{1}{3}$'s from $\frac{1}{3}$ to 4.

a. $5^2 + \sqrt{100}$ **b.** $1000 - 875$ **c.** $\$6.99 \times 5$

d. Double $125.00.   **e.** $12.5 \div 100$ **f.** 20×42

g. $3 \times 4, \div 2, \times 3, + 2, \times 2, + 2, \div 2, \div 3$

Problem Solving:

Copy this problem and fill in the missing digits. No two digits in the problem may be alike.

$$\begin{array}{r} -\ -\ - \\ \times\ \ \ \ 7 \\ \hline 9\ -\ - \end{array}$$

NEW CONCEPT

We have found the area of a rectangle by multiplying the length of the rectangle by its width. This procedure can be described with the following formula:

$$A = lw$$

The letter A stands for the area of the rectangle. The letters l and w stand for the length and width of the rectangle. Written side by side, lw means that we multiply the length by the width. The table below lists formulas for the perimeter and area of squares, rectangles, parallelograms, and triangles.

Figure	Perimeter	Area
Square	$P = 4s$	$A = s^2$
Rectangle	$P = 2l + 2w$	$A = lw$
Parallelogram	$P = 2b + 2s$	$A = bh$
Triangle	$P = s_1 + s_2 + s_3$	$A = \frac{1}{2}bh$

The letters *P* and *A* are abbreviations for *perimeter* and *area*. Other abbreviations are illustrated below:

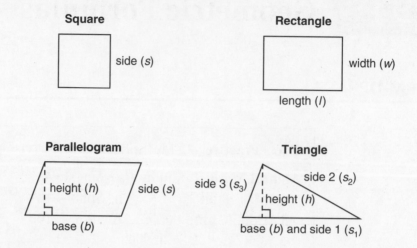

Since squares and rectangles are also parallelograms, the formulas for the perimeter and area of parallelograms may also be used for squares and rectangles.

To use a formula, we substitute each known measure in place of the appropriate letter in the formula. When substituting a number in place of a letter, it is a good practice to write the number in parentheses.

Example   Write the formula for the perimeter of a rectangle. Then substitute 8 cm for the length and 5 cm for the width. Solve the equation to find *P*.

Solution   The formula for the perimeter of a rectangle is

$$P = 2l + 2w$$

We rewrite the equation, substituting 8 cm for *l* and 5 cm for *w*. We write these measurements in parentheses.

$$P = 2(8 \text{ cm}) + 2(5 \text{ cm})$$

We multiply 2 by 8 cm and 2 by 5 cm.

$$P = 16 \text{ cm} + 10 \text{ cm}$$

Now we add 16 cm and 10 cm.

$$P = 26 \text{ cm}$$

The perimeter of the rectangle is **26 cm.**

We summarize the steps below to show how your work should look.

$$P = 2l + 2w$$
$$P = 2(8 \text{ cm}) + 2(5 \text{ cm})$$
$$P = 16 \text{ cm} + 10 \text{ cm}$$
$$P = 26 \text{ cm}$$

LESSON PRACTICE

Practice set

a. Write the formula for the area of a rectangle. Then substitute 8 cm for the length and 5 cm for the width. Solve the equation to find the area of the rectangle.

b. Write the formula for the perimeter of a parallelogram. Then substitute 10 cm for the base and 6 cm for the side. Solve the equation to find the perimeter of the parallelogram.

MIXED PRACTICE

Problem set

1. What is the ratio of prime numbers to composite numbers
(23, 65) in this list?

10, 11, 12, 13, 14, 15, 16, 17, 18, 19, 20, 21

2. Sunrise was at 6:15 a.m., and sunset was at 5:45 p.m. How
(32) many hours and minutes were there from sunrise to sunset?

3. Draw a ratio box for this problem. Then solve the
(88) problem using a proportion.

The ratio of leapers to duckers in the dodgeball game was 3 to 2. If there were 12 leapers, how many duckers were there?

4. A rectangular prism has how many more faces than a
(Inv. 6) triangular prism?

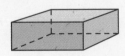 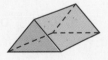

5. John will toss a penny, a nickel, and a dime. Create a
(Inv. 9) table that shows the eight possible outcomes of the tosses.

6. How many degrees does the minute hand of a clock turn
(90) in 45 minutes?

7. A pyramid with a triangular base is
(Inv. 6) shown at right.

(a) How many faces does it have?

(b) How many edges does it have?

(c) How many vertices does it have?

8. (a) What is the perimeter of this
(71) parallelogram?

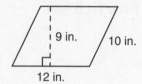

9 in.   10 in.

12 in.

(b) What is the area of this
parallelogram?

9. (a) Write $\frac{7}{20}$ as a decimal number.
(74, 75)

(b) Write $\frac{7}{20}$ as a percent.

10. $P = 2l + 2w$ is a formula for the perimeter of a rectangle.
(91) Find P when l is 5 ft and w is 3 ft.

11. $6\frac{2}{3} + 1\frac{3}{4}$ **12.** $5 - 1\frac{2}{5}$ **13.** $4\frac{1}{4} - 3\frac{5}{8}$
(59)   (63)   (63)

14. $3 \times \frac{3}{4} \times 2\frac{2}{3}$ **15.** $6\frac{2}{3} \div 100$ **16.** $2\frac{1}{2} \div 3\frac{3}{4}$
(72)   (68)   (68)

17. Compare: $\frac{9}{20}$ ◯ 50%
(75)

18. (a) What fraction of this group is
(75)   shaded?

(b) What percent of this group is shaded?

19. If $\frac{5}{6}$ of the 300 seeds sprouted, how many seeds did not
(77) sprout?

20. $6y = 10$ **21.** $\frac{w}{20} = \frac{12}{15}$
(87)   (85)

22. What is the area of this triangle?
(79)

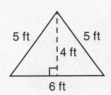

23. The illustration below shows a cube with edges 1 foot
(82) long. (Thus, the edges are also 12 inches long.) What is
the volume of the cube in cubic inches?

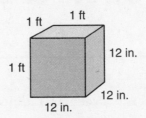

24. Write the prime factorization of 225 using exponents.
(73)

25. The length of segment AC is 56 mm. The length of segment
(69) BC is 26 mm. How long is segment AB?

26. On a number line, $\sqrt{60}$ is between which two consecutive
(89) whole numbers?

27. Which whole number equals $\sqrt{225}$?
(89)

28. A square has vertices at the coordinates (2, 0), (0, −2),
(Inv. 7) (−2, 0), and (0, 2). Draw the points on graph paper, and
draw segments from point to point in the order given.
To complete the square, draw a fourth segment from
(0, 2) to (2, 0).

29. The square in problem 28 encloses some whole squares
(Inv. 7) and some half squares on the graph paper.

(a) How many whole squares are enclosed by the square?

(b) How many half squares are enclosed by the square?

(c) Counting two half squares as a whole square,
calculate the area of the entire square.

30. Write the formula for the area of a parallelogram as given
(91) in this lesson. Then substitute 15 cm for the base and 4 cm
for the height. Solve the equation to find the area of the
parallelogram.

LESSON

92

Expanded Notation with Exponents • Order of Operations with Exponents • Powers of Fractions

WARM-UP

Facts Practice: Liquid Measurement (Test L)

Mental Math: Count up and down by 25's between –150 and 150.

a. 30 · 50
b. 486 + 50
c. 50% of 24 *(Think $\frac{1}{2}$ of 24.)*
d. $20.00 – $14.75
e. 100 × 1.25
f. $\frac{600}{30}$
g. $\sqrt{36}$, + 4, × 3, + 2, ÷ 4, + 1, $\sqrt{\ }$ †

Problem Solving:

Jorge's average grade on four tests is 88%. What grade does he need on his fifth test to have a five-test average of 90%?

†Read $\sqrt{\ }$ as "find the square root."

NEW CONCEPTS

Expanded notation with exponents

In Lesson 32 we began writing whole numbers in expanded notation. Here we show 365 in expanded notation:

$$365 = (3 \times 100) + (6 \times 10) + (5 \times 1)$$

When writing numbers in expanded notation, we may write the powers of 10 with exponents.

$$365 = (3 \times 10^2) + (6 \times 10^1) + (5 \times 10^0)$$

Notice that 10^0 equals 1. The table below shows whole-number place values using powers of 10:

Trillions			Billions			Millions			Thousands			Ones		
hundreds	tens	ones	hundreds	tens	ones	hundreds	tens	ones	hundreds	tens	ones	hundreds	tens	ones
10^{14}	10^{13}	10^{12}	10^{11}	10^{10}	10^9	10^8	10^7	10^6	10^5	10^4	10^3	10^2	10^1	10^0

Example 1 The speed of light is about 186,000 miles per second. Write 186,000 in expanded notation using exponents.

Solution We write each nonzero digit (1, 8, and 6) multiplied by its place value.

$$186{,}000 = (1 \times 10^5) + (8 \times 10^4) + (6 \times 10^3)$$

Order of operations with exponents

In the order of operations, we simplify expressions with exponents or roots before we multiply or divide.

ORDER OF OPERATIONS

1. Simplify within parentheses.
2. Simplify powers and roots.
3. Multiply and divide from left to right.
4. Add and subtract from left to right.

Some students remember the order of operations by using this memory aid:

Please

Excuse

My **D**ear

Aunt **S**ally

The first letter of each word is meant to remind us of the order of operations.

Parentheses

Exponents

Multiplication **D**ivision

Addition **S**ubtraction

Example 2 Simplify: $5 - (8 + 8) \div \sqrt{16} + 3^2 \times 2$

Solution We follow the order of operations.

$5 - (8 + 8) \div \sqrt{16} + 3^2 \times 2$	original problem
$5 - 16 \div \sqrt{16} + 3^2 \times 2$	simplified in parentheses
$5 - 16 \div 4 + 9 \times 2$	simplified powers and roots
$5 - 4 + 18$	multiplied and divided
19	added and subtracted

Powers of   We may use exponents with fractions and with decimals.
fractions   With fractions, parentheses help clarify that an exponent
applies to the whole fraction, not just its numerator.

$$\left(\frac{1}{2}\right)^3 \text{ means } \frac{1}{2} \cdot \frac{1}{2} \cdot \frac{1}{2}$$

$$(0.1)^2 \text{ means } 0.1 \times 0.1$$

Example 3   Simplify: $\left(\frac{2}{3}\right)^2$

Solution   We write $\frac{2}{3}$ as a factor twice and then multiply.

$$\frac{2}{3} \cdot \frac{2}{3} = \frac{4}{9}$$

LESSON PRACTICE

Practice set*   **a.** Write 2,500,000 in expanded notation using exponents.

b. Write this number in standard notation:

$$(5 \times 10^9) + (2 \times 10^8)$$

Simplify:

c. $10 + 2^3 \times 3 - (7 + 2) \div \sqrt{9}$

d. $\left(\frac{1}{2}\right)^3$ **e.** $(0.1)^2$ **f.** $\left(1\frac{1}{2}\right)^2$

g. $(2 + 3)^2 - (2^2 + 3^2)$

MIXED PRACTICE

Problem set   **1.** The weather forecast stated that the chance of rain for
   (58)   Wednesday is 40%. Does this forecast mean that it is
more likely to rain or not to rain? Why?

2. What is the probability of drawing a heart from a normal
(58)   deck of 52 cards?

3. If the sum of three numbers is 144, what is the average of
(18)   the three numbers?

4. "All quadrilaterals are polygons." True or false?
(60)

5. $\sqrt{441}$
(89)

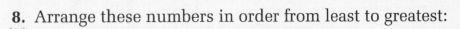

6. $2 \cdot 3^2 - \sqrt{9} + (3 - 1)^3$
(92)

7. Write the formula for the perimeter of a rectangle. Then
(91) substitute 12 in. for the length and 6 in. for the width.
Solve the equation to find the perimeter of the rectangle.

8. Arrange these numbers in order from least to greatest:
(44)

$$1, 0, 0.1, -1$$

9. If $\frac{5}{6}$ of the 30 members were present, how many members
(77) were absent?

10. Reduce before multiplying or dividing: $\dfrac{(24)(36)}{48}$
(70)

11. $\dfrac{\sqrt{100}}{\sqrt{25}}$ **12.** $12\frac{5}{6} + 15\frac{1}{3}$ **13.** $100 - 9.9$
(92)   (59)   (38)

14. $\dfrac{4}{7} \times 100$ **15.** $\dfrac{5}{8} = \dfrac{w}{48}$ **16.** $0.25 \times \$4.60$
(29)   (42)   (39)

17. The diameter of a circular saucepan is 6 inches. What is
(86) the area of the circular base of the pan? Round the answer
to the nearest square inch. (Use 3.14 for π.)

18. Write $3\frac{3}{4}$ as a decimal number, and subtract that number
(74) from 7.4.

19. What percent of the first ten letters of the alphabet are
(75) vowels?

20. Bobby rode his bike north. At Grand Avenue he turned left
(90) 90°. When he reached Arden Road, he turned left 90°. In
what direction was Bobby riding on Arden Road?

21. Estimate the product of 6.95 and 12.1 to the nearest
(51) whole number.

22. Write and solve a proportion for this statement: 16 is to
(85) 10 as what number is to 25?

23. What is the area of the triangle below?
(79)

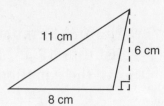

11 cm

6 cm

8 cm

24. This figure is a rectangular prism.
(Inv. 6)
    (a) How many faces does it have?

    (b) How many edges does it have?

Use a ruler to find the length and width of this rectangle to the nearest quarter of an inch. Then refer to the rectangle to answer problems 25 and 26.

25. What is the perimeter of the rectangle?
(91)

26. What is the area of the rectangle?
(91)

27. Each term in this sequence is $\frac{1}{16}$ more than the previous
(17) term. What are the next four terms in the sequence?

$$\frac{1}{16}, \frac{1}{8}, \frac{3}{16}, \frac{1}{4}, \underline{\hspace{1cm}}, \underline{\hspace{1cm}}, \underline{\hspace{1cm}}, \underline{\hspace{1cm}}, \ldots$$

28. The coordinates of the vertices of a parallelogram are
(Inv. 7) (4, 3), (−2, 3), (0, −2), and (−6, −2). What is the area of the parallelogram?

29. Simplify:
(81)
    (a) (12 cm)(8 cm)   (b) $\dfrac{36 \text{ ft}^2}{4 \text{ ft}}$

30. Fernando poured water from one-pint bottles into a
(78) three-gallon bucket. How many pints of water could the bucket hold?

LESSON
93 Classifying Triangles

WARM-UP

Facts Practice: 28 Improper Fractions to Simplify (Test I)

Mental Math: Count up and down by $\frac{1}{8}$'s between $\frac{1}{8}$ and 3.

a. $40 \cdot 60$

b. $234 - 50$

c. 25% of 24 *(Think $\frac{1}{4}$ of 24.)*

d. $5.99 + $2.47

e. $1.2 \div 100$

f. 30×25

g. $8 \times 9, + 3, \div 3, \sqrt{}, \times 6, + 3, \div 3, - 10$

Problem Solving:

Compare the following. Then describe how you performed the comparison.

$$6.142 \times 9.065 \bigcirc 54$$

NEW CONCEPT

All three-sided polygons are triangles, but not all triangles are alike. We distinguish between different types of triangles by using the lengths of their sides and the measures of their angles. We will first classify triangles based on the lengths of their sides.

Triangles Classified by Their Sides

Name	Example	Description
Equilateral triangle	△	All three sides are equal in length.
Isosceles triangle	△	At least two of the three sides are equal in length.
Scalene triangle	△	All three sides have different lengths.

An **equilateral triangle** has three equal sides and three equal angles.

An **isosceles triangle** has at least two equal sides and two equal angles. A **scalene triangle** has three unequal sides and three unequal angles.

Next we consider triangles classified by their angles. In Lesson 28 we learned the names of three different kinds of angles: **acute, right,** and **obtuse.** We can also use these words to describe triangles.

Triangles Classified by Their Angles

Name	Example	Description
Acute triangle		All three angles are acute.
Right triangle		One angle is a right angle.
Obtuse triangle		One angle is an obtuse angle.

Each angle of an equilateral triangle measures 60°, so an equilateral triangle is also an acute triangle. An isosceles triangle may be an acute triangle, a right triangle, or an obtuse triangle. A scalene triangle may also be an acute triangle, a right triangle, or an obtuse triangle.

LESSON PRACTICE

Practice set

a. One side of an equilateral triangle measures 15 cm. What is the perimeter of the triangle?

b. "An equilateral triangle is also an acute triangle." True or false?

c. "All acute triangles are equilateral triangles." True or false?

d. Two sides of a triangle measure 3 inches and 4 inches. If the perimeter is 10 inches, what type of triangle is it?

e. "Every right triangle is a scalene triangle." True or false?

MIXED PRACTICE

Problem set

1. Draw a ratio box for this problem. Then solve the
(88) problem using a proportion.

The ratio of the length to the width of the rectangular lot was 5 to 2. If the lot was 60 ft wide, how long was the lot?

2. On a multiple-choice test, Mitch does not know the
(Inv. 9) correct answers to two questions, so he will make guesses.
The answer choices for each question are labeled A, B, C,
and D. How many different combinations of two guesses
can Mitch make?

3. If the sum of four numbers is 144, what is the average of
(18) the four numbers?

4. The rectangular prism shown below is constructed of
(82) 1-cubic-centimeter blocks. What is the volume of the
prism?

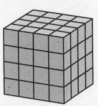

5. Write $\frac{9}{25}$ as a decimal number and as a percent.
(74, 75)

6. Write $3\frac{1}{5}$ as a decimal number and add it to 3.5. What is
(74) the sum?

7. What number is 45% of 80?
(41)

8. $(0.3)^3$ **9.** $\left(2\frac{1}{2}\right)^2$ **10.** $\sqrt{9} \cdot \sqrt{100}$
(73)   (73)   (92)

11. Twenty of the two dozen members voted yes. What
(77) fraction of the members voted yes?

12. If the rest of the members in problem 11 voted no, then
(23) what was the ratio of "no" votes to "yes" votes?

Find each missing number:

13. $w + 4\frac{3}{4} = 9\frac{1}{3}$ **14.** $\frac{6}{5} = \frac{m}{30}$
(63)   (85)

15. In what type of triangle are all three sides the same length?
(93)

16. What mixed number is $\frac{3}{8}$ of 100?
(42)

17. $10 + 6^2 \div 3 - \sqrt{9} \times 3$
(92)

18. A triangular prism has how many faces?
(Inv. 6)

19. How many quarts of milk is $2\frac{1}{2}$ gallons of milk?
(78)

20. Use a factor tree to find the prime factors of 800. Then
(65, 73) write the prime factorization of 800 using exponents.

21. Round the decimal number one hundred twenty-five
(51) thousandths to the nearest tenth.

22. $0.08n = \$1.20$
(87)

23. The diagonal segment through this rectangle divides the
(79) rectangle into two congruent right triangles. What is the
area of one of the triangles?

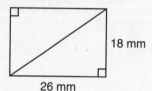

18 mm

26 mm

24. Write $\frac{17}{20}$ as a percent.
(75)

25. On this number line the arrow could be pointing to
(89) which of the following?

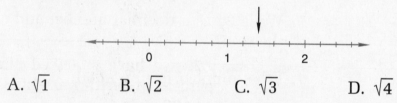

        0   1   2

  A. $\sqrt{1}$ B. $\sqrt{2}$ C. $\sqrt{3}$ D. $\sqrt{4}$

26. Write this number in standard notation:
(92)

$$(7 \times 10^9) + (2 \times 10^8) + (5 \times 10^7)$$

27. (a) What is the probability of rolling a 6 with a single roll
(58) of a number cube?

    (b) What is the probability of rolling a number less than 6
with a single roll of a number cube?

28. The coordinates of the four vertices of a quadrilateral are
(64, Inv. 7) $(-3, -2)$, $(0, 2)$, $(3, 2)$, and $(5, -2)$. What is the name for
this type of quadrilateral?

29. The formula for the area of a triangle is
(91)

$$A = \frac{bh}{2}$$

If the base measures 20 cm and the height measures
15 cm, then what is the area of the triangle?

30. What are the next four numbers in this sequence?
(10, 17)

$$\frac{1}{16}, \frac{1}{8}, \frac{3}{16}, \frac{1}{4}, \frac{5}{16}, \frac{3}{8}, \underline{\quad}, \underline{\quad}, \underline{\quad}, \underline{\quad}, \dots$$

LESSON
94

Writing Fractions and Decimals as Percents, Part 2

WARM-UP

Facts Practice: Linear Measurement (Test K)

Mental Math: Count up and down by 12's between 12 and 144.

  a. $50 \cdot 70$ **b.** $572 + 150$ **c.** 50% of 80

  d. $10.00 - $6.36   **e.** 100×0.02 **f.** $\frac{640}{20}$

  g. $4 \times 5, + 1, \div 3, \times 8, - 1, \div 5, \times 4, - 2, \div 2$

Problem Solving:

The perimeter of this right triangle is 24 cm. What is the area of the triangle?

10 cm   6 cm

NEW CONCEPT

Since Lesson 75 we have practiced changing a fraction or decimal to a percent by writing an equivalent fraction with a denominator of 100.

$$\frac{3}{5} = \frac{60}{100} = 60\%$$

$$0.4 = 0.40 = \frac{40}{100} = 40\%$$

In this lesson we will practice another method of changing a fraction to a percent. Since 100% equals 1, we can multiply a fraction by 100% to form an equivalent number. Here we multiply $\frac{3}{5}$ by 100%:

$$\frac{3}{5} \times \frac{100\%}{1} = \frac{300\%}{5}$$

Then we simplify and find that $\frac{3}{5}$ equals 60%.

$$\frac{300\%}{5} = 60\%$$

We can use the same procedure to change decimals to percents. Here we multiply 0.375 by 100%.

$$0.375 \times 100\% = 37.5\%$$

To change a number to a percent, multiply the number by 100%.

Example 1   Change $\frac{1}{3}$ to a percent.

Solution   We multiply $\frac{1}{3}$ by 100%.

$$\frac{1}{3} \times \frac{100\%}{1} = \frac{100\%}{3}$$

To simplify, we divide 100% by 3 and write the quotient as a mixed number.

$$
\begin{array}{r}
33\frac{1}{3}\% \\
3\overline{)100\%} \\
\underline{9} \\
10 \\
\underline{9} \\
1
\end{array}
$$

Example 2   Write 1.2 as a percent.

Solution   We multiply 1.2 by 100%.

$$1.2 \times 100\% = \mathbf{120\%}$$

In some applications a percent may be greater than 100%. If the number we are changing to a percent is greater than 1, then the percent is greater than 100%.

Example 3   Write $2\frac{1}{4}$ as a percent.

Solution   We show two methods below.

Method 1: We split the whole number and fraction. The mixed number $2\frac{1}{4}$ means "$2 + \frac{1}{4}$." We change each part to a percent and then add.

$$2 + \frac{1}{4}$$

$$200\% + 25\% = \mathbf{225\%}$$

Method 2: We change the mixed number to an improper fraction. The mixed number $2\frac{1}{4}$ equals the improper fraction $\frac{9}{4}$. We then change $\frac{9}{4}$ to a percent.

$$\frac{9}{\underset{1}{\cancel{4}}} \times \frac{\overset{25}{\cancel{100\%}}}{1} = \mathbf{225\%}$$

Example 4   Write $2\frac{1}{6}$ as a percent.

Solution   Method 1 shown in example 3 is quick, if we can recall the percent equivalent of a fraction. Method 2 is easier if the percent equivalent does not readily come to mind. We will use Method 2 in this example. We write the mixed number $2\frac{1}{6}$ as the improper fraction $\frac{13}{6}$ and multiply by 100%.

$$\frac{13}{6} \times \frac{100\%}{1} = \frac{1300\%}{6}$$

Now we divide 1300% by 6 and write the quotient as a mixed number.

$$\frac{1300\%}{6} = \mathbf{216\frac{2}{3}\%}$$

Example 5   Twenty of the thirty people on the bus were reading. What percent of people on the bus were reading?

Solution   We first find the fraction of people who were reading. Then we convert the fraction to a percent.

Of the people on the bus, $\frac{20}{30}$ $\left(\text{or } \frac{2}{3}\right)$ were reading. Now we multiply the fraction by 100%, which is the percent name for 1. We can use either $\frac{20}{30}$ or $\frac{2}{3}$, as we show below.

$$\frac{20}{\underset{3}{30}} \times \overset{10}{100}\% = \frac{200\%}{3} = 66\frac{2}{3}\%$$

$$\frac{2}{3} \times 100\% = \frac{200\%}{3} = 66\frac{2}{3}\%$$

We find that $\mathbf{66\frac{2}{3}\%}$ of people on the bus were reading.

LESSON PRACTICE

Practice set*   Change each decimal number to a percent by multiplying by 100%:

    a. 0.5   **b.** 0.06   **c.** 0.125

    d. 0.45   **e.** 1.3   **f.** 0.025

    g. 0.09   **h.** 1.25   **i.** 0.625

Change each fraction or mixed number to a percent by multiplying by 100%:

    j. $\frac{2}{3}$ **k.** $\frac{1}{6}$ **l.** $\frac{1}{8}$

    m. $1\frac{1}{4}$ **n.** $2\frac{4}{5}$ **o.** $1\frac{1}{3}$

p. What percent of this rectangle is shaded?

q. What percent of a yard is a foot?

MIXED PRACTICE

Problem set

1. Ten of the thirty people on the bus were not reading.
 (94) What percent of people on the bus were not reading?

2. On the Celsius scale water freezes at 0°C and boils at
 (18) 100°C. What temperature is halfway between the freezing and boiling temperatures of water?

3. If the length of segment *AB* is $\frac{1}{3}$ the length of segment *AC*,
 (69) and if segment *AC* is 12 cm long, then how long is segment *BC*?

4. What percent of this group is
 (94) shaded?

5. Change $1\frac{2}{3}$ to a percent by multiplying $1\frac{2}{3}$ by 100%.
 (94)

6. Change 1.5 to a percent by multiplying 1.5 by 100%.
 (94)

7. $6.4 - 6\frac{1}{4}$ (Begin by writing $6\frac{1}{4}$ as a decimal number.)
 (74)

8. $10^4 - 10^3$ **9.** How much is $\frac{3}{4}$ of 360?
 (92)   (70)

Tommy placed a cylindrical can of spaghetti sauce on the counter. He measured the diameter of the can and found that it was about 8 cm. Use this information to answer problems 10 and 11.

Use 3.14 for π.

10. The label wraps around the circumference of the can.
 (47) How long does the label need to be?

11. How many square centimeters of countertop does the
 (86) can occupy?

12. $3\frac{1}{2} + 1\frac{3}{4} + 4\frac{5}{8}$ **13.** $\frac{9}{10} \cdot \frac{5}{6} \cdot \frac{8}{9}$
 (61)   (72)

14. Write 250,000 in expanded notation using exponents.
 (92)

15. $8.47 + 95¢ + $12
(1)

16. 37.5 ÷ 100
(51)

17. $\frac{3}{7} = \frac{21}{x}$
(85)

18. $33\frac{1}{3} ÷ 100$
(68)

19. If ninety percent of the answers were correct, then what
(41) percent were incorrect?

20. Write the decimal number one hundred twenty and three
(35) hundredths.

21. Arrange these numbers in order from least to greatest:
(76)

$$-2.5, \frac{2}{5}, \frac{5}{2}, -5.2$$

22. A pyramid with a square base has how many edges?
(Inv. 6)

23. What is the area of this
(71) parallelogram?

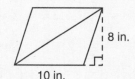

8 in.

10 in.

24. The parallelogram in problem 23 is divided into two
(93) congruent triangles. Both triangles may be described as
which of the following?

A. acute   B. right   C. obtuse

25. During the year, the temperature ranged from −37°F in
(14) winter to 103°F in summer. How many degrees was the
range of temperature for the year?

26. The coordinates of the three vertices of a triangle are (0, 0),
(Inv. 7, 79) (0, −4), and (−4, 0). Graph the triangle and find its area.

27. Margie's first nine test scores are shown below.
(Inv. 5)

21, 25, 22, 19, 22, 24, 20, 22, 24

(a) What is the mode of these scores?

(b) What is the median of these scores?

28. $2^3 + \sqrt{25} \times 3 - 4^2 ÷ \sqrt{4}$
(92)

29. Sandra filled the aquarium with 24 quarts of water. How
(78) many gallons of water did Sandra pour into the aquarium?

30. What is the probability of drawing a red queen from a
(58) normal deck of 52 cards?

LESSON

95 Reducing Units Before Multiplying

WARM-UP

Facts Practice: 30 Fractions to Reduce (Test G)

Mental Math: Count up and down by 20's between −100 and 100.

a. $60 \cdot 80$ b. $437 - 150$ c. 25% of 80

d. $\$3.99 + \$4.28$ e. $17.5 \div 100$ f. 30×55

g. 6×8, $+ 1$, $\sqrt{\ }$, $\times 5$, $+ 1$, $\sqrt{\ }$, $\times 3$, $\div 2$, $\sqrt{\ }$

Problem Solving:

What are the next four numbers in this sequence?

$$\frac{1}{12}, \frac{1}{6}, \frac{1}{4}, \frac{1}{3}, \underline{\quad}, \underline{\quad}, \underline{\quad}, \underline{\quad}, \dots$$

NEW CONCEPT

Since Lesson 70 we have practiced reducing fractions before multiplying. This is sometimes called *canceling*.

$$\frac{\overset{1}{\cancel{3}}}{\underset{2}{\cancel{4}}} \cdot \frac{\overset{1}{\cancel{2}}}{\underset{1}{\cancel{3}}} \cdot \frac{\overset{1}{\cancel{5}}}{\underset{2}{\cancel{6}}} = \frac{1}{4}$$

We can cancel **units** before multiplying just as we cancel numbers.

$$\frac{4 \text{ miles}}{1 \cancel{\text{ hour}}} \times \frac{2 \cancel{\text{ hours}}}{1} = \frac{8 \text{ miles}}{1} = 8 \text{ miles}$$

Example 1 Multiply 55 miles per hour by six hours.

Solution We write 55 miles per hour as the ratio 55 miles over 1 hour, because "per" indicates division. We write six hours as the ratio 6 hours over 1.

$$\frac{55 \text{ miles}}{1 \text{ hour}} \times \frac{6 \text{ hours}}{1}$$

The unit "hour" appears above and below the division line, so we can cancel it.

$$\frac{55 \text{ miles}}{1 \text{ hour}} \times \frac{6 \text{ hours}}{1} = \textbf{330 miles}$$

Can you think of a story to fit this problem?

Example 2 Multiply 5 feet by 12 inches per foot.

Solution We write ratios of 5 feet over 1 and 12 inches over 1 foot. We then cancel units and multiply.

$$\frac{5 \text{ feet}}{1} \cdot \frac{12 \text{ inches}}{1 \text{ foot}} = \textbf{60 inches}$$

LESSON PRACTICE

Practice set When possible, cancel numbers and units before multiplying:

a. $\dfrac{3 \text{ dollars}}{1 \text{ hour}} \times \dfrac{8 \text{ hours}}{1}$

b. $\dfrac{6 \text{ baskets}}{10 \text{ shots}} \times \dfrac{100 \text{ shots}}{1}$

c. $\dfrac{10 \text{ cents}}{1 \text{ kwh}} \times \dfrac{26.3 \text{ kwh}}{1}$

d. $\dfrac{160 \text{ km}}{2 \text{ hours}} \cdot \dfrac{10 \text{ hours}}{1}$

e. Multiply 18 gallons by 29 miles per gallon.

f. Multiply 2.3 meters by 100 centimeters per meter.

MIXED PRACTICE

Problem set 1. What is the total price of a $45.79 item when 7% sales tax
(41)   is added to the price?

2. Jeff is 1.67 meters tall. How many centimeters tall is Jeff?
(95)   (Multiply 1.67 meters by 100 centimeters per meter.)

3. If $\frac{5}{8}$ of the 40 seeds sprouted, how many seeds did not
(77) sprout?

4. Write this number in standard notation:
(46)

$$(5 \times 100) + (6 \times 10) + \left(7 \times \frac{1}{10}\right) + \left(3 \times \frac{1}{100}\right)$$

5. Change $\frac{1}{6}$ to its percent equivalent by multiplying $\frac{1}{6}$ by
(94) 100%.

6. What is the percent equivalent of 2.5?
(94)

7. How much money is 30% of $12.00?
(41)

8. The minute hand of a clock turns 180° in how many
(90) minutes?

9. The circumference of the front tire on Elizabeth's bike is
(27) 6 feet. How many complete turns does the front wheel
make as Elizabeth rides down her 30-foot driveway?

10. Chad built this stack of one-cubic-
(82) foot boxes. What is the volume of
the stack?

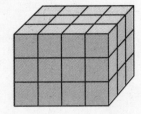

11. $\frac{3}{4} + \frac{3}{5}$
(57)

12. $18\frac{1}{8} - 12\frac{1}{2}$
(63)

13. $3\frac{3}{4} \times 2\frac{2}{3} \times 1\frac{1}{10}$
(72)

14. $\frac{2^5}{2^3}$
(92)

15. How many fourths are in $2\frac{1}{2}$?
(68)

16. $12 + 8.75 + 6.8$
(38)

17. $(1.5)^2$
(38, 39)

18. $6\frac{2}{5} \div 0.8$ (decimal answer)
(74)

19. Estimate the sum of $6\frac{1}{4}$, 4.95, and 8.21 by rounding each
(51) number to the nearest whole number before adding.
Describe how you arrived at your answer.

20. The diameter of a round tabletop is 60 inches.
(86)
    (a) What is the radius of the tabletop?

    (b) What is the area of the tabletop? (Use 3.14 for π.)

21. Arrange these numbers in order from least to greatest:
(75)

$$\tfrac{1}{4}, 4\%, 0.4$$

Find each missing number.

22. $y + 3.4 = 5$
(43)

23. $\dfrac{4}{8} = \dfrac{x}{12}$
(85)

24. A cube has edges that are 6 cm long.
(Inv. 6, 82)
    (a) What is the area of each face of the cube?

    (b) What is the volume of the cube?

25. $\overline{AB}$ is 24 mm long. $\overline{AC}$ is 42 mm long. How long is $\overline{BC}$?
(69)

26. $6^2 \div \sqrt{9} + 2 \times 2^3 - \sqrt{400}$
(89, 92)

27. What is the ratio of a pint of water to a quart of water?
(78)

28. The formula for the area of a parallelogram is $A = bh$. If
(91) the base of a parallelogram is 1.2 m and the height is 0.9 m, what is the area of the parallelogram?

29. Multiply 2.5 liters by 1000 milliliters per liter.
(95)

$$\frac{2.5 \text{ liters}}{1} \times \frac{1000 \text{ milliliters}}{1 \text{ liter}}$$

30. If this spinner is spun once, what is
(58) the probability that the arrow will end up pointing to an even number?

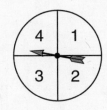

LESSON
96

Functions • Graphing Functions

Facts Practice: 64 Multiplication Facts (Test D)

Mental Math: Count by $\frac{1}{16}$'s from $\frac{1}{16}$ to 1.

  a. 70 · 90   **b.** 364 + 250   **c.** 50% of 60

  d. $5.00 − $0.89   **e.** 100 × 0.015   **f.** $\frac{750}{30}$

  g. 6 × 6, − 1, ÷ 5, × 8, − 1, ÷ 11, × 8, × 2, + 1, $\sqrt{}$

Problem Solving:

Copy this factor tree and fill in the missing numbers:

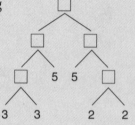

NEW CONCEPTS

Functions Gwen thought of a rule that used the length of a side of a square to find the perimeter of the square. Then she made a table that showed the perimeters that correspond to certain side lengths. In her table *s* stands for the length of a side of a square, and *P* stands for the perimeter of the square.

s	P
5	20
7	28
10	40
15	?

Gwen's rule is "multiply the side length by four to find the perimeter." Another way to state Gwen's rule is shown by this formula:

$$P = 4s$$

Thus, the number that should replace the question mark in the table is 60, because 15 × 4 = 60.

Gwen's rule (or formula) is an example of a **function**. A function pairs one unknown with another unknown. In this case the length of a side is paired with the perimeter.

Because the perimeter of a square depends on the length of the square's sides, we say that the perimeter of a square is a *function* of the side length of the square. If we know the side length of a square, we can find the perimeter by applying the function's rule.

Example 1 Find the rule for this function. Then use the rule to find the value of m when l is 7.

l	m
5	20
7	?
10	25
15	30

m = 4L
25 = 4(10)
m = L + 15
20 = 5 + 15

Solution We study the table to discover the function rule. We see that when l is 5, m is 20. We might guess that the rule is to multiply l by 4. However, when l is 10, m is 25. Since 10×4 does not equal 25, we know that this guess is incorrect. So we look for another rule. We notice that 20 is 15 more than 5 and that 25 is 15 more than 10. Perhaps the rule is to add 15 to l. We see that the values in the bottom row of the table ($l = 15$ and $m = 30$) fit this rule. So the rule is, **to find m, add 15 to l.** To find m when l is 7, we add 15 to 7.

$$7 + 15 = 22$$

The missing number in the table is **22.**

Instead of using the letter m at the top of the table, we could have written the rule. In the table at right, $l + 15$ has replaced m. This means we add 15 to the value of l. We show this type of table in the next example.

l	$l + 15$
5	20
7	?
10	25
15	30

Example 2 Find the missing number in this function table:

x	2	3	4
$3x - 2$	4	7	?

? = 10

Solution This table is arranged horizontally. The rule of the function is stated in the table: multiply the value of x by 3, then subtract 2. To find the missing number in the table, we apply the rule of the function when x is 4.

$$3x - 2$$

$$3(4) - 2 = 10$$

We find that the missing number is **10.**

Graphing functions

Many functions can be graphed on a coordinate plane. Here we show a function table that relates the perimeter of a square to the length of one of its sides. On the coordinate plane we have graphed the number pairs that appear in the table. The coordinate plane's horizontal axis shows the length of a side, and its vertical axis shows the perimeter.

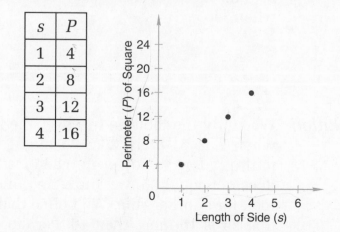

s	P
1	4
2	8
3	12
4	16

We have used different scales on the two axes so that the graph is not too steep. The graphed points show the side length and perimeter of four squares with side lengths of 1, 2, 3, and 4 units. Notice that the graphed points are aligned. Of course, we could graph many more points and represent squares with side lengths of 100 units or more. We could also graph points for squares with side lengths of 0.01 or less. In fact, we can graph points for any side length whatsoever! Such a graph would look like a ray, as shown below.

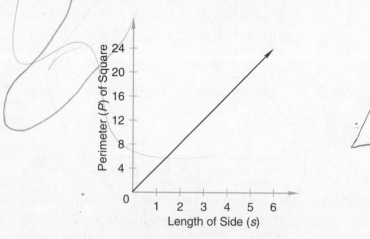

Example 3

The perimeter of an equilateral triangle is a function of the length of its sides. Make a table for this function using side lengths of 1, 2, 3, and 4 units. Then graph the ordered pairs on a coordinate plane. Extend a ray through the points to represent the function for all equilateral triangles.

P = 3s

Solution We create a table of ordered pairs. The letter *s* stands for the length of a side, and *P* stands for the perimeter.

s	P
1	3
2	6
3	9
4	12

Now we graph these points on a coordinate plane with one axis for perimeter and the other axis for side length. Then we draw a ray from the origin through these points.

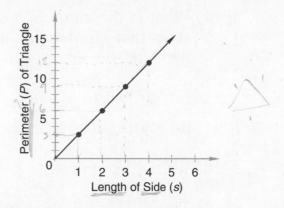

Every point along the ray represents the side length and perimeter of an equilateral triangle.

LESSON PRACTICE

Practice set Find the missing number in each function table:

a.

x	y
3	1
5	3
6	4
10	?

$3 \to \boxed{?} \to 1$

b.

a	b
3	8
5	10
7	12
?	15

c.

x	3	6	8
3x + 1	10	19	?

d.

x	3	4	7
3x − 1	8	?	20

e. The chemist mixed a vat of solution that weighed 2 pounds per quart. Create a table of ordered pairs for this function for 1, 2, 3, and 4 quarts. Then graph the points on a coordinate plane, using the horizontal axis for quarts and the vertical axis for pounds. Would it be appropriate to draw a ray through the points? Why or why not?

MIXED PRACTICE

Problem set

1. When the sum of 2.0 and 2.0 is subtracted from the product of 2.0 and 2.0, what is the difference?
(12, 53)

2. A 4.2-kilogram object weighs the same as how many objects that each weigh 0.42 kilogram?
(49)

3. If the average of 8 numbers is 12, what is the sum of the 8 numbers?
(18)

4. What is the name of a quadrilateral that has one pair of sides that are parallel and one pair of sides that are not parallel?
(64)

5. (a) Write 0.15 as a percent.
(94)

    (b) Write 1.5 as a percent.

6. Write $\frac{5}{6}$ as a percent.
(94)

7. Three of the numbers below are equivalent. Which one is not equivalent to the others?
(41, 76)

   A. 1   B. 100%   C. 0.1   D. $\frac{100}{100}$

8. 11^3
(73)

9. How much is $\frac{5}{6}$ of 360?
(70)

10. Between which two consecutive whole numbers is $\sqrt{89}$?
(89)

11. Silvester ran around the field, turning at each of the three backstops. What was the average number of degrees he turned at each of the three corners?
(90)

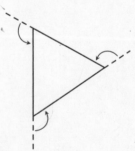

12. Find the missing number in this function table.
(96)

x	4	7	13	15
$2x - 1$	7	13	?	29

13. Factor and reduce: $\dfrac{(45)(54)}{81}$
(67)

14.
(49) $\dfrac{30}{0.08}$

15. $16\dfrac{2}{3} \div 100$
(68)

16. $2\dfrac{1}{2} + 3\dfrac{1}{3} + 4\dfrac{1}{6}$
(61)

17. $6 \times 5\dfrac{1}{3} \times \dfrac{3}{8}$
(72)

18. $\dfrac{2}{5}$ of \$12.00
(22)

19. $0.12 \times \$6.50$
(39)

20. $5.3 - 3\dfrac{3}{4}$ (decimal answer)
(74)

21. What is the ratio of the number of cents in a dime to the
(23) number of cents in a quarter?

Find each missing number:

22. $4n = 6 \cdot 14$
(87)

23. $0.3n = 12$
(87)

24. Draw a segment $1\dfrac{3}{4}$ inches long. Label the endpoints R
(7) and T. Then find and mark the midpoint of $\overline{RT}$. Label the
midpoint S. What are the lengths of $\overline{RS}$ and $\overline{ST}$?

25. Solve this proportion: $\dfrac{6}{9} = \dfrac{36}{w}$
(85)

26. Multiply 4 hours by 6 dollars per hour:
(95)

$$\dfrac{4 \text{ hours}}{1} \times \dfrac{6 \text{ dollars}}{1 \text{ hour}}$$

27. The coordinates of the vertices of a parallelogram are
(Inv. 7, 71) $(0, 0)$, $(6, 0)$, $(4, 4)$, and $(-2, 4)$. What is the area of the
parallelogram?

28. The saying "A pint's a pound the world around" refers to
(78) the fact that a pint of water weighs about one pound.
About how many pounds does a gallon of water weigh?

29. $3^2 + 2^3 - \sqrt{4} \times 5 + 6^2 \div \sqrt{16}$
(92)

30. What is the probability of rolling a prime number with
(58) one roll of a number cube?

LESSON

97   Transversals

Facts Practice: Liquid Measurement (Test L)

Mental Math: Count up and down by $\frac{1}{2}$'s between −3 and 3.

   a. 20 · 50   **b.** 517 − 250   **c.** 25% of 60

   d. $7.99 + $7.58   **e.** 0.1 ÷ 100   **f.** 20 × 75

   g. 5 × 9, − 1, ÷ 2, − 1, ÷ 3, × 10, + 2, ÷ 9, − 2, ÷ 2

Problem Solving:

Chad has taken three tests. His lowest score is 70%. His highest score is 100%. What is Chad's lowest possible three-test average? What is Chad's highest possible three-test average?

NEW CONCEPT

A line that intersects two or more other lines is a **transversal**. In this drawing, line *r* is a transversal of lines *s* and *t*.

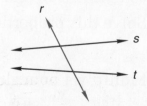

In the drawing, lines *s* and *t* are not parallel. However, in this lesson we will focus on the effects of a transversal intersecting parallel lines.

Below we show parallel lines *m* and *n* intersected by transversal *p*. Notice that eight angles are formed. In this figure there are four acute angles (numbered 2, 4, 6, and 8) and four obtuse angles (numbered 1, 3, 5, and 7).

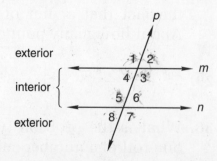

Notice that angle 1, an obtuse angle, and angle 2, an acute angle, together form a straight line. These angles are supplementary, which means their measures total 180°. So if ∠1 measures 110°, then ∠2 measures 70°. Also notice that ∠2 and ∠3 are supplementary. If ∠2 measures 70°, then ∠3 measures 110°. Likewise, ∠3 and ∠4 are supplementary, so ∠4 would measure 70°.

There are names to describe some of the angle pairs. For example, we say that ∠1 and ∠5 are **corresponding angles** because they are in the same relative positions. Notice that ∠1 is the "upper left angle" from line *m,* while ∠5 is the "upper left angle" from line *n.*

Which angle corresponds to ∠2?

Which angle corresponds to ∠7?

Since lines *m* and *n* are parallel, line *p* intersects line *m* at the same angle as it intersects line *n.* So the corresponding angles are congruent. Thus, if we know that ∠1 measures 110°, we can conclude that ∠5 also measures 110°.

The angles between the parallel lines (numbered 3, 4, 5, and 6 in the figure) are **interior angles.** Angle 3 and ∠5 are on opposite sides of the transversal and are called **alternate interior angles.**

Name another pair of alternate interior angles.

Alternate interior angles are congruent if the lines intersected by the transversal are parallel. So if ∠5 measures 110°, then ∠3 also measures 110°.

Angles not between the parallel lines are **exterior angles.** Angle 1 and ∠7, which are on opposite sides of the transversal, are **alternate exterior angles.**

Name another pair of alternate exterior angles.

Alternate exterior angles formed by a transversal intersecting parallel lines are congruent. So if the measure of ∠1 is 110°, then the measure of ∠7 is also 110°.

While we practice the terms for describing angle pairs, it is useful to remember that all the acute angles formed when a transversal intersects parallel lines are equal in measure. Likewise, all the obtuse angles are equal in measure. Thus any acute angle formed will be supplementary to any obtuse angle formed.

Example   Transversal *w* intersects parallel lines *x* and *y*.

(a) Name the pairs of corresponding angles.

(b) Name the pairs of alternate interior angles.

(c) Name the pairs of alternate exterior angles.

(d) If the measure of ∠*a* is 115°, then what are the measures of ∠*e* and ∠*f*?

Solution   (a) **∠*a* and ∠*e*, ∠*b* and ∠*f*, ∠*c* and ∠*g*, ∠*d* and ∠*h***

(b) **∠*d* and ∠*f*, ∠*c* and ∠*e***

(c) **∠*a* and ∠*g*, ∠*b* and ∠*h***

(d) If ∠*a* measures 115°, then ∠*e* also measures **115°** and ∠*f* measures **65°**.

LESSON PRACTICE

Practice set   **a.** Which line in the figure at right is a transversal?

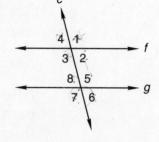

b. Which angle is an alternate interior angle to ∠3?

c. Which angle corresponds to ∠8?

d. Which angle is an alternate exterior angle to ∠7?

e. If the measure of ∠1 is 105°, what is the measure of each of the other angles in the figure?

MIXED PRACTICE

Problem set

1. How many quarter-pound hamburgers can be made from
(49) 100 pounds of ground beef?

2. On the Fahrenheit scale water freezes at 32°F and boils at
(18) 212°F. What temperature is halfway between the freezing
and boiling temperatures of water?

3. This function table shows the relationship between
(96) temperatures measured in degrees Celsius and degrees
Fahrenheit. (To find the Fahrenheit temperature, multiply
the temperature in Celsius by 1.8, then add 32.) Find the
missing number in the table.

C	0	10	20	30
$1.8C + 32$	32	50	68	?

4. Compare: $\dfrac{5}{8}$ 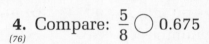 0.675
(76)

5. Write $2\frac{1}{4}$ as a percent.   **6.** Write $1\frac{2}{5}$ as a percent.
(94)   *(94)*

7. Write 0.7 as a percent.   **8.** Write $\frac{7}{8}$ as a percent.
(94)   *(94)*

9. Use division by primes to find the prime factors of 320.
(73) Then write the prime factorization of 320 using exponents.

10. In one minute the second hand of a clock turns 360°.
(90) How many degrees does the minute hand of a clock turn
in one minute?

11. Sammy liked to ride his skateboard around Parallelogram
(90) Park. If he made four turns on each trip around the park,
what was the average number of degrees in each turn?

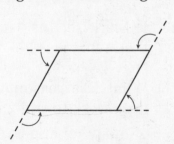

12. $6\frac{3}{4} + 5\frac{7}{8}$ **13.** $6\frac{1}{3} - 2\frac{1}{2}$ **14.** $2\frac{1}{2} \div 100$
(59)   (63)   (68)

15. $6.93 + 8.429 + 12$ **16.** $(1 - 0.1)(1 \div 0.1)$
(38)   (53)

17. $4.2 + \frac{7}{8}$ (decimal answer)
(74)

18. $3\frac{1}{3} - 2.5$ (fraction answer)
(73)

19. If 80% of the 30 moviegoers liked the movie, how many
(41) moviegoers did not like the movie?

20. Compare: $\frac{1}{2} \div \frac{1}{3} \bigcirc \frac{1}{3} \div \frac{1}{2}$
(50)

21. What is the next number in this sequence?
(10)

$$\dots, 1000, 100, 10, 1, \dots$$

Find each missing number:

22. $a + 60 + 70 = 180$ **23.** $\frac{7}{4} = \frac{w}{44}$
(3)   (85)

24. The perimeter of this square is
(79) 48 in. What is the area of one of
the triangles?

Refer to the table below to answer problems 25–27.

Mark's Personal Running Records

Distance	Time (minutes:seconds)
$\frac{1}{4}$ mile	0:58
$\frac{1}{2}$ mile	2:12
1 mile	5:00

25. If Mark set his 1-mile record by keeping a steady pace,
(32) then what was his $\frac{1}{2}$-mile time during the 1-mile run?

26. What is a reasonable expectation for the time it would
(32) take Mark to run 2 miles?

    A. 9:30   B. 11:00   C. 15:00

27. Write a question that relates to this table and answer the
(32) question.

28. Transversal *t* intersects parallel lines *r* and *s*. Angle 2
(97) measures 78°.

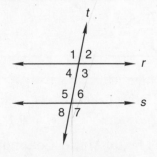

(a) Which angle corresponds to ∠2?

(b) Find the measures of ∠5 and ∠8.

29. $10^2 - \sqrt{49} - (10 + 8) \div 3^2$
(92)

30. What is the probability of rolling a composite number
(58) with one roll of a number cube?

LESSON

98

Sum of the Angle Measures of Triangles and Quadrilaterals

WARM-UP

Facts Practice: 24 Mixed Numbers to Write as Improper Fractions (Test J)

Mental Math: Count up and down by 7's between 7 and 70.

a. 40 · 50
b. 293 + 450
c. 50% of 48
d. $20.00 − $18.72
e. 12.5 × 100
f. $\frac{360}{40}$
g. 8 × 8, − 1, ÷ 9, × 4, + 2, ÷ 2, + 1, $\sqrt{\ }$, $\sqrt{\ }$

Problem Solving:

Use "guess and check" to find this square root: $\sqrt{529}$.

NEW CONCEPT

If we extend a side of a polygon, we form an **exterior angle.** In this figure ∠1 is an exterior angle, and ∠2 is an **interior angle.** Notice that these angles are supplementary. That is, the sum of their measures is 180°.

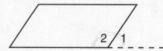

Recall from Lesson 90 that a full turn measures 360°. So if Elizabeth makes three turns to get around a park, she has turned a total of 360°. Likewise, if she makes four turns to get around a park, she has also turned 360°.

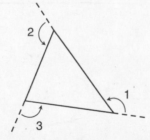

The sum of the measures of angles 1, 2, and 3 is 360°.

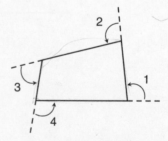

The sum of the measures of angles 1, 2, 3, and 4 is 360°.

If Elizabeth makes three turns to get around the park, then each turn averages 120°.

$$\frac{360°}{3\text{ turns}} = 120° \text{ per turn}$$

If she makes four turns to get around the park, then each turn averages 90°.

$$\frac{360°}{4\text{ turns}} = 90° \text{ per turn}$$

Recall that these turns correspond to exterior angles of the polygons and that the exterior and interior angles at a turn are supplementary. Since the exterior angles of a triangle average 120°, the interior angles must average 60°. A triangle has three interior angles, so the sum of the interior angles is 180° (3 × 60° = 180°).

> **The sum of the interior angles of a triangle is 180°.**

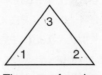

The sum of angles
1, 2, and 3 is 180°.

Since the exterior angles of a quadrilateral average 90°, the interior angles must average 90°. So the sum of the four interior angles of a quadrilateral is 360° (4 × 90° = 360°).

> **The sum of the interior angles of a quadrilateral is 360°.**

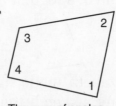

The sum of angles
1, 2, 3, and 4 is 360°.

Example 1   What is m∠A in △ABC?

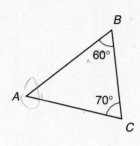

Solution   The measures of the interior angles of a
triangle total 180°.

$$m\angle A + 60° + 70° = 180°$$

Since the measures of ∠B and ∠C total
130°, m∠A is **50°**.

Example 2   What is m∠T in quadrilateral QRST?

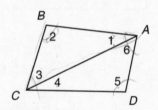

Solution   The measures of the interior angles of a
quadrilateral total 360°.

$$m\angle T + 80° + 80° + 110° = 360°$$

The measures of ∠Q, ∠R, and ∠S total 270°. So m∠T is **90°**.

LESSON PRACTICE

Practice set   Quadrilateral *ABCD* is divided into two triangles by segment *AC*.

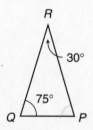

a. What is the sum of m∠1, m∠2, and m∠3?

b. What is the sum of m∠4, m∠5, and m∠6?

c. What is the sum of the measures of the four interior angles
of the quadrilateral?

d. What is m∠P in △PQR?

e. What is the measure of each interior
angle of a regular quadrilateral?

f. Elizabeth made five left turns as she ran around the park.
Draw a sketch that shows the turns in her run around the
park. Then find the average number of degrees in each turn.

MIXED PRACTICE

Problem set

1. When the sum of $\frac{1}{2}$ and $\frac{1}{4}$ is divided by the product of $\frac{1}{2}$
(12, 72)   and $\frac{1}{4}$, what is the quotient?

2. Jenny is $5\frac{1}{2}$ feet tall. She is how many inches tall?
(95)

3. If $\frac{4}{5}$ of the 200 runners finished the race, how many
(77)   runners did not finish the race?

4. Lines p and q are parallel.
(97)

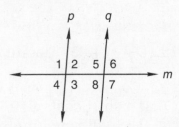

(a) Which angle is an alternate interior angle to $\angle 2$?

(b) If $\angle 2$ measures 85°, what are the measures of $\angle 6$
    and $\angle 7$?

5. The circumference of the earth is about 25,000 miles.
(92)   Write that distance in expanded notation using exponents.

6. Use a ruler to measure the diameter
(17, 27)   of a quarter to the nearest sixteenth
    of an inch.

7. Which of these bicycle wheel parts is the best model of
(27)   the circumference of the wheel?

A. spoke   B. axle   C. tire

8. As this sequence continues, each term equals the sum of
(10)   the two previous terms. What is the next term in this
    sequence?

$$1, 1, 2, 3, 5, 8, 13, \ldots$$

9. If there is a 20% chance of rain, what is the probability
(58)   that it will not rain?

10. Write $1\frac{1}{3}$ as a percent.
(94)

11. $0.08w = \$0.60$
(87)

12. $\dfrac{1 - 0.001}{0.03}$
(49)

13. $\dfrac{3\frac{1}{3}}{100}$
(68)

14. If the volume of each small block is one cubic inch, what is the volume of this rectangular prism?
(82)

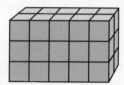

15. $6\frac{1}{2} + 4.95$ (decimal)
(74)

16. $2\frac{1}{6} - 1.5$ (fraction)
(73)

17. If a shirt costs $19.79 and the sales-tax rate is 6%, what is the total price including tax?
(41)

18. What fraction of a foot is 3 inches?
(95)

19. What percent of a meter is 3 centimeters?
(75)

20. The ratio of children to adults in the theater was 5 to 3. If there were 45 children, how many adults were there?
(88)

21. Arrange these numbers in order from least to greatest:
(14, 17)

$$1, -1, 0, \tfrac{1}{2}, -\tfrac{1}{2}$$

22. These two triangles together form a quadrilateral with only one pair of parallel sides. What type of quadrilateral is formed?
(64)

23. Do the triangles in this quadrilateral appear to be congruent or not congruent?
(60)

24. (a) What is the measure of $\angle A$ in $\triangle ABC$?
(98)

(b) What is the measure of the exterior angle marked x?

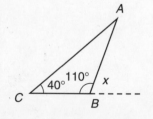

25. Write 40% as a
(33, 74)
    (a) reduced fraction.

    (b) simplified decimal number.

26. The diameter of this circle is 20 mm.
(86) What is the area of the circle? (Use 3.14 for π.)

20 mm

27. $2^3 + \sqrt{81} \div 3^2 + \left(\dfrac{1}{2}\right)^2$
(92)

28. Multiply 120 inches by 1 foot per 12 inches.
(95)

$$\frac{120 \text{ in.}}{1} \times \frac{1 \text{ ft}}{12 \text{ in.}}$$

29. A bag contains 20 red marbles and 15 blue marbles.
(23, 58)
    (a) What is the ratio of red marbles to blue marbles?

    (b) If one marble is drawn from the bag, what is the probability that the marble will be blue?

30. An architect drew a set of plans
(93) for a house. In the plans the roof is supported by a triangular framework. When the house is built, two sides of the framework will be 19 feet long and the base will be 33 feet long. Classified by side length, what type of triangle will be formed?

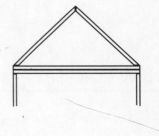

LESSON

99 Fraction-Decimal-Percent Equivalents

WARM-UP

Facts Practice: Linear Measurement (Test K)

Mental Math: Count by $\frac{1}{3}$'s from $\frac{1}{3}$ to 4.

a. 60 · 50   b. 741 − 450   c. 25% of 48
d. $12.99 + $4.75   e. 37.5 ÷ 100   f. 30 × 15
g. 7 × 7, + 1, ÷ 2, $\sqrt{\ }$, × 4, − 2, ÷ 3, × 5, + 3, ÷ 3

Problem Solving:

This quadrilateral is a trapezoid, so the 6-cm segment is parallel to the 9-cm segment. In the figure the 4-cm segment is perpendicular to the parallel segments. The figure is divided into two triangles. What is the area of each triangle? *Hint:* Turn the book upside down to view the upper triangle.

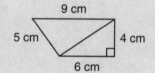

NEW CONCEPT

Fractions, decimals, and percents are three ways to express parts of a whole. An important skill is being able to change from one form to another. In the following problem sets you will be asked to complete tables that show equivalent fractions, decimals, and percents.

Example Complete the table.

FRACTION	DECIMAL	PERCENT
$\frac{1}{2}$	(a)	(b)
(c)	0.3	(d)
(e)	(f)	40%

Solution   The numbers in each row should be equivalent. For $\frac{1}{2}$ we write a decimal and a percent. For 0.3 we write a fraction and a percent. For 40% we write a fraction and a decimal.

(a) $\frac{1}{2} = 2\overline{)1.0}^{\,0.5}$

(b) $\frac{1}{2} \times \frac{100\%}{1} = \mathbf{50\%}$

(c) $0.3 = \frac{3}{10}$

(d) $0.3 \times 100\% = \mathbf{30\%}$

(e) $40\% = \frac{40}{100} = \frac{2}{5}$

(f) $40\% = 0.40 = \mathbf{0.4}$

LESSON PRACTICE

Practice set*   Complete the table.

FRACTION	DECIMAL	PERCENT
$\frac{3}{5}$	a.	b.
c.	0.8	d.
e.	f.	20%
$\frac{3}{4}$	g.	h.
i.	0.12	j.
k.	l.	5%

MIXED PRACTICE

Problem set   **1.** A foot-long hot dog can be cut into how many $1\frac{1}{2}$-inch
(68)   lengths?

2. A can of beans is the shape of what geometric solid?
(Inv. 6)

3. If $\frac{3}{8}$ of the group voted yes and $\frac{3}{8}$ voted no, then what
(77)   fraction of the group did not vote?

4. Nine months is
(29, 75)
(a) what fraction of a year?

(b) what percent of a year?

5. One-cubic-foot boxes were stacked
(82) as shown. What was the volume of
the stack of boxes?

6. Tom was facing east. Then he turned counterclockwise
(90) 270°. After the turn, what direction was Tom facing?

7. If $\frac{1}{5}$ of the pie was eaten, what percent of the pie was left?
(75)

8. Write the percent form of $\frac{1}{7}$.
(94)

9. Solve this proportion: $\frac{15}{20} = \frac{24}{n}$
(85)

10. $5 \cdot 4 \cdot 3 \cdot 2 \cdot 1 \cdot 0$
(5)

11. $\dfrac{4.5}{0.18}$ **12.** $\sqrt{1600}$
(49)   (89)

13. $\sqrt{64} + 5^2 - \sqrt{25} \times (2 + 3)$
(92)

14. $6\frac{3}{4} - 6.2$ (decimal answer)
(74)

15. $12\frac{1}{2} \times 1\frac{3}{5} \times 5$ **16.** $(4.2 \times 0.05) \div 7$
(72)   (53)

17. If the sales-tax rate is 7%, what is the tax on a $111.11
(41) purchase?

18. Perry earned these scores on his first seven tests:
(Inv. 5)
$$80\%,\ 85\%,\ 100\%,\ 80\%,\ 90\%,\ 80\%,\ 95\%$$

(a) Which of the scores did Perry earn most often? That
is, what is the mode of the scores?

(b) If the scores were arranged from lowest to highest,
which score would be the middle score? That is, what
is the median score?

19. Write the prime factorization of 900 using exponents.
(73)

20. Think of two different prime numbers, and write them on
(20) your paper. Then write the greatest common factor (GCF)
of the two prime numbers.

21. The perimeter of a square is 2 meters. How many
(7, 8) centimeters long is each side?

22. (a) What is the area of this triangle?
(79, 93)

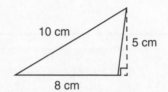

10 cm   5 cm

8 cm

(b) Is this an acute, right, or obtuse triangle?

23. (a) What is the measure of ∠B in
(98)   quadrilateral *ABCD*?

(b) What is the measure of the
exterior angle at *D*?

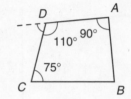

D   A

110° 90°

75°

C   B

Complete the table to answer problems 24–26.

	FRACTION	DECIMAL	PERCENT
24. (99)	(a)	0.6	(b)
25. (99)	(a)	(b)	15%
26. (99)	$\frac{3}{10}$	(a)	(b)

27. Draw $\overline{AC}$ $1\frac{1}{4}$ inches long. Find and mark the midpoint
(17) of $\overline{AC}$, and label the midpoint *B*. What are the lengths
of $\overline{AB}$ and $\overline{BC}$?

28. What is the chance of drawing a diamond from a normal
(58) deck of 52 cards?

29. Compare: 1 gallon ◯ 4 liters
(78)

30. This function table shows the
(27, 96) relationship between the radius (*r*)
and diameter (*d*) of a circle. Find
the missing number.

r	d
1.2	2.4
0.7	1.4
?	5
15	30

LESSON

100 Algebraic Addition of Integers

WARM-UP

Facts Practice: 28 Improper Fractions to Simplify (Test I)

Mental Math: Count up and down by 25's between −150 and 150.

a. 50 · 80   b. 380 + 550   c. 50% of 100

d. $40.00 − $21.89   e. 0.8 × 100   f. $\frac{750}{25}$

g. 5 + 5, × 10, − 1, ÷ 9, + 1, ÷ 3, × 7, + 2, ÷ 2

Problem Solving:

One state uses a license plate that contains one letter followed by five digits. How many different license plates are possible if all of the letters and digits are used?

NEW CONCEPT

The dots on this number line mark the integers from negative five to positive five (−5 to +5).

Recall that the integers consist of the counting numbers (1, 2, 3, ...), their negatives (−1, −2, −3, ...), and the number 0. All other numbers that fall between these numbers are not integers. In this lesson we will practice adding integers.

If we consider a rise in temperature of five degrees as a positive five (+5) and a fall in temperature of five degrees as a negative five (−5), we can use the scale on a thermometer to keep track of the addition.

Imagine that the temperature is 0°F. If the temperature falls five degrees (−5) and then falls another five degrees (−5), the resulting temperature is ten degrees below zero (−10°F). When we add two negative numbers, the sum is negative.

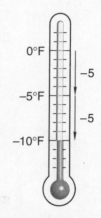

$$-5 + -5 = -10$$

Imagine a different situation. We will again start with a temperature of 0°F. First the temperature falls five degrees (−5). Then the temperature rises five degrees (+5). This brings the temperature back to 0°F. The numbers −5 and +5 are opposites. When we add opposites, the sum is zero.

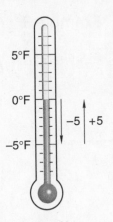

$$-5 + +5 = 0$$

Starting from 0°F, if the temperature rises five degrees (+5) and then falls ten degrees (−10), the temperature will fall through zero to −5°F. The sum is less than zero because the temperature fell more than it rose.

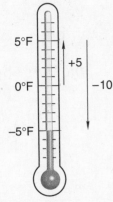

$$+5 + -10 = -5$$

Example 1   Add: +8 + −5

Solution   We will illustrate this addition on a number line. We begin at zero and move eight units in the positive direction (to the right). From +8 we move five units in the negative direction (to the left) to +3.

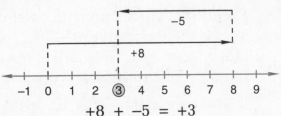

$$+8 + -5 = +3$$

The sum is **+3**, which we write as 3.

Example 2   Add: −5 + −3

Solution   Again using a number line, we start at zero and move in the negative direction, or to the left, five units to −5. From −5 we continue moving left three units to −8.

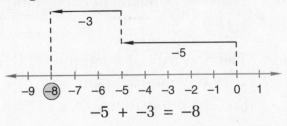

$$-5 + -3 = -8$$

The sum is **−8.**

Example 3   Add: −6 + +6

Solution   We start at zero and move six units to the left. Then we move six units to the right, returning to **zero.**

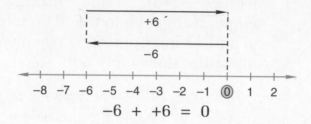

$$-6 + +6 = 0$$

Example 4   Add: (+6) + (−6)

Solution   Sometimes positive and negative numbers are written with parentheses. The parentheses help us see that the positive or negative sign is the sign of the number and not an addition or subtraction operation.

$$(+6) + (−6) = 0$$

Negative 6 and positive 6 are **opposites.** Opposites are numbers that can be written with the same digits but with opposite signs. The opposite of 3 is −3, and the opposite of −5 is 5 (which can be written as +5).

On a number line, we can see that any two opposites lie equal distances from zero. However, they lie on opposite sides of zero from each other.

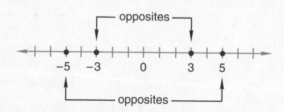

If opposites are added, the sum is zero.

$$-3 + +3 = 0 \qquad -5 + +5 = 0$$

Example 5   Find the opposite of each number:

(a) −7   (b) 10

Solution   The opposite of a number is written with the same digits but with the opposite sign.

(a) The opposite of −7 is **+7,** which is usually written as 7.

(b) The opposite of 10 (which is positive) is **−10.**

Using opposites allows us to change any subtraction problem into an addition problem. Consider this subtraction problem:

$$10 - 6$$

Instead of subtracting 6 from 10, we can add the opposite of 6 to 10. The opposite of 6 is −6.

$$10 + -6$$

In both problems the answer is 4. Adding the opposite of a number to subtract is called **algebraic addition**. Instead of subtracting, we will use algebraic addition to solve subtraction problems.

Example 6 Simplify: −10 − −6

Solution This problem directs us to subtract a negative six from negative ten. Instead, we may add the opposite of negative six to negative ten.

$$-10 - -6$$
$$\downarrow \quad \downarrow$$
$$-10 + +6 = \mathbf{-4}$$

Example 7 Simplify: (−3) − (+5)

Solution Instead of subtracting a positive five, we add a negative five.

$$(-3) - (+5)$$
$$\downarrow \quad \downarrow$$
$$(-3) + (-5) = \mathbf{-8}$$

LESSON PRACTICE

Practice set* Find each sum. Draw a number line to show the addition for problems **a** and **b**. Solve problems **c–h** mentally.

a. −3 + +4   b. −3 + −4

c. −3 + +3   d. +4 + −3

e. (+3) + (−4)   f. (+10) + (−5)

g. (−10) + (−5)   h. (−10) + (+5)

Find the opposite of each number:

i. −8　　　　　　　　　**j.** 4　　　　　　　　　**k.** 0

Solve each subtraction problem using algebraic addition:

l. −3 − −4　　　　　　　　　　**m.** −4 − +2

n. (+3) − (−6)　　　　　　　　　**o.** (−2) − (−4)

MIXED PRACTICE

Problem set

1. If 0.6 is the divisor and 1.2 is the quotient, what is the dividend?
(39)

2. If a number is twelve less than fifty, then it is how much more than twenty?
(12)

3. If the sum of four numbers is 14.8, what is the average of the four numbers?
(18)

4. Illustrate this problem on a number line:
(100)

$$-3 + +5$$

5. Find each sum mentally:
(100)

(a) −4 + +4　　　　　　　　　(b) −2 + −3

(c) −5 + +3　　　　　　　　　(d) +5 + −10

6. Solve each subtraction problem using algebraic addition:
(100)

(a) −2 + +5　　　　　　　　　(b) −3 − −3

(c) +2 − −3　　　　　　　　　(d) −2 − +3

7. What is the measure of each angle of an equilateral triangle?
(93, 98)

8. Quadrilateral *ABCD* is a parallelogram. If angle *A* measures 70°, what are the measures of angles *B*, *C*, and *D*?
(71)

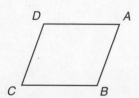

9. (a) If the spinner is spun once, what is the probability that it will stop in a sector with a number 2?
(58)

(b) If the spinner is spun 30 times, about how many times would it be expected to stop in the sector with the number 3?

10. Find the volume of the rectangular prism at right.
(82)

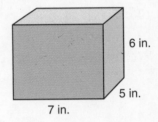

6 in.

5 in.

7 in.

11. Twelve of the 27 people in the room are boys. What is the ratio of girls to boys in the room?
(23)

12. $10^2 + (5^2 - 11) \div \sqrt{49} - 3^3$
(92)

13. The fraction $\frac{2}{3}$ is equal to what percent?
(94)

14. If 20% of the flowers were tulips, then what fraction of the flowers were not tulips?
(33)

15. $\dfrac{4^2}{2^4}$ **16.** $5\dfrac{7}{8} + 4\dfrac{3}{4}$ **17.** $1\dfrac{1}{2} \div 2\dfrac{1}{2}$
(92)   (59)   (68)

18. $5 - (3.2 + 0.4)$
(38)

19. If the diameter of a circular plastic swimming pool is 6 feet, then the area of the bottom of the pool is about how many square feet? Round to the nearest square foot. (Use 3.14 for π.)
(86)

20. We use squares to measure the area of a rectangle. Why do we use cubes instead of squares to measure the volume of a rectangular prism?
(82)

21. Solve this proportion: $\dfrac{9}{12} = \dfrac{15}{x}$
(85)

Rectangle *ABCD* is 8 cm long and 6 cm wide. Segment *AC* is 10 cm long. Use this information to answer problems 22 and 23.

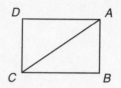

22. What is the area of triangle *ABC*?
(79)

23. What is the perimeter of triangle *ABC*?
(8)

24. Measure the diameter of a nickel to
(27) the nearest millimeter.

25. Calculate the circumference of a
(47) nickel. Round to the nearest millimeter. (Use 3.14 for π.)

26. A bag contains 12 marbles. Eight of the marbles are red
(58) and 4 are blue. If you draw a marble from the bag without looking, what is the probability that the marble will be blue?

Complete the table to answer problems 27–29.

	FRACTION	DECIMAL	PERCENT
27. (99)	$\frac{9}{10}$	(a)	(b)
28. (99)	(a)	1.5	(b)
29. (99)	(a)	(b)	4%

30. A full one-gallon container of milk was used to fill two
(78) one-pint containers. How many quarts of milk were left in the one-gallon container?

INVESTIGATION 10

Focus on

Compound Experiments

Some experiments whose outcomes are determined by chance contain more than one part. Such experiments are called **compound experiments.** In this investigation we will consider compound experiments that consist of two parts performed in order. Here are three examples:

1. A spinner with sectors A, B, and C is spun; then a marble is drawn from a bag that contains 4 red marbles and 2 white marbles.

2. A marble is drawn from a bag with 4 red marbles and 2 white marbles; then, without the first marble being replaced, a second marble is drawn.

3. A number cube is rolled; then a coin is flipped.

The second experiment is actually a way to look at drawing two marbles from the bag at once. We estimated probabilities for this compound experiment in Investigation 9.

A compound experiment can be visualized by a **tree diagram.** Here is a tree diagram for compound experiment 1:

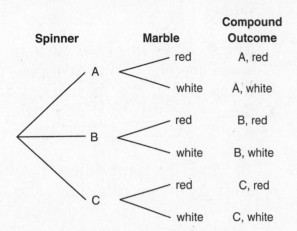

Each **branch** of the tree corresponds to a possible outcome. There are three possible spinner outcomes. For each spinner outcome, there are two possible marble outcomes. To find the total number of **compound outcomes,** we multiply the number of branches in the first part of the experiment by the number of branches in the second part of the experiment.

There are 3 × 2 = 6 branches, so there are six possible compound outcomes. In the column titled "Compound Outcome," we list the outcome for each branch. "A, red" means that the spinner stopped on A, then the marble drawn was red. Although there are six different outcomes, not all the outcomes are equally likely. We need to determine the probability of each part of the experiment in order to find the probability for each compound outcome. To do this, we will use the multiplication principle for compound probability:

> **The probability of a compound outcome is the product of the probabilities of each part of the outcome.**

We will use this principle to calculate the probability of the first branch of the spinner-marble experiment, which corresponds to the compound outcome "A, red."

The first part of the outcome is that the spinner stops in sector A. The probability of this outcome is $\frac{1}{2}$, since sector A occupies half the area of the circle.

The second part of the outcome is that a red marble is drawn from the bag. Since four of the six marbles are red, the probability of this outcome is $\frac{4}{6}$, which reduces to $\frac{2}{3}$.

To find the probability of the compound outcome, we multiply the probabilities of each part.

The probability of "A, red" is $\frac{1}{2} \cdot \frac{2}{3}$, which equals $\frac{1}{3}$.

Notice that although "A, red" is one of six possible outcomes, the probability of "A, red" is greater than $\frac{1}{6}$. This is because "A" is the most likely of the three possible spinner outcomes, and "red" is the more likely of the two possible marble outcomes.

For problems 1–6, copy the table below and calculate the probability of each possible outcome. For the last row, find the sum of the probabilities of the six possible outcomes.

Outcome	Probability
A, red	$\frac{1}{2} \cdot \frac{2}{3} = \frac{1}{3}$
A, white	**1.**
B, red	**2.**
B, white	**3.**
C, red	**4.**
C, white	**5.**
sum of probabilities	**6.**

For problems 7–9 we will consider compound experiment 2, which involves two draws from a bag of marbles that contains four red marbles and two white marbles. The first part of the experiment is that one marble is drawn from the bag and is not replaced. The second part is that a second marble is drawn from the marbles remaining in the bag.

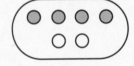

7. Copy and complete this tree diagram showing all possible outcomes of the compound experiment:

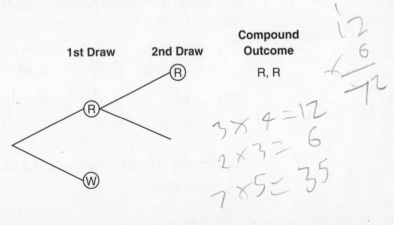

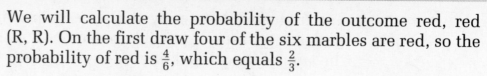

We will calculate the probability of the outcome red, red (R, R). On the first draw four of the six marbles are red, so the probability of red is $\frac{4}{6}$, which equals $\frac{2}{3}$.

If the first marble drawn is red, then three red marbles and two white marbles remain (see picture at right). So the probability of drawing a red marble on the second draw is $\frac{3}{5}$. Therefore, the probability of the outcome red, red is $\frac{2}{3} \cdot \frac{3}{5} = \frac{2}{5}$.

8. Copy and complete this table to show the probability of each remaining possible outcome and the sum of the probabilties of all outcomes. Remember that the first draw changes the collection of marbles in the bag for the second draw.

Outcome	Probability
red, red	$\frac{4}{6} \cdot \frac{3}{5} = \frac{2}{5}$
sum of probabilities	

9. Suppose we draw three marbles from the bag, one at a time and without replacement. What is the probability of drawing three white marbles? What is the probability of drawing three red marbles?

For problems 10–14, consider a compound experiment in which a nickel is flipped and then a quarter is flipped.

10. Create a tree diagram that shows all of the possible outcomes of the compound experiment.

11. Make a table that shows the probability of each possible outcome.

Use the table you made in problem 11 to answer problems 12–14.

12. What is the probability that one of the coins shows "heads" and the other coin shows "tails"?

13. What is the probability that at least one of the coins shows "heads"?

14. What is the probability that the nickel shows "heads" and the quarter shows "tails"?

Extensions   For extensions **a** and **b,** consider the compound experiment in which a bag contains 4 red marbles and 2 white marbles. One marble is drawn from the bag and not replaced, and then a second marble is drawn.

   a. Find the probability that the two marbles drawn from the bag are different colors.

   b. Find the probability that the two marbles drawn from the bag are the same color.

For extensions **c** and **d,** consider the compound experiment involving spinning the spinner and then drawing a marble.

   c. Find the probability that the compound outcome will *not* be "A, red."

   d. Find the probability that the compound outcome will *not* include "A" and will *not* include "red."

For extensions **e** and **f,** consider a compound experiment consisting of rolling a number cube and then flipping a quarter.

   e. Draw a tree diagram showing all possible outcomes for the experiment.

   f. Find the probability of each compound outcome.

LESSON

101 Ratio Problems Involving Totals

WARM-UP

Facts Practice: 72 Multiplication and Division Facts (Test H)

Mental Math: Count by $\frac{1}{16}$'s from $\frac{1}{16}$ to 1.

a. 20 · 300   **b.** 920 − 550   **c.** 25% of 100
d. $18.99 + $5.30   **e.** 3.75 ÷ 100   **f.** 40 × 25
g. Find half of 100, − 1, $\sqrt{\ }$, × 5, + 1, $\sqrt{\ }$, × 3, + 2, ÷ 2
h. What number is represented by the Roman numeral XXVI?[†]

Problem Solving:

Copy this problem and fill in the missing digits:

$$\begin{array}{r} \_1 \\ \times\ \_1 \\ \hline \_1 \\ \_\_\_ \\ \hline 1\,0\,0\,1 \end{array}$$

NEW CONCEPT

In some ratio problems a total is used as part of the calculation. Consider this problem:

> *The ratio of parents to children at the picnic was 5 to 4. If there were 27 picnickers, how many of them were children?*

We begin by drawing a ratio box. In addition to the categories of parents and children, we make a third row for the total number of picnickers. We will use the letters p and c to represent the actual counts of parents and children.

	Ratio	Actual Count
Parents	5	p
Children	4	c
Total	9	27

In the ratio column we add the ratio numbers for parents and children and get the ratio number 9 for the total. We were given 27 as the actual count of picnickers. We will use two of

---

[†]In Lessons 101–120, Mental Math problem **h** reviews concepts from Appendix Topic A. We suggest you complete Appendix Topic A before beginning this lesson.

the three rows from the ratio box to write a proportion. **We use the row we want to complete and the row that is already complete.** Since we are asked to find the actual number of children, we will use the "children" row. And since we know both "total" numbers, we will also use the "total" row. We solve the proportion below.

	Ratio	Actual Count
Parents	5	p
Children	4	c
Total	9	27

$$\frac{4}{9} = \frac{c}{27}$$

$$9c = 4 \cdot 27$$

$$c = 12$$

We find that there were 12 children at the picnic. If we had wanted to find the number of parents, we would have used the "parents" row along with the "total" row to write a proportion.

Example The ratio of football players to band members on the football field was 2 to 5. Altogether, there were 175 football players and band members on the football field. How many football players were on the field?

Solution We use the information in the problem to make a table. We include a row for the total. The ratio number for the total is 7.

	Ratio	Actual Count
Football players	2	f
Band members	5	b
Total	7	175

Next we write a proportion using two rows of the table. We are asked to find the number of football players, so we use the "football players" row. We know both totals, so we also use the "total" row. Then we solve the proportion.

	Ratio	Actual Count
Football players	2	f
Band members	5	b
Total	7	175

$$\frac{2}{7} = \frac{f}{175}$$

$$7f = 2 \cdot 175$$

$$f = 50$$

We find that there were **50 football players** on the field.

LESSON PRACTICE

Practice set   Use ratio boxes to solve problems **a** and **b**.

a. Sparrows and crows perched on the wire in the ratio of 5 to 3. If the total number of sparrows and crows on the wire was 72, how many were crows?

b. Raisins and nuts were mixed by weight in a ratio of 2 to 3. If 60 ounces of mix were prepared, how many ounces of raisins were used?

MIXED PRACTICE

Problem set

1. Draw a ratio box for this problem. Then solve the
(101) problem using a proportion.

> *The ratio of boys to girls in the contest was 3 to 2. If there were 30 contestants, how many girls were there?*

2. A shoe box is the shape of what geometric solid?
(Inv. 6)

3. If the average of six numbers is 12, what is the sum of the
(18) six numbers?

4. If the diameter of a circle is $1\frac{1}{2}$ inches, what is the radius
(27, 68) of the circle?

5. What is the cost of 2.6 pounds of meat priced at $1.65
(95) per pound?

6. If $\overline{AB}$ is $\frac{1}{4}$ the length of $\overline{AC}$, and if $\overline{AC}$ is 12 cm long, then
(69) how long is $\overline{BC}$?

7. Find each sum mentally:
(100)
    (a) −3 + −4   (b) +5 + −5

    (c) −6 + +3   (d) +6 + −3

8. Solve each subtraction problem using algebraic addition:
(100)
    (a) −3 − −4   (b) +5 − −5

    (c) −6 − +3   (d) −6 + +6

9. Two coins are tossed.
(Inv. 10)

(a) What is the probability that both coins will land heads up?

(b) What is the probability that one of the coins will be heads and the other tails?

Complete the table to answer problems 10–12.

	FRACTION	DECIMAL	PERCENT
10. *(99)*	$\frac{3}{4}$	(a)	(b)
11. *(99)*	(a)	1.6	(b)
12. *(99)*	(a)	(b)	5%

13. $1\frac{1}{2} \times 4$
(66)

14. $6 \div 1\frac{1}{2}$
(68)

15. $(0.4)^2 \div 2^3$
(92)

Find each missing number:

16. $x + 2\frac{1}{2} = 5$
(43)

17. $\frac{8}{5} = \frac{40}{x}$
(42)

18. $0.06n = \$0.15$
(49)

19. $6n = 21 \cdot 4$
(87)

20. Nia's garage is 20 feet long, 20 feet wide, and 8 feet high.
(82)

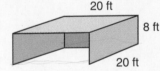

(a) How many 1-by-1-by-1-foot boxes can she fit on the floor (bottom layer) of her garage?

(b) Altogether, how many boxes can Nia fit in her garage if she stacks the boxes 8 feet high?

21. If a roll of tape has a diameter of $2\frac{1}{2}$ inches, then removing one full turn of tape removes about how many inches? Choose the closest answer.
(47)

A. $2\frac{1}{2}$ in.   B. 5 in.   C. $7\frac{3}{4}$ in.   D. $9\frac{1}{4}$ in.

22. $9^2 - \sqrt{9} \times 10 - 2^4 \times 2$
(92)

23. Together, these three triangles form
(60) what kind of polygon?

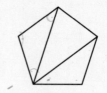

24. What is the sum of the measures of
(98) the angles of each triangle?

25. At 6 a.m. the temperature was −8°F. By noon the
(14, 100) temperature was 15°F. The temperature had risen how
many degrees?

26. To what decimal number is the arrow pointing on the
(50) number line below?

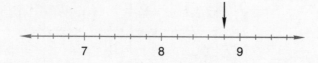

27. What is the probability of rolling a perfect square with one
(38, 58) roll of a number cube?

28. What is the area of a triangle with vertices located at
(Inv. 7, 79) (4, 0), (0, −3), and (0, 0)?

29. Multiply 18 feet by 1 yard per 3 feet:
(95)

$$\frac{18 \text{ ft}}{1} \times \frac{1 \text{ yd}}{3 \text{ ft}}$$

30. If a gallon of milk costs $2.80, what is the cost per quart?
(78)

LESSON

102

Mass and Weight

WARM-UP

Facts Practice: 24 Percent-Fraction-Decimal Equivalents (Test M)

Mental Math: Count up and down by 12's between 12 and 144.

a. 30 · 400
b. 462 + 150
c. 50% of 40

d. $100.00 − $47.50
e. 0.06 × 100
f. 50 × 15

g. 12 + 12, + 1, $\sqrt{\ }$, × 3, + 1, $\sqrt{\ }$, × 2, + 2, × 5

h. What number is represented by the Roman numeral XXXIV?

Problem Solving:

A loop of string was arranged to form a square with sides 9 inches long. If the same loop of string is arranged to form a regular hexagon, how long will each side of the hexagon be?

NEW CONCEPT

Physical objects are composed of matter. The amount of matter in an object is its **mass.** In the metric system we measure the mass of objects in milligrams (mg), grams (g), and kilograms (kg).

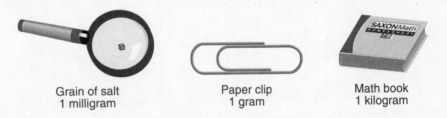

Grain of salt
1 milligram

Paper clip
1 gram

Math book
1 kilogram

1000 mg = 1 g   1000 g = 1 kg

A particular object has the same mass on Earth as it has on the Moon, in orbit, or anywhere else in the universe. In other words, the mass of an object does not change with changes in the force of gravity. However, the **weight** of an object does change with changes in the force of gravity. For example, astronauts who are in orbit feel no gravitational force, so they experience weightlessness. An astronaut who weighs 154 pounds on Earth weighs zero pounds in weightless conditions. Although the weight of the astronaut has changed, his or her mass has not changed.

In the U.S. Customary System we measure the weight of objects in ounces (oz), pounds (lb), or tons (tn). On Earth an object with a mass of 1 kilogram weighs about 2.2 pounds.

Envelope and letter	Shoe	Small car
1 ounce	1 pound	1 ton

16 ounces = 1 pound   2000 pounds = 1 ton

Example 1   Two kilograms is how many grams?

Solution   One kilogram is 1000 grams. So 2 kilograms equals **2000 grams.**

Some measures are given using a mix of units. For example, Sam might finish a facts practice test in 2 minutes 34 seconds. His sister may have weighed 7 pounds 12 ounces when she was born. The following example shows how to add and subtract measures in pounds and ounces.

Example 2   (a) Add: 7 lb 12 oz   (b) Subtract: 9 lb 10 oz
                        + 2 lb   6 oz   − 7 lb 12 oz

Solution   (a) The sum of 12 oz and 6 oz is 18 oz, which is 1 lb 2 oz. We record the 2 oz and then add the pound to 7 lb and 2 lb.

$$\begin{array}{r} \overset{1}{} \\ 7 \text{ lb } 12 \text{ oz} \\ + 2 \text{ lb } 6 \text{ oz} \\ \hline \mathbf{10} \text{ lb } \mathbf{2} \text{ oz} \end{array}$$

(b) Before we can subtract ounces, we convert 9 pounds to 8 pounds plus 16 ounces. We combine the 16 ounces and the 10 ounces to get 26 ounces. Then we subtract.

$$\begin{array}{r} \overset{8}{\cancel{9}} \text{ lb } \overset{26}{\cancel{10}} \text{ oz} \\ - 7 \text{ lb } 12 \text{ oz} \\ \hline \mathbf{1} \text{ lb } \mathbf{14} \text{ oz} \end{array}$$

LESSON PRACTICE

Practice set   **a.** Half of a kilogram is how many grams?

b. The mass of a liter of water is 1 kilogram. So the mass of 2 liters of beverage is about how many grams?

c.   5 lb 10 oz
   + 1 lb   9 oz

d.   9 lb   8 oz
    – 6 lb 10 oz

e. A half-ton pickup truck can haul a half-ton load. Half of a ton is how many pounds?

MIXED PRACTICE

Problem set   On his first six tests, Chris had scores of 90%, 92%, 96%, 92%, 84%, and 92%. Use this information to answer problems 1 and 2.

1. (a) Which score occurred most frequently? That is, what is the mode of the scores?
(Inv. 5)

(b) The difference between Chris's highest score and his lowest score is how many percentage points? That is, what is the range of the scores?

2. What was Chris's average score for the six tests? That is, what is the mean of the scores?
(18)

3. In basketball there are one-point baskets, two-point baskets, and three-point baskets. If a team scored 96 points and made 18 one-point baskets and 6 three-point baskets, how many two-point baskets did the team make?
(87)

4. Which ratio forms a proportion with $\frac{4}{7}$?
(83)

A. $\frac{7}{4}$ B. $\frac{14}{17}$ C. $\frac{12}{21}$ D. $\frac{2}{3}$

5. Complete this proportion: Four is to five as what number is to twenty?
(85)

6. Arrange these numbers in order from least to greatest:
(50)

$$-1, 1, 0.1, -0.1, 0$$

7. The product of $10^3 \cdot 10^2$ equals which of the following?
(92)

A. 10^9 B. 10^6 C. 10^5 D. 10

8. The area of the square in this figure is 100 mm².
(86)

(a) What is the radius of the circle?

(b) What is the diameter of the circle?

(c) What is the area of the circle? (Use 3.14 for π.)

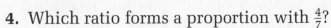

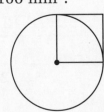

Complete the table to answer problems 9–11.

	FRACTION	DECIMAL	PERCENT
9. (99)	$\frac{4}{25}$	(a)	(b)
10. (99)	(a)	0.01	(b)
11. (99)	(a)	(b)	90%

12. $1\frac{2}{3} + 3\frac{1}{2} + 4\frac{1}{6}$ **13.** $\frac{5}{6} \times \frac{3}{10} \times 4$
(61)   (72)

14. $6\frac{1}{4} \div 100$ **15.** $6.437 + 12.8 + 7$
(68)   (38)

16. Convert $\frac{1}{7}$ to a decimal number by dividing 1 by 7. Stop
(74) dividing after three decimal places, and round your answer to two decimal places.

17. An octagon has how many more sides than a pentagon?
(60)

18. $4 \times 5^2 - 50 \div \sqrt{4} + (3^2 - 2^3)$
(92)

19. Sector 2 on this spinner is a 90°
(Inv. 10) sector. If the spinner is spun twice, what is the probability that it will stop in sector 2 both times?

20. If the spinner is spun 100 times, about how many times
(58) would it be expected to stop in sector 1?

21. How many cubes 1 inch on each
(82) edge would be needed to build this larger cube?

4 in.

22. The average of four numbers is 5. What is their sum?
(18)

23. When Andy was born, he weighed 8 pounds 4 ounces.
(102) Three weeks later he weighed 10 pounds 1 ounce. How many pounds and ounces had he gained in three weeks?

24. Lines *s* and *t* are parallel.
(97)

(a) Which angle is an alternate interior angle to ∠5?

(b) If the measure of ∠5 is 76°, what are the measures of ∠1 and ∠2?

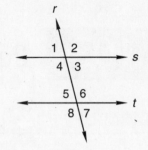

25. Find the missing number in this function table:
(96)

x	1	2	4	5
3*x* − 5	−2	1	7	?

26. What is the perimeter of this hexagon? Dimensions are in centimeters.
(8)

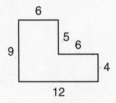

27. (a) What is the area of the parallelogram at right?
(71, 79)

(b) What is the area of the triangle?

(c) What is the combined area of the parallelogram and triangle?

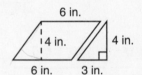

28. How many milligrams is half of a gram?
(102)

29. The coordinates of the endpoints of a line segment are (3, −1) and (3, 5). The midpoint of the segment is the point halfway between the endpoints. What are the coordinates of the midpoint?
(Inv. 7)

30. Tania took 10 steps to walk across the tetherball circle and 31 steps to walk around the tetherball circle. Use this information to find the approximate number of diameters in the circumference of the tetherball circle.
(47)

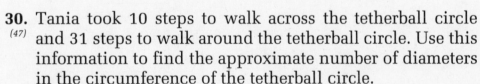
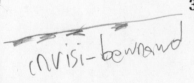
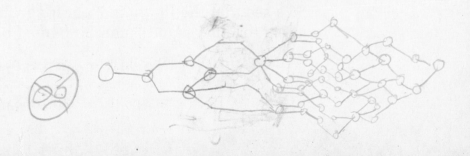

LESSON
103 Perimeter of Complex Shapes

WARM-UP

Facts Practice: 24 Percent-Fraction-Decimal Equivalents (Test M)

Mental Math: Count by 3's from −30 to 30.

a. 50 · 60   b. 543 − 250   c. 25% of 40
d. $5.65 + $3.99   e. 87.5 ÷ 100   f. $\frac{500}{20}$
g. 6 × 6, − 1, ÷ 5, × 6, − 2, ÷ 5, × 4, − 2, × 3
h. What number is represented by the Roman numeral XXIX?

Problem Solving:

Laura has nickels, dimes, and quarters in her pocket. She has half as many dimes as nickels and half as many quarters as dimes. If Laura has four dimes, then how much money does she have in her pocket?

NEW CONCEPT

In this lesson we will practice finding the perimeters of complex shapes. The figure below is an example of a complex shape. Notice that the lengths of two of the sides are not given. We will first find the lengths of these sides; then we will find the perimeter of the shape. (In this book, assume that corners that look square are square.)

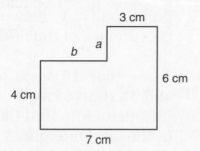

We see that the figure is 7 cm long. The sides marked b and 3 cm together equal 7 cm. So b must be 4 cm.

$$b + 3 \text{ cm} = 7 \text{ cm}$$

$$b = 4 \text{ cm}$$

The width of the figure is 6 cm. The sides marked 4 cm and a together equal 6 cm. So a must equal 2 cm.

$$4 \text{ cm} + a = 6 \text{ cm}$$

$$a = 2 \text{ cm}$$

We have found that b is 4 cm and a is 2 cm.

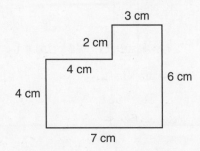

We add the lengths of all the sides and find that the perimeter is 26 cm.

$$6 \text{ cm} + 7 \text{ cm} + 4 \text{ cm} + 4 \text{ cm} + 2 \text{ cm} + 3 \text{ cm} = 26 \text{ cm}$$

Example Find the perimeter of this figure.

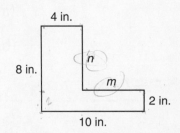

Solution To find the perimeter, we add the lengths of the six sides. The lengths of two sides are not given in the illustration. We will write two equations to find the lengths of these sides. The length of the figure is 10 inches. The sides parallel to the 10-inch side have lengths of 4 inches and m inches. Their combined length is 10 inches, so m must equal 6 inches.

$$4 \text{ in.} + m = 10 \text{ in.}$$

$$m = 6 \text{ in.}$$

The width of the figure is 8 inches. The sides parallel to the 8-inch side have lengths of n inches and 2 inches. Their combined measures equal 8 inches, so n must equal 6 inches.

$$n + 2 \text{ in.} = 8 \text{ in.}$$

$$n = 6 \text{ in.}$$

We add the lengths of the six sides to find the perimeter of the complex shape.

$$10 \text{ in.} + 8 \text{ in.} + 4 \text{ in.} + 6 \text{ in.} + 6 \text{ in.} + 2 \text{ in.} = \textbf{36 in.}$$

LESSON PRACTICE

Practice set   Find the perimeter of each complex shape:

a.
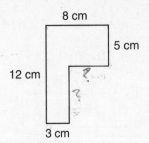
8 cm

5 cm

12 cm

3 cm

b.
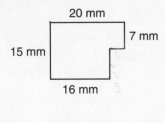
20 mm

7 mm

15 mm

16 mm

MIXED PRACTICE

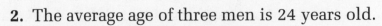

Problem set   **1.** When the sum of $\frac{1}{2}$ and $\frac{1}{3}$ is divided by the product of
$^{(12,\ 72)}$ $\frac{1}{2}$ and $\frac{1}{3}$, what is the quotient?

2. The average age of three men is 24 years old.
$^{(18)}$

(a) What is the sum of their ages?

(b) If two of the men are 22 years old, how old is the third?

3. A string one yard long is formed into the shape of a square.
$^{(38)}$

(a) How many inches long is each side of the square?

(b) How many square inches is the area of the square?

4. Complete this proportion: Five is to three as thirty is to
$^{(85)}$ what number?

5. In a group of 30 players, there are 14 boys. What is the
$^{(23)}$ boy-girl ratio of the group?

6. In another group of 33 players, the ratio of boys to girls is
$^{(101)}$ 4 to 7. How many girls are in that group?

7. $100 \div 10^2 + 3 \times (2^3 - \sqrt{16})$
$^{(92)}$

8. Robert complained that he had a "ton" of homework.
$^{(58,\ 102)}$

(a) How many pounds is a ton?

(b) What is the probability that Robert would literally have a ton of homework?

Complete the table to answer problems 9–11.

	FRACTION	DECIMAL	PERCENT
9. (99)	$\frac{1}{100}$	(a)	(b)
10. (99)	(a)	0.4	(b)
11. (99)	(a)	(b)	8%

12. (68) $10\frac{1}{2} \div 3\frac{1}{2}$

13. (53) $(6 + 2.4) \div 0.04$

Find each missing number:

14. (61) $7\frac{1}{2} + 6\frac{3}{4} + n = 15\frac{3}{8}$

15. (63) $x - 1\frac{3}{4} = 7\frac{1}{2}$

16. (43) Instead of dividing $10\frac{1}{2}$ by $3\frac{1}{2}$, Guadalupe doubled both numbers before dividing. What was Guadalupe's division problem and its quotient?

17. (47) Mariabella used a tape measure to find the circumference and the diameter of a plate. The circumference was about 35 inches, and the diameter was about 11 inches. Find the approximate number of diameters in the circumference. Round to the nearest tenth.

18. (92) Write twenty million, five hundred thousand in expanded notation using exponents.

19. (19) List the prime numbers between 40 and 50.

20. (100) Calculate mentally:

(a) $-3 + -8$

(b) $-3 + +8$

(c) $-8 + +3$

(d) $-8 - +3$

21. (98) In $\triangle ABC$ the measure of $\angle A$ is $40°$. Angles B and C are congruent. What is the measure of $\angle C$?

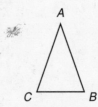

22. (8, 79) (a) What is the perimeter of this triangle?

(b) What is the area of this triangle?

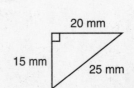

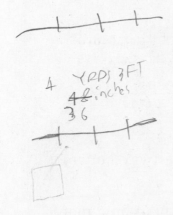

23. The Simpsons rented a trailer that was 8 feet long and
(82) 5 feet wide. If they load the trailer with 1-by-1-by-1-foot boxes to a height of 3 feet, how many boxes can be loaded onto the trailer?

24. What is the probability of drawing the queen of spades
(58) from a normal deck of 52 cards?

25. (a) What temperature is shown on
(10, 100)   this thermometer?
       (b) If the temperature rises 12°F, what will the temperature be?

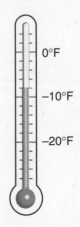

26. Find the perimeter of the figure below.
(103)

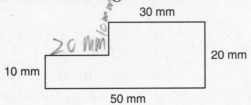

27. (a) What is the area of the shaded
(31)   rectangle?
       (b) What is the area of the unshaded rectangle?
       (c) What is the combined area of the two rectangles?

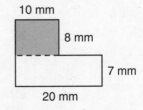

28. What are the coordinates of the point halfway between
(Inv. 7) (−3, −2) and (5, −2)?

29. A pint of milk weighs about 16 ounces. About how many
(78, 102) pounds does a half gallon of milk weigh?

30. Tad walked around a building whose perimeter was
(90) shaped like a regular pentagon.
       (a) At each corner of the building, Tad turned about how many degrees?
       (b) What is the measure of each interior angle of the regular pentagon?

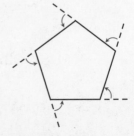

LESSON

104 Algebraic Addition Activity

WARM-UP

Note: Because the New Concept in this lesson takes more time than usual, today's Warm-Up has been omitted.

NEW CONCEPT

One model for the addition of signed numbers is the number line. Another model for the addition of signed numbers is the electrical-charge model, which is used in the Sign Wars game. In this model signed numbers are represented by positive and negative charges that can neutralize each other.

Activity: *Sign Wars Game*

In Sign Wars positives "battle" negatives. After each battle we ask ourselves, "Who survived?" and then write our answer. There are four skill levels to the game. Be sure you are successful at one level before moving to the next level.

Level 1   Positive and negative signs are placed randomly on a "screen." For the battle we neutralize positive and negative pairs by crossing out the signs as shown. (Appropriate sound effects strengthen the experience!)

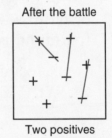

After the battle we count the remaining positives or negatives to determine who survived. In the battle shown above, there are two positive survivors. See whether you can determine the number and type of survivors for the following practice screens:

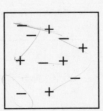
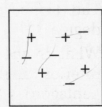
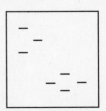

Level 2 Positives and negatives are displayed in counted clusters. The suggested strategy is to group forces before the battle. So +3 combines with +1 to form +4, and −5 combines with −2 to form −7.

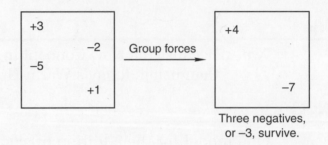

Three negatives,
or −3, survive.

In this battle there were three more negatives than positives, so −3 survived. See whether you can determine the number and type of survivors for the following practice screens:

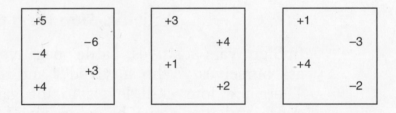

Level 3 Positive and negative clusters can be displayed with two signs, one sign, or no sign. Clusters appear "in disguise" by taking on an additional sign or by dropping a sign. The first step is to remove the disguise. A cluster with no sign, with "− −," or with "+ +" is a positive cluster. A cluster with "+ −" or with "− +" is a negative cluster. If a cluster has a "shield" (parentheses), look through the shield to see the sign.

Examples of Positives   Examples of Negatives

$$-(-3) = +3$$

$$--2 = +2$$

$$4 = +4$$

$$++1 = +1$$

$$-(+2) = -2$$

$$+(-3) = -3$$

$$+-1 = -1$$

$$-+4 = -4$$

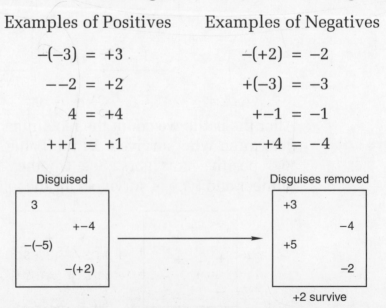

+2 survive

See whether you can determine the number and type of survivors for the following practice screens:

--3 +(−5) −+6 +(+4)	+(−3) −2 −(+4)	−(+6) --3 +4 −(+2) +−6

Level 4   Extend Level 3 to a line of clusters without using a screen. Determine the survivors for this battle:

$$-3 + (-4) - (-5) - (+2) + (+6)$$

Use the following steps to find the answer:

Step 1: Remove the disguises: $-3 - 4 + 5 - 2 + 6$
Step 2: Group forces: $-9 + 11$
Step 3: Who survived? $+2$

LESSON PRACTICE

Practice set   Simplify:

a. $-2 + -3 - -4 + -5$

b. $-3 + (+2) - (+5) - (-6)$

c. $+3 + -4 - +6 + +7 - -1$

d. $2 + (-3) - (-9) - (+7) + (+1)$

e. $3 - -5 + -4 - +2 + +8$

f. $(-10) - (+20) - (-30) + (-40)$

MIXED PRACTICE

Problem set   **1.** A pyramid with a square base has how many more edges
(Inv. 6)   than vertices?

2. Becki weighed 7 lb 8 oz when she was born and 12 lb 6 oz
(102)   at 3 months. How many pounds and ounces did Becki gain in 3 months?

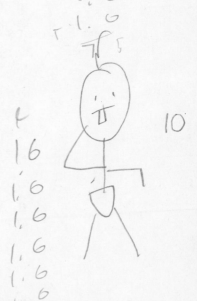

3. The team won 6 games and lost 10. What was its win-
(23) loss ratio?

4. Another team's win-loss ratio was 3 to 2. If the team
(101) had played 20 games without a tie, how many games
had it won?

5. If Molly tosses a coin and rolls a number cube, what is
(Inv. 10) the probability of the coin landing heads up and the
number cube stopping with a 6 on top?

6. (a) What is the perimeter of this
(71) parallelogram?

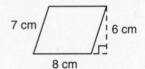

    (b) What is the area of this
parallelogram?

7. If each acute angle of a parallelogram measures 59°, then
(71) what is the measure of each obtuse angle?

8. The center of this circle is the origin.
(86) The circle passes through (2, 0).

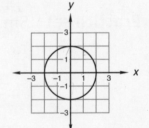

    (a) Estimate the area of the circle
in square units by counting
squares.

    (b) Calculate the area of the circle
by using 3.14 for π.

9. Which ratio forms a proportion with $\frac{2}{3}$?
(83)

    A. $\frac{2}{4}$ B. $\frac{3}{4}$ C. $\frac{4}{6}$ D. $\frac{3}{2}$

10. Complete this proportion: $\dfrac{6}{8} = \dfrac{a}{12}$
(85)

11. What is the perimeter of the
(103) hexagon at right? Dimensions are
in centimeters.

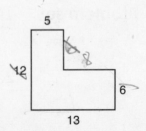

Complete the table to answer problems 12–14.

	FRACTION	DECIMAL	PERCENT
12. (99)	$\frac{3}{20}$	(a)	(b)
13. (99)	(a)	1.2	(b)
14. (99)	(a)	(b)	10%

15. Sharon bought a notebook for 40% off the regular price of
(41) $6.95. What was the sale price of the notebook?

16. Between which two consecutive whole numbers is $\sqrt{200}$?
(89)

17. Compare:
(92, Inv. 10)

   $\left(\frac{1}{2}\right)^3$ ◯ the probability of 3 consecutive "heads" coin tosses

18. Divide 0.624 by 0.05 and round the quotient to the
(49) nearest whole number.

19. The average of three numbers is 20. What is the sum of
(18) the three numbers?

20. Write the prime factorization of 450 using exponents.
(73)

21. −3 + −5 − −4 − +2
(104)

22. $3^4 + 5^2 \times 4 - \sqrt{100} \times 2^3$
(92)

23. How many blocks 1 inch on each edge would it take to
(82) fill a shoe box that is 12 inches long, 6 inches wide, and
5 inches tall?

24. Three fourths of the 60 athletes played in the game. How
(77) many athletes did not play?

25. The distance a car travels can be found by multiplying
(95) the **speed** of the car by the amount of **time** the car travels
at that speed. How far would a car travel in 4 hours at 88
kilometers per hour?

$$\frac{88 \text{ km}}{1 \text{ hr}} \times \frac{4 \text{ hr}}{1}$$

26. (a) What is the area of the shaded rectangle?
(31)

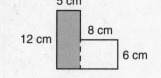

(b) What is the area of the unshaded rectangle?

(c) What is the combined area of the two rectangles?

27. Kobe measured the circumference and diameter of four
(18, 51) circles. Then he divided the circumference by the diameter of each circle to find the number of diameters in a circumference. Here are his answers:

$$3.12,\ 3.2,\ 3.15,\ 3.1$$

Find the average of Kobe's answers. Round the average to the nearest hundredth.

28. Norton was thinking of a two-digit counting number, and
(58) he asked Simon to guess the number. Describe how you can find the probability that Simon will guess correctly on the first try.

29. The coordinates of three vertices of a triangle are (3, 5),
(Inv. 7, 79) (–1, 5), and (–1, –3). What is the area of the triangle?

30. $\dfrac{2\ \text{gal}}{1} \times \dfrac{4\ \text{qt}}{1\ \text{gal}} \times \dfrac{2\ \text{pt}}{1\ \text{qt}}$
(95)

LESSON

105 Using Proportions to Solve Percent Problems

WARM-UP

Facts Practice: 24 Percent-Fraction-Decimal Equivalents (Test M)

Mental Math: Count by $\frac{1}{4}$'s from −2 to 2.

  a. 200 · 40   **b.** 567 − 150   **c.** 50% of 200

  d. $17.20 + $2.99   **e.** 7.5 ÷ 100   **f.** $\frac{440}{20}$

  g. 6 × 8, + 1, $\sqrt{}$, × 3, − 1, ÷ 2, × 10, − 1, ÷ 9

  h. What number is represented by the Roman numeral LXIV?

Problem Solving:

The numbers in these boxes form number patterns. What one number should be placed in both empty boxes to complete the patterns?

1	2	3
2	4	
3		9

NEW CONCEPT

We know that a percent can be expressed as a fraction with a denominator of 100.

$$30\% = \frac{30}{100}$$

A percent can also be regarded as a ratio in which 100 represents the total number in the group, as we show in the following example.

Example 1 Thirty percent of the paintings were portraits. If there were twelve portraits, how many paintings were there in all?

Solution We construct a ratio box. The ratio numbers we are given are 30 and 100. (We know from the word *percent* that 100 represents the ratio total.) The actual count we are given is 12. Our categories are "portraits" and "not portraits."

	Percent	Actual Count
Portraits	30	12
Not portraits		
Total	100	

Since the ratio total is 100, we calculate that the ratio number for "not portraits" is 70. We use n to stand for "not portraits" and t for "total" in the actual-count column. We use two rows from the table to write a proportion. Since we know both numbers in the "portraits" row, we use the numbers in the "portraits" row for the proportion. Since we want to find the total number of paintings, we also use the numbers from the "total" row. We will then solve the proportion using cross products.

	Percent	Actual Count
Portraits	30	12
Not portraits	70	n
Total	100	t

$$\frac{30}{100} = \frac{12}{t}$$

$$30t = 12 \cdot 100$$

$$t = \frac{\overset{4}{\cancel{12}} \cdot \overset{10}{\cancel{100}}}{\underset{\underset{1}{3}}{\cancel{30}}}$$

$$t = 40$$

We find that there were **40 paintings** in all.

In the above problem we did not need to use the 70% that were "not portraits." In the next example we will need to use the "not" percent in order to solve the problem.

Example 2 Only 40% of the team members played in the game. If 24 team members did not play, then how many did play?

Solution We construct a ratio box. The categories are "played," "did not play," and "total." Since 40% played, we calculate that 60% did not play. We are asked for the actual count of those who played. So we use the "played" row and the "did not play" row (because we know both numbers in that row) to write the proportion.

	Percent	Actual Count
Played	40	p
Did not play	60	24
Total	100	t

$$\frac{40}{60} = \frac{p}{24}$$

$$60p = 40 \cdot 24$$

$$p = \frac{\overset{4}{\cancel{40}} \cdot \overset{4}{\cancel{24}}}{\underset{\underset{1}{6}}{\cancel{60}}}$$

$$p = 16$$

We find that **16 team members** played in the game.

LESSON PRACTICE

Practice set Solve these percent problems using proportions. Make a ratio box for each problem.

a. Forty percent of the paintings were portraits. If 24 paintings were not portraits, how many paintings were there in all?

b. Seventy percent of the team members played in the game. If 21 team members played, how many team members did not play?

c. Referring to problem **b,** what proportion would we use to find the number of members on the team?

MIXED PRACTICE

Problem set **1.** How far would a car travel in $2\frac{1}{2}$ hours at 50 miles per hour?
(95)

$$\frac{50 \text{ mi}}{1 \text{ hr}} \times \frac{2\frac{1}{2} \text{ hr}}{1}$$

2. A map is drawn with the scale 1 inch = 2 miles. How
(95) many miles apart are two towns that are 3 inches apart on the map?

3. The ratio of humpback whales to killer whales was 2 to 7.
(88) If there were 28 killer whales, how many humpback whales were there?

4. When Robert measured a half-
(82) gallon box of ice cream, he found it had the dimensions shown in the illustration. What was the volume of the box in cubic inches?

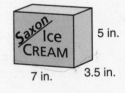

5 in.

7 in.   3.5 in.

5. Calculate mentally:
(100, 104)

(a) +10 + −10   (b) −10 − −10   (c) +6 + −5 − −4

6. On Earth a 1-kilogram object weighs about 2.2 pounds.
(102) Sid weighs 50 kilograms. About how many pounds does Sid weigh?

7. Sonia has only dimes and nickels in her coin jar; they
(101) are in a ratio of 3 to 5. If she has 120 coins in the jar, how many are dimes?

8. The airline sold 25% of the seats on the plane at
(105) discount. If 45 seats were sold at a discount, how many
seats were on the plane?

Complete the table to answer problems 9–11.

	Fraction	Decimal	Percent
9. *(99)*	$\frac{3}{50}$	(a)	(b)
10. *(99)*	(a)	0.04	(b)
11. *(99)*	(a)	(b)	150%

12. $4\frac{1}{12} + 5\frac{1}{6} + 2\frac{1}{4}$ **13.** $\frac{4}{5} \times 3\frac{1}{3} \times 3$
(61)   *(72)*

14. 0.125×80 **15.** $(1 + 0.5) \div (1 - 0.5)$
(39)   *(53)*

16. Solve: $\dfrac{c}{12} = \dfrac{3}{4}$
(85)

17. What is the total cost of an $8.75 purchase after 8% sales
(41) tax is added?

18. Write the decimal number one hundred five and five
(35) hundredths.

19. The measure of $\angle A$ in quadrilateral
(98) *ABCD* is 115°. What are the
measures of $\angle B$ and $\angle C$?

20. Write the prime factorization of 500 using exponents.
(73)

21. A quart is a little less than a liter, so a gallon is a little less
(78) than how many liters?

22. Diane will spin the spinner twice.
(Inv. 10) What is the probability that it will
stop in sector 2 both times?

23. The perimeter of this isosceles
(8) triangle is 18 cm. What is the
length of its longest side?

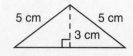

24. What is the area of the triangle in problem 23?
(79)

25. The temperature was –5°F at 6:00 a.m. By noon the
(14, 100) temperature had risen 12 degrees. What was the noontime temperature?

26. The weather report stated that the chance of rain is 30%.
(58) Use a decimal number to express the probability that it will not rain.

27. Find the perimeter of this figure.
(103) Dimensions are in inches.

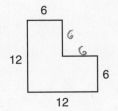

28. Study this function table and
(96) describe the rule that helps you find *y* if you know *x*.

x	y
$\frac{1}{2}$	$1\frac{1}{2}$
1	3
$1\frac{1}{2}$	$4\frac{1}{2}$
2	6

29. A room is 15 feet long and 12 feet wide.
(7, 31)

(a) The room is how many **yards** long and wide?

(b) What is the area of the room in square yards?

30. Ned rolled a dot cube and it turned up 6. If he rolls the
(58) dot cube again, what is the probability that it will turn up 6? Explain your answer.

LESSON

106   Two-Step Equations

WARM-UP

Facts Practice: Linear Measurement (Test K)

Mental Math: Count by 25's from −200 to 200.

    a. 40 · 600   **b.** 429 + 350   **c.** 25% of 200

    d. $60.00 − $59.45   **e.** 1.2 × 100   **f.** 60 × 12

    g. Square 5, − 1, ÷ 4, × 5, + 2, ÷ 4, × 3, + 1, $\sqrt{}$

    h. What number is represented by the Roman numeral XLIX?

Problem Solving:

Copy this problem and fill in the missing digits. (Assume the addends are reduced.)

$$\frac{\square}{4} + \frac{\square}{6} = \frac{11}{12}$$

NEW CONCEPT

Since Lessons 3 and 4 we have solved one-step equations in which we look for a missing number in addition, subtraction, multiplication, or division. In this lesson we will begin solving two-step equations that involve more than one operation.

Example   Solve: $3n - 1 = 20$

Solution   Let us think about what this equation means. When 1 is subtracted from $3n$, the result is 20. So $3n$ equals 21.

$$3n = 21$$

Since $3n$ means "3 times n" and $3n$ equals 21, we know that n equals 7.

$$n = 7$$

We show our work this way:

$$3n - 1 = 20$$
$$3n = 21$$
$$n = 7$$

We check our answer this way:

$$3(7) - 1 = 20$$

$$21 - 1 = 20$$

$$20 = 20$$

LESSON PRACTICE

Practice set   Solve each equation, showing the steps of the solution. Then check your answer.

    a. $3n + 1 = 16$ **b.** $2x - 1 = 9$

    c. $3y - 2 = 22$ **d.** $5m + 3 = 33$

    e. $4w - 1 = 35$ **f.** $7a + 4 = 25$

MIXED PRACTICE

Problem set   **1.** The average of three numbers is 20. If the greatest is 28
    (18) and the least is 15, what is the third number?

    2. A map is drawn to the scale of 1 inch = 10 miles. How
    (95) many miles apart are two points that are $2\frac{1}{2}$ inches apart
    on the map?

$$\frac{2\frac{1}{2} \text{ in.}}{1} \times \frac{10 \text{ mi}}{1 \text{ in.}}$$

    3. What number is one fourth of 360?
    (29)

    4. What percent of a quarter is a nickel?
    (75)

    5. If you draw one card from a normal deck of 52 cards,
    (58) what is the probability that the card will be a face card
    (jack, queen, or king)?

    Solve and check:

    6. $8x + 1 = 25$ **7.** $3w - 5 = 25$
    (106)   *(106)*

    8. Calculate mentally:
    (100, 104)
    (a) $-15 + +20$

    (b) $-15 - +20$

    (c) $(-3) + (-2) - (-1)$

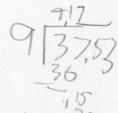

9. A sign in the elevator says that the maximum load is
(102) 4000 pounds. How many tons is 4000 pounds?

10. One gallon minus one quart equals how many pints?
(78)

11. The ratio of kangaroos to koalas was 9 to 5. If there were
(88) 414 kangaroos, how many koalas were there?

Complete the table to answer problems 12–14.

	FRACTION	DECIMAL	PERCENT
12. (99)	$\frac{1}{8}$	(a)	(b)
13. (99)	(a)	1.8	(b)
14. (99)	(a)	(b)	3%

15. $8\frac{1}{3} - 3\frac{1}{2}$ **16.** $2\frac{1}{2} \div 100$
(63)   (68)

17. $0.014 \div 0.5$
(49)

18. Write the standard notation for the following:
(92)

$$(6 \times 10^4) + (9 \times 10^2) + (7 \times 10^0)$$

19. The prime factorization of one hundred is $2^2 \cdot 5^2$. The
(73) prime factorization of one thousand is $2^3 \cdot 5^3$. Write the
prime factorization of one million using exponents.

20. A 1-foot ruler broke into two pieces so that one piece was
(63) $5\frac{1}{4}$ inches long. How long was the other piece?

21. $6 + 3^2(5 - \sqrt{4})$
(92)

22. If each small block has a volume of
(82) one cubic centimeter, what is the
volume of this rectangular prism?

23. Three inches is what percent of a foot?
(75)

24. (a) What is the area of the shaded rectangle?
⁽³¹⁾

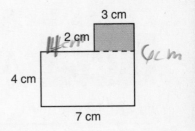

(b) What is the area of the unshaded rectangle?

(c) What is the combined area of the two rectangles?

25. What is the perimeter of the hexagon in problem 24?
⁽¹⁰³⁾

26. The diameter of each tire on Jan's bike is two feet. The circumference of each tire is closest to which of the following? (Use 3.14 for π.)
⁽⁴⁷⁾

A. 6 ft   B. 6 ft 3 in.   C. 6 ft 8 in.   D. 7 ft

27. What is the area of this triangle?
⁽⁷⁹⁾

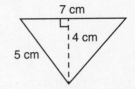

This table shows the number of miles Celina rode her bike each day during the week. Use this information to answer problems 28–30.

28. If the data were rearranged in order of distance (with 3 miles listed first and 10 miles listed last), then which distance would be in the middle of the list?
^(Inv. 5)

Miles of Bike Riding for the Week

Day	Miles
Sunday	7
Monday	3
Tuesday	6
Wednesday	10
Thursday	5
Friday	4
Saturday	7

29. What was the average number of miles Celina rode each day?
⁽¹⁸⁾

30. Write a comparison question that relates to the table, and then answer the question.
⁽¹³⁾

LESSON

107   Area of Complex Shapes

WARM-UP

Facts Practice: 24 Percent-Fraction-Decimal Equivalents (Test M)

Mental Math: Count by $\frac{1}{3}$'s from −2 to 2.

   a. 100 · 100   **b.** 376 − 150   **c.** 10% of 200

   d. $12.89 + $9.99   **e.** 6.0 ÷ 100   **f.** $\frac{360}{60}$

   g. 10 × 6, + 4, $\sqrt{\ }$, × 3, + 1, $\sqrt{\ }$, × 7, + 1, $\sqrt{\ }$

   h. What number is represented by the Roman numeral CCL?

Problem Solving:

On the first two tests Jason's average score was 80. On the next three tests his average score was 90. What was Jason's average score on his first five tests?

NEW CONCEPT

In Lesson 103 we found the perimeter of complex shapes. In this lesson we will practice finding the area of complex shapes. One way to find the area of a complex shape is to divide the shape into two or more parts, find the area of each part, and then add the areas. Think of how the shape below could be divided into two rectangles.

Example   Find the area of this figure.

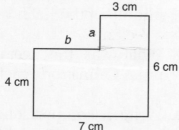

Solution   We will show two ways to divide this shape into two rectangles. We use the skills we learned in Lesson 103 to find that side *a* is 2 cm and side *b* is 4 cm. We extend side *b* with a dashed line segment to divide the figure into two rectangles.

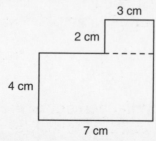

The length and width of the smaller rectangle are 3 cm and 2 cm, so its area is 6 cm^2. The larger rectangle is 7 cm by 4 cm, so its area is 28 cm^2. We find the combined area of the two rectangles by adding.

$$6 \text{ cm}^2 + 28 \text{ cm}^2 = \textbf{34 cm}^2$$

A second way to divide the figure into two rectangles is to extend side a.

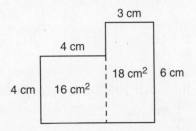

Extending side a forms a 4-cm by 4-cm rectangle and a 3-cm by 6-cm rectangle. Again we find the combined area of the two rectangles by adding.

$$16 \text{ cm}^2 + 18 \text{ cm}^2 = 34 \text{ cm}^2$$

Either way we divide the figure, we find that its area is 34 cm^2.

LESSON PRACTICE

Practice set*   **a.** Draw two ways to divide this figure into two rectangles. Then find the area of the figure each way.

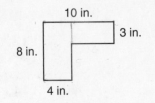

b. This trapezoid can be divided into a rectangle and a triangle. Find the area of the trapezoid.

MIXED PRACTICE

Problem set   **1.** If the divisor is eight tenths and the dividend is forty-
(49)   eight hundredths, what is the quotient?

2. The plans for the clubhouse were drawn so that 1 inch
(95)   equaled 2 feet. In the plans the clubhouse was 4 inches tall. The actual clubhouse will be how tall?

3. If all the king's horses total 600 and all the king's men
(23)   total 800, then what is the ratio of men to horses?

4. What percent of the perimeter of a regular pentagon is the
(8, 60)   length of one side?

5. The mass of a dollar bill is about one gram. A gram is
(102) what fraction of a kilogram?

6. Calculate mentally:
(100, 104)

(a) $+15 + -10$

(b) $-15 - -10$

(c) $(+3) + (-5) - (-2) - (+4)$

7. $10^3 - (10^2 - \sqrt{100}) - 10^3 \div 10^2$
(92)

8. Complete this proportion: $\dfrac{6}{n} = \dfrac{8}{12}$
(85)

Complete the table to answer problems 9–11.

	Fraction	Decimal	Percent
9. (99)	$1\frac{1}{10}$	(a)	(b)
10. (99)	(a)	0.45	(b)
11. (99)	(a)	(b)	80%

12. $5\dfrac{3}{8} + 4\dfrac{1}{4} + 3\dfrac{1}{2}$ **13.** $\dfrac{8}{3} \cdot \dfrac{5}{12} \cdot \dfrac{9}{10}$
(61)   (72)

14. $64.8 + 8.42 + 24$
(38)

15. If one acute angle of a right triangle measures 55°, then
(93, 98) what is the measure of the other acute angle?

16. How many ounces is one half of a pint?
(78)

17. Solve and check: $3m + 8 = 44$
(106)

18. Write one hundred ten million in expanded notation
(92) using exponents.

19. What is the greatest common factor of 30 and 45?
(20)

20. A square with sides 1 inch long is
(29, 31) divided into $\frac{1}{2}$-by-$\frac{1}{4}$-inch rectangles.

(a) What is the area of each $\frac{1}{2}$-by-$\frac{1}{4}$-in. rectangle?

(b) What fraction of the square is shaded?

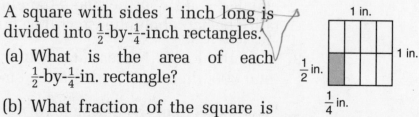

21. How many blocks with edges 1 foot long would be needed
(82) to fill a cubical box with edges 1 yard long?

22. $0.3n = \$6.39$
(49)

23. What is the perimeter of this
(103) hexagon?

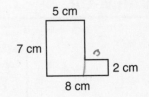

24. Divide the hexagon at right into two
(107) rectangles. What is the combined
area of the two rectangles?

25. This trapezoid has been divided
(107) into two triangles. Find the area of
the trapezoid by adding the areas
of the two triangles.

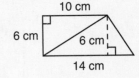

Alberto earned the following scores on his quizzes. Use this
information to answer problems 26 and 27.

$$8, 9, 7, 8, 10, 9, 8, 7, 10$$

26. (a) Which score did Alberto earn most often? That is,
(Inv. 5)   what is the mode of the scores?

(b) If the scores were arranged in order from lowest to
highest, what would be the middle score? That is,
what is the median score?

27. Find Alberto's average quiz score. Round the answer to
(18, 51) the nearest tenth.

28.   12 lb 3 oz
(102)   − 8 lb 7 oz

29. Use a compass to draw a circle that has a diameter of 10 cm.
(47)
(a) What is the radius of the circle?

(b) Calculate the circumference of the circle. (Use 3.14
for π.)

30. $\dfrac{10 \text{ gallons}}{1} \times \dfrac{31.5 \text{ miles}}{1 \text{ gallon}}$
(95)

LESSON

108 Transformations

WARM-UP

Facts Practice: 24 Percent-Fraction-Decimal Equivalents (Test M)

Mental Math: Count by $\frac{1}{16}$'s from $\frac{1}{16}$ to 1.

a. $70 \cdot 70$ **b.** $296 - 150$ **c.** 25% of $20

d. $8.23 + $8.99   **e.** $75 \div 100$ **f.** $\frac{800}{40}$

g. $8 \times 8, -4, \div 2, +5, \div 5, \times 8, -1, \div 5, \times 2, -1, \div 3$

h. What number is represented by the Roman numeral CCXXII?

Problem Solving:

These two triangles are congruent. The perimeter of each triangle is 12 cm. What is the area of each triangle?

NEW CONCEPT

Recall from Lesson 60 that two figures are **congruent** if one figure has the same shape and size as the other figure. One way to determine whether two figures are congruent is to position one figure "on top of" the other. The two triangles below are congruent. As we will see below, triangle *ABC* can be positioned "on top of" triangle *XYZ*, illustrating that it is congruent to triangle *XYZ*.

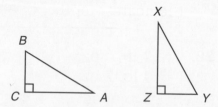

To position triangle *ABC* on triangle *XYZ*, we make three different kinds of moves. First, we **rotate** (turn) triangle *ABC* 90° counterclockwise.

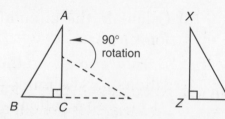

Second, we **translate** (slide) triangle *ABC* to the right so that side *AC* aligns with side *XZ*.

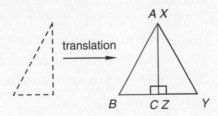

Third, we **reflect** (flip) triangle *ABC* so that angle *B* is positioned on top of angle *Y*.

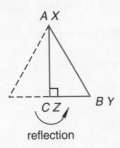

The three different kinds of moves we made are called **transformations**. We list them in the following table:

Transformations

Name	Movement
Rotation	turning a figure about a certain point
Translation	sliding a figure in one direction without turning the figure
Reflection	reflecting a figure as in a mirror or "flipping" a figure over a certain line

Activity: *Transformations*

Materials needed:

- scissors
- pencil and paper

Cut out a pair of congruent triangles as follows:

Step 1: Fold a piece of paper in half.

Step 2: Draw a triangle on the folded paper.

Step 3: While the paper is folded, cut out the triangle so that two triangles are cut out at the same time.

Place the two triangles on a desk or table so that the triangles are apart and in different orientations. Move one of the triangles until it is positioned on top of the other triangle. The moves permitted are rotation, translation, and reflection. Take the moves one at a time and describe them as you go. After successfully aligning the triangles, place the triangles in different starting orientations and repeat the procedure.

LESSON PRACTICE

Practice set   For problems **a–e,** what transformation(s) could be used to position triangle I on triangle II?

a.

b.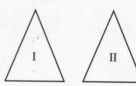

c.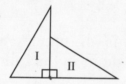

d.

e.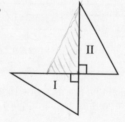

MIXED PRACTICE

Problem set

1. What is the sum of the first five positive even numbers?
(10)

2. The team's win-loss ratio is 4 to 3. If the team has won 12 games, how many games has the team lost?
(88)

3. Ginger answered 80% of the questions correctly. If she missed 5 questions, how many questions did she answer correctly?
(105)

4. Pete had to clean the inside of every porthole on the ship. If the diameter of each porthole was 12 inches, then what was the area of glass Pete had to clean on each porthole? Round the answer to the nearest square inch. (Use 3.14 for π.)
(86)

5. Three eighths of the 48 band members played woodwinds.
(77) How many woodwind players were in the band?

6. What is the least common multiple (LCM) of 6, 8, and 12?
(30)

7. Triangles I and II are congruent. Describe the
(108) transformations that would position triangle I on triangle II.

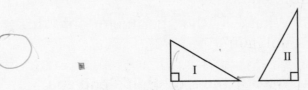

8. Complete this proportion: $\dfrac{7}{20} = \dfrac{n}{100}$
(85)

Complete the table to answer problems 9–11.

	FRACTION	DECIMAL	PERCENT
9. (99)	$1\frac{1}{10}$	(a)	(b)
10. (99)	(a)	0.45	(b)
11. (99)	(a)	(b)	80%

12. $4\dfrac{3}{4} + \left(2\dfrac{1}{4} - \dfrac{7}{8}\right)$
(63)

13. $1\dfrac{1}{5} \div \left(2 \div 1\dfrac{2}{3}\right)$
(68)

14. $6.2 + (9 - 2.79)$
(38)

15. $-3 + +7 + -8 - -1$
(104)

16. Find 6% of $2.89. Round the product to the nearest cent.
(41)

17. What fraction of a meter is a millimeter?
(95)

18. Arrange these numbers in order from least to greatest:
(44)

$$0.3,\ 0.31,\ 0.305$$

19. If each edge of a cube is 10 centimeters long, then its
(82) volume is how many cubic centimeters?

20. $2^5 - 5^2 + \sqrt{25} \times 2$
(92)

21. Solve and check: $8a - 4 = 60$
(106)

22. Acute angle *a* is one third of a
(28, Inv. 3) right angle. What is the measure
of angle *a*?

Refer to the figure at right to answer
problems 23 and 24. Dimensions are in
millimeters.

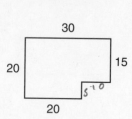

23. What is the perimeter of this
(103) polygon?

24. What is the area of this polygon?
(107)

25. A pint of water weighs about one pound. About how
(78) much does a two-gallon bucket of water weigh?
(Disregard the weight of the bucket.)

26. Recall that a trapezoid is a
(107) quadrilateral with two parallel
sides. The parallel sides of this
trapezoid are 10 mm apart. The
trapezoid is divided into two
triangles. What is the area of the
trapezoid?

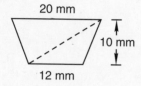

27. A cubic container that is 10 cm
(78) long, 10 cm wide, and 10 cm deep
can contain one liter of water. One
liter is how many milliliters?

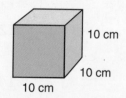

28. A bag contains 6 red marbles and 4 blue marbles. If Delia
(Inv. 10) draws one marble from the bag and then draws another
marble without replacing the first, what is the probability
that both marbles will be red?

29. One and one half kilometers is how many meters?
(95)

30. On a coordinate plane draw a segment from (−3, 4) to
(28, Inv. 7) (1, −1). Draw another segment from (1, −1) to (5, −1).
Together, the segments form what type of angle?

LESSON
109
Corresponding Parts • Similar Triangles

Facts Practice: Measurement Facts (Test N)

Mental Math: Count down by 25's from 200 to −200.

    a. 400 · 30   **b.** 687 + 250   **c.** 10% of $20

    d. $10.00 − $6.87   **e.** 0.5 × 100   **f.** 70 × 300

    g. Square 7, + 1, ÷ 2, × 3, − 3, ÷ 8, $\sqrt{}$

    h. What number is represented by the Roman numeral CXC?

Problem Solving:

One state uses a license plate that contains two letters followed by four digits. How many license plates are possible if all of the letters and digits are used?

NEW CONCEPTS

Corresponding parts

The two triangles below are congruent. Each triangle has three angles and three sides. The angles and sides of triangle *ABC* **correspond** to the angles and sides of triangle *XYZ*.

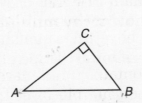 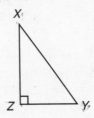

By rotating, translating, and reflecting triangle *ABC*, we could position it on top of triangle *XYZ*. Then their **corresponding parts** would be in the same place.

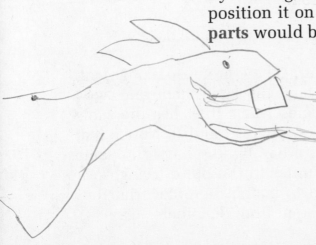

                   ∠*A* corresponds to ∠*X*.

                   ∠*B* corresponds to ∠*Y*.

                   ∠*C* corresponds to ∠*Z*.

                   $\overline{AB}$ corresponds to $\overline{XY}$.

                   $\overline{BC}$ corresponds to $\overline{YZ}$.

                   $\overline{AC}$ corresponds to $\overline{XZ}$.

If two figures are congruent, their corresponding parts are congruent. So the measures of the corresponding parts are equal.

Example 1 These triangles are congruent. What is the perimeter of each?

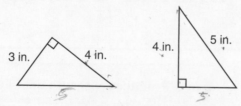

Solution We will rotate the triangle on the left so that the corresponding parts are easier to see.

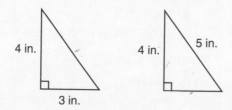

Now we can more easily see that the unmarked side on the left-hand triangle corresponds to the 5-inch side on the right-hand triangle. Since the triangles are congruent, the measures of the corresponding parts are equal. So each triangle has sides that measure 3 inches, 4 inches, and 5 inches. Adding, we find that the perimeter of each triangle is **12 inches.**

$$3 \text{ in.} + 4 \text{ in.} + 5 \text{ in.} = 12 \text{ in.}$$

Similar triangles Figures that have the same shape but are not necessarily the same size are **similar.** Three of these four triangles are similar:

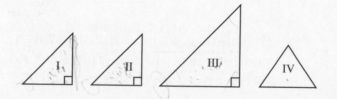

Triangles I and II are similar. They are also congruent. (Remember, congruent figures have the same shape *and* size.) Triangle III is similar to triangles I and II. It has the same shape but not the same size as triangles I and II. Notice that the corresponding angles of similar figures have the same measure. Triangle IV is not similar to the other triangles. It cannot be reduced or enlarged to match the other triangles. The shape of triangle IV is different from the others.

Example 2   The two triangles below are similar. What is the measure of angle *A*?

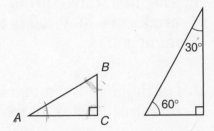

Solution   We will rotate and reflect triangle *ABC* so that the corresponding angles are easier to see.

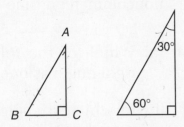

We see that angle *A* in triangle *ABC* corresponds to the 30° angle in the similar triangle. Since corresponding angles of similar triangles have the same measure, the measure of angle *A* is **30°**.

LESSON PRACTICE

Practice set   **a.** "All squares are similar." True or false?

b. "All similar triangles are congruent." True or false?

c. "If two polygons are similar, then their corresponding angles are equal in measure." True or false?

d. These two triangles are congruent. Which side of triangle *PQR* is the same length as $\overline{AB}$?

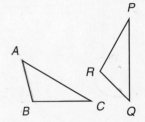

e. Which of these two triangles appear to be similar?

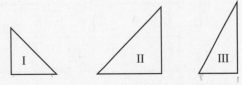

MIXED PRACTICE

Problem set

1. The first three prime numbers are 2, 3, and 5. Their
(19) product is 30. What is the product of the next three prime
numbers?

2. On the map 2 cm equals 1 km. What is the actual length
(95) of a street that is 10 cm long on the map?

3. Between 8 p.m. and 9 p.m. the station broadcast 8 minutes
(23) of commercials. What was the ratio of commercial time to
noncommercial time during that hour?

4. (a) Which of the following triangles appears to have a
(93)   right angle as one of its angles?

(b) In which triangle do all three sides appear to be the
same length?

A.   B.    C.

5. If the two acute angles of a right triangle are congruent,
(28, 98) then what is the measure of each acute angle?

6. What is the probability of drawing a king or a queen from
(58) a regular deck of 52 cards?

Solve:

7. $7w - 3 = 60$
(106)

8. $\dfrac{8}{n} = \dfrac{4}{25}$
(85)

Complete the table to answer problems 9–11.

	FRACTION	DECIMAL	PERCENT
9. (99)	$\frac{5}{8}$	(a)	(b)
10. (99)	(a)	1.25	(b)
11. (99)	(a)	(b)	70%

12. (a) If the spinner is spun once, what
(58)   is the probability that it will stop
       on a number less than 4?

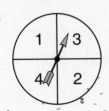

(b) If the spinner is spun 100 times,
    how many times would it be
    expected to stop on a prime
    number?

13. Convert 200 centimeters to meters by completing this
(95) multiplication:

$$\frac{200 \text{ cm}}{1} \cdot \frac{1 \text{ m}}{100 \text{ cm}}$$

14. $(6.2 + 9) - 2.79$ **15.** $10^3 \div 10^2 - 10^1$
(38)   (92)

16. Create a function table for x and y. In your table, record
(96) four pairs of numbers that follow this rule: y is twice x.

17. Write the fraction $\frac{2}{3}$ as a decimal number rounded to the
(74) hundredths place.

18. The Zamoras rent a storage room that is 10 feet wide,
(82) 12 feet long, and 8 feet high. How many cube-shaped boxes
       1 foot on each edge can the Zamoras store in the room?

19. Explain why figures that are congruent are also similar.
(109)

20. $0.12m = \$4.20$
(49)

21. Calculate mentally:
(100)   (a) $+7 + -8$ (b) $-7 + +8$

        (c) $-7 - +8$ (d) $-7 - -8$

Triangles I and II below are congruent. Refer to the triangles
to answer problems 22 and 23.

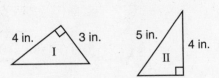

22. What is the area of each triangle?
(109)

23. Name the transformations that would position triangle I
(108) on triangle II.

24. This trapezoid has been divided
(107) into a rectangle and a triangle.
What is the area of the trapezoid?

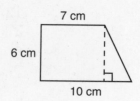

6 cm

25. Write the length of the segment below in
(7, 50) (a) millimeters.

(b) centimeters.

26. The triangles below are similar. What is the measure of
(109) angle *A*?

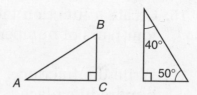

27. The ratio of peanuts to cashews in the mix was 9 to 2.
(88) Horace counted 36 cashews in all. How many peanuts
were there?

28. A soup label must be long enough to wrap around a can. If
(47) the diameter of the can is 7 cm, then the label must be at
least how long? Round up to the nearest whole number.

29. $6\frac{2}{3} \div 100$
(68)

30. Compare: $\left(\frac{1}{10}\right)^2 \bigcirc 0.01$
(92)

LESSON

110 Symmetry

WARM-UP

Facts Practice: 24 Percent-Fraction-Decimal Equivalents (Test M)

Mental Math: Count up and down by $\frac{1}{8}$'s between $\frac{1}{8}$ and 2.

a. 90 · 90
b. 726 − 250
c. 50% of $50
d. $7.62 + $3.98
e. 8 ÷ 100
f. $\frac{350}{50}$
g. Square 10, − 1, ÷ 9, × 3, − 1, ÷ 4, × 7, + 4, ÷ 3
h. What number is represented by the Roman numeral CXCIX?

Problem Solving:

Copy this problem and fill in the missing digits. Use only zeros or ones in the blanks.

$$
\begin{array}{r}
9\_\ \\
\_\_\overline{)\_\_\_\_} \\
99 \\
\_\_ \\
== \\
\_
\end{array}
$$

NEW CONCEPT

A figure is **symmetrical** if it can be divided in half so that the halves are mirror images of each other. We can observe symmetry in nature. For example, butterflies, leaves, and most types of fish are symmetrical. In many respects our bodies are also symmetrical. Manufactured items are sometimes designed with symmetry (as we see in many lamps, chairs, and kitchen sinks, for example). Two-dimensional figures can also be symmetrical. A two-dimensional figure is symmetrical if a line can divide the figure into two mirror images. Line r divides the triangle below into two mirror images. Thus the triangle is symmetrical, and line r is called a **line of symmetry.**

Example 1 This rectangle has how many lines of symmetry?

Solution There are two ways to divide the rectangle into mirror images:

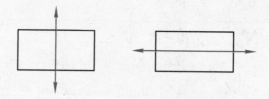

We see that the rectangle has **2 lines of symmetry.**

Example 2 Which of these triangles does not appear to be symmetrical?

A.   B.   C.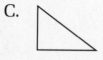

Solution We check each triangle to see whether we can find a line of symmetry. In choice A all three sides of the triangle are the same length. We can find three lines of symmetry in the triangle.

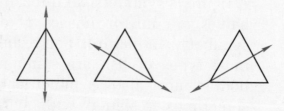

In choice B two sides of the triangle are the same length. The triangle in choice B has one line of symmetry.

In choice C each side of the triangle is a different length. The triangle has no line of symmetry, so the triangle is not symmetrical. The answer is **C.**

LESSON PRACTICE

Practice set   **a.** Draw four squares. Then draw a different line of symmetry for each square.

b. All but one of these letters can be drawn to have a line of symmetry. Which letter is that?

A   B   C   D   E   F

MIXED PRACTICE

Problem set   **1.** When the greatest four-digit number is divided by the
   (2)   greatest two-digit number, what is the quotient?

2. The ratio of the length to the width of a rectangle is 3 to 2.
   (88)   If the width is 60 mm, what is the length?

3. A box of saltine crackers in the
   (82)   shape of a square prism had a
   length, width, and height of
   4 inches, 4 inches, and 10 inches
   respectively. How many cubic
   inches was the volume of the box?

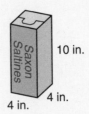

10 in.

4 in.   4 in.

4. A full turn is 360°. How many degrees is $\frac{1}{6}$ of a turn?
   (90)

Refer to these triangles to answer problems 5–7:

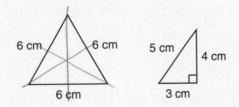

6 cm   6 cm   5 cm   4 cm

6 cm   3 cm

5. The three sides of an equilateral triangle are equal in
   (110)   length. The equilateral triangle has how many lines of
   symmetry?

6. What is the perimeter of the equilateral triangle?
(93)

7. What is the area of the right triangle?
(93)

8. Draw a ratio box for this problem. Then solve the
(105) problem using a proportion.

> *Ms. Smith noted that there were 12 girls among her piano students and that 40% of the students were boys. How many piano students did Ms. Smith teach?*

Complete the table to answer problems 9–11.

	FRACTION	DECIMAL	PERCENT
9. (99)	$2\frac{3}{4}$	(a)	(b)
10. (99)	(a)	1.1	(b)
11. (99)	(a)	(b)	64%

12. $24\frac{1}{6} + 23\frac{1}{3} + 22\frac{1}{2}$
(61)

13. $\left(1\frac{1}{5} \div 2\right) \div 1\frac{2}{3}$
(68)

14. $9 - (6.2 + 2.79)$
(38)

15. $0.36m = \$63.00$
(49)

16. Find 6.5% of $24.89 by multiplying 0.065 by $24.89.
(51) Round the product to the nearest cent.

17. Round the quotient to the nearest thousandth:
(51)

$$0.065 \div 4$$

18. Write the prime factorization of 1000 using exponents.
(73)

19. "All squares are similar." True or false?
(109)

20. $3^3 - 3^2 \div 3 - 3 \times 3$
(92)

21. What is the perimeter of this
(103) polygon?

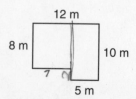

22. What is the area of this polygon?
(107)

Triangles I and II are congruent. Refer to these triangles to answer problems 23 and 24.

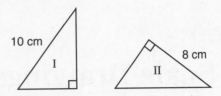

23. Name the transformations that would position triangle I
(108) on triangle II.

24. The perimeter of each triangle is 24 cm. What is the
(8, 109) length of the shortest side of each triangle?

25. Mae-Ying rolled an old automobile tire down the road. If
(47) the tire had a diameter of two feet, how far down the road would the tire roll on each complete turn? (Use 3.14 for π.)

26. Use a ruler to draw $\overline{AB}$ $1\frac{3}{4}$ inches long. Then draw a dot at
(7) the midpoint of $\overline{AB}$, and label the point M. How long is $\overline{AM}$?

27. Use a compass to draw a circle on a coordinate plane.
(27, Inv. 7) Make the center of the circle the origin, and make the radius five units. At which two points does the circle cross the x-axis?

28. What is the area of the circle in problem 27? (Use 3.14 for π.)
(86)

29. $-3 + -4 - -5 - +7$
(104)

30. If Freddy tosses a coin four times, what is the
(Inv. 10) probability that the coin will turn up heads, tails, heads, tails in that order?

INVESTIGATION 11

Focus on

Scale Drawings and Models

Before a building is constructed, architects create **scale drawings** to guide the construction. Sometimes, a **scale model** of the building is also constructed to show the appearance of the finished project.

Scale drawings, such as architectural plans, are two-dimensional representations of larger objects. Scale models, such as model cars and action figures, are three-dimensional representations of larger objects. In some cases, however, a scale drawing or model represents an object smaller than the model itself. For example, we might want to construct a large model of a bee in order to more easily portray its anatomy.

In scale drawings and models the **legend** gives the relationship between a unit of length in the drawing and the actual measurement that the unit represents. The drawing below shows the floorplan of Angelo's studio apartment. The legend for this scale drawing is $\frac{1}{2}$ inch = 5 feet.

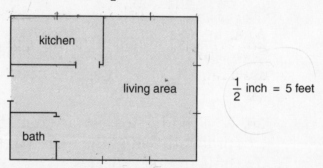

If we measure the scale drawing above, we find that it is 2 inches long and $1\frac{1}{2}$ inches wide. Using these measurements, we can determine the actual dimensions of Angelo's apartment. Below, we show some relationships that are based on the scale drawing's legend.

$$\frac{1}{2} \text{ inch} = 5 \text{ feet (given)}$$
$$1 \text{ inch} = 10 \text{ feet}$$
$$1\frac{1}{2} \text{ inches} = 15 \text{ feet}$$
$$2 \text{ inches} = 20 \text{ feet}$$

Since the scale drawing is 2 inches long by $1\frac{1}{2}$ inches wide, we find that Angelo's apartment is 20 feet long by 15 feet wide.

1. What are the actual length and width of Angelo's kitchen?

2. In the scale drawing each doorway measures $\frac{1}{4}$ inch wide. Since $\frac{1}{4}$ inch is half of $\frac{1}{2}$ inch, what is the actual width of each doorway in Angelo's apartment?

3. A dollhouse was built as a scale model of an actual house using 1 in. to represent 1.5 ft. What are the dimensions of a room in the actual house if the corresponding dollhouse room measures 8 in. by 10 in.? What is the area of the room in the actual house?

4. A scale model of an airplane is built using 1 inch to represent 2 feet. The wingspan of the model airplane is 24 inches. What is the wingspan of the actual airplane in feet?

The lengths of corresponding parts of scale drawings or models and the objects they represent are proportional. For example, the distance between New York City and Washington, D.C., is 197 miles. Suppose that on a map the distance between those two cities is 5 inches. Also suppose that the map distance between New York City and Cleveland measures 10 inches. What is the actual distance between New York City and Cleveland? Since the relationships are proportional, we can use a ratio box to sort out the numbers.

	Map Distance	Actual Distance
New York to Washington	5	197
New York to Cleveland	10	m

Boring

(very, Boring)

From the box we write this proportion:

$$\frac{5}{10} = \frac{197}{m}$$

We will solve this proportion using cross products. We first notice that the ratio $\frac{5}{10}$ reduces to $\frac{1}{2}$, so we can solve this proportion:

$$\frac{1}{2} = \frac{197}{m}$$

$$m = 2(197)$$

$$m = 394$$

So the actual distance between New York City and Cleveland is 394 miles.

To answer problems 5–8, first set up a proportion. Then solve for the unknown measurement either by using cross products or by writing an equivalent ratio.

5. A scale model of a sports car is 7 inches long. The car itself is 14 feet long. If the model is 3 inches wide, how wide is the actual car?

6. For the sports car in problem 5, suppose the actual height is 4 feet. What is the height of the model?

7. The femur is the large bone that runs from the knee to the hip. In a scale drawing of a human skeleton the length of the femur measures 3 cm, while the full skeleton measures 12 cm. If the scale drawing represents a 6-ft-tall person, what is the actual length of the person's femur?

8. The humerus is the bone that runs from the elbow to the shoulder. Suppose the humerus of a 6-ft-tall person is 1 ft long. How long should the humerus be on the scale drawing of the skeleton in problem 7?

9. A scale drawing of a room addition that measures 28 ft by 16 ft is shown below. The scale drawing measures 7 cm by 4 cm.

   (a) Complete this legend for the scale drawing:

   $$1 \text{ cm} = \underline{\quad\quad} \text{ ft}$$

   (b) Estimate the actual length and width of the bathroom, rounding to the nearest foot.

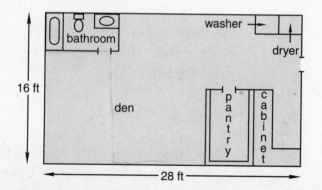

10. A natural history museum contains a 44-inch-long scale model of a *Stegosaurus* dinosaur. The actual length of the *Stegosaurus* was 22 feet. What should be the legend for the scale model of the dinosaur?

$$1 \text{ inch } = \underline{\hspace{2cm}} \text{ feet}$$

Maps, blueprints, and models are called *renderings*. Scale drawings are two-dimensional renderings, and scale models are three-dimensional renderings. If a rendering is smaller than the actual object it represents, then the dimensions of the rendering are a fraction of the dimensions of the actual object. This fraction is called the **scale** of the rendering.

To determine the scale of a rendering, we form a fraction with a dimension of the model as the numerator and the corresponding dimension of the actual object as the denominator. Then we reduce. (*Note:* To determine scale, we must use the same units for both the model and the actual object.)

$$\text{scale} = \frac{\text{dimension of rendering}}{\text{dimension of object}}$$

In the case of the *Stegosaurus* in problem 10, the corresponding lengths are 44 inches and 22 feet. Before reducing the fraction, we will convert 22 feet to inches.

$$\frac{22 \text{ feet}}{1} \times \frac{12 \text{ inches}}{1 \text{ foot}} = 264 \text{ inches}$$

Now we form a fraction, using the length of the model as the numerator and the dinosaur's actual length as the denominator.

$$\text{scale} = \frac{44 \text{ inches}}{22 \text{ feet}} = \frac{44 \text{ inches}}{264 \text{ inches}} = \frac{1}{6}$$

So the model is a $\frac{1}{6}$ scale model. The reciprocal of the scale is the **scale factor.** So the scale factor from the model to the actual *Stegosaurus* is 6. This means we can multiply any dimension of the model by 6 to determine the corresponding dimension of the actual object.

11. What is the scale of the model car in problem 5? What is the scale factor?

12. A scale may be written as a ratio that uses a colon. For example, we can write the scale of the *Stegosaurus* model as 1:6. Suppose that a toy company makes action figures of sports stars using a scale of 1:10. How many inches tall will a figure of a 6-ft-8-in. basketball player be?

13. In a scale drawing of a wall mural, the scale factor is 6. If the scale drawing is 3 feet long by 1.5 feet wide, what are the dimensions of the actual mural?

Extensions

a. Make a scale drawing of your bedroom's floor plan, where 1 in. = 3 ft. Include in your drawing the locations of doors and windows as well as major pieces of furniture. What is the scale factor you used?

b. Cut out and assemble the pieces from Activity Sheets 17 and 18 (available in *Saxon Math 7/6—Homeschool Tests and Worksheets*) to make a scale model of the *Freedom 7* spacecraft. This spacecraft was piloted by Alan B. Shepard, the first American to go into space. With *Freedom 7* sitting atop a rocket, Shepard blasted off from Cape Canaveral, Florida, on May 5, 1961. Because he did not orbit (circle) the earth, the trip lasted only 15 minutes before splashdown in the Atlantic Ocean. Shepard was one of six astronauts to fly Mercury spacecraft like *Freedom 7*. Each Mercury spacecraft could carry only one astronaut, because the rockets available in the early 1960's were not powerful enough to lift heavier loads.

Your completed *Freedom 7* model will have a scale of 1:24. After you have constructed the model, measure its length. Use this information to determine the length of an actual Mercury spacecraft.

c. Find a 3-by-5-inch photograph and a picture frame that is larger than the photograph. To what size should the photograph be enlarged so that the enlargement is at least as large as the opening in the frame? How much must be cut from the enlargement so that it fits the opening exactly?

d. Using plastic straws (or wooden skewers), scissors, and glue, make a scale model of a triangular prism whose base has edges of 3 ft, 4 ft, and 5 ft and whose height is 5 ft. For the model, let $\frac{3}{4}$ in. = 1 ft. What is the scale factor?

LESSON
111   Applications Using Division

WARM-UP

Facts Practice: Measurement Facts (Test N)

Mental Math: Count up and down by $\frac{1}{2}$'s between −3 and 3.

a. $\frac{2}{3}$ of 24   b. 7 × 35   c. 50% of $48

d. $10.00 − $8.59   e. 0.5 × 100   f. $\frac{1600}{400}$

g. 8 × 8, − 1, ÷ 9, × 4, + 2, ÷ 2, ÷ 3, × 5

h. What number is represented by the Roman numeral MM?

Problem Solving:

Sonya, Sid, and Sinéad met at the gym on Monday. Sonya goes to the gym every two days. The next day she will be at the gym is Wednesday. Sid goes to the gym every three days. The next day Sid will be at the gym is Thursday. Sinéad goes to the gym every four days. She will next be at the gym on Friday. What will be the next day that Sonya, Sid, and Sinéad are at the gym on the same day?

NEW CONCEPT

When a division problem has a remainder, there are several ways to write the answer: with a remainder, as a mixed number, or as a decimal number.

$$4\overline{)15} \;\; 3\text{ R }3 \qquad 4\overline{)15} \;\; 3\tfrac{3}{4} \qquad 4\overline{)15.00} \;\; 3.75$$

How a division answer should be written depends upon the question to be answered. In real-world applications we sometimes need to round an answer up, and we sometimes need to round an answer down. The quotient of 15 ÷ 4 rounds up to 4 and rounds down to 3.

Example 1   One hundred campers are to be assigned to 3 campsites. How many campers should be assigned to each campsite so that the numbers are as balanced as possible?

Solution   Dividing 100 by 3 gives us 33 R 1. Assigning 33 campers per campsite totals 99 campers. We assign the remaining camper to one of the campsites, giving that campsite 34 campers. We write the answer **33, 33,** and **34.**

Example 2   Movie tickets cost $4. Jim has $15. How many tickets can he buy?

Solution   We divide 15 dollars by 4 dollars per ticket. The quotient is $3\frac{3}{4}$ tickets.

$$\frac{15 \text{ dollars}}{4 \text{ dollars per ticket}} = 3\frac{3}{4} \text{ tickets}$$

Jim cannot buy $\frac{3}{4}$ of a ticket, so we round down to the nearest whole number. Jim can buy **3 tickets.**

Example 3   Fifteen children need a ride to the fair. Each car can transport 4 children. How many cars are needed to transport 15 children?

Solution   We divide 15 children by 4 children per car. The quotient is $3\frac{3}{4}$ cars.

$$\frac{15 \text{ children}}{4 \text{ children per car}} = 3\frac{3}{4} \text{ cars}$$

Three cars are not enough. Four cars will be needed. One of the cars will be $\frac{3}{4}$ full. We round $3\frac{3}{4}$ cars up to **4 cars.**

Example 4   Four workers were paid a total of $50 for cleaning a yard. If the workers divide the money equally, how much money will each worker receive?

Solution   We divide $50 by 4 workers. This time we write the quotient as a decimal number.

$$\frac{\$50.00}{4 \text{ workers}} = \$12.50 \text{ per worker}$$

Each worker will receive **$12.50.**

LESSON PRACTICE

Practice set*   **a.** Ninety campers were assigned to four campsites as equally as possible. How many campers were assigned to each of the four campsites?

b. Movie tickets cost $6.00. Aluna has $20.00. How many movie tickets can she buy?

c. Twenty-eight children need a ride to the fair. Each van can carry six children. How many vans are needed?

d. Four workers were paid a total of $45 for cleaning a yard. If the workers divide the money equally, how much money will each worker receive?

MIXED PRACTICE

Problem set   **1.** Eighty campers are to be assigned to three campsites. How
(111)   many campers should be assigned to each campsite so that
the numbers are as balanced as possible? (Write the
numbers.)

2
37
4
× 148

2. Round ten and eighty-six thousandths to the nearest
(51)   hundredth.

3. Shauna bought a sheet of 37¢ stamps at the post office for
(49) $14.80. How many stamps were in the sheet?

4. Eight sugar cubes were used to build
(82)   this 2-by-2-by-2 cube. How many
sugar cubes are needed to build a
cube that has three sugar cubes
along each edge?

2
37
× 40
‾‾‾‾‾
1980

5. Write the standard notation for the following:
(92)

$$(5 \times 10^3) + (4 \times 10^1) + (3 \times 10^0)$$

6. Use a centimeter ruler and an inch ruler to answer this
(7)   question. Twelve inches is closest to how many
centimeters? Round the answer to the nearest centimeter.

7. The legend on a scale drawing of a house indicates that
(Inv. 11)   1 inch = 2 feet.

(a) What is the scale of the drawing?

(b) What is the scale factor?

8. If two angles of a triangle measure 70° and 80°, then what
(98)   is the measure of the third angle?

Complete the table to answer problems 9–11.

	FRACTION	DECIMAL	PERCENT
9. (99)	$\frac{11}{20}$	(a)	(b)
10. (99)	(a)	1.5	(b)
11. (99)	(a)	(b)	1%

12. Calculate mentally:
_(100, 104)

(a) $-6 + -12$ (b) $-6 - -12$

(c) $-12 + +6$ (d) $-12 - +6$

13. $6\frac{1}{4} \div 100$ **14.** $0.3m = \$4.41$
₍₆₈₎   ₍₄₉₎

15. Kim scored 15 points, which was 30% of the team's total.
₍₁₀₅₎ How many points did the team score in all?

16. Andrea received the following scores from the judges:
₍₁₈₎

6.7	7.6	6.6	6.7	6.5	6.7	6.8

The highest score and the lowest score are not counted. What is the average of the remaining scores?

17. Refer to problem 16 to write a comparison question about
₍₁₃₎ the scores Andrea received from the judges. Then answer the question.

18. What is the area of the quadrilateral below?
₍₁₀₇₎

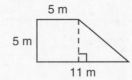

19. What is the ratio of vertices to edges on a pyramid with a
_(Inv. 6) square base?

20. Line *r* is called a line of symmetry
₍₁₁₀₎ because it divides the equilateral triangle into two mirror images. Which other line is also a line of symmetry?

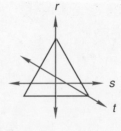

21. Solve and check: $3m + 1 = 100$
₍₁₀₆₎

22. Write the prime factorization of 600 using exponents.
₍₇₃₎

23. The candy bar has 10 sections. If
₍₁₅₎ one section weighs 12 grams, what does the whole candy bar weigh?

24. The price of an item is 89¢. The sales-tax rate is 7%.
$^{(41)}$ What is the total for the item, including tax?

25. The probability of winning a prize in the drawing is one
$^{(58)}$ in a million. What is the probability of not winning a
prize in the drawing?

Triangles *ABC* and *CDA* are congruent. Refer to this figure to
answer problems 26 and 27.

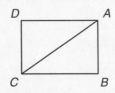

26. Which angle in triangle *ABC* corresponds to angle *D* in
$^{(109)}$ triangle *CDA?*

27. Which transformations would position triangle *CDA* on
$^{(108)}$ triangle *ABC?*

28. Ted used a compass to draw a circle with a radius of
$^{(47)}$ 5 centimeters. What was the circumference of the circle?
(Use 3.14 for π.)

29. Solve this proportion: $\dfrac{10}{16} = \dfrac{25}{y}$
$_{(85)}$

30. The formula $d = rt$ shows that the distance traveled (d)
$^{(95)}$ equals the rate (r) times the time (t) spent traveling at that
rate. (Here, *rate* means "speed.") This function table
shows the relationship between distance and time when
the rate is 50.

t	1	2	3	4
d	50	100	150	200

Find the value of d in $d = rt$ when r is $\frac{50 \text{ mi}}{1 \text{ hr}}$ and t is 5 hr.

good fruit names!
pulp
Juice
citrus
rind

LESSON

112 Multiplying and Dividing Integers

WARM-UP

Facts Practice: 24 Percent-Fraction-Decimal Equivalents (Test M)

Mental Math: Count up and down by 5's between −25 and 25.

a. $\frac{3}{4}$ of 24   b. 6 × 48   c. 25% of $48

d. $4.98 + $2.49   e. 0.5 ÷ 10   f. 500 · 30

g. 11 × 4, + 1, ÷ 5, $\sqrt{}$, × 4, − 2, × 5, − 1, $\sqrt{}$

h. What number is represented by the Roman numeral MCM?

Problem Solving:

An auditorium is shaped to have 8 seats in the first row, 10 seats in the second row, 12 seats in the third row, and so on. Altogether, how many seats are in the first 8 rows?

NEW CONCEPT

We know that when we multiply two positive numbers the product is positive.

$$(+3)(+4) = +12$$

positive × positive = positive

Notice that when we write (+3)(+4) there is no + or − sign between the sets of parentheses.

When we multiply a positive number and a negative number, the product is negative. We show an example on this number line by multiplying 3 and −4.

$$3 \times -4 \text{ means } (-4) + (-4) + (-4)$$

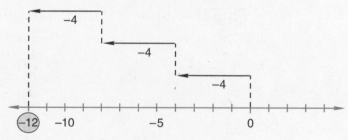

We write the multiplication this way:

$$(+3)(-4) = -12$$

Positive three times *negative* four equals *negative* 12.

positive × negative = negative

When we multiply two negative numbers, the product is positive. Consider this sequence of equations:

1. Three times 4 is 12 (3 × 4 = 12).
2. Three times the opposite of 4 is the opposite of 12 (3 × −4 = −12).
3. The opposite of 3 times the opposite of 4 is the opposite of the opposite of 12 (−3 × −4 = +12).

negative × negative = positive

Recall that we can rearrange the numbers of a multiplication fact to make two division facts.

Multiplication Facts	Division Facts	
(+3)(+4) = +12	$\frac{+12}{+3} = +4$	$\frac{+12}{+4} = +3$
(+3)(−4) = −12	$\frac{-12}{+3} = -4$	$\frac{-12}{-4} = +3$
(−3)(−4) = +12	$\frac{+12}{-3} = -4$	$\frac{+12}{-4} = -3$

Studying these nine facts, we can summarize the results in two rules:

1. If the two numbers in a multiplication or division problem have the **same sign,** the answer is positive.
2. If the two numbers in a multiplication or division problem have **different signs,** the answer is negative.

Examples Calculate mentally:

(a) (+8)(+4)   (b) (+8) ÷ (+4)

(c) (+8)(−4)   (d) (+8) ÷ (−4)

(e) (−8)(+4)   (f) (−8) ÷ (+4)

(g) (−8)(−4)   (h) (−8) ÷ (−4)

Solutions (a) **+32**   (b) **+2**

(c) **−32**   (d) **−2**

(e) **−32**   (f) **−2**

(g) **+32**   (h) **+2**

LESSON PRACTICE

Practice set*   a. (−5)(+4)   b. (−5)(−4)

c. (+5)(+4)   d. (+5)(−4)

e. $\dfrac{+12}{-2}$ f. $\dfrac{+12}{+2}$ g. $\dfrac{-12}{+2}$ h. $\dfrac{-12}{-2}$

MIXED PRACTICE

Problem set   **1.** Two hundred people are traveling by bus on a guided
(111)   tour. The maximum number of people allowed on each
bus is 84. How many buses are needed for the trip?

2. The wingspan on the model plane was 12 inches, while
(Inv. 11)   the wingspan on the actual plane was 60 feet.

(a) What was the scale of the model?

(b) If the model of the plane is 10 inches long, then how
long is the actual plane?

3. Calculate mentally:
(112)

(a) (−2)(−6)   (b) $\dfrac{+6}{-2}$

(c) $\dfrac{-6}{-6}$ (d) (−2)(+6)

4. Calculate mentally:
(100)   (a) −2 + −6   (b) −2 − −6

(c) +2 + −6   (d) +2 − −6

5. Rafael correctly answered 27 questions, which was 90%
(105)   of the questions on the test. How many questions did he
not answer correctly?

6. Write twenty million, five hundred ten thousand in
(92)   expanded notation using exponents.

7. Find 8% of $3.65 and round the product to the nearest cent.
(41)

8. $\left(\dfrac{1}{2}\right)^2 + \dfrac{1}{8} \div \dfrac{1}{2}$
(92)

Complete the table to answer problems 9–11.

	FRACTION	DECIMAL	PERCENT
9. (99)	$1\frac{4}{5}$	(a)	(b)
10. (99)	(a)	0.6	(b)
11. (99)	(a)	(b)	2%

Solve:

12. (63) $5\frac{1}{2} - m = 2\frac{5}{6}$

13. (85) $\frac{6}{10} = \frac{9}{n}$

14. (106) $9x - 7 = 92$

15. (49) $0.05w = 8$

16. (15) All eight books in the stack are the same weight. Three books weigh a total of six pounds.

(a) How much does each book weigh?

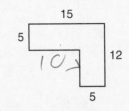
6 lb

(b) How much do all eight books weigh?

17. (91) Find the volume of a rectangular prism using the formula $V = lwh$ when the length is 8 cm, the width is 5 cm, and the height is 2 cm.

18. (95) How many millimeters is 1.2 meters (1 m = 1000 mm)?

19. (103) What is the perimeter of the polygon at right? (Dimensions are in millimeters.)

15
5
12
5

20. (107) What is the area of the polygon in problem 19?

21. (Inv. 6) If the pattern shown below were cut out and folded on the dotted lines, would it form a cube, a pyramid, or a cylinder?

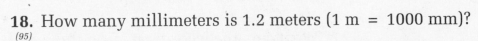
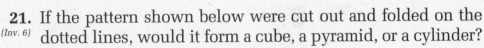

22. Which one of these numbers is not a composite number?
(65)

    A. 34   B. 35   C. 36   D. 37

23. Debbie wants to decorate a cylindrical wastebasket by
(47) wrapping it with wallpaper. The diameter of the wastebasket is 12 inches. The length of the wallpaper should be at least how many inches? Round up to the next inch.

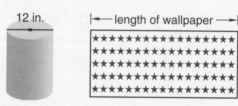

24. Which one of these letters has two lines of symmetry?
(110)

<h1 style="text-align:center">H   A   V   E</h1>

25. Which arrow is pointing to $-\frac{1}{2}$?
(17)

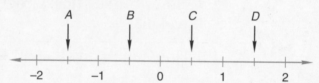

26. The ratio of good guys to bad guys in the movie was 2 to 3.
(101) If the total number of good guys and bad guys was 30, how many good guys were there?

27. What are the coordinates of the point that is halfway
(Inv. 7) between (−2, −3) and (6, −3)?

28. A heart is drawn from a normal deck of 52 cards and is
(Inv. 10) not replaced. A second card is drawn from the remaining 51 cards. What is the probability that the second card will be a heart?

29. Combine the areas of the two
(107) triangles to find the area of this trapezoid.

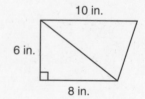

30. $3^2 + 2 \times 5^2 - 50 \div \sqrt{25}$
(92)

LESSON

113 Adding and Subtracting Mixed Measures • Multiplying by Powers of Ten

WARM-UP

Facts Practice: Measurement Facts (Test N)

Mental Math: Count up and down by $\frac{1}{8}$'s between $\frac{1}{8}$ and 2.

a. $\frac{3}{10}$ of 40   b. 4×38 c. 25% of $200

d. $100.00 − $9.50   e. $0.12 \div 10$ f. $\frac{2000}{500}$

g. $6 \times 8, + 2, \times 2, - 1, \div 3, - 1, \div 4, + 2, \div 10, - 1$

h. What number is represented by the Roman numeral MMIV?

Problem Solving:

Copy this problem and fill in the missing digits:

$$\begin{array}{r} \text{\_\,\_\_,\_\_\_} \\ \times \qquad\quad 7 \\ \hline 9\,9\,9,9\,9\,9 \end{array}$$

NEW CONCEPTS

Adding and subtracting mixed measures

Measurements that include more than one unit of measurement are mixed measures. If we say that a movie is an hour and 40 minutes long, we have used a mixed measure that includes hours and minutes. When adding or subtracting mixed measures, we may need to convert from one unit to another unit. In Lesson 102 we added and subtracted mixed measures involving pounds and ounces. In this lesson we will consider other mixed measures.

Example 1 Add: 1 hr 40 min
           + 1 hr 50 min

Solution We add 40 minutes and 50 minutes to get 90 minutes. Sixty minutes equals 1 hour, so 90 minutes equals 1 hour 30 minutes.

$$\begin{array}{r} 1 \text{ hr } 40 \text{ min} \\ + 1 \text{ hr } 50 \text{ min} \\ \hline 90 \text{ min} \end{array}$$ (which is 1 hr 30 min)

We change 90 minutes to 1 hour 30 minutes. We write "30 minutes" in the minutes column and add the 1 hour to the hours column. Then we add the hours.

$$
\begin{array}{r}
\overset{1}{1} \text{ hr } 40 \text{ min} \\
+ \ 1 \text{ hr } 50 \text{ min} \\
\hline
3 \text{ hr } \underset{30}{\cancel{90}} \text{ min}
\end{array}
$$

The sum is **3 hours 30 minutes.**

Example 2 Subtract: 6 ft 5 in.
       − 4 ft 8 in.

Solution Before we can subtract inches, we rename 6 feet as 5 feet plus 12 inches. The 12 inches combine with the 5 inches to make 17 inches. Then we subtract.

$$
\begin{array}{r}
\overset{5}{\cancel{6}} \text{ ft } \overset{17}{\cancel{5}} \text{ in.} \\
- \ 4 \text{ ft } 8 \text{ in.} \\
\hline
1 \text{ ft } 9 \text{ in.}
\end{array}
$$

The difference is **1 foot 9 inches.**

Multiplying by powers of ten We can multiply by powers of ten very easily. Multiplying by powers of ten does not change the digits, only the place value of the digits. We can change the place value by moving the decimal point the number of places shown by the exponent. To write 1.2×10^3 in standard notation, we simply move the decimal point three places to the right and fill the empty places with zeros.

$$1.2 \times 10^3 = 1200. = 1200$$

Example 3 Write 6.2×10^2 in standard notation.

Solution To multiply by a power of ten, simply move the decimal point the number of places shown by the exponent. In this case, we move the decimal point two places to the right.

$$6.2 \times 10^2 = 620. = \textbf{620}$$

Sometimes powers of ten are named with words instead of numbers. For example, we might read that the population of Hong Kong is 7.2 million. The number 7.2 million means $7.2 \times 1{,}000{,}000$. We can write this number by shifting the decimal point of 7.2 six places to the right, which gives us 7,200,000.

Example 4 Write $\frac{1}{2}$ billion in standard notation.

Solution The expression $\frac{1}{2}$ billion means "one half of one billion." First we write $\frac{1}{2}$ as the decimal number 0.5. Then we multiply by one billion, which shifts the decimal point nine places.

$$\frac{1}{2} \text{ billion} = 0.5 \times 1{,}000{,}000{,}000 = \textbf{500{,}000{,}000}$$

LESSON PRACTICE

Practice set* Find each sum or difference:

a. 6 ft 5 in.
+ 4 ft 8 in.

b. 3 hr 15 min
− 1 hr 40 min

Write the standard notation for each of the following numbers. Change fractions and mixed numbers to decimal numbers before multiplying.

c. 1.2×10^4

d. 1.5 million

e. $2\frac{1}{2}$ billion

f. $\frac{1}{4}$ million

MIXED PRACTICE

Problem set

1. For cleaning the yard, four teenagers were paid a total of
(111) $35.00. If they divide the money equally, how much money will each teenager receive?

2. Which of the following is the best estimate of the length
(7) of a bicycle?

A. 0.5 m   B. 2 m   C. 6 m   D. 36 m

3. If the chance of rain is 80%, what is the probability that it
(58) will not rain? Express the answer as a decimal.

4. The team's win-loss ratio was 5 to 3. If the team played
(101) 120 games and did not tie any games, then how many games did the team win?

5. Write 4.5×10^6 as a standard numeral.
(113)

6. Calculate mentally:
(112)

(a) (−12)(+3)   (b) (−12)(−3)

(c) $\dfrac{-12}{+3}$ (d) $\dfrac{-12}{-3}$

7. Calculate mentally:
(100)

(a) −12 + −3   (b) −12 − −3

(c) +3 + −12   (d) +3 − −12

8. Describe a method for arranging these fractions from least
(76) to greatest:

$$\frac{3}{4}, \frac{3}{5}, \frac{4}{5}$$

Complete the table to answer problems 9–11.

	FRACTION	DECIMAL	PERCENT
9. (99)	$\frac{1}{50}$	(a)	(b)
10. (99)	(a)	1.75	(b)
11. (99)	(a)	(b)	25%

12. $12\frac{1}{4}$ in. − $3\frac{5}{8}$ in.   **13.** $3\frac{1}{3}$ ft × $2\frac{1}{4}$ ft
(63)   (66)

14. (3 cm)(3 cm)(3 cm)   **15.** 0.6 m × 0.5 m
(81)   (81)

16. $5^2 + 2^5$
(92)

17. Find the area of this trapezoid by
(107) combining the area of the triangle
and the area of the rectangle.

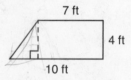

18. 2 feet 3 inches − 1 foot 9 inches
(113)

19. Which line in this figure is not a line of symmetry?
(110)

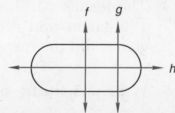

20. How many cubes one centimeter
(82) on each edge would be needed to
fill this box?

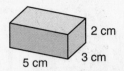

2 cm
3 cm
5 cm

21. Elizabeth worked for three days and earned $240. At that
(88) rate, how much would she earn in ten days?

22. Seventy is the product of which three prime numbers?
(65)

23. Saturn is about 900 million miles from the Sun. Write that
(12) distance in standard notation.

24. Recall that a rhombus is a parallelogram in which all four
(71) sides are equal in length.

(a) What is the perimeter of this
rhombus?

(b) What is the area of this
rhombus?

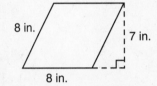

8 in.

7 in.

8 in.

(c) If an acute angle of this rhombus measures 61°, then
what is the measure of each obtuse angle?

25. The ratio of quarters to dimes in the soda machine was
(88) 5 to 8. If there were 120 quarters, how many dimes
were there?

26. The coordinates of the three vertices of a triangle are
(Inv. 7, 79) (0, 0), (0, 4), and (4, 4). What is the area of the triangle?

The following list shows the ages of the children attending a
party. Use this information to answer problems 27 and 28.

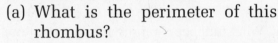

8, 9, 8, 8, 7, 9, 12, 12, 11, 16

27. What was the median age of the children attending the
(Inv. 5) party?

28. What was the mean age of the children at the party?
(Inv. 5)

29. The diameter of a playground ball is
(47) 10 inches. What is the circumference
of the ball? (Use 3.14 for π.)

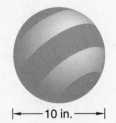

|— 10 in. —|

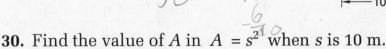

30. Find the value of A in $A = s^2$ when s is 10 m.
(91)

114   Unit Multipliers

WARM-UP

Facts Practice: 24 Percent-Fraction-Decimal Equivalents (Test M)

Mental Math: Count by $\frac{1}{4}$'s from −1 to 1.

a. $\frac{7}{10}$ of 40 **b.** 6 × 480 **c.** 10% of $500

d. $4.99 + 65¢ **e.** 0.125 × 1000 **f.** 40 · 900

g. 5 × 7, + 1, ÷ 4, $\sqrt{\ }$, × 7, − 1, × 3, − 10, × 2, $\sqrt{\ }$

h. What number is represented by the Roman numeral MD?

Problem Solving:

How many different license plates can be made if the pattern is three letters followed by three digits and if all letter and digit combinations are used?

NEW CONCEPT

A **unit multiplier** is a fraction that equals 1 and that is written with two different units of measure. Recall that when the numerator and denominator of a fraction are equal (and are not zero), the fraction equals 1. Since 1 foot equals 12 inches, we can form two unit multipliers with the measures 1 foot and 12 inches.

$$\frac{1 \text{ ft}}{12 \text{ in.}} \qquad \frac{12 \text{ in.}}{1 \text{ ft}}$$

Each of these fractions equals 1 because the numerator and denominator of each fraction are equal.

We can use unit multipliers to help us convert from one unit of measure to another. If we want to convert 60 inches to feet, we can multiply 60 inches by the unit multiplier $\frac{1 \text{ ft}}{12 \text{ in.}}$.

$$\frac{\overset{5}{\cancel{60} \text{ in.}}}{1} \times \frac{1 \text{ ft}}{\underset{1}{\cancel{12} \text{ in.}}} = 5 \text{ ft}$$

Example 1 (a) Write two unit multipliers using these equivalent measures:

$$3 \text{ ft} = 1 \text{ yd}$$

(b) Which unit multiplier would you use to convert 30 yards to feet?

Solution (a) We use the equivalent measures to write two fractions equal to 1.

$$\frac{3 \text{ ft}}{1 \text{ yd}} \qquad \frac{1 \text{ yd}}{3 \text{ ft}}$$

(b) We want the units we are changing **from** to appear in the denominator and the units we are changing **to** to appear in the numerator. To convert 30 yards to feet, we use the unit multiplier that has yards in the denominator and feet in the numerator.

$$\frac{30 \text{ yd}}{1} \times \frac{3 \text{ ft}}{1 \text{ yd}}$$

Here we show the work. Notice that the yards "cancel," and the product is expressed in feet.

$$\frac{30 \cancel{\text{ yd}}}{1} \times \frac{3 \text{ ft}}{1 \cancel{\text{ yd}}} = 90 \text{ ft}$$

Example 2 Convert 30 feet to yards using a unit multiplier.

Solution We can form two unit multipliers.

$$\frac{1 \text{ yd}}{3 \text{ ft}} \quad \text{and} \quad \frac{3 \text{ ft}}{1 \text{ yd}}$$

We are asked to convert from feet to yards, so we use the unit multiplier that has feet in the denominator and yards in the numerator.

$$\frac{\overset{10}{\cancel{30 \text{ ft}}}}{1} \times \frac{1 \text{ yd}}{\cancel{3 \text{ ft}}} = 10 \text{ yd}$$

Thirty feet converts to **10 yards**.

LESSON PRACTICE

Practice set   **a.** Write two unit multipliers for these equivalent measures:

$$1 \text{ gal } = 4 \text{ qt}$$

b. Which unit multiplier from problem **a** would you use to convert 12 gallons to quarts?

c. Write two unit multipliers for these equivalent measures:

$$1 \text{ m } = 100 \text{ cm}$$

d. Which unit multiplier from problem **c** would you use to convert 200 centimeters to meters?

e. Use a unit multiplier to convert 12 quarts to gallons.

f. Use a unit multiplier to convert 200 meters to centimeters.

g. Use a unit multiplier to convert 60 feet to yards (1 yd = 3 ft).

MIXED PRACTICE

Problem set   **1.** Tickets to the matinee are $6 each. How many tickets can
(111)   Maela buy with $20?

2. Maria ran four laps of the track at an even pace. If it took
(88)   6 minutes to run the first three laps, how long did it take to run all four laps?

3. Fifteen of the 25 members played in the game. What
(77)   fraction of the members did not play?

4. Two fifths of the thirty plants in the garden are tomato
(77)   plants. How many other plants are in the garden?

5. Which digit in 94,763,581 is in the ten-thousands place?
(12)

6. (a) Write two unit multipliers for these equivalent
(114)   measures:

$$1 \text{ gallon } = 4 \text{ quarts}$$

(b) Which of the two unit multipliers from part (a) would you use to convert 8 gallons to quarts?

7. Estimate the sum of $36.43, $41.92, and $26.70 to the
(51) nearest dollar.

8. $4 + 4^2 \div \sqrt{4} - \dfrac{4}{4}$
(92)

9. $3\dfrac{1}{4}$ in. $+ 2\dfrac{1}{2}$ in. $+ 4\dfrac{5}{8}$ in.
(61)

Complete the table to answer problems 10–12.

	FRACTION	DECIMAL	PERCENT
10. (99)	$\dfrac{1}{8}$	(a)	(b)
11. (99)	(a)	0.9	(b)
12. (99)	(a)	(b)	60%

13. $3.25 \div \dfrac{2}{3}$ (fraction answer)
(73)

Solve:

14. $3m - 10 = 80$
(106)

15. $\dfrac{3}{2} = \dfrac{18}{m}$
(85)

16. Calculate mentally:
(112)

(a) $(-5)(-20)$

(b) $(-5)(+20)$

(c) $\dfrac{-20}{+5}$

(d) $\dfrac{-20}{-5}$

17. If a car travels 6 hours at an average speed of 55 miles per
(95) hour, how far will it travel?

$$\dfrac{6 \text{ hours}}{1} \times \dfrac{55 \text{ miles}}{1 \text{ hour}}$$

18. What is the area of this polygon?
(107)

19. What is the perimeter of this
(103) polygon?

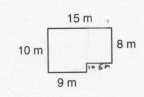

15 m

10 m

8 m

9 m

20. Calculate mentally:
(100)

(a) $-5 + -20$

(b) $-20 - -5$

(c) $-5 - -5$

(d) $+5 - -20$

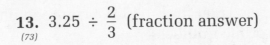

21. Transversal *t* intersects parallel lines *q* and *r*. Angle 1 is
(97) half the measure of a right angle.

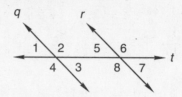

(a) Which angle corresponds to ∠1?

(b) What is the measure of each obtuse angle?

22. Fifty people responded to the survey, a number that
(105) represented 5% of the surveys mailed. How many surveys
were mailed?

23. Think of two different prime numbers, and write them on
(19, 30) your paper. Then find the least common multiple (LCM) of
the two prime numbers.

24. Write 1.5×10^6 as a standard number.
(113)

25. A room that is 30 feet long, 30 feet wide, and 10 feet high
(82) has a volume of how many cubic feet?

26. Convert 8 quarts to gallons using a unit multiplier.
(114)

27. A circle was drawn on a coordinate plane. The
(Inv. 7, 86) coordinates of the center of the circle were (1, 1). One
point on the circle was (1, −3).

(a) What was the radius of the circle?

(b) What was the area of the circle? (Use 3.14 for π.)

28. On the driving exam the range of scores was 35 points. If
(Inv. 5) the highest score was 95, what was the lowest score?

29. 4 ft 3 in. − 2 ft 9 in.
(113)

30. Study this function and describe a
(96) rule for finding *A* when *s* is known.

s	A
1	1
2	4
3	9
4	16

LESSON
115 Writing Percents as Fractions, Part 2

WARM-UP

Facts Practice: Measurement Facts (Test N)

Mental Math: Count up and down by 5's between −25 and 25.

a. $\frac{3}{4}$ of 16   **b.** 9 × 507   **c.** 10% of $2.50

d. $10.00 − $9.59   **e.** 0.5 ÷ 100   **f.** $\frac{2400}{300}$

g. 10 × 9, − 10, ÷ 2, + 2, ÷ 6, × 10, + 2, ÷ 9, − 9

h. What number is represented by the Roman numeral MDCC?

Problem Solving:

When Sarah said 7, Tom said 15. When Perry said 11, Tom said 23. When Hector said 3, Tom said 7. Figure out the rule Tom used to change the numbers he was given. What number would Tom say if Simon says 5?

NEW CONCEPT

Recall that a percent is a fraction with a denominator of 100. We can write a percent in fraction form by removing the percent sign and writing the denominator 100.

$$50\% = \frac{50}{100}$$

We then reduce the fraction to lowest terms. If the percent includes a fraction, we actually divide by 100 to simplify the fraction.

$$33\tfrac{1}{3}\% = \frac{33\tfrac{1}{3}}{100}$$

In this case we divide $33\frac{1}{3}$ by 100. We have performed division problems similar to this in the problem sets.

$$33\frac{1}{3} \div 100 = \frac{\overset{1}{\cancel{100}}}{3} \times \frac{1}{\underset{1}{\cancel{100}}} = \frac{1}{3}$$

We see that $33\frac{1}{3}\%$ equals $\frac{1}{3}$.

Example Convert $3\frac{1}{3}\%$ to a fraction.

Solution We remove the percent sign and write the denominator 100.

$$3\frac{1}{3}\% = \frac{3\frac{1}{3}}{100}$$

We perform the division.

$$3\frac{1}{3} \div 100 = \frac{\overset{1}{\cancel{10}}}{3} \times \frac{1}{\underset{10}{\cancel{100}}} = \frac{1}{30}$$

We find that $3\frac{1}{3}\%$ equals $\frac{1}{30}$.

LESSON PRACTICE

Practice set **a.** Convert $66\frac{2}{3}\%$ to a fraction.

b. Convert $6\frac{2}{3}\%$ to a fraction.

c. Convert $12\frac{1}{2}\%$ to a fraction.

d. Write $14\frac{2}{7}\%$ as a fraction.

e. Write $83\frac{1}{3}\%$ as a fraction.

MIXED PRACTICE

Problem set **1.** What is the total cost of a $12.60 item plus 7% sales tax?
(41)

2. Which is the greatest weight?
(44)

A. 6.24 lb   B. 6.4 lb   C. 6.345 lb

3. Draw a ratio box for this problem. Then solve the problem
(105) using a proportion.

> *Ines missed three questions on the test but answered 90% of the questions correctly. How many questions were on the test?*

4. Sound travels about 331 meters per second in air. How
(95) far will it travel in 60 seconds?

$$\frac{331 \text{ m}}{1 \text{ s}} \cdot \frac{60 \text{ s}}{1}$$

5. Write the standard number for $(5 \times 10^4) + (6 \times 10^2)$.
(92)

6. If the radius of a circle is seventy-five hundredths of a
(27, 37) meter, what is the diameter?

7. Round the product of $3\frac{2}{3}$ and $2\frac{2}{3}$ to the nearest whole
(66) number.

8. In a bag are three red marbles and three white marbles. If
(Inv. 10) two marbles are taken from the bag at the same time,
what is the probability that both marbles will be red?

Complete the table to answer problems 9 and 10.

	FRACTION	DECIMAL	PERCENT
9. (99)	$2\frac{2}{5}$	(a)	(b)
10. (99)	(a)	0.85	(b)

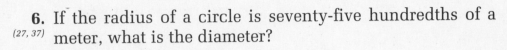

Solve:

11. $7x - 3 = 39$
(106)

12. $\frac{x}{7} = \frac{35}{5}$
(85)

13. Calculate mentally:
(112)

  (a) $(-3)(-15)$

  (b) $\frac{-15}{+3}$

  (c) $\frac{-15}{-3}$

  (d) $(+3)(-15)$

14. $-6 + -7 + +5 - -8$
(104)

15. $0.12 \div (12 \div 0.4)$
(49)

16. Write $\frac{22}{7}$ as a decimal rounded to the hundredths place.
(74)

17. What whole number multiplied by itself equals 10,000?
(89)

18. What is the area of this hexagon?
(107)

19. What is the perimeter of this
(103) hexagon?

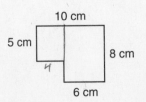

20. What is the volume of this cube?
(82)

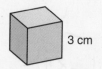

3 cm

21. What is the mode of the number of days in the twelve
(Inv. 5) months of the year?

22. If seven of the containers can hold a
(88) total of 84 ounces, then how many
ounces can 10 containers hold?

84 ounces

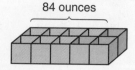

23. Write the standard number for $4\frac{1}{2}$ million.
(113)

24. Round 58,697,284 to the nearest million.
(16)

25. Which arrow is pointing to 0.4?
(50)

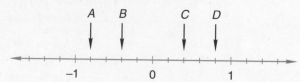

26. When Rosita was born, she weighed 7 pounds 9 ounces.
(102, 113) Two months later she weighed 9 pounds 7 ounces. How
much weight did she gain in two months?

27. The coordinates of the vertices of a parallelogram are
(Inv. 7, 71) (0, 0), (5, 0), (6, 3), and (1, 3). What is the area of the
parallelogram?

28. Convert $16\frac{2}{3}\%$ to a fraction.
(115)

29.   2 gal 2 qt 1 pt
(113) + 2 gal 2 qt 1 pt

30. Gilbert started the trip with a full tank of gas. He drove
(81) 323.4 miles and then refilled the tank with 14.2 gallons of
gas. How can Gilbert calculate the average number of miles
he traveled on each gallon of gas?

LESSON
116 Compound Interest

WARM-UP

Facts Practice: 24 Percent-Fraction-Decimal Equivalents (Test M)

Mental Math: Count down by 2's from 10 to −10.

a. $\frac{3}{8}$ of 16   b. 4 × 560   c. 50% of $2.50

d. $8.98 + 49¢   e. 0.375 × 100   f. 50 · 400

g. 11 × 6, − 2, $\sqrt{\ }$, × 3, + 1, $\sqrt{\ }$, × 10, − 1, $\sqrt{\ }$

h. What number is represented by the Roman numeral MDCXX?

Problem Solving:

Copy this problem and fill in the missing digits. The product should be a perfect square.

$$
\begin{array}{r}
-\ - \\
\times\ \underline{-\ -} \\
-\ -\ - \\
\hline
=\ = \\
3\ \_\ 1
\end{array}
$$

NEW CONCEPT

When you deposit money in a bank, the bank uses a portion of that money to make loans and investments that earn money for the bank. To attract deposits, banks offer to pay **interest,** a percentage of the money deposited. The amount deposited is called the **principal.**

There is a difference between **simple interest** and **compound interest.** Simple interest is paid on the principal only and not paid on any accumulated interest. For instance, if you deposited $100 in an account that pays 6% simple interest, you would be paid 6% of $100 ($6) each year your $100 was on deposit. If you take your money out after three years, you would have a total of $118.

Simple Interest

$100.00	principal
$6.00	first-year interest
$6.00	second-year interest
+ $6.00	third-year interest
$118.00	total

Most interest-bearing accounts, however, are compound-interest accounts. In a compound-interest account, interest is paid on accumulated interest as well as on the principal. If you deposited $100 in an account with 6% annual percentage rate, the amount of interest you would be paid each year increases if the earned interest is left in the account. After three years you would have a total of $119.10.

Compound Interest

$100.00	principal
$6.00	first-year interest (6% of $100.00)
$106.00	total after one year
$6.36	second-year interest (6% of $106.00)
$112.36	total after two years
$6.74	third-year interest (6% of $112.36)
$119.10	total after three years

Example 1 Samantha deposited $2000 in an account that pays 10% interest compounded annually. If she does not withdraw any money from the account, how much money will be in the account (a) after one year, (b) after two years, and (c) after three years?

Solution We calculate the total amount of money in the account at the end of each year by finding the interest earned in that year and adding it to the amount of money in the account at the start of that year.

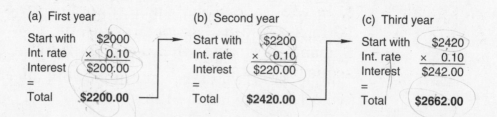

Notice that the amount of interest earned increased each year, even though the interest rate stayed the same. This increase occurred because interest is paid on interest earned in prior years as well as on the original principal. The effect of compound interest becomes more dramatic as the number of years increases.

In our solution above we multiplied each starting amount by 10% (0.10) to find the amount of interest earned, then we added the interest to the starting amount. Instead of multiplying by 10% and adding, we can multiply by 110% (1.10) to find the total amount in the account after each year.

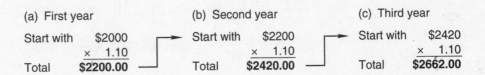

(a) First year	(b) Second year	(c) Third year
Start with $2000	Start with $2200	Start with $2420
× 1.10	× 1.10	× 1.10
Total **$2200.00**	Total **$2420.00**	Total **$2662.00**

We will use this second method with a calculator to find the amount of money in Samantha's account after one, two, and three years. We use 1.1 for 110% and follow this keystroke sequence:[†]

Display

2 0 0 0 × 1 . 1 = 2200 (1st yr)

× 1 . 1 = 2420 (2nd yr)

× 1 . 1 = 2662 (3rd yr)

We can use the calculator memory to reduce the number of keystrokes. (*Note:* Abbreviations on memory keys vary. In the keystroke sequences below, we will use →M for "enter memory" and MR for "memory recall.") Instead of entering 1.1 for every year, we enter 1.1 into the memory with these keystrokes:

1 . 1 →M

Now we can use the "memory recall" key instead of 1.1 to perform the calculations. We find the amount of money in Samantha's account after one, two, and three years with this sequence of keystrokes:

Display

2 0 0 0 × MR = 2200 (1st yr)

× MR = 2420 (2nd yr)

× MR = 2662 (3rd yr)

[†]If this keystroke sequence does not produce the indicated result, consult the manual for your calculator for the appropriate keystroke sequence.

Example 2   Kayla deposited $2000 in an account that pays 7% interest compounded annually. Use the memory keys to find the amount of money in Kayla's account (a) after three years, (b) after ten years, and (c) after twenty years.

Solution   We will repeatedly multiply the deposit and earned interest by 107%. So we enter 1.07 into the memory.

Now we can use memory recall instead of entering 1.07 to perform the calculations. Here we show the keystroke sequence for the first two years.

We continue the keystroke pattern and record the results after the third, tenth, and twentieth years.

(a) **$2450.09**   (b) **$3934.30**   (c) **$7739.37**

LESSON PRACTICE

Practice set   **a.** After the third year, $2662 was in Samantha's account. If the account continues to earn 10% interest, how much money will be in the account (1) after the tenth year and (2) after the twentieth year? Round answers to the nearest cent.

b. Nelson deposited $2000 in an account that pays 9% interest per year. If he does not withdraw any money from the account, how much will be in the account (1) after three years, (2) after 10 years, and (3) after 20 years? (Multiply by 1.09. Round answers to the nearest cent.)

c. How much more money will be in Samantha's account than in Nelson's account (1) after three years, (2) after 10 years, and (3) after 20 years?

Extension   Find an advertisement that gives a bank's or other saving institution's interest rate for savings accounts. Write a problem based on the advertisement. Then find the answer to your problem.

MIXED PRACTICE

Problem set **1.** The 306 people were assigned to ten groups so that there
(111) were 30 or 31 people in each group. How many groups
had exactly 30 people?

2. If 5 feet of ribbon costs $1.20, then 10 feet of ribbon
(95) would cost how much?

3. (a) Six is what fraction of 15?
(22, 75)
    (b) Six is what percent of 15?

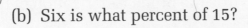

4. The multiple-choice question has four choices. Raidon
(58) knows that one of the choices must be correct, but he has no
idea which one. If Raidon simply guesses, what is the
chance that he will guess the correct answer?

5. If $\frac{2}{5}$ of the 30 full-grown horses in the stable are stallions
(77) (males), then what is the ratio of stallions to mares
(females) in the stable?

6. Write 1.2×10^9 as a standard number.
(113)

7. The cost (c) of corn is related to its price per pound (p)
(95) and its weight (w) by this formula:

$$c = pw$$

Find the cost when p is $\frac{\$0.65}{1 \text{ pound}}$ and w is 5 pounds.

8. Arrange these numbers in order from least to greatest:
(44)

$$9.9, \ 9.95, \ 9.925, \ 9.09$$

Complete the table to answer problems 9 and 10.

	Fraction	Decimal	Percent
9. (99)	$3\frac{3}{8}$	(a)	(b)
10. (99)	(a)	(b)	15%

Solve and check:

11. $9x + 17 = 80$ **12.** $\frac{x}{3} = \frac{16}{12}$
(106)   (85)

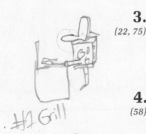

13. $-6 + -4 - +3 - -8$
₍₁₀₄₎

14. $6 + 3\frac{3}{4} + 4.6$ (decimal answer)
₍₇₄₎

15. The Gateway Arch in St. Louis, Missouri, is approximately
₍₁₁₄₎ 210 yards tall. How tall is it in feet?

16. Use division by primes to find the prime factors of 648.
₍₇₃₎ Then write the prime factorization of 648 using exponents.

17. If a 32-ounce box of cereal costs \$3.84, what is the cost
₍₁₅₎ per ounce?

18. Find the area of this trapezoid by
₍₁₀₇₎ combining the area of the rectangle
and the area of the triangle.

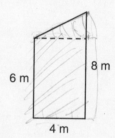

19. The radius of a circle is 10 cm. Use 3.14 for π to calculate the
_(47, 86)

  (a) circumference of the circle.

  (b) area of the circle.

20. The volume of the pyramid is $\frac{1}{3}$ the
₍₈₂₎ volume of the cube. What is the
volume of the pyramid?

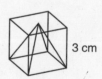

21. Solve: $0.6y = 54$
₍₄₉₎

22. Calculate mentally:
₍₁₁₂₎

  (a) $(-8)(-2)$ (b) $(+8)(-2)$

  (c) $\dfrac{+8}{-2}$ (d) $\dfrac{-8}{-2}$

23. John drew a right triangle with sides 6 inches, 8 inches, and
_(79, 93) 10 inches long. What was the area of the triangle?

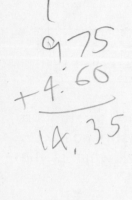

24. Two angles of a triangle measure 40° and 110°.
(98)

(a) What is the measure of the third angle?

(b) Make a rough sketch of the triangle.

25. If Anthony spins the spinner 60
(58) times, about how many times should he expect the arrow to stop in sector 3?

A. 60 times   B. 40 times

C. 20 times   D. 10 times

26. An equilateral triangle and a
(8, 38) square share a common side. If the area of the square is 100 mm², then what is the perimeter of the equilateral triangle?

27. Write $11\frac{1}{9}\%$ as a reduced fraction.
(115)

28. The heights of the five starters on the basketball team are
(Inv. 5) listed below. Find the mean, median, and range of these measures.

181 cm, 177 cm, 189 cm, 158 cm, 195 cm

29. George bought two bunches of bananas. The smaller
(102, 113) bunch weighed 2 lb 12 oz. The larger bunch weighed 3 lb 8 oz. What was the total weight of the two bunches of bananas?

30. Which type of triangle has no lines of symmetry?
(93, 110)

A. equilateral   B. isosceles   C. scalene

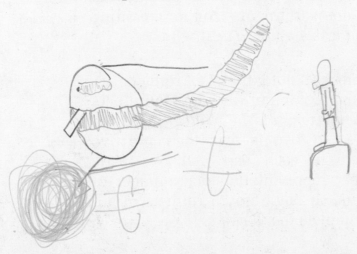

LESSON

117

Finding a Whole When a Fraction Is Known

WARM-UP

Facts Practice: Measurement Facts (Test N)

Mental Math:

a. $\frac{7}{8}$ of 16   b. 3 × 760   c. 25% of $80
d. 5 is what % of 20? e. 0.6 × 40   f. 60 · 700
g. 8 × 8, − 1, ÷ 7, $\sqrt{}$, × 4, ÷ 2, × 3, ÷ 2
h. What number is represented by the Roman numeral MCM?

Problem Solving:

Coins from the vending machine were put into coin rolls. Forty nickels fill a roll, 50 dimes fill a roll, and 40 quarters fill a roll. The coins in the machine filled four quarter rolls, two dime rolls, and three nickel rolls. All nine rolls equaled how much money?

NEW CONCEPT

Consider the following fractional-parts problem:

> *Two fifths of the runners in the race are warming up. If there are ten runners warming up, how many runners are in the race?*

A diagram can help us understand and solve this problem. We have drawn a rectangle to represent all of the runners. The problem states that two fifths are warming up, so we divide the rectangle into five parts. Two of the parts are warming up, so the remaining three parts are not warming up.

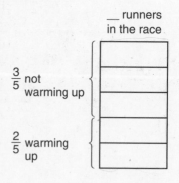

We are also told that ten runners are warming up. In our diagram ten runners make up two of the parts. Since ten divided by two is five, there are five runners in each part. All five parts together represent the total number of runners, so there are 25 runners in all. We complete the diagram.

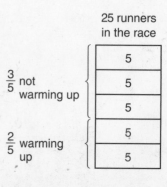

Example 1 Three eighths of the townspeople voted. If 120 of the townspeople voted, how many people live in the town?

Solution We are told that $\frac{3}{8}$ of the town voted, so we divide the whole into eight parts and mark off three of the parts. We are told that these three parts total 120 people. Since the three parts total 120, each part must be 40 (120 ÷ 3 = 40). Each part is 40, so all eight parts must be 8 times 40, which is **320 people.**

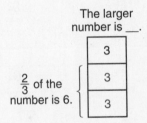

Example 2 Six is $\frac{2}{3}$ of what number?

Solution A larger number has been divided into three parts. Six is the total of two of the three parts. So each part equals three, and all three parts together equal **9.**

LESSON PRACTICE

Practice set* Solve. Draw a diagram for each problem.

   a. Eight is $\frac{1}{5}$ of what number?

   b. Eight is $\frac{2}{5}$ of what number?

   c. Nine is $\frac{3}{4}$ of what number?

   d. Sixty is $\frac{3}{8}$ of what number?

   e. Three fifths of the bunches of grapes were red grapes. If there were 18 bunches of red grapes, how many bunches of grapes were there altogether?

MIXED PRACTICE

Problem set   **1.** Three fifths of the townspeople voted. If 120 of the
(117)   townspeople voted, how many people live in the town?

2. If 130 children are separated as equally as possible into
(111)   four groups, how many will be in each group? (Write four
numbers, one for each of the four groups.)

3. If the parking lot charges $1.25 per half hour, what is the
(32, 95)   cost of parking a car from 11:15 a.m. to 2:45 p.m.?

4. If the area of the square is 400 m²,
(86)   then what is the area of the circle?
(Use 3.14 for π.)

5. Jackson correctly answered 46 of the 50 questions. What
(94)   percent of the questions did Jackson answer correctly?

6. The coordinates of the vertices of a triangle are (3, 6), (5, 0),
(Inv. 7, 79)   and (0, 0). What is the area of the triangle?

7. Write one hundred five thousandths as a decimal number.
(35)

8. Round the quotient of $7.00 ÷ 9 to the nearest cent.
(51)

9. Arrange in order from least to greatest:
(99)

$$81\%, \frac{4}{5}, 0.815$$

Solve and check:

10. $6x - 12 = 60$
(106)

11. $\frac{9}{15} = \frac{m}{25}$
(85)

12. Six is $\frac{2}{5}$ of what number?
(117)

13. $\left(5 - 1\frac{2}{3}\right) - 1\frac{1}{2}$
(63)

14. $2\frac{2}{5} \div 1\frac{1}{2}$
(68)

15. 0.625 × 2.4
(39)

16. −5 + −5 + −5
(104)

17. The prime factorization of 24 is 2 · 2 · 2 · 3, which we
(73)   can write as 2^3 · 3. Write the prime factorization of 36
       using exponents.

18. What is the total price of a $12.50 item plus 6% sales tax?
(41)

19. What is the area of this pentagon?
(107)

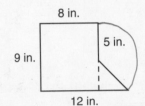
8 in.

5 in.

9 in.

12 in.

20. Write the standard numeral for 6×10^5.
(113)

21. Calculate mentally:
(112)

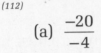

(a) $\dfrac{-20}{-4}$

(b) $\dfrac{-36}{6}$

(c) (−3)(8)

(d) (−4)(−9)

22. If each small cube has a volume of
(82)   one cubic inch, then what is the
       volume of this rectangular solid?

23. Draw a triangle in which each angle measures less than
(93)   90°. What type of triangle did you draw?

24. The mean, median, and mode of Austin's test scores were
(Inv. 5) 89, 87, and 92 respectively. About half of the test scores
       were what score or higher?

25. A bank offers an annual percentage rate (APR) of 6.5%.
(116)

(a) By what decimal number do we multiply a deposit to
    find the total amount in an account after one year at
    this rate?

(b) Maria deposited $1000 into an account at this rate.
    How much money was in the account after three years?
    (Assume that the account earns compound interest.)

26. If the spinner is spun twice, what
(Inv. 10) is the probability that the arrow
will stop on a number greater than
1 on both spins?

27. By rotation and translation, these
(108) two congruent triangles can be
arranged to form a:

A. square   B. parallelogram   C. octagon

28. Draw a ratio box for this problem. Then solve the
(101) problem using a proportion.

> *The ratio of cattle to horses on the ranch was 15
> to 2. The combined number of cattle and horses
> was 1020. How many horses were on the ranch?*

29. $\sqrt{100} + 3^2 \times 5 - \sqrt{81} \div 3$
(92)

30. Which of these figures has the greatest number of lines of
(110) symmetry?

A. $\triangle$ B. $\square$ C. $\bigcirc$

LESSON
118   Estimating Area

WARM-UP

Facts Practice: 24 Percent-Fraction-Decimal Equivalents (Test M)

Mental Math:

   a. 6 is $\frac{1}{3}$ of what number?   **b.** $\frac{2}{3}$ of 15

   c. 40% of $20   **d.** 10 is what % of 40?

   e. 0.3 × 20   **f.** 300 · 300

   g. 10 × 10, − 10, ÷ 2, − 1, ÷ 4, × 3, − 1, ÷ 4

   h. What number is represented by the Roman numeral MCMLXIX?

Problem Solving:

When Toshi said 9, Sarah said 80. When Brad said 6, Sarah said 35. When Diana said 7, Sarah said 48. Find the rule Sarah used to change the numbers she was given. What number would Sarah say if Sonia said 10?

NEW CONCEPT

In Lesson 86 we used a grid to estimate the area of a circle. Recall that we counted squares with most of their area within the circle as whole units. We counted squares with about half of their area in the circle as half units. In the figure at right, we have marked these "half squares" with dots.

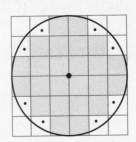

We can use a grid to estimate the areas of shapes whose areas would otherwise be difficult to calculate.

Example A one-acre grid is placed over an aerial photograph of a lake. Estimate the surface area of the lake in acres.

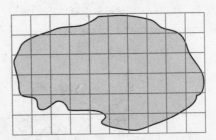

Solution   Each square on the grid represents an area of one acre. The curve is the shoreline of the lake. We count each square that is entirely or mostly within the curve as one acre. We count as half acres those squares that are about halfway within the curve. (Those squares are marked with dots in the figure below.) We ignore bits of squares within the curve because we assume that they balance out the bits of squares we counted that lay outside the curve.

		•	•	1	2	3	•	
•	4	5	6	7	8	9	10	11
12	13	14	15	16	17	18	19	20
•	21	22	23	24	25	26	27	28	29
	•	30	31	32	33	34	35	•
			36	37	•			

We count 37 entire or nearly entire squares within the shoreline. We also count ten squares with about half of their area within the shoreline. Ten half squares is equivalent to five whole squares. So we estimate the surface area of the lake to be about **42 acres.**

LESSON PRACTICE

Practice set   Estimate the area of the paw print shown below.

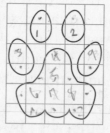

MIXED PRACTICE

Problem set

1. If 52 cards are dealt to 7 people as evenly as possible, how many people will be dealt 8 cards?
(111)

2. About how long is a new pencil?
(7)

A. 1.8 cm   B. 18 cm   C. 180 cm

3. Texas is the second most populous state in the United
(117) States. About 7 million people under the age of 21 lived
in Texas in the year 2000. This number was about $\frac{1}{3}$ of the
total population of the state at that time. About how many
people lived in Texas in 2000?

4. The symbol $\neq$ means "is not equal to." Which statement
(76) is true?

A. $\frac{3}{4} \neq \frac{9}{12}$ B. $\frac{3}{4} \neq \frac{9}{16}$ C. $\frac{3}{4} \neq 0.75$

5. What is the total price, including 7% tax, of a $14.49 item?
(41)

6. As Bryan peered out his window he saw 48 trucks, 84 cars,
(23) and 12 motorcycles go by his home. What was the ratio of
trucks to cars that Bryan saw?

7. What is the mean of 17, 24, 27, and 28?
$(Inv. 5)$

8. Arrange in order from least to greatest:
(74)

$$6.1, \sqrt{36}, 6\frac{1}{4}$$

9. Nine cookies were left in the package. That was $\frac{3}{10}$ of the
(117) original number of cookies. How many were in the
package originally?

10. Buz measured the circumference of the trunk of the old
(47) oak tree. How can Buz calculate the approximate
diameter of the tree?

11. Twelve is $\frac{3}{4}$ of what number?
(117)

12. $2\frac{2}{3} + \left(5\frac{1}{3} - 2\frac{1}{2}\right)$ **13.** $6\frac{2}{3} \div 4\frac{1}{6}$
(72) (68)

14. $4\frac{1}{4} + 3.2$ (decimal answer)
(74)

15. $1 - (0.1)^2$ **16.** $\sqrt{441}$
(92) (89)

17. Use a unit multiplier to convert 2.5 pounds to ounces.
(114)

18. The quadrilateral at right is a
(71) parallelogram.

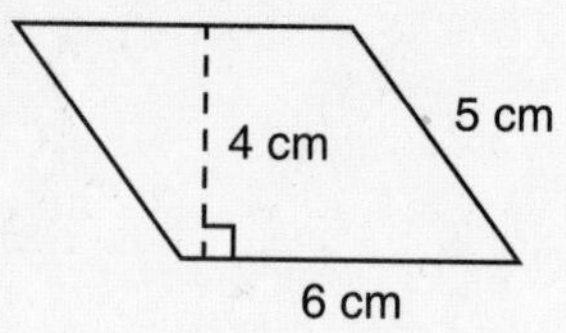

(a) What is the area of the parallelogram?

(b) If each obtuse angle measures 127°, then what is the measure of each acute angle?

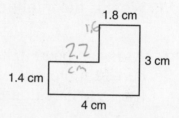

19. What is the perimeter of this hexagon?
(103)

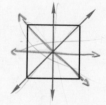

20. We show two lines of symmetry for
(110) this square. A square has a total of how many lines of symmetry?

21. Each edge of a cube measures 4 feet. What is the volume
(82) of the cube?

22. Complete the proportion: $\dfrac{f}{12} = \dfrac{12}{16}$
(85)

23. Find the rule for this function.
(96) Then use the rule to find the missing number.

x	y
2	10
3	15
5	25
?	40

24. Write the standard number for 1.25×10^4.
(113)

25. $-5 + {}^{+}2 - {}^{+}3 - {}^{-}4 + {}^{-}1$
(104)

(handwritten in left margin)
2
1.6
2.2
1.8
3.0
4.0
+1.4
─────
14.0

Mastermind

26. Chad deposited $4000 in an account that pays $7\frac{1}{2}\%$ interest
(116) compounded annually. How much money was in the
account after two years?

27. Convert $7\frac{1}{2}\%$ to a fraction.
(115)

28. At noon the temperature was −3°F. By sunset the
(14) temperature had dropped another five degrees. What was
the temperature at sunset?

29. There were three red marbles, three white marbles, and
(Inv. 10) three blue marbles in a bag. Luis drew a white marble out
of the bag and held it. If he draws another marble out of
the bag, what is the probability that the second marble
also will be white?

30. The area of square $ABCD$ is 16 units². Estimate the area of
(118) square $QRST$.

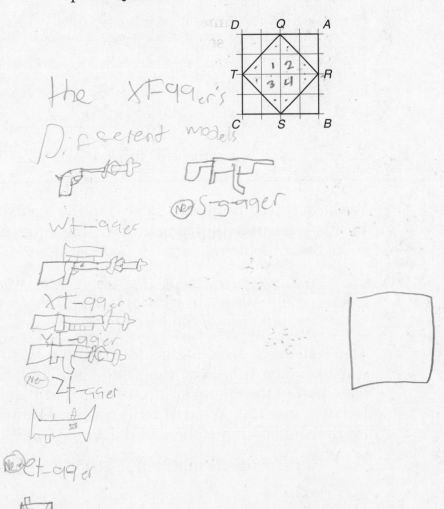

LESSON
119

Finding a Whole When a Percent Is Known

WARM-UP

Facts Practice: Measurement Facts (Test N)

Mental Math:

a. $\frac{2}{3}$ of what number is 6?   **b.** $\frac{7}{10}$ of 60

c. 70% of $50   **d.** 3 is what % of 6?

e. 0.8 × 70   **f.** $\frac{5000}{25}$

g. $\frac{1}{2}$ of 50, $\sqrt{}$, × 6, + 2, ÷ 4, × 3, + 1, × 4, $\sqrt{}$

h. What number is represented by the Roman numeral MDCCLXXVI?

Problem Solving:

At one o'clock the hour and minute hand of a clock form an angle that measures how many degrees?

NEW CONCEPT

We have solved problems like the following using a ratio box. In this lesson we will practice writing equations to help us solve these problems.

> *Thirty percent of the tribe were warriors. There were 150 warriors in all. What was the population of the tribe?*

The statement above tells us that 30% of the tribe were warriors. The tribe was the whole group, and the warriors were part of the whole group. We are told that the number of warriors was 150. We will write an equation using *t* to stand for the number of members in the whole tribe.

30% of the members of the tribe were warriors.

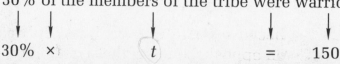

$$30\% \times \qquad t \qquad = \qquad 150$$

Now we change 30% to a fraction or to a decimal. For this problem we choose to write 30% as a decimal.

$$0.3t = 150$$

Now we find t by dividing 150 by three tenths.

$$\overset{500.}{03.\overline{)1500.}}$$

We find that there were 500 members in the entire tribe.

Example 1 Thirty percent of what number is 120?

Solution To translate the question into an equation, we translate the word *of* into a multiplication sign and the word *is* into an equal sign. For the words *what number* we write the letter *n*.

Thirty percent of what number is 120?

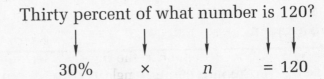

$$30\% \quad \times \quad n \quad = 120$$

We may choose to change 30% to a fraction or a decimal number. We choose the decimal form.

$$0.3n = 120$$

Now we find n by dividing 120 by 0.3.

$$\overset{400.}{03.\overline{)1200.}}$$

Thirty percent of **400** is 120.

Example 2 Sixteen is 25% of what number?

Solution We translate the question into an equation, using an equal sign for *is*, a multiplication sign for *of*, and a letter for *what number*.

Sixteen is 25% of what number?

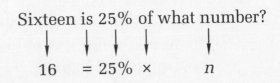

$$16 \quad = 25\% \times \quad n$$

Because of the way the question was asked, the numbers are on opposite sides of the equal sign as compared to example 1.

We can solve the equation in this form, or we can rearrange the equation. Either form of the equation may be used.

$$16 = 25\% \times n$$

$$25\% \times n = 16$$

We will use the first form of the equation and change 25% to the fraction $\frac{1}{4}$.

$$16 = 25\% \times n$$

$$16 = \frac{1}{4}n$$

We find n by dividing 16 by $\frac{1}{4}$.

$$16 \div \frac{1}{4} = \frac{16}{1} \times \frac{4}{1} = 64$$

Sixteen is 25% of **64**.

LESSON PRACTICE

Practice set* Translate each question into an equation and solve:

a. Twenty percent of what number is 120?

b. Fifty percent of what number is 30?

c. Twenty-five percent of what number is 12?

d. Twenty is 10% of what number?

e. Twelve is 100% of what number?

f. Fifteen is 15% of what number?

MIXED PRACTICE

Problem set **1.** Divide 555 by 12 and write the quotient
(25)   (a) with a remainder.

(b) as a mixed number.

2. The six gymnasts scored 9.75, 9.8, 9.9, 9.4, 9.9, and 9.95.
(37)   The lowest score was not counted. What was the sum of the five highest scores?

3. Cantara said that the six trumpet players made up 10% of
(105)   the band. The band had how many members?

4. Eight is $\frac{2}{3}$ of what number?
(117)

5. Write the standard number for the following:
(92)

$$(1 \times 10^5) + (8 \times 10^4) + (6 \times 10^3)$$

6. On Rob's scale drawing, each inch represents 8 feet. One
(Inv. 11) of the rooms in his drawing is $2\frac{1}{2}$ inches long. How long is
the actual room?

7. Two angles of a triangle each measure 45°.
(98) (a) What is the measure of the third angle?

(b) Make a rough sketch of the triangle.

8. Convert $8\frac{1}{3}$% to a fraction.
(115)

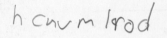

9. Nine dollars is what percent of $12?
(75)

10. Twenty percent of what number is 12?
(119)

11. Three tenths of what number is 9?
(117)

12. $(-5) - (+6) + (-7)$ **13.** $(-15)(-6)$
(104)   (112)

14. Reduce: $\dfrac{60}{84}$ **15.** $2\frac{1}{2} - 1\frac{2}{3}$
(29)   (63)

16. Stephen competes in a two-event race made up of biking
(101) and running. The ratio of the length of the distance run to
the length of the bike ride is 2 to 5. If the distance run was
10 kilometers, then what was the total length of the two-
event race?

17. The area of the shaded triangle is 2.8 cm². What is the
(79) area of the parallelogram?

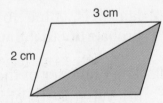

18. The figurine was packed in a box that was 10 in. long, 3 in.
(82) wide, and 4 in. deep. What was the volume of the box?

19. A rectangle that is not a square has
(110) a total of how many lines of
symmetry?

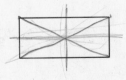

20. If this shape were cut out and
(Inv. 6) folded on the dotted lines, would it
form a cube, a pyramid, or a cone?

21. $3m - 5 = 25$
(106)

22. Find the rule for this function. Then use the rule to find the
(96) missing number.

x	3	4	6	?
y	12	16	24	32

23. How many pounds is 10 tons?
(102)

24. Which of these polygons is not a quadrilateral?
(64)

    A. parallelogram   B. pentagon   C. trapezoid

25. Compare: area of the square ◯ area of the circle
(86)

26. The coordinates of three points that are on the same line
(Inv. 7) are (–3, –2), (0, 0), and (x, 4). What number should
replace x in the third set of coordinates?

27. Robert flipped a coin. It landed heads up. He flipped the
(58) coin a second time. It landed heads up. If he flips the coin
a third time, what is the probability that it will land
heads up?

28. James is going to flip a coin three times. What is the
(Inv. 10) probability that the coin will land heads up all three times?

29. The diameter of the circle is 10 cm.
(38, 86)
    (a) What is the area of the square?

    (b) What is the area of the circle?

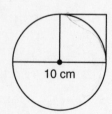

10 cm

30. What is the mode and the range of this set of numbers?
(Inv. 5)

4, 7, 6, 4, 5, 3, 2, 6, 7, 9, 7, 4, 10, 7, 9

LESSON

120 Volume of a Cylinder

Facts Practice: 24 Percent-Fraction-Decimal Equivalents (Test M)

Mental Math:

a. $\frac{2}{3}$ of 27

b. 80% of $60

c. $\frac{3}{4}$ of what number is 9?

d. 5 is what % of 5?

e. 0.8 × 400

f. 10 · 20 · 30

g. $\frac{1}{4}$ of 24, × 5, + 5, ÷ 7, × 8, + 2, ÷ 7, + 1, ÷ 7

h. What number is represented by the Roman numeral MCDXCII?

Problem Solving:

The first three prime numbers are 2, 3, and 5. These numbers can be arranged to form a three-digit number that is also a prime number. What is that three-digit prime number?

NEW CONCEPT

Imagine pressing a quarter down into a block of soft clay.

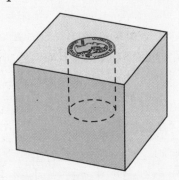

As the quarter is pressed into the block, it creates a hole in the clay. The quarter sweeps out a cylinder as it moves through the clay. We can calculate the volume of the cylinder by multiplying the area of the circular face of the quarter by the distance it moved through the clay. The distance the quarter moved is the **height** of the cylinder.

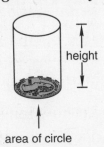

height

area of circle

Example   The diameter of this cylinder is 20 cm. Its height is 10 cm. What is its volume?

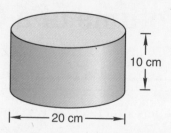

Solution   To calculate the volume of a cylinder, we find the area of a circular end of the cylinder and multiply that area by the height of the cylinder—the distance between the circular ends.

Since the diameter of the cylinder is 20 cm, the radius is 10 cm. A square with a side the length of the radius has an area of 100 cm². So the area of the circle is about 3.14 times 100 cm², which is 314 cm².

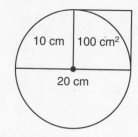

Now we multiply the area of the circular end of the cylinder by the height of the cylinder.

$$314 \text{ cm}^2 \times 10 \text{ cm} = 3140 \text{ cm}^3$$

We find that the volume of the cylinder is approximately **3140 cm³**.

LESSON PRACTICE

Practice set   A large can of soup has a diameter of about 8 cm and a height of about 12 cm. The volume of the can is about how many cubic centimeters? Round your answer to the nearest hundred cubic centimeters.

MIXED PRACTICE

Problem set   **1.** Write the prime factorization of 750 using exponents.
(73)

2. About how long is your little finger?
(7)
   A. 0.5 mm   B. 5 mm   C. 50 mm   D. 500 mm

3. If 3 parts weigh 24 grams, how much do 8 parts weigh?
(88)

24 grams

4. Complete the proportion: $\frac{3}{24} = \frac{8}{w}$
(85)

5. Write the standard number for $(7 \times 10^3) + (4 \times 10^0)$.
(92)

6. Use digits to write two hundred five million, fifty-six thousand.
(12)

7. The mean of four numbers is 25. Three of the numbers are 17, 23, and 25.
(Inv. 5)

(a) What is the fourth number?

(b) What is the range of the four numbers?

8. Calculate mentally:
(100, 112)

(a) −6 − −4   (b) −10 + −15   (c) (−10)(−10)

9. Write $16\frac{2}{3}\%$ as a reduced fraction.
(115)

10. Twenty-four guests came to the party. This was $\frac{4}{5}$ of those who were invited. How many guests were invited?
(117)

11. $1\frac{1}{3} + 3\frac{3}{4} + 1\frac{1}{6}$
(61)

12. $\frac{5}{6} \times 3 \times 2\frac{2}{3}$
(72)

13. 5.62 + 0.8 + 4
(38)

14. 0.08 ÷ (1 ÷ 0.4)
(49)

15. (−2) + (−2) + (−2)
(104)

16. $\sqrt{2500} + \sqrt{25}$
(89)

17. At $1.12 per pound, what is the price per ounce (1 pound = 16 ounces)?
(114)

18. The children held hands and stood in a circle. The diameter of the circle was 10 m. What was the circumference of the circle? (Use 3.14 for π.)
(47)

19. If the area of a square is 36 cm², what is the perimeter of the square?
(38)

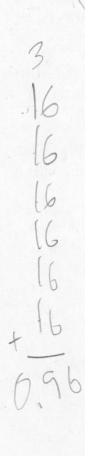

20. If each small cube has a volume of
(82) 1 cm³, then what is the volume of
this rectangular solid?

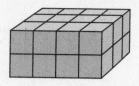

21. Sixty percent of the votes were cast for Shayla. If Shayla
(105) received 18 votes, how many votes were cast in all?

22. Rodric does not know the answers to three multiple-
(Inv. 10) choice questions. If the options are A, B, C, and D, and if
he cannot eliminate any choices, what is the probability
that he will correctly guess all three answers?

23. On a 20-question multiple-choice test with four choices
(58) for each question, how many questions are likely to be
answered correctly by simply guessing each answer?

24. (a) $(-8) - (+7)$ (b) $(-8) - (-7)$
(100)

25. $+3 + -5 - -7 - +9 + +11 + -7$
(104)

26. The three vertices of a triangle have the coordinates
(Inv. 7, 79) $(0, 0)$, $(-8, 0)$, and $(-8, -8)$. What is the area of the
triangle?

27. Jan tossed a coin and it landed heads up. What is the
(Inv. 10) probability that her next two tosses of the coin will also
land heads up?

28. The inside diameter of a mug is
(120) 8 cm. The height of the mug is
7 cm. What is the capacity of the
mug in cubic centimeters? (Think
of the capacity of the mug as the
volume of a cylinder with the
given dimensions.)

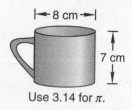

Use 3.14 for π.

29. A cubic centimeter of liquid is a milliliter of liquid. The
(78) mug in problem 28 will hold how many milliliters of hot
chocolate? Round to the nearest ten milliliters.

30. Draw a ratio box for this problem. Then solve the
(105) problem using a proportion.

*Ricardo correctly answered 90% of the questions on
the test. If he incorrectly answered four questions,
how many questions did he answer correctly?*

INVESTIGATION 12

Focus on

Platonic Solids

Recall from Lesson 60 that polygons are closed, two-dimensional figures with straight sides. If every face of a solid figure is a polygon, then the solid figure is called a **polyhedron.** Thus polyhedrons do not have any curved surfaces. So rectangular prisms and pyramids are polyhedrons, but spheres and circular cylinders are not.

Remember also that regular polygons have sides of equal length and angles of equal measure. Just as there are regular polygons, so there are regular polyhedrons. A cube is one example of a regular polyhedron. All the edges of a cube are of equal length, and all the angles are of equal measure, so all the faces are congruent regular polygons.

There are five regular polyhedrons. These polyhedrons are known as the **Platonic solids,** named after the ancient Greek philosopher Plato. We illustrate the five Platonic solids below.

   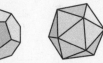

tetrahedron   cube   octahedron dodecahedron icosahedron

In this activity we will construct models of four of the Platonic solids.

Activity: *Platonic Solids*

Materials needed:

- Activity Sheets 19 and 20 (available in *Saxon Math 7/6—Homeschool Tests and Worksheets*)
- ruler
- scissors
- glue or tape

Working with your teacher or a partner is helpful. Sometimes more than two hands are needed to fold and glue.

Beginning with the tetrahedron pattern, cut around the border of the pattern. The line segments in the pattern are fold lines. Do not cut these. The folds will become the edges of the polyhedron. The triangles marked with "T" are tabs for gluing. These tabs are tucked inside the polyhedron and are hidden from view when the polyhedron is finished.

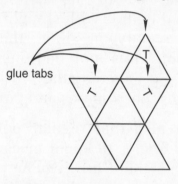

glue tabs

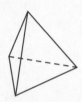

tetrahedron

Fold the pattern to make a pyramid with four faces. Glue the tabs or tape across the joining edges to hold the pattern in place.

When you have completed the tetrahedron, select another pattern to cut, fold, and form. All tabs are hidden when the pattern is properly folded, but all other polygons should be fully visible. When you have completed the models, copy this table and fill in the missing information by studying your models.

Platonic Solid	Each face is what polygon?	How many faces?	How many vertices?	How many edges?
tetrahedron	equilateral triangle	4		
cube	square	6	8	12
octahedron				
dodecahedron				

Extensions   **a.** This arrangement of four equilateral triangles was folded to make a model of a tetrahedron. Draw another arrangement of four adjoining equilateral triangles that can be folded to make a tetrahedron model. (Omit tabs.)

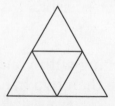

b. This arrangement of six squares was folded to make a model of a cube. How many other different patterns of six adjacent squares can you draw that can be folded to make a model of a cube? (Omit tabs.)

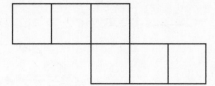

c. Using scissors and glue, cut out and construct the icosahedron model on Activity Sheet 21 (available in *Saxon Math 7/6—Homeschool Tests and Worksheets*). Working with your teacher or a partner is helpful. We suggest pre-folding the pattern before making the cuts to separate the tabs. Remember that the triangles marked with a "T" are tabs and should be hidden from view when the model is finished.

Once you have constructed the model, hold it lightly between your thumb and forefinger. You should be able to rotate the icosahedron while it is in this position. Since your icosahedron is a regular polyhedron, you also should be able to reposition the model so that your fingers touch two different vertices but the appearance of the figure remains unchanged.

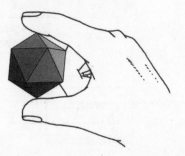

Holding the model as shown, your thumb and finger each touch a vertex. As you rotate the icosahedron, you can count the vertices. How many vertices are there in all? How many faces are there in all? What is the shape of each face?

New "Bernardepedia"

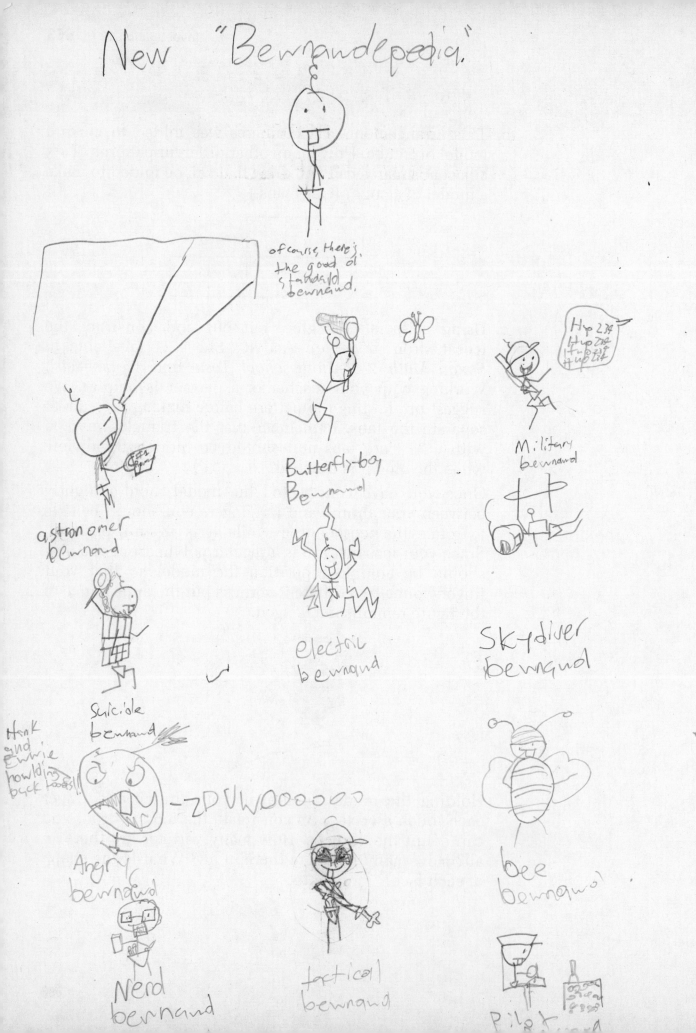

of course, there's the good ol' standard bernard.

astronomer bernard

Butterfly-boy Bernard

Military bernard

electric bernard

Skydiver bernard

Suicide bernard

Hank and Ennie howlding back loose!!

-ZPUWOOOOOO

Angry bernard

bee bernard

Nerd bernard

tactical bernard

Pilot

Additional Topics and Supplemental Practice

the bewnawd's friend,

T X-T 99er in 12 different
colors

Scope available in 4 models
foresight detatchable
takes 9-volt

$50 buck toofs

See page 679

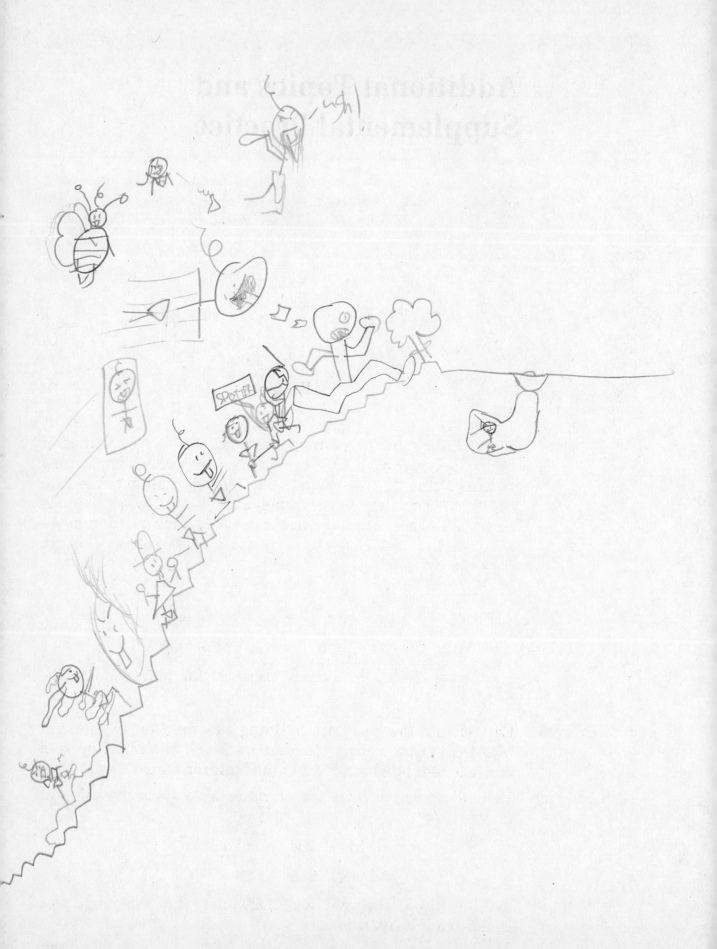

TOPIC

A   Roman Numerals

NEW CONCEPT

Not all number systems use place value. The value of a Roman numeral is the same whatever its place. Here are the values of the Roman numerals we will consider in this book:

I	1
V	5
X	10
L	50
C	100
D	500
M	1000

To find the value of a Roman numeral, we add the values of the numerals. So MCLXII equals 1000 plus 100 plus 50 plus 10 plus 1 plus 1, which equals 1162. An exception to the rule of adding the values occurs when a numeral of lesser value is to the left of a numeral of greater value. In such a situation we subtract the lesser value from the greater value. The six possible combinations are these:

IV = 4	IX = 9
XL = 40	XC = 90
CD = 400	CM = 900

So, for example, the Roman numeral for 999 is CMXCIX, not IM.

Example   Carved into the base of a building was the Roman numeral MCMXXIV, indicating the year in which the building was constructed. In what year was the building constructed?

Solution   We will spread out the Roman numeral to show the value of its parts.

M	CM	XX	IV
1000	900	20	4

Adding these values, we find that the building was constructed in **1924.**

LESSON PRACTICE

Practice set   Find the value of each Roman numeral:[†]

   a. XXXIX   **b.** LXIV

   c. MCMXIX   **d.** MMII

Bilbo
Bewncwd
MUST
DEFEAT
SMAUG
FRIZZLEPOOP!

[†]Roman numerals are practiced in Warm-ups 101–120.

Supplemental Practice Problems for Selected Lessons

This appendix contains additional practice problems for concepts presented in selected lessons. It is very important that no problems in the regular problem sets be skipped to make room for these problems. Saxon math is designed to produce long-term retention through repeated exposure to concepts in the problem sets. The problem sets provide enough exposure to concepts for most students. However, the problems in this appendix can be completed to provide additional exposure as necessary.

Lesson 2   Multiply:

Set A

1. 576 × 8   **2.** $3.08 × 7

3. 784 × 6   **4.** 4306 × 9

5. 42¢ × 30   **6.** 56 × 40

7. 60 × 78   **8.** 70¢ × 64

9. 90 × 70   **10.** 300 × 60

11. 400 × 50   **12.** 80 × 500

13. 37 × 43   **14.** 62¢ × 74

15. 86 × 27   **16.** 94 × 63

17. $4.08 × 24   **18.** 507 × 37

19. 62 × 409   **20.** 84 × $3.06

21. 520 × 36   **22.** 940 × 42

23. $7.90 × 86   **24.** 243 × 67

Set B   Divide. Write uneven division answers with a remainder.

1. 357 ÷ 5   **2.** $2.44 ÷ 4

3. 892 ÷ 7   **4.** 143 ÷ 4

5. $4.12 ÷ 4   **6.** 423 ÷ 6

7. $7.28 ÷ 7   **8.** 812 ÷ 9

9. 1206 ÷ 6   **10.** $8.24 ÷ 4

11. 906 ÷ 9   **12.** 3492 ÷ 7

13. $1.44 ÷ 12   **14.** 472 ÷ 10

15. 893 ÷ 15   **16.** 762 ÷ 25

17. 432 ÷ 20   **18.** 986 ÷ 50

19. 1427 ÷ 25   **20.** 3819 ÷ 19

21. 4126 ÷ 32   **22.** 968 ÷ 24

23. 377 ÷ 18   **24.** 566 ÷ 42

Lesson 3   Find the missing number in each problem:

1.
$$\begin{array}{r} 12 \\ + \ A \\ \hline 27 \end{array}$$

2.
$$\begin{array}{r} B \\ + \ 15 \\ \hline 40 \end{array}$$

3.
$$\begin{array}{r} 8 \\ + \ C \\ \hline 33 \end{array}$$

4.
$$\begin{array}{r} D \\ + \ 21 \\ \hline 50 \end{array}$$

5.
$$\begin{array}{r} 16 \\ + \ E \\ \hline 45 \end{array}$$

6.
$$\begin{array}{r} F \\ + \ 26 \\ \hline 72 \end{array}$$

7.
$$\begin{array}{r} 36 \\ + \ G \\ \hline 64 \end{array}$$

8.
$$\begin{array}{r} H \\ + \ 55 \\ \hline 81 \end{array}$$

9.
$$\begin{array}{r} 25 \\ - \ J \\ \hline 12 \end{array}$$

10.
$$\begin{array}{r} K \\ - \ 36 \\ \hline 24 \end{array}$$

11.
$$\begin{array}{r} 40 \\ - \ L \\ \hline 17 \end{array}$$

12.
$$\begin{array}{r} M \\ - \ 17 \\ \hline 32 \end{array}$$

13.
$$\begin{array}{r} 38 \\ - \ N \\ \hline 16 \end{array}$$

14.
$$\begin{array}{r} P \\ - \ 43 \\ \hline 19 \end{array}$$

15.
$$\begin{array}{r} 63 \\ - \ Q \\ \hline 48 \end{array}$$

16.
$$\begin{array}{r} R \\ - \ 24 \\ \hline 39 \end{array}$$

17. $S + 26 = 62$

18. $T - 26 = 43$

19. $35 + U = 57$

20. $42 - V = 15$

21. $W + 54 = 80$

22. $X - 38 = 14$

23. $27 + Y = 72$

24. $60 - Z = 32$

Lesson 4   Find the missing number in each problem:

1.
$$\begin{array}{r} A \\ \times \ 5 \\ \hline 30 \end{array}$$

2.
$$\begin{array}{r} 7 \\ \times \ B \\ \hline 63 \end{array}$$

3.
$$\begin{array}{r} C \\ \times \ 8 \\ \hline 120 \end{array}$$

4.
$$\begin{array}{r} 9 \\ \times \ D \\ \hline 180 \end{array}$$

5.
$$\begin{array}{r} E \\ \times \ 6 \\ \hline 150 \end{array}$$

6.
$$\begin{array}{r} 4 \\ \times \ F \\ \hline 56 \end{array}$$

7.
$$\begin{array}{r} G \\ \times \ 9 \\ \hline 234 \end{array}$$

8.
$$\begin{array}{r} 8 \\ \times \ H \\ \hline 176 \end{array}$$

9. $J\overline{)96}$ with 8 above

10. $8\overline{)K}$ with 7 above

11. $L\overline{)105}$ with 7 above

12. $12\overline{)M}$ with 15 above

13. $\dfrac{N}{5} = 14$

14. $\dfrac{90}{P} = 6$

15. $\dfrac{Q}{25} = 20$

16. $\dfrac{84}{R} = 7$

17. $6S = 90$

18. $T \div 15 = 9$

19. $7U = 126$

20. $152 \div V = 8$

21. $128 = 8W$

22. $X \div 16 = 8$

23. $153 = 9Y$

24. $144 \div Z = 9$

Lesson 12   Use digits to write each number:

1. five thousand

2. two hundred eight

3. one thousand, two hundred

4. six thousand, fifty

5. nine hundred forty-three

6. eight thousand, one hundred ten

7. ten thousand

8. twenty-one thousand

9. forty thousand, nine hundred

10. one thousand, ten

11. fifteen thousand, twenty-one

12. nineteen thousand, eight hundred

13. one hundred thousand

14. two hundred ten thousand

15. four hundred five thousand

16. three hundred twenty-five thousand

17. one million

18. one million, two hundred thousand

19. ten million, one hundred fifty thousand

20. five hundred million

21. two million, fifty thousand

22. twenty-five million, seven hundred fifty thousand

23. five billion

24. one billion, two hundred fifty million

25. twenty-one billion, five hundred ten million

26. two hundred billion

27. one trillion

28. ten trillion

29. two trillion, five hundred billion

30. two hundred trillion

Lesson 16 Round each number to the nearest ten:

1. 678 **2.** 83 **3.** 575 **4.** 909

5. 99 **6.** 1492 **7.** 104 **8.** 1321

Round each number to the nearest hundred:

9. 678 **10.** 437 **11.** 846 **12.** 1587

13. 1023 **14.** 987 **15.** 3679 **16.** 4981

Round each number to the nearest thousand:

17. 1986 **18.** 2317 **19.** 1484

20. 3675 **21.** 5280 **22.** 1760

23. 36,102 **24.** 57,843 **25.** 375,874

Lesson 18 Find the average of each set of numbers:

1. 15, 18, 21 **2.** 16, 18, 20, 22

3. 5, 6, 7, 8, 9 **4.** 2, 4, 6, 8, 10, 12

5. 100, 200, 300, 400 **6.** 20, 30, 40, 50, 60

7. 23, 35, 32 **8.** 136, 140, 141

9. 94, 94, 98, 98 **10.** 68, 72, 68, 76, 76

11. 6847, 6951 **12.** 86, 86, 86, 86, 86

13. 562, 437, 381 **14.** 6, 6, 3, 9, 3, 6, 9

What number is halfway between each pair of numbers?

15. 47 and 91

16. 56 and 88

17. 75 and 57

18. 1 and 101

19. 92 and 136

20. 253 and 325

21. 548 and 752

22. 1776 and 1986

Lesson 20 List the factors of each number:

1. 30   **2.** 40   **3.** 50   **4.** 60

5. 35   **6.** 36   **7.** 37

8. 38   **9.** 39   **10.** 49

Find the greatest common factor (GCF) of each set of numbers:

11. 14, 28   **12.** 12, 20   **13.** 15, 16

14. 15, 25   **15.** 25, 50   **16.** 40, 70

17. 24, 42   **18.** 12, 21   **19.** 22, 55

20. 12, 30   **21.** 4, 3, 2   **22.** 2, 4, 6

23. 4, 8, 12   **24.** 6, 12, 20   **25.** 24, 30, 12, 15

Lesson 22 Solve:

1. $\frac{1}{2}$ of 42

2. $\frac{1}{3}$ of \$42

3. $\frac{2}{3}$ of 42

4. $\frac{1}{4}$ of \$60

5. $\frac{3}{4}$ of 60

6. $\frac{2}{3}$ of \$60

7. $\frac{1}{5} \times 60$

8. $\frac{2}{5} \times \$60$

9. $\frac{3}{5} \times 20$

10. $\frac{3}{8} \times \$24$

11. $\frac{5}{6} \times 24$

12. $\frac{3}{10} \times \$100$

13. What number is $\frac{2}{3}$ of 48?

14. How much money is $\frac{1}{5}$ of $90?

15. What number is $\frac{5}{8}$ of 40?

16. How much money is $\frac{1}{10}$ of $200?

17. What number is $\frac{9}{10}$ of 60?

18. How much money is $\frac{3}{4}$ of $28?

19. One third of 27 is what number?

20. Two thirds of 36 is what number?

21. Three fourths of 24 is what number?

22. Four fifths of 35 is what number?

23. Two ninths of 36 is what number?

24. Seven tenths of 30 is what number?

25. Five twelfths of 24 is what number?

Lesson 25 Write each improper fraction as a mixed number:

1. $\frac{3}{3}$ **2.** $\frac{5}{4}$ **3.** $\frac{7}{3}$ **4.** $\frac{17}{10}$ **5.** $\frac{24}{6}$

6. $\frac{24}{5}$ **7.** $\frac{24}{4}$ **8.** $\frac{32}{15}$ **9.** $\frac{32}{16}$

10. $\frac{27}{5}$ **11.** $\frac{36}{7}$ **12.** $\frac{25}{6}$ **13.** $\frac{35}{5}$

14. $\frac{12}{5}$ **15.** $\frac{31}{10}$ **16.** $1\frac{5}{2}$ **17.** $3\frac{6}{3}$

18. $7\frac{9}{4}$ **19.** $6\frac{8}{2}$ **20.** $4\frac{5}{3}$ **21.** $11\frac{6}{5}$

22. $4\frac{11}{10}$ **23.** $2\frac{13}{12}$ **24.** $1\frac{10}{3}$ **25.** $23\frac{7}{2}$

Lesson 29   Reduce each fraction to lowest terms:
Set A

1. $\dfrac{2}{6}$ 2. $\dfrac{3}{6}$ 3. $\dfrac{4}{6}$ 4. $\dfrac{2}{8}$ 5. $\dfrac{4}{8}$

6. $\dfrac{6}{8}$ 7. $\dfrac{3}{9}$ 8. $\dfrac{2}{10}$ 9. $\dfrac{4}{10}$

10. $\dfrac{5}{10}$ 11. $\dfrac{8}{10}$ 12. $\dfrac{2}{12}$ 13. $\dfrac{3}{12}$

14. $\dfrac{4}{12}$ 15. $\dfrac{6}{12}$ 16. $\dfrac{8}{12}$ 17. $\dfrac{9}{12}$

18. $3\dfrac{10}{12}$ 19. $4\dfrac{6}{15}$ 20. $1\dfrac{18}{24}$ 21. $2\dfrac{15}{18}$

22. $6\dfrac{16}{24}$ 23. $8\dfrac{12}{24}$ 24. $9\dfrac{8}{24}$ 25. $10\dfrac{10}{24}$

Set B   Find each sum or difference. Simplify the answer when possible.

1. $\dfrac{5}{8} + \dfrac{2}{8}$ 2. $\dfrac{5}{8} - \dfrac{2}{8}$ 3. $\dfrac{3}{6} + \dfrac{2}{6}$

4. $\dfrac{3}{6} - \dfrac{2}{6}$ 5. $\dfrac{1}{3} + \dfrac{1}{3}$ 6. $\dfrac{1}{3} - \dfrac{1}{3}$

7. $\dfrac{4}{9} + \dfrac{1}{9}$ 8. $\dfrac{4}{9} - \dfrac{2}{9}$ 9. $\dfrac{1}{4} + \dfrac{1}{4} + \dfrac{1}{4}$

10. $\dfrac{1}{7} + \dfrac{2}{7} + \dfrac{3}{7}$ 11. $\dfrac{3}{4} + \dfrac{2}{4}$ 12. $\dfrac{3}{4} - \dfrac{1}{4}$

13. $\dfrac{2}{3} + \dfrac{2}{3}$ 14. $\dfrac{3}{8} - \dfrac{1}{8}$ 15. $\dfrac{4}{5} + \dfrac{4}{5}$

16. $\dfrac{6}{5} - \dfrac{1}{5}$ 17. $\dfrac{5}{8} + \dfrac{3}{8}$ 18. $\dfrac{5}{8} - \dfrac{1}{8}$

19. $\dfrac{3}{10} + \dfrac{2}{10}$ 20. $\dfrac{9}{10} - \dfrac{1}{10}$ 21. $\dfrac{5}{12} + \dfrac{5}{12}$

22. $\dfrac{5}{12} - \dfrac{1}{12}$ 23. $\dfrac{3}{10} + \dfrac{7}{10}$ 24. $\dfrac{7}{10} - \dfrac{3}{10}$

Lesson 30 Find the least common multiple (LCM) of each set of numbers:

1. 3, 4   **2.** 3, 5   **3.** 3, 6   **4.** 4, 6

5. 6, 8   **6.** 4, 8   **7.** 3, 8   **8.** 2, 8

9. 3, 9   **10.** 6, 9   **11.** 6, 10   **12.** 4, 10

13. 8, 12   **14.** 9, 12   **15.** 10, 12   **16.** 2, 5, 10

17. 2, 3, 4   **18.** 2, 3, 6   **19.** 2, 4, 8   **20.** 2, 4, 6

Investigation 3

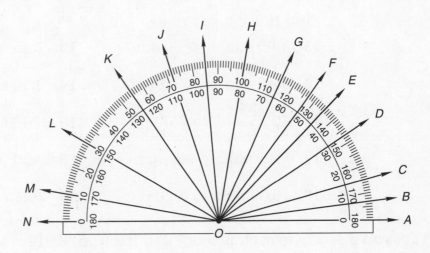

Using the protractor, find the measure of each of the following angles:

1. ∠AOB   **2.** ∠AOE   **3.** ∠AOH   **4.** ∠AOK

5. ∠AON   **6.** ∠AOD   **7.** ∠AOG   **8.** ∠AOJ

9. ∠AOM   **10.** ∠AOC   **11.** ∠AOF   **12.** ∠AOI

13. ∠AOL   **14.** ∠NOL   **15.** ∠NOI   **16.** ∠NOF

17. ∠NOC   **18.** ∠NOK   **19.** ∠NOH   **20.** ∠NOE

21. ∠NOM   **22.** ∠NOJ   **23.** ∠NOG   **24.** ∠NOD

Lesson 32 Write each number in standard form:

1. $(6 \times 100) + (7 \times 10)$

2. $(5 \times 1000) + (4 \times 100)$

3. $(7 \times 100) + (3 \times 1)$

4. $(8 \times 10) + (1 \times 1)$

5. $(9 \times 1000) + (5 \times 10)$

6. $(7 \times 100) + (3 \times 10)$

Write each number in expanded notation:

7. 560 8. 5600 9. 706 10. 5280

Find the elapsed time:

11. 3:00 a.m. to 8:30 a.m. 12. 7:15 a.m. to 10:00 a.m.

13. 8:45 a.m. to 11:20 a.m. 14. 9:30 a.m. to 1:15 p.m.

15. 10:25 a.m. to 2:00 p.m. 16. 9:40 a.m. to 2:30 p.m.

17. 5:15 p.m. to 9:05 p.m. 18. 2:50 p.m. to 6:15 p.m.

19. 10:35 a.m. to 6:20 p.m. 20. 10:53 a.m. to 2:27 p.m.

Lesson 33 Write each percent as a fraction. Reduce when possible.

1. 50% 2. 25% 3. 75% 4. 10%

5. 20% 6. 90% 7. 30% 8. 60%

9. 80% 10. 1% 11. 5% 12. 99%

13. 2% 14. 35% 15. 45% 16. 4%

17. 14% 18. 24% 19. 40% 20. 70%

Lesson 35 Write each number as a decimal number:

1. five tenths

2. three hundredths

3. eleven hundredths

4. one thousandth

5. twenty-five thousandths

6. one and two tenths

7. ten and four tenths

8. two and one hundredth

9. five and twelve hundredths

10. one hundred twenty thousandths

11. two hundred five thousandths

12. six and fifteen hundredths

13. ten and one hundred thousandths

14. twelve and six hundredths

15. ten and twenty-two thousandths

Write each fraction or mixed number as a decimal number:

16. $\dfrac{5}{100}$ **17.** $\dfrac{12}{1000}$ **18.** $\dfrac{3}{10}$ **19.** $\dfrac{1}{10}$

20. $\dfrac{23}{100}$ **21.** $\dfrac{124}{1000}$ **22.** $\dfrac{1}{1000}$

23. $\dfrac{45}{100}$ **24.** $\dfrac{3}{100}$ **25.** $\dfrac{52}{1000}$

Lesson 36   Subtract:

1. $1 - \dfrac{2}{3}$ **2.** $2 - 1\dfrac{1}{3}$ **3.** $6 - 3\dfrac{3}{4}$

4. $1 - \dfrac{2}{5}$ **5.** $3 - 1\dfrac{4}{5}$ **6.** $7 - 3\dfrac{1}{2}$

7. $1 - \dfrac{3}{8}$ **8.** $4 - 1\dfrac{1}{4}$ **9.** $8 - 3\dfrac{1}{3}$

10. $1 - \dfrac{7}{10}$ **11.** $3 - 1\dfrac{1}{2}$ **12.** $10 - 4\dfrac{5}{8}$

13. $1 - \dfrac{3}{4}$ **14.** $5 - 3\dfrac{3}{8}$ **15.** $12 - 4\dfrac{3}{10}$

Lesson 38 Simplify:

    1. $\sqrt{36}$ **2.** $\sqrt{4}$ **3.** $\sqrt{81}$ **4.** $\sqrt{49}$

    5. $\sqrt{121}$ **6.** $\sqrt{16}$ **7.** $\sqrt{64}$ **8.** $\sqrt{100}$

    9. $\sqrt{25}$ **10.** $\sqrt{9}$ **11.** $\sqrt{144}$ **12.** $\sqrt{1}$

Lesson 40 Find each sum or difference:
Set A

    1. 0.62 + 0.4   **2.** 0.62 − 0.4

    3. 1.5 + 0.15   **4.** 1.2 − 0.15

    5. 0.5 + 0.41   **6.** 0.5 − 0.41

    7. 0.23 + 0.6 + 1.4   **8.** 5.3 − 4.29

    9. 3.6 + 2 + 0.75   **10.** 1 − 0.3

    11. 4.75 + 3 + 12.5   **12.** 15.4 − 15.40

    13. 0.3 + 0.4 + 0.5   **14.** 5 − 1.25

    15. 0.36 + 0.4 + 0.575   **16.** 0.3 − 0.036

    17. 1 + 0.2 + 3.456   **18.** 10 − 0.7

    19. 0.6 + 0.7 + 0.8   **20.** 1 − 0.21

Set B Multiply:

    1. 0.3 × 4   **2.** 0.4 × 0.6   **3.** 0.3 × 0.2

    4. 0.4 × 0.3   **5.** 7 × 0.21   **6.** 0.6 × 1.24

    7. 0.36 × 0.4   **8.** 0.012 × 10   **9.** 1.2 × 8

    10. 6.2 × 0.07   **11.** 1.2 × 0.12   **12.** 1.25 × 10

    13. 3.6 × 1.2   **14.** 4.5 × 9   **15.** 0.015 × 0.03

    16. 6.75 × 0.1   **17.** 0.01 × 3.75   **18.** 1.5 × 1.5

    19. 0.25 × 0.25   **20.** 6.3 × 0.24   **21.** 4.2 × 100

Lesson 41 Complete this table:

Set A

	PERCENT	FRACTION	DECIMAL
1.	10%	(a)	(b)
2.	20%	(a)	(b)
3.	30%	(a)	(b)
4.	40%	(a)	(b)
5.	50%	(a)	(b)
6.	15%	(a)	(b)
7.	25%	(a)	(b)
8.	45%	(a)	(b)
9.	75%	(a)	(b)
10.	1%	(a)	(b)
11.	2%	(a)	(b)
12.	4%	(a)	(b)
13.	5%	(a)	(b)
14.	6%	(a)	(b)
15.	12%	(a)	(b)
16.	24%	(a)	(b)
17.	90%	(a)	(b)
18.	95%	(a)	(b)
19.	36%	(a)	(b)
20.	80%	(a)	(b)

Set B 21. What is 25% of 100?   22. What is 25% of 200?

23. What is 25% of 400?   24. What is 25% of 40?

25. What is 25% of 20?   26. What is 20% of 100?

27. What is 20% of 50?   28. What is 20% of 5?

29. What is 50% of 200?   30. What is 50% of 100?

31. What is 50% of 50?   32. What is 50% of 12?

33. What is 60% of 100?   **34.** What is 60% of 200?

35. What is 60% of 50?   **36.** What is 60% of 25?

37. What is 75% of 100?   **38.** What is 75% of 400?

39. What is 75% of 40?   **40.** What is 75% of 4?

Lesson 45   Divide:

1. $4.8 \div 6$ **2.** $0.48 \div 4$ **3.** $0.48 \div 8$

4. $0.125 \div 5$ **5.** $1.44 \div 6$ **6.** $0.018 \div 3$

7. $0.24 \div 12$ **8.** $5.6 \div 8$ **9.** $17.1 \div 9$

10. $3.65 \div 5$ **11.** $42.80 \div 10$ **12.** $3.10 \div 10$

13. $0.190 \div 5$ **14.** $0.234 \div 9$ **15.** $5.00 \div 4$

16. $0.7 \div 5$ **17.** $0.4 \div 4$ **18.** $0.5 \div 4$

19. $3.6 \div 10$ **20.** $0.24 \div 10$ **21.** $0.12 \div 8$

22. $0.9 \div 4$ **23.** $1.1 \div 8$ **24.** $0.51 \div 10$

Lesson 46   Write each number in expanded notation:

1. 3.5   **2.** 0.26

3. 4.08   **4.** 3.14

5. 0.015   **6.** 0.09

7. 12.5   **8.** 0.405

Write each number in decimal form:

9. $(6 \times 1) + \left(5 \times \dfrac{1}{10}\right)$

10. $\left(7 \times \dfrac{1}{10}\right) + \left(5 \times \dfrac{1}{100}\right)$

11. $(5 \times 10) + \left(5 \times \dfrac{1}{10}\right)$

12. $\left(8 \times \dfrac{1}{100}\right)$

13. $(7 \times 1) + \left(5 \times \dfrac{1}{100}\right)$

14. $\left(3 \times \dfrac{1}{100}\right) + \left(9 \times \dfrac{1}{1000}\right)$

15. $(8 \times 10) + (3 \times 1) + \left(2 \times \dfrac{1}{10}\right)$

16. $(7 \times 10) + \left(8 \times \dfrac{1}{10}\right) + \left(1 \times \dfrac{1}{100}\right)$

Lesson 48   Subtract:

1. $1 - \dfrac{1}{5}$ **2.** $1 - \dfrac{3}{8}$ **3.** $2 - \dfrac{1}{2}$

4. $2 - \dfrac{1}{3}$ **5.** $2 - 1\dfrac{1}{4}$ **6.** $3 - 1\dfrac{3}{8}$

7. $3 - 2\dfrac{5}{8}$ **8.** $8 - 3\dfrac{3}{4}$ **9.** $8 - 5\dfrac{1}{8}$

10. $10 - 4\dfrac{2}{5}$ **11.** $4\dfrac{1}{3} - 1\dfrac{2}{3}$ **12.** $4\dfrac{1}{5} - 1\dfrac{3}{5}$

13. $6\dfrac{1}{10} - 3\dfrac{4}{10}$ **14.** $5\dfrac{3}{8} - 3\dfrac{6}{8}$ **15.** $2\dfrac{1}{4} - 1\dfrac{2}{4}$

16. $5\dfrac{2}{5} - 3\dfrac{4}{5}$ **17.** $7\dfrac{3}{8} - 4\dfrac{4}{8}$ **18.** $9\dfrac{1}{10} - 3\dfrac{5}{10}$

19. $4\dfrac{2}{4} - 1\dfrac{3}{4}$ **20.** $8\dfrac{3}{5} - 1\dfrac{4}{5}$ **21.** $4\dfrac{1}{4} - 1\dfrac{3}{4}$

22. $4\dfrac{3}{8} - 1\dfrac{5}{8}$ **23.** $3\dfrac{1}{6} - 1\dfrac{5}{6}$ **24.** $6\dfrac{1}{10} - 4\dfrac{3}{10}$

Lesson 49   Change each problem so that the divisor is a whole number. Then divide.

    1. $5.2 \div 0.4$ **2.** $0.144 \div 0.8$ **3.** $3.21 \div 0.3$

    4. $1.00 \div 0.4$ **5.** $0.525 \div 0.05$ **6.** $8.1 \div 0.09$

    7. $1.2 \div 0.003$ **8.** $0.54 \div 0.006$ **9.** $1.2 \div 0.12$

  10. $0.12 \div 1.2$ **11.** $0.5 \div 0.04$ **12.** $3.6 \div 0.5$

  13. $0.12 \div 10$ **14.** $6.4 \div 100$ **15.** $3.5 \div 0.08$

  16. $8 \div 0.5$ **17.** $4 \div 0.4$ **18.** $12 \div 0.06$

  19. $18 \div 0.20$ **20.** $16 \div 0.008$ **21.** $5 \div 0.25$

  22. $4.44 \div 0.06$ **23.** $16 \div 0.25$ **24.** $0.3 \div 0.4$

Investigation 5   Find the (a) mean, (b) median, (c) mode, and (d) range for each set of data.

    1. Quiz scores: 10, 8, 7, 8, 6, 9, 8, 4, 9, 8

    2. Best times in 50 m race: 8.7 s, 9.3 s, 7.9 s, 9.8 s, 8.7 s, 9.6 s

    3. Ages of children in club: 6 yr, 8 yr, 9 yr, 10 yr, 10 yr

    4. Number of poems in each collection: 24, 23, 21, 24, 28, 25, 30, 27, 32

Lesson 51   Round each number to the nearest tenth:

    1. 0.48   **2.** 0.133   **3.** 0.375   **4.** 4.28

    5. 62.84   **6.** 0.0984   **7.** 6.25   **8.** 1.97

Round each number to the nearest hundredth (or cent):

    9. 0.8181   **10.** 0.6666   **11.** 1.333   **12.** 4.321

  13. $0.2345   **14.** $7.675   **15.** $0.166   **16.** $3.422

Round each number to the nearest whole number:

  17. 12.34   **18.** 4.567   **19.** 91.66   **20.** 142.8

Lesson 52   Solve mentally:

1. 4.2×10
2. 0.35×10

3. 0.178×10
4. 3.65×10

5. 4.21×100
6. 0.375×100

7. 6.5×100
8. 4.323×100

9. 7.275×1000
10. 6.4×1000

11. 0.86×1000
12. 0.01625×1000

13. $4.2 \div 10$
14. $0.42 \div 10$

15. $42.1 \div 10$
16. $6 \div 10$

17. $87.5 \div 100$
18. $6.5 \div 100$

19. $0.4 \div 100$
20. $372.8 \div 100$

21. $123.4 \div 1000$
22. $42.5 \div 1000$

23. $7.6 \div 1000$
24. $4 \div 1000$

Lesson 57   Find each sum or difference:

1. $\dfrac{1}{2} + \dfrac{1}{8}$
2. $\dfrac{1}{2} - \dfrac{1}{8}$
3. $\dfrac{3}{4} + \dfrac{1}{8}$

4. $\dfrac{3}{4} - \dfrac{1}{8}$
5. $\dfrac{2}{3} + \dfrac{1}{6}$
6. $\dfrac{2}{3} - \dfrac{1}{6}$

7. $\dfrac{1}{3} + \dfrac{1}{4}$
8. $\dfrac{1}{3} - \dfrac{1}{4}$
9. $\dfrac{3}{4} + \dfrac{2}{3}$

10. $\dfrac{3}{4} - \dfrac{2}{3}$
11. $\dfrac{1}{2} + \dfrac{1}{10}$
12. $\dfrac{1}{2} - \dfrac{1}{10}$

13. $\dfrac{3}{4} + \dfrac{3}{8}$
14. $\dfrac{3}{4} - \dfrac{3}{8}$
15. $\dfrac{2}{3} + \dfrac{1}{2}$

16. $\dfrac{2}{3} - \dfrac{1}{2}$
17. $\dfrac{7}{10} + \dfrac{1}{2}$
18. $\dfrac{7}{10} - \dfrac{1}{2}$

19. $\dfrac{1}{4} + \dfrac{1}{5}$
20. $\dfrac{1}{4} - \dfrac{1}{5}$
21. $\dfrac{3}{5} + \dfrac{1}{2}$

Lesson 63 Find each sum or difference:

1. $3\frac{1}{8} + 2\frac{1}{4}$

2. $3\frac{1}{4} - 2\frac{1}{8}$

3. $1\frac{1}{6} + 1\frac{1}{3}$

4. $2\frac{1}{3} - 1\frac{1}{6}$

5. $3\frac{3}{4} + 4\frac{1}{8}$

6. $4\frac{3}{4} - 3\frac{1}{8}$

7. $5\frac{3}{5} + 1\frac{3}{10}$

8. $5\frac{3}{5} - 1\frac{3}{10}$

9. $4\frac{1}{2} + 2\frac{1}{12}$

10. $4\frac{1}{2} - 2\frac{1}{12}$

11. $6\frac{1}{2} + 2\frac{1}{3}$

12. $4\frac{1}{2} - 2\frac{1}{3}$

13. $5\frac{1}{2} + 1\frac{2}{3}$

14. $5\frac{1}{2} - 1\frac{2}{3}$

15. $1\frac{1}{2} + \frac{3}{4}$

16. $1\frac{1}{2} - \frac{3}{4}$

17. $6\frac{3}{5} + 1\frac{1}{2}$

18. $6\frac{3}{5} - 1\frac{1}{2}$

19. $2\frac{7}{10} + 1\frac{1}{5}$

20. $2\frac{7}{10} - 1\frac{1}{5}$

21. $8\frac{2}{3} + 1\frac{3}{4}$

22. $8\frac{2}{3} - 1\frac{3}{4}$

23. $3\frac{1}{2} + 1\frac{1}{3}$

24. $5\frac{1}{4} + 3\frac{1}{2}$

Lesson 65 Write the prime factorization of each number:

1. 6

2. 8

3. 9

4. 10

5. 12

6. 14

7. 15

8. 16

9. 18

10. 20

11. 21

12. 24

13. 30

14. 36

15. 39

16. 40

17. 42

18. 48

19. 60

20. 100

Lesson 66   Multiply:

1. $3 \times 1\frac{1}{4}$ **2.** $1\frac{1}{2} \times 3$ **3.** $1\frac{1}{2} \times 1\frac{1}{4}$

4. $1\frac{2}{3} \times 2\frac{1}{2}$ **5.** $3\frac{1}{2} \times 5$ **6.** $1\frac{3}{4} \times 1\frac{1}{2}$

7. $3\frac{1}{3} \times 1\frac{2}{3}$ **8.** $7\frac{1}{2} \times 2$ **9.** $\frac{4}{5} \times 1\frac{1}{5}$

10. $\frac{5}{6} \times 1\frac{1}{5}$ **11.** $1\frac{1}{2} \times 1\frac{1}{3}$ **12.** $1\frac{1}{2} \times 1\frac{2}{3}$

13. $1\frac{1}{4} \times 2\frac{2}{5}$ **14.** $3\frac{2}{3} \times 3$ **15.** $4 \times 3\frac{1}{2}$

16. $\frac{5}{6} \times 3\frac{3}{5}$ **17.** $3\frac{1}{3} \times 2\frac{1}{10}$ **18.** $5\frac{1}{3} \times 1\frac{1}{8}$

19. $2\frac{1}{2} \times 1\frac{1}{3}$ **20.** $\frac{7}{8} \times 2\frac{2}{3}$ **21.** $1\frac{2}{5} \times 2\frac{1}{2}$

Lesson 68   Divide:

1. $1\frac{1}{2} \div 3$ **2.** $3 \div 1\frac{1}{2}$ **3.** $1\frac{2}{3} \div 2$

4. $2 \div 1\frac{2}{3}$ **5.** $\frac{3}{4} \div 1\frac{1}{2}$ **6.** $1\frac{1}{2} \div \frac{3}{4}$

7. $1\frac{2}{3} \div 1\frac{1}{2}$ **8.** $1\frac{1}{2} \div 1\frac{2}{3}$ **9.** $\frac{3}{8} \div 2$

10. $2 \div \frac{3}{8}$ **11.** $1\frac{3}{5} \div 2\frac{1}{3}$ **12.** $2\frac{1}{3} \div 1\frac{3}{5}$

13. $4\frac{1}{2} \div 2\frac{1}{4}$ **14.** $2\frac{1}{4} \div 4\frac{1}{2}$ **15.** $5 \div 1\frac{1}{4}$

16. $1\frac{1}{4} \div 5$ **17.** $2\frac{2}{3} \div 2$ **18.** $2 \div 2\frac{2}{3}$

19. $2\frac{1}{2} \div 1\frac{3}{4}$ **20.** $1\frac{3}{4} \div 2\frac{1}{2}$ **21.** $\frac{3}{4} \div 2\frac{1}{4}$

Investigation   Refer to the coordinate plane below to answer problems 1–25.
7

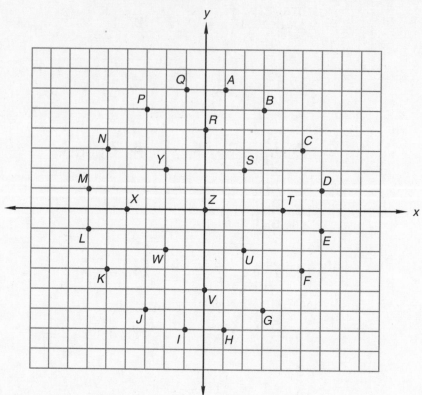

1. What is the name of the vertical axis?

2. What is the name of the horizontal axis?

3. What is the name given to the point at which the x-axis and y-axis cross?

4. What name do we use to describe the numbers that give the location of a point?

Which points have the following coordinates?

5. (3, 5)   6. (0, 4)   7. (−2, −2)   8. (−4, 0)

9. (2, −2)   10. (5, −3)   11. (−1, 6)   12. (−3, −5)

What are the coordinates of the following points?

13. L   14. A   15. C

16. D   17. G   18. I

19. K   20. V   21. M

22. P   23. Y   24. S   25. Z

Lesson 73 Convert each decimal number to a reduced fraction or mixed number:

1. 0.1   **2.** 1.2   **3.** 0.4   **4.** 2.5

5. 0.8   **6.** 3.9   **7.** 0.12   **8.** 4.15

9. 0.25   **10.** 3.75   **11.** 0.025   **12.** 0.005

Write each answer as a fraction:

13. $0.5 + \dfrac{1}{3}$ **14.** $0.8 - \dfrac{2}{5}$ **15.** $\dfrac{3}{4} - 0.1$

16. $\dfrac{1}{3} \times 0.6$ **17.** $0.2 \times \dfrac{1}{2}$ **18.** $0.75 \div \dfrac{3}{4}$

19. $\dfrac{3}{4} \div 0.25$ **20.** $\dfrac{7}{10} + 0.3$ **21.** $\dfrac{1}{10} + 0.3$

22. $0.4 - \dfrac{3}{10}$ **23.** $\dfrac{2}{3} \times 0.75$ **24.** $\dfrac{1}{5} \div 0.6$

Lesson 74 Convert each fraction or mixed number to a decimal number:

1. $\dfrac{1}{2}$ **2.** $\dfrac{1}{4}$ **3.** $\dfrac{1}{8}$ **4.** $\dfrac{1}{10}$

5. $\dfrac{3}{4}$ **6.** $1\dfrac{3}{8}$ **7.** $2\dfrac{3}{10}$

8. $4\dfrac{3}{5}$ **9.** $7\dfrac{5}{8}$ **10.** $3\dfrac{7}{10}$

Write each answer as a decimal:

11. $3.6 + \dfrac{1}{2}$ **12.** $\dfrac{1}{4} - 0.2$ **13.** $1.2 \times \dfrac{1}{5}$

14. $\dfrac{3}{5} \div 0.3$ **15.** $4.4 + \dfrac{3}{5}$ **16.** $\dfrac{3}{4} - 0.15$

17. $1.2 \times \dfrac{3}{4}$ **18.** $\dfrac{3}{10} \div 0.3$ **19.** $1 + \dfrac{1}{2}$

20. $\dfrac{3}{8} - 0.3$ **21.** $0.8 \times \dfrac{1}{8}$ **22.** $\dfrac{1}{2} \div 0.2$

Lesson 75 Translate each question into an equation and solve:

    1. Six is what percent of 12?

    2. Six is what percent of 10?

    3. Six is what percent of 8?

    4. Six is what percent of 6?

    5. Twelve is what percent of 120?

    6. Twelve is what percent of 60?

    7. Twelve is what percent of 48?

    8. Twelve is what percent of 24?

    9. Twelve is what percent of 12?

    10. Twelve is what percent of 16?

    11. Twelve is what percent of 15?

    12. Twelve is what percent of 100?

    13. What percent of 100 is 8?

    14. What percent of 10 is 8?

    15. What percent of 16 is 8?

    16. What percent of 8 is 8?

    17. What percent of 80 is 8?

    18. What percent of 100 is 20?

    19. What percent of 80 is 20?

    20. What percent of 40 is 20?

    21. What percent of 25 is 20?

    22. What percent of 20 is 20?

    23. What percent of 15 is 3?

    24. What percent of 100 is 30?

    25. What percent of 40 is 30?

Lesson 77   Answer the questions that follow each sentence.

Two fifths of the 25 team members played.

1. Into how many parts was the team divided?

2. How many were in each part?

3. How many parts played?

4. How many team members played?

5. How many parts did not play?

6. How many team members did not play?

Five sixths of the 300 members paid their dues.

7. Into how many parts was the group divided?

8. How many members were in each part?

9. How many parts paid dues?

10. How many members paid dues?

11. How many parts did not pay dues?

12. How many members did not pay dues?

Three tenths of the $6000 in prize money went to pay taxes.

13. Into how many parts was the prize money divided?

14. How much money was in each part?

15. How many parts went to pay taxes?

16. How much money went to pay taxes?

17. How many parts did not go to pay taxes?

18. How much money did not go to pay taxes?

Out of 800 rooms in the hotel, $\frac{27}{100}$ were still available.

19. Into how many parts were the rooms divided?

20. How many rooms were in each part?

21. How many parts were still available?

22. How many rooms were still available?

23. How many parts were not available?

24. How many rooms were not available?

Lesson 79   Find the area of each triangle or shaded triangle:

1.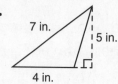

2.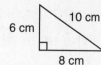

3.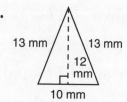

4.

5.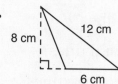

6.

7.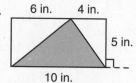

8.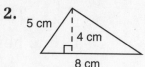

9.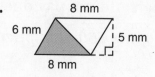

10.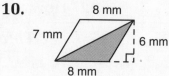

11.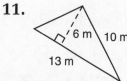

12.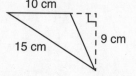

Lesson 84   Simplify:

1. $3 + 3 \div 3 + 3 \times 3$

2. $3 \times 3 - 3 - 3 \div 3$

3. $3 + 3 \times 3 \div 3 - 3$

4. $3 - 3 \div 3 + 3 \times 3$

5. $4 + 4 \times 4 + 4 \div 4$

6. $4 \div 4 + 4 \times 4 - 4$

7. $5 - 4 \div 2 + 3 \times 6$

8. $6 - 3 \times 2 + 5 - 4$

9. $5 \times 5 - 5 \div 5 - 5$

10. $6 + 5 \times 4 - 3 - 2 \div 1$

11. $4(5) - 4 + 5(4)$

12. $3(2 + 1) - 2 + 3 \times 2$

13. $4(5 - 4) + 5 - 4$

14. $32 + 1.8 \times 10$

15. $2(5 + 4) - 2(4 + 3)$ **16.** $6 \times 4 \div 2 - 6 \div 2 \times 4$

17. $(5 + 3)(5 - 3)$ **18.** $32 + 1.8(50)$

19. $10 + 10 \times 10 - 10 \div 10$ **20.** $2(5) + 2(4) - 2(5 + 4)$

Lesson 85 Use cross products to complete each proportion:

1. $\frac{3}{4} = \frac{a}{20}$ **2.** $\frac{4}{6} = \frac{12}{b}$ **3.** $\frac{c}{6} = \frac{8}{12}$

4. $\frac{4}{d} = \frac{24}{60}$ **5.** $\frac{8}{10} = \frac{12}{e}$ **6.** $\frac{9}{10} = \frac{f}{1000}$

7. $\frac{g}{4} = \frac{27}{36}$ **8.** $\frac{15}{h} = \frac{30}{42}$ **9.** $\frac{12}{15} = \frac{i}{25}$

10. $\frac{7}{8} = \frac{700}{j}$ **11.** $\frac{k}{60} = \frac{4}{5}$ **12.** $\frac{12}{l} = \frac{60}{100}$

13. $\frac{3}{25} = \frac{m}{100}$ **14.** $\frac{3}{8} = \frac{9}{n}$ **15.** $\frac{p}{20} = \frac{35}{100}$

16. $\frac{12}{q} = \frac{10}{20}$ **17.** $\frac{27}{50} = \frac{r}{100}$ **18.** $\frac{6}{18} = \frac{7}{s}$

19. $\frac{t}{4} = \frac{75}{100}$ **20.** $\frac{27}{u} = \frac{15}{20}$

Lesson 86 Complete the chart for each circle. (Use $\pi = 3.14$.)

RADIUS	DIAMETER	CIRCUMFERENCE	AREA
1 in.	**1.**	**2.**	**3.**
4.	4 ft	**5.**	**6.**
3 cm	**7.**	**8.**	**9.**
10.	8 m	**11.**	**12.**
5 mm	**13.**	**14.**	**15.**
16.	12 yd	**17.**	**18.**
10 km	**19.**	**20.**	**21.**
22.	100 mi	**23.**	**24.**

25. What word is used to describe the perimeter of a circle?

Lesson 89 Each square root is between which two consecutive whole numbers? Find the answer without the aid of a calculator.

1. $\sqrt{10}$ **2.** $\sqrt{30}$ **3.** $\sqrt{40}$ **4.** $\sqrt{50}$ **5.** $\sqrt{60}$

6. $\sqrt{70}$ **7.** $\sqrt{80}$ **8.** $\sqrt{90}$ **9.** $\sqrt{5}$ **10.** $\sqrt{2}$

Lesson 92 Write each number in expanded notation using exponents:

1. 450,000   **2.** 25,000,000   **3.** 16,000,000,000

Write each number in standard notation:

4. 5×10^6

5. $(3 \times 10^4) + (6 \times 10^3)$

6. $(1 \times 10^9) + (5 \times 10^8)$

Find each power:

7. $\left(\frac{1}{3}\right)^3$ **8.** $\left(2\frac{1}{2}\right)^2$ **9.** $(0.1)^3$ **10.** $\left(1\frac{1}{2}\right)^3$

Simplify each expression using the proper order of operations:

11. $20 + 3^2 \times 2 - (8 + 2) \div \sqrt{25}$

12. $(2 + 3)^2 - 5(4) + \sqrt{100} \div \sqrt{25}$

13. $2 \times 3^2 - 2^3 + \sqrt{64} - 2(3)$

14. $\sqrt{9} + 5^2 - \sqrt{36} \div 3 - 2 \times 1^4$

15. $4^2 \div \sqrt{16} \times (2 + 1)^2 \div \sqrt{36} - 3 \times 2$

Lesson 94 Write each fraction as a percent:

1. $\frac{1}{10}$ **2.** $\frac{9}{10}$ **3.** $\frac{1}{5}$ **4.** $\frac{3}{4}$ **5.** $\frac{3}{20}$

6. $\frac{3}{25}$ **7.** $\frac{3}{50}$ **8.** $\frac{3}{100}$ **9.** $\frac{11}{100}$

10. $\frac{11}{50}$ **11.** $\frac{11}{25}$ **12.** $\frac{11}{20}$ **13.** $\frac{1}{3}$

14. $\frac{2}{3}$ **15.** $\frac{1}{8}$ **16.** $\frac{3}{8}$ **17.** $\frac{1}{9}$

18. $\frac{5}{4}$ **19.** $\frac{1}{6}$ **20.** $\frac{5}{8}$ **21.** $\frac{1}{7}$

22. $\frac{5}{2}$ **23.** $\frac{5}{6}$ **24.** $\frac{7}{9}$ **25.** $\frac{5}{12}$

Write each decimal as a percent:

26. 0.6   **27.** 3.4   **28.** 0.01   **29.** 1.2   **30.** 0.5

31. 1.0   **32.** 0.37   **33.** 4.5   **34.** 2.0

35. 0.1   **36.** 1.05   **37.** 0.6   **38.** 3.0

Lesson 99 Complete this table:

	FRACTION	DECIMAL	PERCENT
1.	$\frac{1}{100}$	(a)	(b)
2.	(a)	0.8	(b)
3.	(a)	(b)	25%
4.	$\frac{3}{4}$	(a)	(b)
5.	(a)	0.7	(b)
6.	(a)	(b)	90%
7.	$\frac{1}{20}$	(a)	(b)
8.	(a)	0.5	(b)
9.	(a)	(b)	4%
10.	$\frac{1}{50}$	(a)	(b)
11.	(a)	0.45	(b)
12.	(a)	(b)	23%
13.	$1\frac{1}{2}$	(a)	(b)
14.	(a)	0.15	(b)
15.	(a)	(b)	10%
16.	$\frac{1}{8}$	(a)	(b)
17.	(a)	0.2	(b)
18.	(a)	(b)	35%

Lesson 100 Simplify:

1. −3 + (−4)

2. (−5) + (−8)

3. +3 + (−5)

4. 3 + (−8

5. −3 + 4

6. (−4) + (+3)

7. −7 + 6

8. (−6) + (+6)

9. 5 + (−11)

10. (+5) + (+7)

11. (−12) + (−12)

12. −12 + 12

13. (−12) + (+15)

14. (+8) + (−1)

15. (−8) + (+1)

16. −5 + 8

17. +15 + (−18)

18. −25 + (−30)

19. (+15) + (−20)

20. +8 + (−16)

21. −9 + 15

22. (−6) + (+20)

23. (−20) + (−12)

24. (+6) + (−4)

25. +8 + (−18)

26. 3 − (−5)

27. −3 − 5

28. −3 − (−5)

29. (−5) − (−3)

30. (−5) − (+3)

31. 5 − (−6)

32. 7 − (−12)

33. −7 − (−12)

34. (−7) − (+12)

35. −12 − 7

36. −12 − (−7)

37. 12 − (−7)

38. (−12) − (+7)

39. −6 − (−6)

40. (+6) − (−6)

41. (−10) − (+5)

42. −10 − (−5)

43. −5 − 10

44. −5 − (−10)

45. $10 - (-5)$ **46.** $-12 - (-8)$

47. $(-12) - (+8)$ **48.** $-8 - 12$

49. $-8 - (-12)$ **50.** $8 - (-12)$

Lesson 107   Find the perimeter and area of each polygon. Assume that angles that appear to be right angles are right angles.

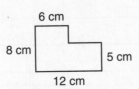

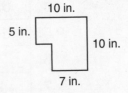

1. perimeter

2. area

3. perimeter

4. area

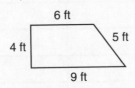

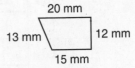

5. perimeter

6. area

7. perimeter

8. area

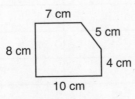

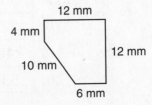

9. perimeter

10. area

11. perimeter

12. area

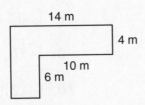

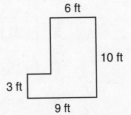

13. perimeter

14. area

15. perimeter

16. area

Lesson 111   Write each quotient in three forms:

	PROBLEM	WITH REMAINDER	WITH FRACTION	AS A DECIMAL
1.	100 ÷ 8			
2.	50 ÷ 4			
3.	56 ÷ 10			
4.	65 ÷ 10			
5.	63 ÷ 12			
6.	72 ÷ 5			
7.	49 ÷ 4			
8.	38 ÷ 8			
9.	47 ÷ 4			
10.	146 ÷ 8			
11.	390 ÷ 20			
12.	625 ÷ 10			
13.	432 ÷ 5			
14.	650 ÷ 8			
15.	325 ÷ 20			
16.	562 ÷ 8			
17.	530 ÷ 40			
18.	375 ÷ 50			
19.	240 ÷ 100			
20.	534 ÷ 10			

Lesson 112   Find each product or quotient:

1. $(-3)(+4)$

2. $(-8)(-12)$

3. $(+12)(-15)$

4. $(-18)(-20)$

5. $(+21)(-7)$

6. $(-8)(+17)$

7. $(-25)(-15)$

8. $(+18)(+5)$

9. $(-7)(+43)$

10. $(-12)(-24)$

11. $(+6)(-18)$

12. $(-9)(-25)$

13. $(-24) \div (+2)$

14. $(-24) \div (-3)$

15. −72 ÷ 8

16. 400 ÷ (−5)

17. (−234) ÷ (−6)

18. (−144) ÷ (+12)

19. −125 ÷ 25

20. (−375) ÷ (−15)

Lesson 113   Simplify:

1.   2 hr 45 min
       + 3 hr 25 min

2.   10 min 15 s
       − 5 min 50 s

3.   1 yd 2 ft 8 in.
       + 2 yd 1 ft 9 in.

4.   6 ft 3 in.
       − 5 ft 7 in.

5.   6 lb 10 oz
       + 3 lb   8 oz

6.   8 lb 3 oz
       − 6 lb 9 oz

7.   2 gal 3 qt
       + 3 gal 2 qt

8.   5 gal 2 qt
       − 1 gal 3 qt

Write each number in standard notation:

9. $\frac{1}{2}$ million

10. 2.5×10^7

11. 7×10^5

12. $1\frac{1}{2}$ billion

13. 1.25×10^9

14. 15 million

15. 3.5×10^6

16. $\frac{1}{2}$ billion

Lesson 117   Solve:

1. Fifty is $\frac{1}{2}$ of what?

2. Forty is $\frac{1}{4}$ of what?

3. Thirty is $\frac{1}{5}$ of what?

4. Twenty is $\frac{1}{10}$ of what?

5. Ten is $\frac{2}{5}$ of what?

6. Twenty is $\frac{2}{3}$ of what?

7. Thirty is $\frac{3}{4}$ of what?

8. Forty is $\frac{2}{5}$ of what?

9. Fifty is $\frac{5}{8}$ of what?

10. Twelve is $\frac{1}{4}$ of what?

11. Twelve is $\frac{3}{4}$ of what?

12. Twelve is $\frac{1}{3}$ of what?

13. Twelve is $\frac{3}{5}$ of what?   **14.** Sixty is $\frac{1}{10}$ of what?

15. Sixty is $\frac{1}{5}$ of what?   **16.** Sixty is $\frac{3}{10}$ of what?

17. Sixty is $\frac{2}{5}$ of what?   **18.** Sixty is $\frac{1}{2}$ of what?

19. Sixty is $\frac{3}{5}$ of what?   **20.** Thirty-five is $\frac{7}{10}$ of what?

21. Forty is $\frac{4}{5}$ of what?   **22.** Forty-five is $\frac{9}{10}$ of what?

23. Thirty is $\frac{3}{4}$ of what?   **24.** Twenty is $\frac{1}{4}$ of what?

Lesson 119 Translate each question into an equation and solve:

1. Ten is 10% of what number?

2. Ten is 20% of what number?

3. Ten is 40% of what number?

4. Ten is 25% of what number?

5. Ten is 5% of what number?

6. Ten is 2% of what number?

7. One hundred is 100% of what number?

8. Twenty is 20% of what number?

9. Twenty is 10% of what number?

10. Twenty is 1% of what number?

11. Six is 10% of what number?

12. Six is 25% of what number?

13. Six is 1% of what number?

14. Five is 100% of what number?

15. Fifty is 50% of what number?

16. Fifty is 100% of what number?

17. Fifty is 10% of what number?

18. Six is 75% of what number?

19. Ten is 100% of what number?

20. Fifty is 25% of what number?

21. Forty is 100% of what number?

22. Forty is 40% of what number?

23. Forty is 50% of what number?

24. Forty is 25% of what number?

25. Twenty is 25% of what number?

introducing, the one-of-a-kind-Mario Bennend!

acute angle An angle whose measure is more than 0° and less than 90°.

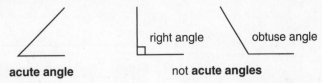

acute angle   not **acute angles**

*An **acute angle** is smaller than both a right angle and an obtuse angle.*

acute triangle A triangle whose largest angle measures more than 0° and less than 90°.

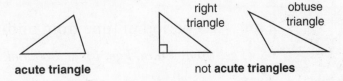

acute triangle   not **acute triangles**

addend One of two or more numbers that are added to find a sum.

$7 + 3 = 10$ *The **addends** in this problem are 7 and 3.*

algebraic addition The combining of positive and negative numbers to form a sum.

*We use **algebraic addition** to find the sum of –3, +2, and –11:*
$$(-3) + (+2) + (-11) = -12$$

algorithm Any process for solving a mathematical problem.

*In the addition **algorithm** we add the ones first, then the tens, and then the hundreds.*

alternate exterior angles A special pair of angles formed when a transversal intersects two lines. Alternate exterior angles lie on opposite sides of the transversal and are outside the two intersected lines.

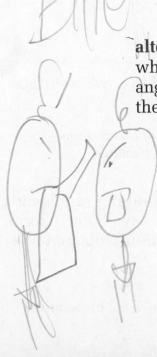

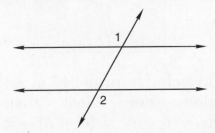

*∠1 and ∠2 are **alternate exterior angles**. When a transversal intersects parallel lines, as in this figure, **alternate exterior angles** have the same measure.*

693

alternate interior angles   A special pair of angles formed when a transversal intersects two lines. Alternate interior angles lie on opposite sides of the transversal and are inside the two intersected lines.

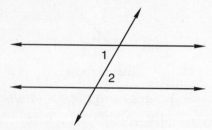

∠1 *and* ∠2 *are* **alternate interior angles.** *When a transversal intersects parallel lines, as in this figure,* **alternate interior angles** *have the same measure.*

a.m.   The period of time from midnight to just before noon.

I get up at 7 **a.m.** *I get up at 7 o'clock in the morning.*

angle   The opening that is formed when two lines, rays, or segments intersect.

These rays form an **angle.**

angle bisector   A line, ray, or segment that divides an angle into two congruent parts.

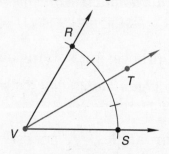

$\overrightarrow{VT}$ *is an* **angle bisector.**
It divides ∠RVS *in half.*

area   The number of square units needed to cover a surface.

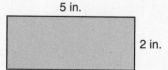

The **area** *of this rectangle is 10 square inches.*

associative property of addition   The grouping of addends does not affect their sum. In symbolic form, $a + (b + c) = (a + b) + c$. Unlike addition, subtraction is not associative.

$(8 + 4) + 2 = 8 + (4 + 2)$ $(8 - 4) - 2 \neq 8 - (4 - 2)$

Addition is **associative.**   *Subtraction is not* **associative.**

associative property of multiplication The grouping of factors does not affect their product. In symbolic form, $a \times (b \times c) = (a \times b) \times c$. Unlike multiplication, division is not associative.

$(8 \times 4) \times 2 = 8 \times (4 \times 2)$ $(8 \div 4) \div 2 \neq 8 \div (4 \div 2)$

Multiplication is **associative**.   Division is not **associative**.

average The number found when the sum of two or more numbers is divided by the number of addends in the sum; also called *mean*.

To find the **average** of the numbers 5, 6, and 10, first add.

$5 + 6 + 10 = 21$

Then, since there were three addends, divide the sum by 3.

$21 \div 3 = 7$

The **average** of 5, 6, and 10 is 7.

base (1) A designated side or face of a geometric figure.

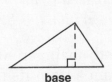 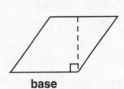

base   base   base

(2) The lower number in an exponential expression.

$$base \longrightarrow 5^3 \longleftarrow exponent$$

5^3 means $5 \times 5 \times 5$, and its value is 125.

bimodal Having two modes.

The numbers 5 and 7 are the modes of the data at right.   5, 1, 44, 5, 7, 13, 9, 7
This set of data is **bimodal**.

bisect To divide a segment or angle into two equal halves.

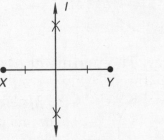   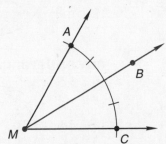

Line l **bisects** $\overline{XY}$.   Ray MB **bisects** $\angle AMC$.

capacity   The amount of liquid a container can hold.

*Cups, gallons, and liters are units of **capacity**.*

Celsius scale   A scale used on some thermometers to measure temperature.

*On the **Celsius scale**, water freezes at 0°C and boils at 100°C.*

center   The point inside a circle from which all points on the circle are equally distant.

*The **center** of circle A is 2 inches from every point on the circle.*

chance   A way of expressing the likelihood of an event; the probability of an event expressed as a percentage.

*The **chance** of snow is 10%. It is not likely to snow.*
*There is an 80% **chance** of rain. It is likely to rain.*

circle   A closed, curved shape in which all points on the shape are the same distance from its center.

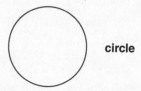

circle

circle graph   A method of displaying data, often used to show information about percentages or parts of a whole. A circle graph is made of a circle divided into sectors.

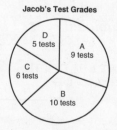

*This **circle graph** shows data for Jacob's test grades.*

circumference   The perimeter of a circle.

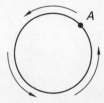

*If the distance from point A around to point A is 3 inches, then the **circumference** of the circle is 3 inches.*

closed-option survey A survey in which the possible responses are limited.

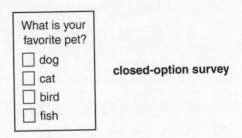

closed-option survey

common denominator A number that is the denominator of two or more fractions.

The fractions $\frac{2}{5}$ and $\frac{3}{5}$ have **common denominators.**

commutative property of addition Changing the order of addends does not change their sum. In symbolic form, $a + b = b + a$. Unlike addition, subtraction is not commutative.

$8 + 2 = 2 + 8$ $8 - 2 \neq 2 - 8$

Addition is **commutative.**   *Subtraction is not* **commutative.**

commutative property of multiplication Changing the order of factors does not change their product. In symbolic form, $a \times b = b \times a$. Unlike multiplication, division is not commutative.

$8 \times 2 = 2 \times 8$ $8 \div 2 \neq 2 \div 8$

Multiplication is **commutative.**   *Division is not* **commutative.**

compass A tool used to draw circles and arcs.

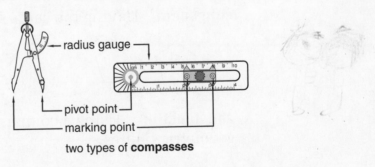

two types of **compasses**

complementary angles Two angles whose sum is 90°.

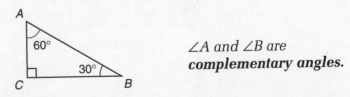

∠A and ∠B are
complementary angles.

composite number   A counting number greater than 1 that is divisible by a number other than itself and 1. Every composite number has three or more factors.

> *9 is divisible by 1, 3, and 9. It is **composite**.*
>
> *11 is divisible by 1 and 11. It is not **composite**.*

compound interest   Interest that pays on previously earned interest.

	Compound Interest			Simple Interest
$100.00	*principal*		$100.00	*principal*
+ $6.00	*first-year interest (6% of $100)*		$6.00	*first-year interest (6% of $100)*
$106.00	*total after one year*	+	$6.00	*second-year interest (6% of $100)*
+ $6.36	*second-year interest (6% of $106)*		$112.00	*total after two years*
$112.36	*total after two years*			

concentric circles   Two or more circles with a common center.

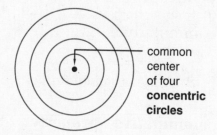

common
center
of four
**concentric
circles**

cone   A three-dimensional solid with a circular base and a single vertex.

cone

congruent   Having the same size and shape.

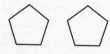

*These polygons are **congruent**. They have the same size and shape.*

construction   Using a compass and/or a straightedge to draw geometric figures.

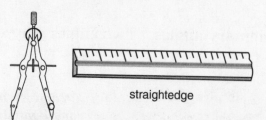

straightedge

*These tools are used in **construction**.*

compass

coordinate(s) (1) A number used to locate a point on a number line.

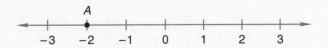

*The **coordinate** of point A is –2*

(2) An ordered pair of numbers used to locate a point in a coordinate plane.

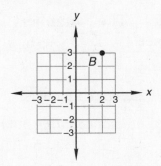

*The **coordinates** of point B are (2, 3). The x-coordinate is listed first, the y-coordinate second.*

coordinate plane A grid on which any point can be identified by an ordered pair of numbers.

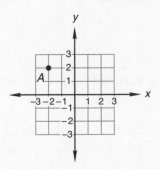

*Point A is located at (–2, 2) on this **coordinate plane.***

corresponding angles A special pair of angles formed when a transversal intersects two lines. Corresponding angles lie on the same side of the transversal and are in the same position relative to the two intersected lines.

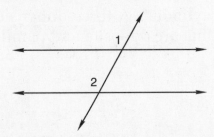

*∠1 and ∠2 are **corresponding angles**. When a transversal intersects parallel lines, as in this figure, **corresponding angles** have the same measure.*

corresponding parts Sides or angles that occupy the same relative positions in similar polygons.

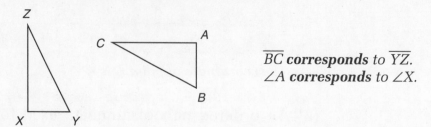

$\overline{BC}$ **corresponds** to $\overline{YZ}$.
$\angle A$ **corresponds** to $\angle X$.

counting numbers The numbers used to count; the members of the set {1, 2, 3, 4, 5, ...}. Also called *natural numbers*.

1, 24, and 108 are **counting numbers.**

–2, 3.14, 0, and $2\frac{7}{9}$ are not **counting numbers.**

cross product The product of the numerator of one fraction and the denominator of another.

$$5 \times 16 = 80 \qquad 20 \times 4 = 80$$
$$\frac{16}{20} = \frac{4}{5}$$

The **cross products** *of these two fractions are equal.*

cube A three-dimensional solid with six square faces. Adjacent faces are perpendicular and opposite faces are parallel.

cube

cylinder A three-dimensional solid with two circular bases that are opposite and parallel to each other.

cylinder

decimal number A numeral that contains a decimal point.

23.94 is a **decimal number** *because it contains a decimal point.*

decimal places Places to the right of a decimal point.

*5.47 has two **decimal places**.*

*6.3 has one **decimal place**.*

*8 has no **decimal places**.*

decimal point The symbol in a decimal number used as a reference point for place value.

34.15

↑

decimal point

degree (°) (1) A unit for measuring angles.

*There are 90 **degrees** (90°) in a right angle.*

*There are 360 **degrees** (360°) in a circle.*

(2) A unit for measuring temperature.

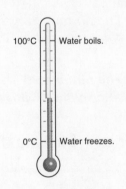

*There are 100 **degrees** between the freezing and boiling points of water on the Celsius scale.*

denominator The bottom term of a fraction.

$$\frac{5}{9}$$ ← numerator
← **denominator**

diameter The distance across a circle through its center.

3 in.

*The **diameter** of this circle is 3 inches.*

difference   The result of subtraction.

> $12 - 8 = 4$ *The **difference** in this problem is 4.*

digit   Any of the symbols used to write numbers: 0, 1, 2, 3, 4, 5, 6, 7, 8, 9.

> *The last **digit** in the number 7862 is 2.*

distributive property   A number times the sum of two addends is equal to the sum of that same number times each individual addend: $a \times (b + c) = (a \times b) + (a \times c)$.

> $8 \times (2 + 3) = (8 \times 2) + (8 \times 3)$ *Multiplication is **distributive** over addition.*

dividend   A number that is divided.

> $12 \div 3 = 4$ $3\overline{)12}$ (with 4 above) $\dfrac{12}{3} = 4$ *The **dividend** is 12 in each of these problems.*

divisible   Able to be divided by a whole number without a remainder.

> $4\overline{)20}$ (with 5 above)   *The number 20 is **divisible** by 4, since $20 \div 4$ has no remainder.*

> $3\overline{)20}$ (with 6 R 2 above)   *The number 20 is not **divisible** by 3, since $20 \div 3$ has a remainder.*

divisor   (1) A number by which another number is divided.

> $12 \div 3 = 4$ $3\overline{)12}$ (with 4 above) $\dfrac{12}{3} = 4$ *The **divisor** is 3 in each of these problems.*

(2) A factor of a number.

> *2 and 5 are divisors of 10.*

edge   A line segment formed where two faces of a polyhedron intersect.

*One **edge** of this cube is colored. A cube has 12 **edges**.*

endpoint   A point at which a segment ends.

A •————————————• B

*Points A and B are the **endpoints** of segment AB.*

equation   A statement that uses the symbol "=" to show that two quantities are equal.

$x = 3$ $3 + 7 = 10$ $4 + 1$ $x < 7$

equations   not **equations**

equilateral triangle   A triangle in which all sides are the same length.

*This is an **equilateral triangle**.
All of its sides are the same length.*

equivalent fractions   Different fractions that name the same amount.

$\frac{1}{2}$ [shaded bar diagram] = [shaded bar diagram] $\frac{2}{4}$

*$\frac{1}{2}$ and $\frac{2}{4}$ are **equivalent fractions**.*

estimate   To determine an approximate value.

*We **estimate** that the sum of 199 and 205 is about 400.*

evaluate   To find the value of an expression.

*To **evaluate** $a + b$ for $a = 7$ and $b = 13$, we replace a with 7 and b with 13:*

$$7 + 13 = 20$$

even numbers   Numbers that can be divided by 2 without a remainder; the members of the set {..., −4, −2, 0, 2, 4, ...}.

***Even numbers** have 0, 2, 4, 6, or 8 in the ones place.*

expanded notation   A way of writing a number as the sum of the products of the digits and the place values of the digits.

*In **expanded notation** 6753 is written*

$$(6 \times 1000) + (7 \times 100) + (5 \times 10) + (3 \times 1).$$

experimental probability   The probability of an event occurring as determined by experimentation.

*If we roll a number cube 100 times and get 22 threes, the **experimental probability** of getting three is $\frac{22}{100}$, or $\frac{11}{50}$.*

exponent   The upper number in an exponential expression; it shows how many times the base is to be used as a factor.

$$base \longrightarrow 5^3 \longleftarrow exponent$$

5^3 means $5 \times 5 \times 5$, and its value is 125.

exponential expression   An expression that indicates that the base is to be used as a factor the number of times shown by the exponent.

$$4^3 = 4 \times 4 \times 4 = 64$$

*The **exponential expression** 4^3 is evaluated by using 4 as a factor 3 times. Its value is 64.*

exterior angle   In a polygon, the supplementary angle of an interior angle.

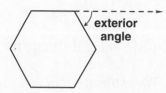

exterior angle

face   A flat surface of a geometric solid.

*One **face** of the cube is shaded. A cube has six **faces**.*

fact family   A group of three numbers related by addition and subtraction or by multiplication and division.

*The numbers 3, 4, and 7 are a **fact family**. They make these four facts:*

$$3 + 4 = 7 \qquad 4 + 3 = 7 \qquad 7 - 3 = 4 \qquad 7 - 4 = 3$$

factor (1) Noun: One of two or more numbers that are multiplied.

> $3 \times 5 = 15$ The **factors** in this problem are 3 and 5.

(2) Noun: A whole number that divides another whole number without a remainder.

> The numbers 3 and 5 are **factors** of 15.

(3) Verb: To write as a product of factors.

> We can **factor** the number 15 by writing it as 3×5.

Fahrenheit scale A scale used on some thermometers to measure temperature.

> On the **Fahrenheit scale,** water freezes at 32°F and boils at 212°F.

fraction A number that names part of a whole.

> $\frac{1}{4}$ of the circle is shaded.
>
> $\frac{1}{4}$ is a **fraction.**

function A rule for using one number (an input) to calculate another number (an output). Each input produces only one output.

$y = 3x$

x	y
3	9
5	15
7	21
10	30

> There is exactly one resulting number for every number we multiply by 3. Thus, $y = 3x$ is a **function.**

geometric solid A three-dimensional geometric figure.

geometric solids		not geometric solids		
cube	cylinder	circle	rectangle	hexagon

geometry A major branch of mathematics that deals with shapes, sizes, and other properties of figures.

Some figures we study in **geometry** *are angles, circles, and polygons.*

graph (1) Noun: A diagram, such as a bar graph, a circle graph (pie chart), or a line graph, that displays quantitative information.

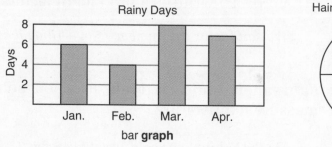

bar **graph**

circle **graph**

(2) Noun: A point, line, or curve on a coordinate plane.

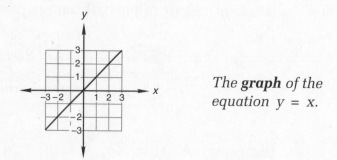

The **graph** *of the equation* y = x.

(3) Verb: To draw a point, line, or curve on a coordinate plane.

greatest common factor (GCF) The largest whole number that is a factor of two or more given numbers.

The factors of 12 are 1, 2, 3, 4, 6, and 12.

The factors of 18 are 1, 2, 3, 6, 9, and 18.

The **greatest common factor** *of 12 and 18 is 6.*

height The perpendicular distance from the base to the opposite side of a parallelogram or trapezoid; from the base to the opposite face of a prism or cylinder; or from the base to the opposite vertex of a triangle, pyramid, or cone.

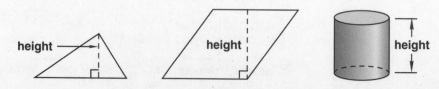

histogram A method of displaying a range of data. A histogram is a special type of bar graph that displays data in intervals of equal size with no space between bars.

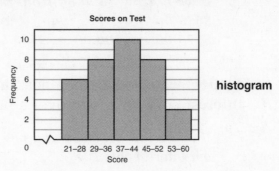

horizontal Parallel to the horizon; perpendicular to vertical.

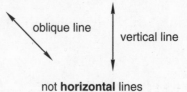

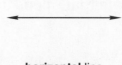

horizontal line   not **horizontal** lines

identity property of addition The sum of any number and 0 is equal to the initial number. In symbolic form, $a + 0 = a$. The number 0 is referred to as the *additive identity*.

> The **identity property of addition** is shown by this statement:
>
> $$13 + 0 = 13$$

identity property of multiplication The product of any number and 1 is equal to the initial number. In symbolic form, $a \times 1 = a$. The number 1 is referred to as the *multiplicative identity*.

> The **identity property of multiplication** is shown by this statement:
>
> $$94 \times 1 = 94$$

improper fraction A fraction with a numerator equal to or greater than the denominator.

> $\frac{12}{12}$, $\frac{57}{3}$, and $2\frac{15}{2}$ are **improper fractions**.
> All **improper fractions** are greater than or equal to 1.

integers The set of counting numbers, their opposites, and zero; the members of the set $\{..., -2, -1, 0, 1, 2, ...\}$.

> -57 and 4 are **integers**. $\frac{15}{8}$ and -0.98 are not **integers**.

interest An amount added to a loan, account, or fund, usually based on a percentage of the principal.

*If we borrow $500.00 from the bank and repay the bank $575.00 for the loan, the **interest** on the loan is $575.00 − $500.00 = $75.00.*

interest rate A percentage that determines the amount of interest paid on a loan over a period of time.

*If we borrow $1000.00 and pay back $1100.00 after one year, our **interest rate** is:*

$$\frac{\$1100.00 - \$1000.00}{\$1000.00} \times 100\% = 10\% \text{ per year}$$

interior angle An angle that opens to the inside of a polygon.

*This hexagon has six **interior angles**.*

International System *See* **metric system.**

intersect To share a point or points.

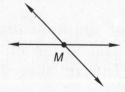

*These two lines **intersect**. They share the point M.*

inverse operations Operations that "undo" one another.

$a + b - b = a$ *Addition and subtraction are*
$a - b + b = a$ ***inverse operations.***

$a \times b \div b = a \quad (b \neq 0)$ *Multiplication and division are*
$a \div b \times b = a \quad (b \neq 0)$ ***inverse operations.***

$\sqrt{a^2} = a \quad (a \geq 0)$ *Squaring and finding square*
$(\sqrt{a})^2 = a \quad (a \geq 0)$ *roots are **inverse operations**.*

invert To switch the numerator and denominator of a fraction.

*If we **invert** $\frac{7}{8}$, we get $\frac{8}{7}$.*

irrational numbers   Numbers that cannot be expressed as a ratio of two integers. Their decimal expansions are nonending and nonrepeating.

*π and √3 are **irrational numbers.***

isosceles triangle   A triangle with at least two sides of equal length.

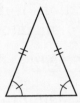

*Two of the sides of this **isosceles triangle** have equal lengths.*

least common multiple (LCM)   The smallest whole number that is a multiple of two or more given numbers.

Multiples of 6 are 6, 12, 18, 24, 30, 36, ...

Multiples of 8 are 8, 16, 24, 32, 40, 48, ...

*The **least common multiple** of 6 and 8 is 24.*

legend   A notation on a map, graph, or diagram that describes the meaning of the symbols and/or the scale used.

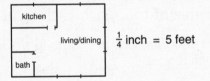

$\frac{1}{4}$ inch = 5 feet

*The **legend** of this scale drawing shows that $\frac{1}{4}$ inch represents 5 feet.*

line   A straight collection of points extending in opposite directions without end.

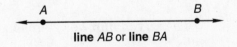

line *AB* or **line** *BA*

line of symmetry   A line that divides a figure into two halves that are mirror images of each other.

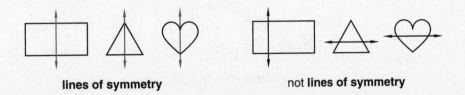

lines of symmetry   **not lines of symmetry**

line plot A method of plotting a set of numbers by placing a mark above a number on a number line each time it occurs in the set.

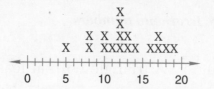

*This is a **line plot** of the numbers 5, 8, 8, 10, 10, 11, 12, 12, 12, 12, 13, 13, 14, 16, 17, 17, 18, and 19.*

lowest terms A fraction is in *lowest terms* if the only common factor of the numerator and denominator is 1.

*When written in **lowest terms**, the fraction $\frac{8}{20}$ becomes $\frac{2}{5}$.*

mass The amount of matter in an object.

*Grams and kilograms are units of **mass**.*

mean *See* **average.**

median The middle number (or the average of the two central numbers) of a list of data when the numbers are arranged in order from the least to the greatest.

*In the data at right, 7 is the **median**.*   *1, 1, 2, 5, 6, 7, 9, 15, 24, 36, 44*

metric system An international system of measurement based on multiples of ten. Also called *International System.*

*Centimeters and kilograms are units in the **metric system**.*

minuend A number from which another number is subtracted.

12 − 8 = 4   *The **minuend** in this problem is 12.*

mixed number A whole number and a fraction together.

*The **mixed number** $2\frac{1}{3}$ means "two and one third."*

mode The number or numbers that appear most often in a list of data.

*In the data at right, 5 is the **mode**.*   *5, 12, 32, 5, 16, 5, 7, 12*

multiple A product of a counting number and another number.

*The **multiples** of 3 include 3, 6, 9, and 12.*

negative numbers Numbers less than zero.

*–15 and –2.86 are **negative numbers**.*

*19 and 0.74 are not **negative numbers**.*

number line A line for representing and graphing numbers. Each point on the line corresponds to a number.

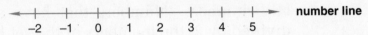

numeral A symbol or group of symbols that represents a number.

*4, 72, and $\frac{1}{2}$ are examples of **numerals**. "Four," "seventy-two," and "one half" are words that name numbers but are not **numerals**.*

numerator The top term of a fraction.

$$\frac{9}{10} \begin{matrix} \leftarrow \text{ numerator} \\ \leftarrow \text{ denominator} \end{matrix}$$

oblique line(s) (1) A line that is neither horizontal nor vertical.

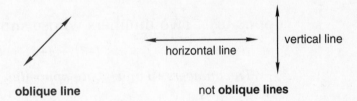

(2) Lines in the same plane that are neither parallel nor perpendicular.

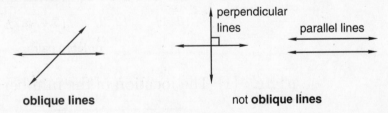

obtuse angle An angle whose measure is more than 90° and less than 180°.

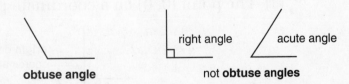

*An **obtuse angle** is larger than both a right angle and an acute angle.*

obtuse triangle A triangle whose largest angle measures more than 90° and less than 180°.

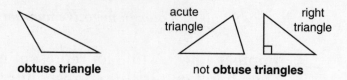

obtuse triangle   **not obtuse triangles**

odd numbers Numbers that have a remainder of 1 when divided by 2; the members of the set {..., −3, −1, 1, 3, ...}.

Odd numbers have 1, 3, 5, 7, or 9 in the ones place.

open-option survey A survey that does not limit the possible responses.

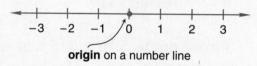

open-option survey

operations of arithmetic The four basic mathematical operations: addition, subtraction, multiplication, and division.

$$1 + 9 \qquad 21 - 8 \qquad 6 \times 22 \qquad 3 \div 1$$
the **operations of arithmetic**

opposites Two numbers whose sum is 0.

$$(-3) + (+3) = 0$$

*The numbers +3 and −3 are **opposites**.*

ordered pair A pair of numbers, written in a specific order, that are used to designate the position of a point on a coordinate plane. *See also* **coordinate(s).**

$$(0, 1) \qquad (2, 3) \qquad (3.4, 5.7) \qquad \left(\tfrac{1}{2}, -\tfrac{1}{2}\right)$$
ordered pairs

origin (1) The location of the number 0 on a number line.

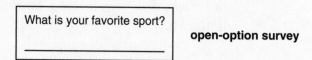

origin on a number line

(2) The point (0, 0) on a coordinate plane.

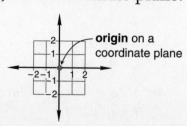

outlier   A number that is distant from most of the other numbers in a list of data.

*In the data at right, the number 28 is an **outlier** because it is distant from the other numbers in the list.*   *1, 5, 4, 3, 6, 28, 8, 2*

parallel lines   Lines in the same plane that do not intersect.

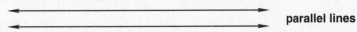

parallel lines

parallelogram   A quadrilateral that has two pairs of parallel sides.

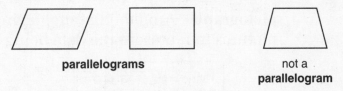

parallelograms   not a
                              parallelogram

percent   A fraction whose denominator of 100 is expressed as a percent sign (%).

$$\frac{99}{100} = 99\% = 99 \text{ \textbf{percent}}$$

perfect square   The product when a whole number is multiplied by itself.

*The number 9 is a **perfect square** because 3 × 3 = 9.*

perimeter   The distance around a closed, flat shape.

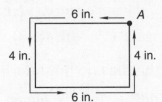

*The **perimeter** of this rectangle (from point A around to point A) is 20 inches.*

perpendicular bisector   A line, ray, or segment that intersects a segment at its midpoint at a right angle, thereby dividing the segment into two congruent parts.

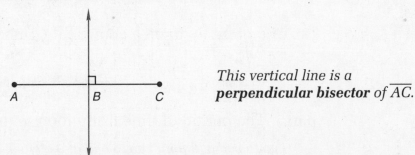

*This vertical line is a **perpendicular bisector** of $\overline{AC}$.*

perpendicular lines   Two lines that intersect at right angles.

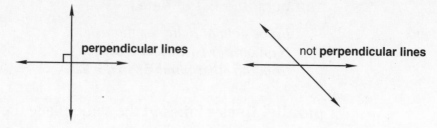

perpendicular lines   not **perpendicular lines**

pi (π)   The number of diameters equal to the circumference of a circle.

*Approximate values of **pi** are 3.14 and $\frac{22}{7}$.*

pictograph   A method of displaying data that involves using pictures to represent the data being counted.

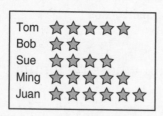

*This is a **pictograph**.*
It shows how many stars
each person saw.

pie graph   *See* **circle graph**.

place value   The value of a digit based on its position within a number.

$$\begin{array}{r} 341 \\ 23 \\ +7 \\ \hline 371 \end{array}$$

***Place value** tells us that 4 in 341 is worth four tens.*
In addition and subtraction problems we align
*digits with the same **place value**.*

plane   A flat surface that has no boundaries.

*The flat surface of a desk is part of a **plane**.*

Platonic solid   Any one of the five regular polyhedrons: tetrahedron, cube, octahedron, dodecahedron, and icosahedron.

Platonic solids

tetrahedron   cube   octahedron   dodecahedron   icosahedron

p.m.   The period of time from noon to just before midnight.

*I go to bed at 9 **p.m.** I go to bed at 9 o'clock at night.*

point An exact position on a line, on a plane, or in space.

$\bullet_A$ *This dot represents **point** A.*

polygon A closed, flat shape with straight sides.

polygons   not **polygons**

polyhedron A geometric solid whose faces are polygons.

polyhedrons   not **polyhedrons**

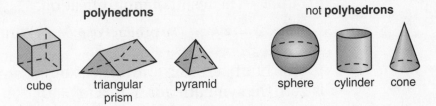

cube   triangular   pyramid   sphere   cylinder   cone
       prism

positive numbers Numbers greater than zero.

*0.25 and 157 are **positive numbers.***

*−40 and 0 are not **positive numbers.***

power (1) The value of an exponential expression.

*16 is the fourth **power** of 2 because $2^4 = 16$.*

(2) An exponent.

*The expression 2^4 is read "two to the fourth **power.**"*

prime factorization The expression of a composite number as a product of its prime factors.

*The **prime factorization** of 60 is $2 \times 2 \times 3 \times 5$.*

prime number A counting number greater than 1 whose only two factors are the number 1 and itself.

*7 is a **prime number.** Its only factors are 1 and 7.*

*10 is not a **prime number.** Its factors are 1, 2, 5, and 10.*

principal The amount of money borrowed in a loan, deposited in an account that earns interest, or invested in a fund.

*If we borrow $750.00, the **principal** is $750.00.*

prism   A polyhedron with two congruent parallel bases.

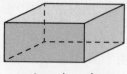 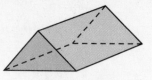

rectangular **prism**   triangular **prism**

probability   A way of describing the likelihood of an event; the ratio of favorable outcomes to all possible outcomes.

*The **probability** of rolling a 3 with a standard number cube is $\frac{1}{6}$.*

product   The result of multiplication.

$5 \times 4 = 20$ *The **product** of 5 and 4 is 20.*

property of zero for multiplication   Zero times any number is zero. In symbolic form, $0 \times a = 0$.

*The **property of zero for multiplication** tells us that $89 \times 0 = 0$.*

proportion   A statement that shows two ratios are equal.

$$\frac{6}{10} = \frac{9}{15}$$

*These two ratios are equal, so this is a **proportion**.*

protractor   A tool used to measure and draw angles.

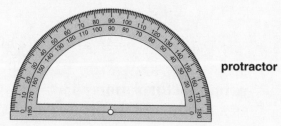

protractor

pyramid   A three-dimensional solid with a polygon as its base and triangular faces that meet at a vertex.

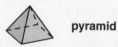

pyramid

quadrilateral   Any four-sided polygon.

*Each of these polygons has 4 sides. They are all **quadrilaterals**.*

qualitative Expressed in or relating to categories rather than quantities or numbers.

> *Qualitative data are categorical. Examples include the month in which someone is born and a person's favorite flavor of ice cream.*

quantitative Expressed in or relating to quantities or numbers.

> *Quantitative data are numerical. Examples include the population of a city, the number of pairs of shoes someone owns, and the number of hours per week someone watches television.*

quotient The result of division.

$$12 \div 3 = 4 \qquad 3)\overline{12}^{\,4} \qquad \frac{12}{3} = 4$$

> *The **quotient** is 4 in each of these problems.*

radius (Plural: *radii*) The distance from the center of a circle to a point on the circle.

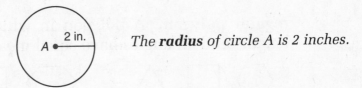

> *The **radius** of circle A is 2 inches.*

range The difference between the largest number and smallest number in a list.

> *To calculate the **range** of the data at right, we subtract the smallest number from the largest number. The **range** of this set of data is 29.*

> 5, 17, 12, 34, 29, 13

ratio A comparison of two numbers by division.

△ △ △
☆ ☆ ☆ ☆ ☆ ☆

> *There are 3 triangles and 6 stars. The **ratio** of triangles to stars is $\frac{3}{6}$ (or $\frac{1}{2}$), which is read as "3 to 6" (or "1 to 2").*

ray A part of a line that begins at a point and continues without end in one direction.

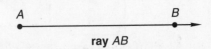

ray AB

reciprocals   Two numbers whose product is 1.

$$\frac{3}{4} \times \frac{4}{3} = \frac{12}{12} = 1$$ *Thus, the fractions $\frac{3}{4}$ and $\frac{4}{3}$ are **reciprocals**.*

rectangle   A quadrilateral that has four right angles.

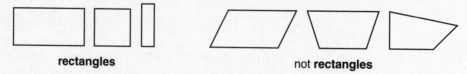

rectangles   not **rectangles**

reduce   To rewrite a fraction in lowest terms.

*If we **reduce** the fraction $\frac{9}{12}$, we get $\frac{3}{4}$.*

reflection   Flipping a figure to produce a mirror image.

reflection

regular polygon   A polygon in which all sides have equal lengths and all angles have equal measures.

regular polygons   not **regular polygons**

remainder   An amount left after division.

$$\begin{array}{r} 7\ R\ 1 \\ 2\overline{)15} \\ \underline{14} \\ 1 \end{array}$$ *When 15 is divided by 2, there is a **remainder** of 1.*

rhombus   A parallelogram with all four sides of equal length.

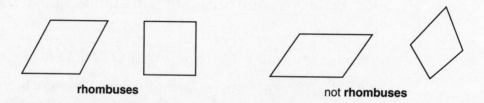

rhombuses   not **rhombuses**

right angle   An angle that forms a square corner and measures 90°. It is often marked with a small square.

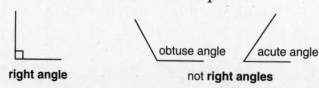

right angle   obtuse angle   acute angle
                        not **right angles**

right triangle   A triangle whose largest angle measures 90°.

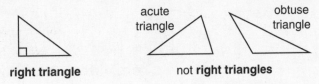

right triangle   not **right triangles**

rotation   Turning a figure about a specified point called the *center of rotation.*

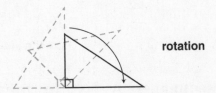

rotation

sales tax   The tax charged on the sale of an item and based upon the item's purchase price.

*If the **sales-tax** rate is 7%, the **sales tax** on a $5.00 item will be $5.00 × 7% = $0.35.*

scale   A ratio that shows the relationship between a scale drawing or model and the actual object.

*If a drawing of the floor plan of a house has the legend 1 inch = 2 feet, the **scale** of the drawing is $\frac{1\,in.}{2\,ft} = \frac{1}{24}$.*

scale drawing   A two-dimensional representation of a larger or smaller object.

*Blueprints and maps are examples of **scale drawings**.*

scale factor   The number that relates corresponding sides of similar geometric figures.

25 mm

10 mm   10 mm

4 mm

*The **scale factor** from the smaller rectangle to the larger rectangle is 2.5.*

scale model   A three-dimensional rendering of a larger or smaller object.

*Globes and model airplanes are examples of **scale models**.*

scalene triangle   A triangle with three sides of different lengths.

*All three sides of this **scalene triangle** have different lengths.*

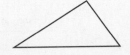

sector   A region bordered by part of a circle and two radii.

*This circle is divided into 3 **sectors.***

segment   A part of a line with two distinct endpoints.

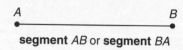

segment *AB* or **segment** *BA*

sequence   A list of numbers arranged according to a certain rule.

*The numbers 2, 4, 6, 8, ... form a **sequence.** The rule is "count up by twos."*

side   A line segment that is part of a polygon.

*This pentagon has 5 **sides.***

similar   Having the same shape but not necessarily the same size. Corresponding angles of similar figures are congruent. Corresponding sides of similar figures are proportional.

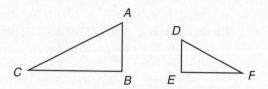

*△ABC and △DEF are **similar.** They have the same shape but not the same size.*

simple interest   Interest calculated as a percentage of the principal only.

Simple Interest		**Compound Interest**	
$100.00	*principal*	$100.00	*principal*
$6.00	*first-year interest (6% of $100)*	+ $6.00	*first-year interest (6% of $100)*
+ $6.00	*second-year interest (6% of $100)*	$106.00	*total after one year*
$112.00	*total after two years*	+ $6.36	*second-year interest (6% of $106)*
		$112.36	*total after two years*

solid   *See* **geometric solid.**

sphere   A round geometric solid in which every point on the surface is at an equal distance from its center.

sphere

square   (1) A rectangle with all four sides of equal length.

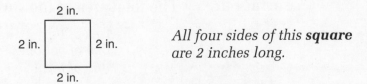

*All four sides of this **square** are 2 inches long.*

(2) The product of a number and itself.

*The **square** of 4 is 16.*

square root   One of two equal factors of a number. The symbol for the principal, or positive, square root of a number is $\sqrt{\ }$.

*A **square root** of 49 is 7 because 7 × 7 = 49.*

stem-and-leaf plot   A method of graphing a collection of numbers by placing the "stem" digits (or initial digits) in one column and the "leaf" digits (or remaining digits) out to the right.

Stem	Leaf
2	1 3 5 6 6 8
3	0 0 2 2 4 5 6 6 8 9
4	0 0 1 1 1 2 3 3 5 7 7 8
5	0 1 1 2 3 5 8

*In this **stem-and-leaf plot,** 3|5 represents 35.*

straight angle   An angle that measures 180° and thus forms a straight line.

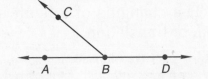

*Angle ABD is a **straight angle.** Angles ABC and CBD are not **straight angles.***

subtrahend   A number that is subtracted.

*12 − 8 = 4   The **subtrahend** in this problem is 8.*

sum The result of addition.

$$7 + 6 = 13 \qquad \text{The \textbf{sum} of 7 and 6 is 13.}$$

supplementary angles Two angles whose sum is 180°.

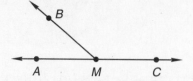

∠*AMB* and ∠*CMB* are
supplementary.

surface area The total area of the surface of a geometric solid.

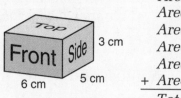

Area of top	= 5 cm × 6 cm =	30 cm²
Area of bottom	= 5 cm × 6 cm =	30 cm²
Area of front	= 3 cm × 6 cm =	18 cm²
Area of back	= 3 cm × 6 cm =	18 cm²
Area of side	= 3 cm × 5 cm =	15 cm²
+ *Area of side*	= 3 cm × 5 cm =	15 cm²
Total **surface area**		= 126 cm²

survey A method of collecting data about a particular population.

*Mya conducted a **survey** by asking each twelve-year-old on her block the name of his or her favorite television show.*

term (1) A number that serves as a numerator or denominator of a fraction.

$$\frac{5}{6} \Big\rangle \; \text{terms}$$

(2) A number in a sequence.

$$1, 3, 5, 7, 9, 11, \ldots$$

*Each number in this sequence is a **term.***

theoretical probability The probability that an event will occur, as determined by analysis rather than by experimentation.

*The **theoretical probability** of rolling a three with a standard number cube is $\frac{1}{6}$.*

transformation The changing of a figure's position through rotation, reflection, or translation.

Transformations

Movement	Name
flip	reflection
slide	translation
turn	rotation

translation Sliding a figure from one position to another without turning or flipping the figure.

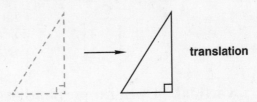

transversal A line that intersects one or more other lines in a plane.

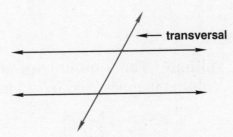

trapezium A quadrilateral with no parallel sides.

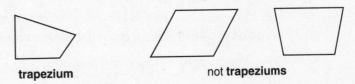

trapezoid A quadrilateral with exactly one pair of parallel sides.

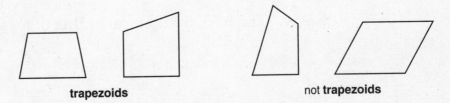

unit multiplier A ratio equal to 1 that is composed of two equivalent measures.

$$\frac{12\ inches}{1\ foot} = 1$$

*We can use this **unit multiplier** to convert feet to inches.*

U.S. Customary System A system of measurement used almost exclusively in the United States.

*Pounds, quarts, and feet are units in the **U.S. Customary System**.*

vertex (Plural: *vertices*) A point of an angle, polygon, or polyhedron where two or more lines, rays, or segments meet.

One **vertex** *of this cube is colored. A cube has eight* **vertices.**

vertical Upright; perpendicular to horizontal.

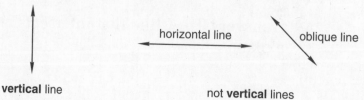

horizontal line   oblique line

vertical line   not **vertical** lines

volume The amount of space a solid shape occupies. Volume is measured in cubic units.

This rectangular prism is 3 units wide, 3 units high, and 4 units deep. Its **volume** *is 3 · 3 · 4 = 36 cubic units.*

weight The measure of how heavy an object is.

The **weight** *of the car was about 1 ton.*

whole numbers The members of the set {0, 1, 2, 3, 4, ...}.

0, 25, and 134 are **whole numbers.**
–3, 0.56, and 100$\frac{3}{4}$ are not **whole numbers.**

x-axis The horizontal number line of a coordinate plane.

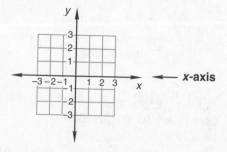

y-axis The vertical number line of a coordinate plane.

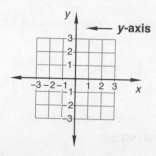

INDEX

Note: Asterisks (*) indicate that the cited topic is covered in a lesson Warm-up.

Page locators followed by the letter "n" are references to footnotes on the indicated pages.

A

Abbreviations
a.m., 169–170
Celsius (C), 44
centimeters (cm), 32
cubic centimeters (cm^3), 326
cubic inches ($in.^3$), 326
Fahrenheit (F), 44
feet (ft), 31
gallons (gal), 415
grams (g), 551
greatest common factor (GCF), 100
inches (in.), 31
kilograms (kg), 551
kilometers (km), 31
least common multiple (LCM), 154
liters (L), 415
meters (m), 31
miles (mi), 31
miles per gallon (mpg), 436–437
miles per hour (mph), 436–437
milligrams (mg), 551
milliliters (mL), 415
millimeters (mm), 31
ounces (oz), 415
pints (pt), 415
p.m., 169–170
pounds (lb), 552
quarts (qt), 415
square centimeters (cm^2), 163
square feet (ft^2), 164
square inches ($in.^2$), 164
square miles (mi^2), 436–437
square (sq.), 164
tons (tn), 552
yards (yd), 31
See also Symbols and signs
Activities
algebraic addition of integers, 561–563
angles
bisectors, 431–433
measuring with protractors, 160–162
sum of angle measures of triangles
and quadrilaterals, 524–526
area
of parallelograms, 380–382
of triangles, 419–421
bisectors
angle, 431–433
perpendicular, 429–431

Activities (cont.)
circles, circumference of, 248–251
compasses, using, 139
coordinate planes, drawing on, 377–378
experimental probability, 486–488
fraction manipulatives, 104–107
length
centimeters, 31–32
inches, 31–32, 83–84
lines of symmetry, 591–592
parallelogram area, 381–382
perimeter, 35
perpendicular bisectors, 429–431
Platonic solids, 651–653
prime numbers, 97
probability, experimental, 486–488
quadrilaterals, sum of angle measures in,
524–526
rulers
centimeter, 31–32
inch, 31–32, 83–84
Sign Wars game, 561–563
transformations, 581–582
triangle area, 419–421
Acute angles, 143–144
formed by transversals, 518–520
in triangles, 500
Acute triangles, 500
Addends, 1
missing, 12–13, 228–229
Addition
activities, 561–563
addends, 1
missing, 12–13, 228–229
addition patterns, 52–55
algebraic, of integers, 534–537, 561–563
associative property of, 23
checking answers, 1
commutative property of, 1
in decimal number chart, 283
of decimals, 193–194, 197–198, 283
estimating sums, 78
fact families, 3–4
of fractions
with common denominators, 123–124,
293–294, 303–304, 327–328
with different denominators, 294–295,
298–299, 303–304
renaming and, 294
steps for, 303–304

Exponents, 198–200, 392
in comparison problems, 393
in expanded notation, 494–495
fractions and, 496
in order of operations, 495
as powers, 392
reading, 392–393
See also Powers of ten
Exterior angles, 524
alternate, 519–520
measuring turns and, 481, 524–526
of polygons, 524–526
transversals and, 342, 519–520

F

Faces, 323–324, 651–652
Fact families
addition and subtraction, 3–4
division and multiplication, 9–10, 607
Factorization, prime. *See* Prime factorization
Factors, 6, 94
of counting numbers, 96–97
divisibility and, 94–95, 97, 108
greatest common factor (GCF), 100–101, 150
missing, 17–19, 465–466
of prime numbers, 96–97
reducing fractions and, 287–288
reversing order of, as multiplication check, 8
Factor trees, 347–348, 379*, 512*
Fahrenheit (F), 44–45
Feet (ft), 31, 436
Feet, square (ft^2), 164, 436
Figures. *See* Geometric figures
Flipping of geometric figures as reflections 581–582, 585
Formulas
for area, 489–490
of circles, 462
of parallelograms, 381, 420, 489–490
of rectangles, 164, 489–490
of squares, 489–490
of triangles, 420–421, 489–490
circumference, 250–251
functions and, 512–513
for perimeter
of parallelograms, 489
of rectangles, 489–491
of squares, 489
of triangles, 489
for surface area of geometric solids, 324
for volume, 441–442, 647
Fractional-parts problems, 410–411, 632–633
Fraction bar, —, 26, 397

Fraction-decimal-percent equivalents, 530–531
Fraction-decimal-percent table, 530
Fraction form, 352–353
Fractions, 26
adding
with common denominators, 123–124, 294, 303–304, 327–328
with different denominators, 294–295, 298–299, 303–304
steps for, 303–304
three or more, 327–328
bar notation, 26, 397
canceling terms of, before multiplying, 369–371, 508–509
common denominators (*See* Common denominators)
comparing, 299–300, 406–407, 445*
converting
to decimals, 397–398, 406–407, 530–531
decimals to, 183–184, 203, 393–394, 530–531
to percents, 401–402, 503–505, 530–531
percents to, 174–175, 217–218, 621–622
and cross products, 454–455
denominator (*See* Denominators)
dividing, 265–266, 288–290
equivalent, 530–531
exponents and, 496
improper (*See* Improper fractions)
manipulatives, 95, 104–107, 123–124, 133–134, 288–290
missing digits in, 572*
missing-number problems with, 228–229
multiplying, 148–150
three or more, 388
on number lines, 82–84
numerator (*See* Numerators)
parts of, 26–27
powers of, 496
rational numbers as, 119
as ratios, 119–120
reciprocals (*See* Reciprocals)
reducing (*See* Reducing fractions)
renaming, 222–223, 294, 298–300
SOS memory aid, 387–388
subtracting
with common denominators, 123–124, 294–295, 303–304
with different denominators, 294–295, 298–299, 303–304
with regrouping, 255–256
from whole numbers, 188–189
terms of, 155

M

Manipulatives
    blocks, 326, 440–442
    calculators (*See* Calculators)
    coins
        combination problems, 26*, 77*,
           94*, 556*
        in compound experiments, 541–545
        probability and, 148*, 310, 541–545
    compasses (*See* Compasses)
    cubes or blocks
        building with, 326, 440–442
        warm-up problems, 119*, 142*,
           174*, 197*, 238*, 264*
    for fractions, 95, 104–107, 123–124,
        133–134, 288–290
    graph/grid paper, 381–382, 460–461,
        637–638
    inch rulers, 31–32, 83–84
    marbles, 310, 486–488, 541–545
    paper (*See* Paper)
    protractors, 159–162
    round objects, 248–251
    ruler, tagboard, 31–32
    spinners, 308–309, 311, 484, 541–543,
        545
    string, 249
    tape measures, 249
Marbles, 310, 486–488, 541–545
Mass versus weight, 551–552
Mean, 88–90, 272
    See also Average; Central tendency
Measurement
    of angles, 159–162, 524–526
        estimation of, 642
    of area (*See* Area)
    arithmetic and, 434–437
    of capacity, 415–416
    of central tendency (*See* Central tendency)
    of circles, 138–139
    of height (*See* Height)
    of length (*See* Length)
    linear (*See* Length)
    of liquid capacity, 415–416
    of mass, 551–552
    of temperature, 44–45
    of time, 169–170, 611–612
    of turns, 479–481, 524–526
    of volume (*See* Volume)
    of weight, 551–552
    See also Metric System; U.S. Customary
        System
Measuring turns, 479–481, 524–526

Median, 272–273
    See also Central tendency
Memory aids
    decimal number chart, 283
    order of operations, 495
    SOS to solve fraction problems, 387–388
Mental math *(A variety of mental math
    skills and strategies are developed
    in the Warm-up activities at the
    beginning of every lesson. With
    few exceptions, these skills and
    strategies have not been indexed.)*
    dividing decimals by ten and by one
        hundred, 279–280
    elapsed-time word problems, 170
    estimating answer by rounding, 78
    multiplying decimals by ten and one
        hundred, 243–244
    rounding, 77–78
Meter, (m), 31–32
Metric system
    converting units of, 32, 415, 552
    units of
        area, 163–165, 199
        capacity, 415–416
        length, 31–32
        liquid measure, 415–416
        mass, 551–552
Miles (mi), 31, 436–437
Miles per gallon (mpg), 437
Miles per hour (mph), 436–437
Milligrams (mg), 551
Milliliters (mL), 415
Millimeters, (mm), 31–32
Minuends, 2
    missing, 13–14, 73, 228–229
Minus sign, –, 2
    See also Negative numbers; Signed
        numbers
Minutes, 169–170, 611–612
Mirror images, 591
Missing-digit problems
    addition, 22*, 43*, 183*, 242*, 298*
    division, 275*, 327*, 352*, 434*, 591*
    fractions, 572*
    multiplication, 154*, 208*, 283*, 460*,
        489*, 546*, 611*, 625*
    subtraction, 73*, 100*, 138*
Missing numbers
    in addition, 12–13, 228–229
    checking answers, 13–14, 17–19, 229
    about combining, 52–54
    in division, 18–19
    about equal groups, 73–74, 89, 113–116
    missing factors, 465–466
    in fractions, 228–229

Diagram
of
Junior

THE
Bennards take
MANHATTAN!